WELL CONTROL FOR COMPLETIONS AND INTERVENTIONS

WELL CONTROL FOR COMPLETIONS AND INTERVENTIONS

HOWARD CRUMPTON
SPE (Society of Petroleum Engineers),
Point Five (Well Services) Ltd. Isle of Skye, Scotland

Gulf Professional Publishing is an imprint of Elsevier
50 Hampshire Street, 5th Floor, Cambridge, MA 02139, United States
The Boulevard, Langford Lane, Kidlington, Oxford, OX5 1GB, United Kingdom

British Library Cataloguing-in-Publication Data
A catalogue record for this book is available from the British Library

Library of Congress Cataloging-in-Publication Data
A catalog record for this book is available from the Library of Congress

ISBN: 978-0-08-100196-7

For Information on all Gulf Professional Publishing publications
visit our website at https://www.elsevier.com/books-and-journals

Publishing Director: Joe Hayton
Senior Acquisition Editor: Katie Hammon
Senior Editorial Project Manager: Kattie Washington
Production Project Manager: Anitha Sivaraj
Cover Designer: Greg Harris

Typeset by MPS Limited, Chennai, India

Transferred to Digital Printing in 2018

CONTENTS

ACKNOWLEDGMENTS

In April 2010 I was teaching a Tubing Stress Analysis course to a group of completion and drilling engineers from BP and Chevron at the Drilling Training Alliance facility in Houston, Texas. On the morning of the 21st I began to hear rumors of a major incident out in the Gulf of Mexico involving the Deepwater Horizon; a semisubmersible drilling rig contracted by BP to drill a well in the Macondo field. Switching on the news I watched the first reports of the Macondo blowout. The BP engineers on my course did not make it in that day, or for the rest of the week. In the aftermath of the Macondo blowout, I became involved in the creation of a Completion and Well Intervention Well Control course for the operating company Shell. It was the writing of a Well Control Manual to accompany that course that gave me the motivation to write this book.

Completions and Interventions covers an extremely wide range of disciplines and techniques; although I have spent almost 40 years working exclusively in this field, I do not know everything and continue to learn. In writing this book I have drawn heavily upon the huge range of literature available on the subject. I have also drawn upon the advice and expertise of many subject matter experts. Thanks are due to Bob Baister and Peter Plummer with whom I shared the challenging, but enjoyable task of creating the advanced well control course for Shell. Thanks, are also due to Jonathan Bellarby. His own experiences in writing the excellent Completion Design book have enabled him to pass on a great deal of very much needed advice and guidance. I would also like to thank Katie Hammon and Kattie Washington at Elsevier for their endless patience; I would originally thought this book would take about a year to write. It is been nearer to 3 years!

This book was written from our home on the Isle of Skye; a wonderful place to live, but far too many distractions. When the weather is fine the urge to be up in the hills or out on the sea is sometimes irresistible. On top of these distractions I have had to juggle writing with my many overseas teaching assignments (Completion design and Intervention courses). My wonderful wife, Anita, has had to put up with me disappearing into my study to write—sometimes after long absences abroad.

Despite this, her support throughout has been unstinting. She has shared her scientific expertise (BSc Hon, PhD, Chemistry) and her advice and guidance have been invaluable. She has read and questioned me on every word I have written, and I could not have finished this book without her.

Howard Crumpton
Isle of Skye. January 2018

CHAPTER ONE

Introduction and Well Control Fundamentals

1.1 INTRODUCTION

Well control is the primary objective of any workover operation.[1]

The primary goal of every completion and workover is to complete the task in a safe and efficient manner.[2]

Well pressure control is the most critical consideration in the planning and performing any well servicing operation.[3]

Three statements from three different manuals, each one dealing with the management of well control during completion and workover operations. Most instructional documents covering intervention well control have similar opinions. Clearly, our industry recognizes the importance of well control during well servicing work. And yet, despite these concerns, accidents and incidents still occur. Well control incidents attributable to completion, workover, and intervention activities account for a significant proportion of the total.

As the Table 1.1 shows, exploration drilling carries the highest risk; this is to be expected. However, completion, workover, and intervention activities account for more well control incidents than development drilling, at more than one third of the total.

Whilst these statistics are the result of a study of one area (Texas and the Gulf Coast), they are symptomatic of a worldwide problem. There are several compelling explanations for why well control problems occur so frequently during completion, workover, and intervention activities.

- Many workover operations are carried out to repair or replace failing equipment. Working on a well where integrity is already compromised increases the risk.

Well Control for Completions and Interventions.
DOI: https://doi.org/10.1016/B978-0-08-100196-7.00001-4

Table 1.1 Operational phase during blowouts[a]
Number of well control incidents by activity and area: 1960–96

Operational phase	Texas	Offshore Continental Shelf (United States)
Exploration drilling	244	45
Development drilling	180	49
Other drilling	14	4
Completion	64	25
Workover	197	23
Wireline	19	5
Production	85	12
Missing data	15	23
Total	817	186

[a]Trends extracted from 1200 Gulf Coast blowouts during 1960–96. Pal Skalle (NTNU, Trondheim, Norway) A.L. Podio, (University of Texas). World Oil, June 1998.

- Completions and workovers are normally carried out with clear, solids free, fluid in the well. The risk of fluid losses is greater than when using mud.
- Interventions are routinely performed with the well live (pressure at surface). Any failure of the pressure control equipment results in an immediate release of hydrocarbons.

Each of these risks can be managed if the crew is experienced and well-trained. However, for many years the only well control training available (and recognized by the industry) was drilling well control. Candidates were (and still are) taught how to recognize a kick, how to shut in the well, and then how circulate out the kick with weighted up mud. Whilst this is a vital skill for anyone working as a member of a drill crew, it ignored many of the well control complexities than can arise during a completion or workover.

In today's oilfield, there is much more emphasis on workover and intervention well control training. However, many would argue that there is still a bias towards drilling. The aim of this book is to redress the balance and provide the reader with a better understanding of well control problems that can arise when completing, working over, or intervening in wells.

1.1.1 Workover or intervention?

Well intervention, completion, and workover are common industry terms. Whilst the term completion is generally unambiguous, the terms "intervention" and "workover" are used differently by operating companies and regulatory bodies.

Workover is, for some operating companies and jurisdictions, an operation that materially alters the structure of the well. Adding perforations, setting bridge plugs in a liner to isolate unwanted water, or any of a range of stimulation treatments are all classified as workover operations. For others, *workover* means a recompletion, the removal and replacement of all the major completion components including the production tubing. This generally means killing the well and using a drilling derrick or hydraulic workover unit to pull and rerun the tubing. For the purposes of clarity, where the term "workover" is used in this book, it will mean recompletion of the well.

Well intervention will mean "through tree" intervention on live wells using wireline, coil tubing, or a workstring run against pressure using a hydraulic workover unit. Well interventions also include pumped treatments, stimulation, and well testing operations.

For nearly all of these interventions, well control is provided by pressure control equipment, for example wireline lubricator and stuffing box, coil tubing stripper, and when using a hydraulic workover unit, stripper rams or annular preventer. Operations using live well pressure control equipment are significantly different from those carried out on a dead well where a fluid barrier is used, consequently, management of well control and well integrity must be viewed differently.

1.1.1.1 Pressure control and well control

The terms well control and pressure control are used extensively throughout this book. For the purposes of clarity, the term "pressure control" is used to describe live well interventions, where pressure retaining equipment is being used to prevent the escape of pressurized fluids at the surface. It is mainly applicable to wireline, coiled tubing, and hydraulic workover operations on live wells. "Well control" is generally used in the context of maintaining a hydrostatic overbalance during operations on a dead well.

1.1.2 Why interventions and workovers are performed

Workovers and Interventions are performed for two reasons:

1. To repair or replace failed equipment.
2. To increase production, either through improving existing production or slowing the rate of decline.

1.1.2.1 Production decline

If production from a well is in decline, the immediate task is to determine the reason why. It may simply be the result of declining reservoir pressure and in line with expectations. However, a declining rate can also be an early indication of a problem in the reservoir or the wellbore. Diagnostic interventions are often performed to try and establish the nature and location of flow restrictions. There are several common production related problems that are routinely managed by intervening in, or working over a well.

1.1.2.1.1 Scale precipitation

Scale precipitation in the formation or perforation tunnels will increase the skin factor and reduce productivity. Scale forming in the wellbore will reduce the tubing ID, and consequently reduce production by choking the well. The two most common oilfield scales are calcium carbonate ($CaCO_3$) and barium sulfate ($BaSO_4$).

Increasing temperature and reducing pressure promote the precipitation of calcium carbonate scale, so it tends to form high up in the tubing where pressure is low. Several intervention techniques are used to manage calcium carbonate scale. It can be prevented or slowed by bullheading a large volume of scale inhibitor into the producing formation. Where already formed, it can be dissolved using hydrochloric acid (HCl). This is normally accomplished using coiled tubing. Small deposits of soft scale can be removed using various wireline deployed tools.

Sulfate scale forms when sulfates, present in some injection water, mix with barium ions in formation water. Consequently, sulfate scale can form anywhere in the producing system, from the reservoir to the process facilities. Unlike carbonate scale, it cannot easily be removed using chemicals. It is normally removed from the wellbore using coiled tubing deployed scale mills or high-pressure jetting.

1.1.2.1.2 Wax deposits

Wax is a long chain alkane hydrocarbon that solidifies at relatively low temperatures. Wax accumulation in the tubing causes a decline in flow rate. It can be removed by circulating hot fluid (hot oiling), mechanical removal using wireline or coiled tubing, or chemical solvents such as xylene or toluene.

1.1.2.1.3 Asphaltines

Asphaltines appear as hard deposits resembling an asphalt road surface. They are organic solids that precipitate from crude oil systems, and are most likely to occur at close to the oil bubble point pressure. Like scale and wax, asphaltine precipitation causes a reduction in flow rate. Asphaltines are difficult to remove from the wellbore, and normally require mechanical removal or hydro-jetting using coiled tubing. They can also be removed chemically using xylene or toluene.

1.1.2.1.4 Water and gas production

There are several sources of water ingress into oil producing wells. Water from nearby injection wells can flow preferentially through high permeability layers or fractures in the formation. Underlying aquifer water can be drawn up towards the producing zone, a problem made worse by high drawdown. Water will also find its way through channels in cement or leaks in the completion. Similarly, overlying gas from the gas cap can be drawn into the well. Water entry will reduce oil production, because of increased hydrostatic head in the tubing and relative permeability changes in the reservoir. Water production also brings corrosion, scaling, and in many cases an increase in sand production.

Water breakthrough and water production are monitored by taking flowline samples and conducting well tests. Logging can reveal where water (or gas) is entering the well, essential information if the water is to be isolated at source. Water and gas shut-off interventions are common, employing several widely used techniques. These include mechanical bridge plugs and straddles, casing patches, remedial cementing, and chemical treatments.

1.1.2.2 Well stimulation

1.1.2.2.1 Hydraulic fracturing (Fracking)

Hydraulic fracturing of sandstone and shale formations can significantly increase productivity from low permeability reservoirs. Fluid is pumped down the wellbore at above the formation fracture pressure, and the resulting fracture is packed with proppant to keep it open. Hydraulic fracturing is often performed through casing before the completion is installed, as is normally the case for shale gas wells. However, many through tubing fracking operations are carried out on existing wells to improve inflow.

1.1.2.2.2 Acid fracturing and acid matrix treatments

Carbonate (limestone and dolomite) formations are fractured using acid, normally HCl. Acid fractures the formation, or enters existing fractures where it dissolves the carbonate material, creating highly permeable pathways into the wellbore. Acid treatments are frequently performed as interventions in existing wells, often through coiled tubing.

1.1.2.3 Artificial lift

Many workovers are performed to install artificial lift. Gas lift is widely used, and works by reducing the density of the fluid in the tubing. All other artificial lift systems use a pump. There are several types of pump used, including electric submersible pumps (ESP), beam pumps, and progressive cavity pumps (PCP).

1.1.2.4 Mechanical repairs

Completion equipment can fail. Some failures occur very early in the life of the well, and are a result of poor design, wrongly specified equipment, or damage during installation. Most failures occur late in well life because of corrosion, erosion, and fatigue. A plot of failure frequency against time resembles a bathtub (colloquially known as a bathtub curve) (Fig. 1.1).

Many completion components are barrier elements (Chapter 6. Well Barriers), and their failure can compromise the primary or secondary well barrier envelope. When this occurs, they must be repaired or replaced. Some components are repairable or replaceable using through tubing interventions with the well still live, e.g., replacement of a failed tubing retrievable safety valve, or the replacement of a gas lift valve when the check valves have failed. Other failures can only be remedied by replacing the completion, a tubing collapse for example.

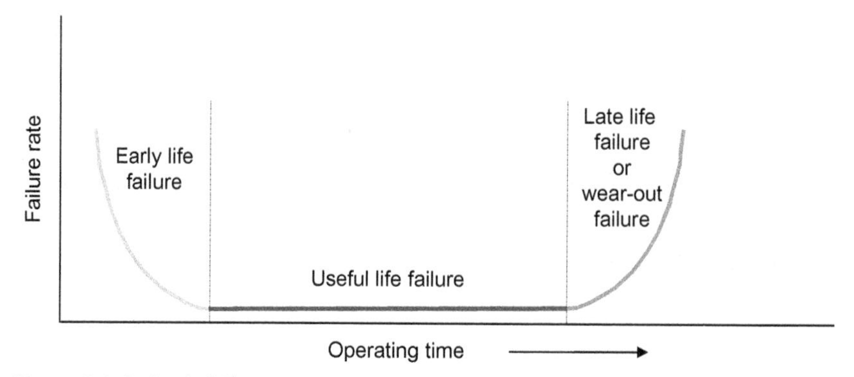

Figure 1.1 Bathtub failure curve.

1.1.3 The geology of hydrocarbon reservoirs

An understanding of some very basic hydrocarbon reservoir properties is useful for anyone working with wells. Permeability, formation fracture pressure, and formation pore pressure all have a direct influence on how well control is managed. With a few notable exceptions, hydrocarbon reservoirs are found in sedimentary rocks.

1.1.3.1 Sedimentary rock

Formed by the deposition of sediments settling in layers over long periods of geological time, sedimentary rocks can be classified as clastics, evaporites, and organics.

Clastics, from the Greek word Klastos meaning "broken" are formed from the compacted fragments of other rocks. Sandstone is a common type of clastic rock. Individual grains of sandstone are fragments of older rocks that have been weathered, crushed, and broken down until they are small enough to be transported by wind and water to a place of deposition. If they remain in place for long enough, the deposited sand grains become covered by layer upon layer of other sediments. The buried sediment is then consolidated by heat, chemical action, and the pressure (overburden) of the overlying formations. Many hydrocarbon accumulations are in sandstone formations (Fig. 1.2).

Figure 1.2 Horizontally bedded sandstone formation: analogous to that found in many hydrocarbon reservoirs. Isle of Skye, Scotland.

Figure 1.3 Limestone outcrops in the Yorkshire Dales (United Kingdom).

Evaporites form when a body of saline water evaporates. As the salinity increases, chemical precipitates build up in layers. Common evaporates are gypsum, anhydrite, and halite (rock salt). Halite is of interest to oilfield geologists, since some oilfields form around salt domes.

Most organic sediment is classified as "carbonate," including limestone and dolomite formations. Organic sediments are the skeletons of dead marine creatures that sink to the bottom of the ocean. Over time, the carbonate material builds into beds that can be many hundreds of feet thick. Carbonate formations are an important reservoir rock, accounting for approximately 60% of the world's oil and gas production (Fig. 1.3).

A few hydrocarbon reservoirs are found in naturally fractured crystalline basement rocks, such as granite or basalt.

1.1.3.2 Hydrocarbon traps

Many theories concerning the origin of oil and gas have been advanced over the years. Current thinking is that hydrocarbons are produced through a complex chemical reaction involving the bacterial decay of phytoplankton and algae. Once formed, hydrocarbons, being less dense than surrounding formation water, migrate upwards from the source rock. If there are no impermeable barriers in the overlying formations, the hydrocarbons will eventually migrate all the way to the surface. Where an impermeable barrier is present, hydrocarbons become trapped. Although the hydrocarbons displace formation water as they migrate, some residual formation water remains in place. Hydrocarbon traps exist where permeable reservoir rocks have overlying low permeability formations. A formation that prevents upward migration of the hydrocarbons is known as a caprock. These are often compacted shales, evaporites, or tightly cemented sandstones. There are two main categories of trap, structural and stratigraphic.

Structural traps hold hydrocarbons because the formation has been folded or faulted in some way. Structural traps include domes, anticlines, and sealing faults. The most common is an anticline, accounting for approximately 75% of all reservoirs traps. Fault traps are rare, making up only about 1% of reservoirs. They form when faults seal the hydrocarbon zone. A salt dome or salt diapir forms when salt of a lower density than the surrounding formation plastically deforms the surrounding formation as it "flows" upward towards the surface. The deformed structure creates a trap for hydrocarbons.

Stratigraphic traps are depositional. The hydrocarbon reservoir forms in place, usually when sandstone or limestone is covered by impermeable shale. In some, both strata and structure will combine for create a formation (Fig. 1.4).

1.1.3.3 Porosity and permeability

To form a commercially viable hydrocarbon reservoir, sedimentary rocks must exhibit two essential characteristics.

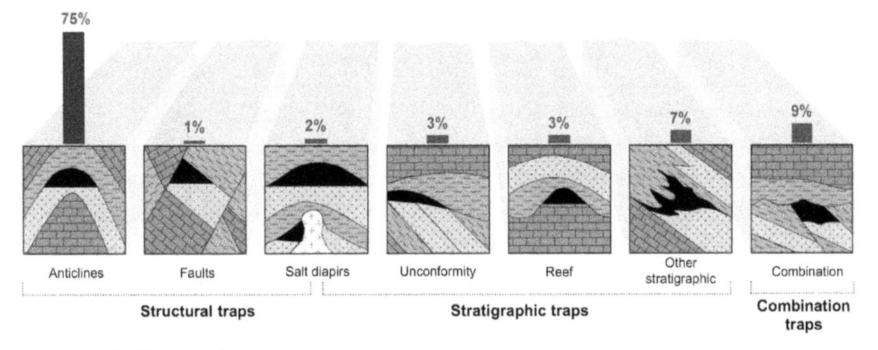

Figure 1.4 Hydrocarbon traps.

Table 1.2 Porosity values for sandstone and carbonate material

Formation	ϕ Range (%)	Typical average (%)	Normal economic limit (%)
Sandstone	8–39	18	8
Carbonates	3–50	8	3–5

1. The capacity for storage (porosity).
2. The transmissibility of fluids (permeability).

Porosity is a measure of the storage capacity of a formation, and is directly related to the volume of void space in the reservoir rock.

Voids occur at intergranular gaps between the grains, and are termed pores. Porosity is defined as the percentage or fraction of the void space to the total bulk volume of the rock, and is normally indicated using the symbol ϕ (phi). Porosity varies enormously (Table 1.2).

Porosity alone is not enough to make a formation commercially viable. Hydrocarbons must be able to flow through the formation and reach the wellbore. This can only happen if there is interconnectivity between pore spaces allowing fluids to flow. Permeability is a measure of flow capacity through a formation, and can only be determined by flow experiments using core from the reservoir. Since permeability depends upon the continuity of pore space, there is no unique relationship between porosity and permeability (Fig. 1.5).

Permeability is represented by the Greek letter k (kappa). Reservoir permeability is most commonly measured in Darcy (D) or millidarcy (mD). The coefficient of permeability (k) is a characteristic of the rock, and is independent of the fluid used for measurement. Rock has a permeability of 1 D, if a pressure gradient of 1 atm/cm induces a flow rate of 1 cm^3/s across a cross-sectional area of 1 cm^2 using a liquid with a viscosity of 1 cP (fresh water). Since most oilfield reservoirs have permeabilities that are less than 1 D, the millidarcy (10^{-3} D) is more commonly used.

To relate this to oilfield applications and hydrocarbon reservoirs, 20/40 mesh gravel, of the type used in propped fracs and sand control completions, has an unstressed permeability of approximately 120 D (120,000 mD). By contrast, unconsolidated well sorted course sandstone formation might have a permeability of, e.g., 5 or 6 D. Compacted, poorly sorted sandstones will have much lower permeability, and must be measured in millidarcy. The shale gas reservoirs currently being exploited in North America, whilst having good porosity, have very poor

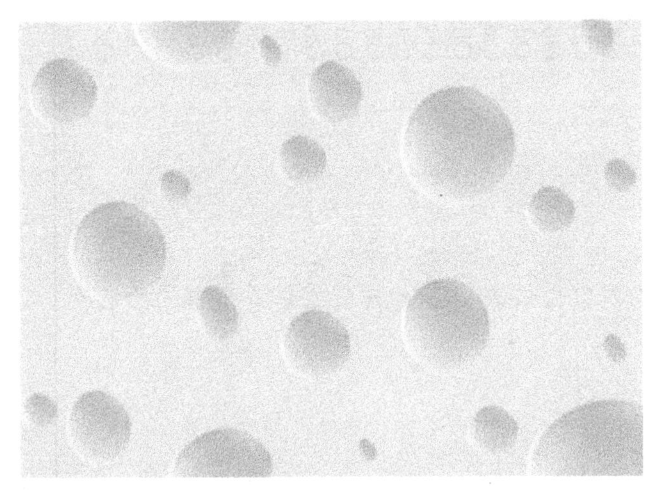

Figure 1.5 Swiss cheese—high porosity but poor permeability!

permeability, measured in micro (10^{-6}), even nano (10^{-9}) Darcy. They will only flow at commercially viable rates after massive hydraulic fracturing operations. Table 1.3 shows porosity and permeability from a sample from the North Sea fields.

Understanding porosity and permeability in the reservoir is essential for managing well control. Losses and kicks are far more likely when working on wells with highly permeable or fractured formations. Conversely, tight formations can create problems if a bullhead is needed to stimulate or kill the well. Porosity data is used to estimate depth of invasion where losses have occurred, and volume required for matrix stimulation treatments. Knowing the size distribution of the pore spaces is also useful for sizing solid lost circulation material, such as calcium carbonate or sized salt.

1.1.4 Formation pressure and reservoir pressure

Hydrocarbon accumulation displaces formation water from the permeable reservoir rock. Unless subsequent tectonic movements completely seal the reservoir, any underlying water (the aquifer) is contiguous. Pressure in the aquifer will be equivalent to native or regional hydrostatic gradient. In the water column, the pressure at any depth is approximated by:

$$P = hG_{\mathrm{w}} \tag{1.1}$$

Table 1.3 Porosity and permeability range from North Sea Fields

Field	Age/Formation	Reservoir	Porosity (%)	Permeability (mD)	Fluid
Alwyn North	Jurassic/Brent	Sandstone	17	500–800	Oil
Alwyn North	Jurassic/Statfjord	Sandstone	14	330	Oil
Auk	Permian/Zechstein	Fractured dolomite	13	53	Oil
Auk	Permian/Rotliegend	Sandstone	19	5	Oil
Brae South	Jurassic/Brae	Sandstone	12	130	Oil
Buchan	Devonian/Old Red	Fractured sandstone	9	38	Oil
Cleeton	Permian/Rotliegend	Sandstone	18	95	Gas
Cyrus	Paleocene/Andrew	Sandstone	20	200	Oil
Ekofisk	Cretaceous/Chalk	Limestone (fractured chalk)	32	>150	Oil
Ekofisk	Cretaceous/Chalk	Limestone (chalk)	30	2	Oil and gas
Forties	Paleocene/Forties	Sandstone	27	30–4000	Oil
Fulmar	Jurassic/Fulmar	Sandstone	23	500	Oil
Frigg	Eocene/Frigg	Sandstone	29	1500	Gas
Heather	Jurassic/Brent	Sandstone	10	20	Oil
Leman	Permian/Rotliegend	Sandstone	13	0.5–15	Gas
Piper	Jurassic/Piper	Sandstone	24	4000	Oil
Ravenspurn South	Permian/Rotliegend	Sandstone	13	55	Gas
South Morecambe	Triassic/Ormskirk	Sandstone	14	150	Gas
Scapa	Cretaceous/Valhall	Sandstone	18	111	Oil
Staffa	Jurassic/Brent	Sandstone	10	10–100	Oil
West Sole	Permian/Rotliegend	Sandstone	12	—	Gas

Table 1.4 Regional variations for pore pressure gradient

Region	Pore pressure gradient	
	psi/ft	**KPa/m**
Anadarko basin	0.433	9.64
California	0.439	9.77
Gulf of Mexico	0.465	10.35
Malaysia	0.442	9.84
North Sea	0.452	10.06
Rocky Mountains	0.436	9.71
West Africa	0.452	9.84
West Texas	0.433	9.64

Where h = the vertical depth (ft or m); G_w = the pressure gradient (psi/ft or kPa/m).

Formation water is normally saline and more dense than fresh water. However, increasing temperature with depth reduces fluid density, so a common "normal" value used is the fresh water gradient (0.433 psi/ft or 9.80 kPa/m). Gradients within the range 0.433–0.5 psi/ft are considered normal (Table 1.4).

Pressure at the top of a hydrocarbon-bearing structure can be expected to be higher than the hydrostatic gradient extrapolated from the hydrocarbon/water contact caused by the reduced pressure gradient in the oil and gas column (Fig. 1.6).

Where pressure in the formation is greater than that caused by a column of formation brine, the pressure is considered "abnormal." Although the term "abnormal" is used, the condition is in fact quite common, and a characteristic of some of the best oil and gas reservoirs. There are several causes.

Where the vertical depth of the water column is more than well depth, the well will be abnormally pressured. Perhaps the best example of this is the artesian well Fig. 1.7.

When drilling a low-lying area of a mountainous region, a relatively short borehole can penetrate a formation that is pressurized by a fluid column that has a higher elevation than wellbore ground level. Balancing fluid needs to be very dense to prevent uncontrolled flow. Similarly, in dipping and folded permeable reservoirs, pressure from the deepest part of the formation can be transmitted to the shallowest part. Whilst pressure at the deepest point may be normal for the depth, pressure at the crest can be significantly higher than "normal". For example, the pore pressure gradient in the North Sea is generally given as 0.452 psi/ft (10.06 KPa/m).

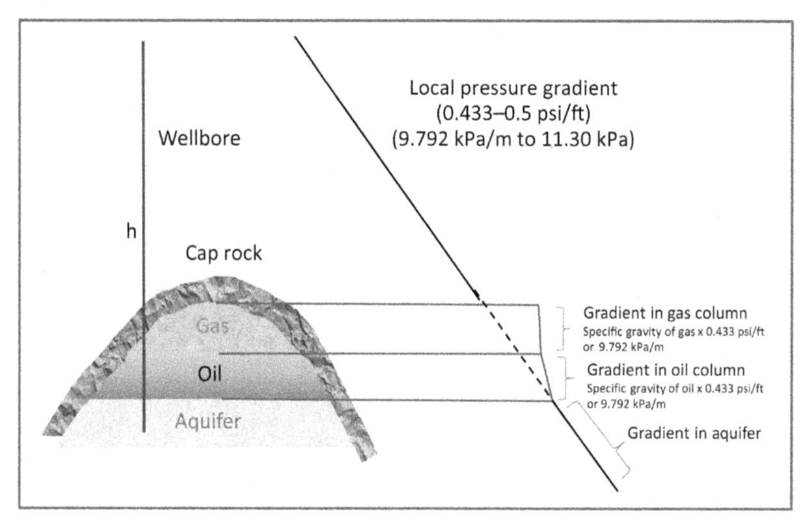

Figure 1.6 "Normal" pressure distribution from the surface through a reservoir structure.

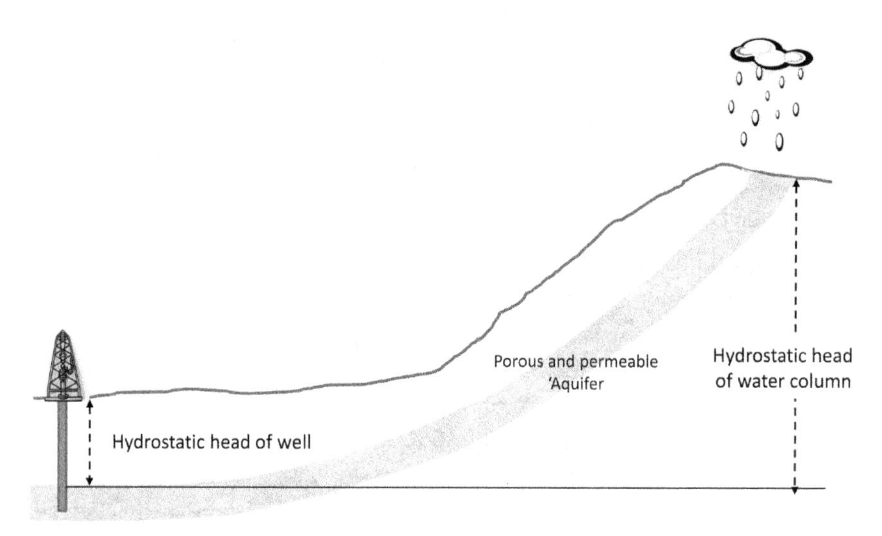

Figure 1.7 Artesian well—abnormally pressured.

Fig. 1.8 shows an anticline structure. The permeable sandstone in the structure is filled with gas with a pressure gradient of 0.1 psi/ft. The surrounding formation is filled with salt water with a pressure gradient of 0.465 psi/ft, normal for the North Sea.

Reservoir pressure at 5000 ft is 2325 psi. The gas bearing formation is 2000 ft thick from crest to base, so the pressure at the top of the

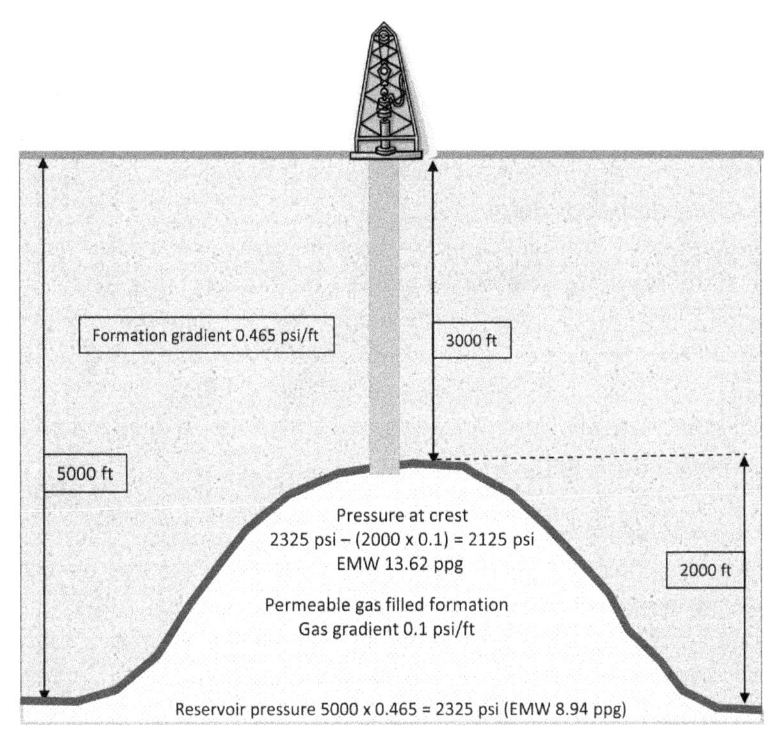

Figure 1.8 Overpressure at reservoir crest.

formation is $2325 - (2000 \times 0.1) = 2125$ psi. At 5000 ft, the formation is normally pressured and has an equivalent mud weight of 8.94 ppg. A well drilled into the crest at 3000 ft would need 13.62 ppg (0.708 psi/ft) mud to balance reservoir pressure.

1.1.4.1 Under-compaction in massive shale beds

Immediately following deposition, shale has high porosity. More than 50% of the total volume of un-compacted muds and clays can be the water in which the solids were deposited. Normally, during compaction, water is squeezed out of the formation, and porosity reduces as the weight of overlying new sediment increases. If the removal of the water is impeded, fluid pressure in the shale increases, since the trapped water has a very low compressibility coefficient. A porous fluid-filled shale supporting heavy overburden weight abnormally is likely to be over-pressured.

1.1.4.2 Salt beds

Deposition of salt can occur over wide areas. Since salt is impermeable to fluids, the underlying formations become over-pressured. Abnormal pressures are frequently found in zones directly below a salt layer.

1.1.4.3 Salt domes or diapirs

Salt domes form when overburden pressure acting on a salt formation causes it to plastically deform and push up through weaknesses in the overlying formations. Upwards movement of salt through the sedimentary strata, and the associated deformation of the formation above, is called "*halokinetics*" or salt tectonics. Movement may continue for several 100 million years. Overpressure can occur because of the folding and faulting of the formation.

1.1.4.4 Tectonic forces

Tectonic movement can give rise to horizontal forces in the formation. In a normally pressured formation, water is squeezed out of clays as they are compacted by increasing overburden. However, if the horizontal force is such that it squeezes the formation laterally, and fluids are prevented from escaping at a rate equal to the reduction in pore volume, an increase in pore pressure will result.

1.1.4.5 Faulting

Formation blocks sometimes contain sealed-in pressure that is normal for the depth of burial. If, however, the formation is uplifted to a shallower depth because of fault movement, the pressure will be abnormal for the new depth (Fig. 1.9).

1.1.4.6 Cross-flow

Some completions allow flow between layers in the reservoir. High pressure zones can cross-flow into lower pressured zones. This is sometimes referred to as an underground blowout. Cross-flow can be very problematic during some intervention activities. It can also complicate and compromise well kill operations.

1.1.5 Formation fracture pressure

It is possible to hydraulically fracture a formation by applying pressure to the wellbore. When a formation fractures, cracks are created within the rock matrix, and fluid in the wellbore will be lost into the

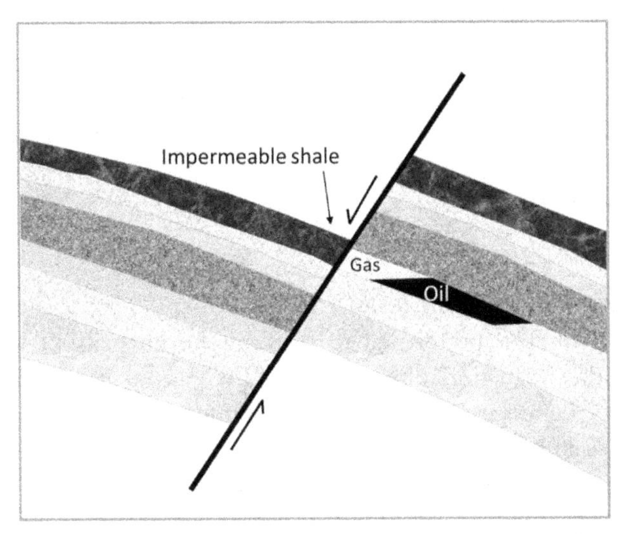

Figure 1.9 A trap where the oil bearing formation has moved up from its original place of deposition.

fractures. The pressure required to create a fracture is termed "fracture pressure."

Fracture pressure is expressed as either:

- A pressure—psi, bar, or kPa.
- A fluid gradient—psi/ft, Bar/m, or kPa/m.
- A fluid weight equivalent—ppg, kg/l, or SG.

Knowing the fracture pressure is essential for workover and intervention operation, as exceeding fracture pressure would lead to severe fluid loss and a consequent loss of the hydrostatic overbalance. Fluid loss to the formation also carries a risk of formation damage, and the severe losses associated with a fractured formation are very damaging. The impact on productivity is likely to be severe. Most operating companies will have policy and procedures in place to ensure that fracture pressure is not accidentally exceeded during completion and workover operations. However, there are occasions when fracturing is a required part of the intervention. Fracture pressure is deliberately exceeded during the installation of frac-pack sand control completions. It is also routinely exceeded during acid fracturing and propped frac stimulation operations.

Fracture pressure is related to the weight of formation matrix (rock and sediments), and the fluid occupying the pore spaces above the zone of interest. These two factors combine to produce what is termed "overburden pressure." Although the density of the overlying formation varies with depth, a rough approximation of fracture pressure can be estimated if it is assumed that average density of the overlying formation and the associated liquids is roughly equivalent to a gradient of 1 psi/ft (22.6 kPa/m). For most completion and intervention activities, fracture pressure will have been determined during the drilling of the well by performing a leak-off test (LOT).

1.1.5.1 Formation leak-off tests

A formation LOT is performed to confirm the integrity of the cement bond, and the formation directly below the casing seat. Normally, the zone directly beneath the casing seat is assumed to be the weakest point during the drilling of the next hole section. Since it is the shallowest part of the next section of formation to be drilled, it will have the lowest overburden pressure.

LOTs are normally carried out at each casing point. After setting, cementing, and testing the new casing string, the shoe track and casing shoe are drilled out, and a few feet of new formation drilled. Normally, this is about 15 ft, to ensure enough formation is exposed. A formation LOT is then performed. A routine formation LOT is typically performed as follows:

- Check pressure gauges are working and have been recently calibrated.
- Condition (circulate) the mud to ensure weight is consistent throughout the system and confirm mud density.
- Ensure the bit is back inside the casing shoe, then close the well (close the BOP annular preventer or pipe rams).
- Start to slowly increase pressure by pumping a small volume at a steady rate (¼–½ bbl/min). Measure and record the pressure increase against volume pumped.
  - Note: Slightly different techniques are used by some operators. Some will increase pressure incrementally, stopping between increments. Others do not like to pump into a closed system, and will circulate whilst increasing back pressure by gradually closing the choke.
- As the formation fractures, mud will start to leak into the formation, and the rate of pressure increase will fall off. Pump rate should be reduced.
- When no further increase in pressure is observed, or pressure begins to fall off, stop pumping.

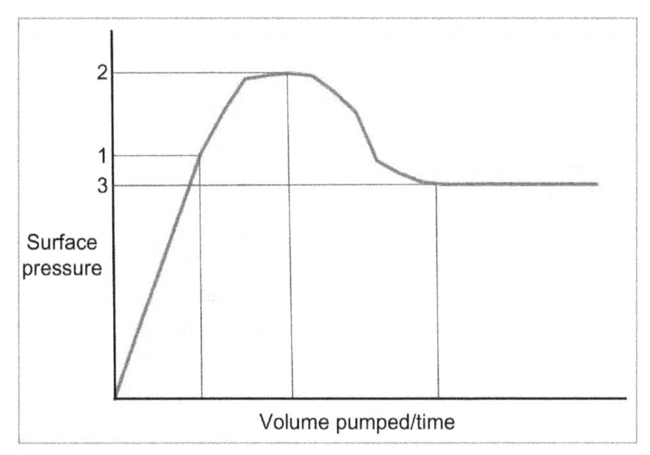

Figure 1.10 LOT plot—Surface pressure vs volume pumped.
1. Leak-Off Pressure—The pressure at which the fracture begins to open and fluid starts to leak off into the formation. This will be seen as a change in the slope of the plot. At this point, the pump rate should be reduced.
2. Rupture Pressure—This is the maximum pressure the formation can sustain before irreversible fracture occurs. This will be determined by a sharp drop in the pressure being applied—pumping should be halted.
3. If no more pressure is applied at this point, most formations will recover to a certain degree, and the Propagation Pressure is determined when the pressure becomes stable again.

- Bleed off and measure the volume of mud returned—record the volume lost to the formation.

As pressure is increased during a LOT, three pressure stages are normally evident, and it is the operator's decision as to which one will be taken as the pressure on which to base subsequent formation integrity calculations (Fig. 1.10).

A major disadvantage of a LOT is that fracturing the formation can weaken it, reducing fracture pressure to below the undisturbed value. Propagation (recovery pressure) is normally lower than the original fracture pressure, so the integrity of the formation during the drilling of the next hole section is compromised to some degree.

During a LOT, there are two forces acting on the formation. Firstly, there is the hydrostatic pressure from the column of mud, secondly, applied pressure at the surface. Fracture pressure is:

- Mud hydrostatic pressure at the casing shoe + applied surface pressure.
- To calculate the fracture pressure as an equivalent mud weight:

$$\text{Fracture pressure EMW} = (P_{\text{surface}}/\text{TVD}/0.052) + \text{Mud weight} \quad (1.2)$$

Note: Always **round down** *to one decimal place when calculating LOT equivalent mud weight, i.e., 15.69 becomes 15.6 ppg.*

When calculating kill fluid weight, always **round up** *to one decimal place, i.e., 15.12 becomes 15.2 ppg.*

1.1.5.2 Formation integrity tests

LOTs are generally restricted to exploration wells, or wells in a development area where there is uncertainty about fracture gradient and formation pressure. Where reliable offset data is available, deliberately fracturing the formation during drilling is normally avoided. The formation is pressure tested, but at below the anticipated fracture pressure. This type of test is called a "Formation Integrity Test" (FIT). The advantage of the FIT is that there is no compromise of the formation fracture pressure.

LOTs and Formation Integrity Tests can only be successfully carried out with drilling mud in the well. Solids in drilling mud allow a filter cake to build up on the bore-hole wall, limiting fluid loss into the formation. With correctly formulated mud in the hole, fluid loss can only be induced if the formation is fractured. If solid-free fluids are in the well (i.e., completion brines), fluid will leak off at above pore pressure, but below fracture pressure. Losses can be controlled using lost circulation material, such as calcium carbonate ($CaCO_3$) or sized salt. Completion fluids, their use and properties are described in Chapter 5.

1.1.5.3 Unit systems

To date, no industry-wide standardization of units has been achieved. Across the world, the oil industry uses a variety of unit systems. Oilfield units are the most widely used, but metric and SI systems are becoming more popular. It is not uncommon to find different disciplines at the same location using different unit systems. For example, drilling teams often work in oilfield units, whilst the process engineers use SI. It is not unusual to find a mixture of unit systems in some locations, e.g., some North Sea operators measure depth and pipe length in meters, but pipe diameter in inches.

It is not surprising that confusion over unit systems and conversion factors leads to mistakes. The safest approach is for an individual to use the unit system they are most comfortable with. Simple arithmetical errors are more likely to be recognized, since the magnitude of the answer will be anticipated.

This book uses a mixture of oilfield and SI units, with the emphasis on oilfield units, since they are still the most widely used. Most people working in the industry will have had some exposure to psi, barrels, feet, etc. SI answers are provided for some of the worked examples.

1.1.6 Hydrostatic pressure calculations

Understanding hydrostatic pressure is fundamental to well control. An ability to calculate hydrostatic pressure at any point in the wellbore is an essential skill.

Mass: Mass is the term for a quantity of matter. The oilfield unit of measurement is the pound, and the SI unit of measurement is the gram or kilogram.

Density: Density is an expression giving the mass of gas, fluid, or solid matter in a given volume, i.e., mass per unit volume. For example, in oilfield units, density is normally reported as pounds (mass) per US gallon (volume), and abbreviated as ppg. In SI units, density is normally recorded in kilograms per m^3, or kilograms per liter (Kg/L).

Liquid density can also be expressed relative to fresh water. Fresh water density will vary with temperature, and there are various figures quoted. However, for consistency, 8.33 lbs/gallon is used for fresh water. This is the value used in nearly all oil industry text books.

Since the mass of 1 (US) gallon of fresh water is known (8.33 ppg), that value is used to calculate the mass of any fluid relative to the mass of fresh water. This is termed specific gravity (SG):

$$\mathrm{SG} = \frac{\mathrm{ppg}}{8.33} \tag{1.3}$$

Example: What is the SG of a 10 ppg brine?

$$\frac{10\mathrm{ppg}}{8.33\mathrm{ppg}} = 1.2004\mathrm{SG}.$$

The equation can be rearranged to find the mass of the fluid (ppg) if the relative density (SG) is known.

$$\mathrm{ppg} = 8.33 \times \mathrm{SG}$$

In SI units, fluid mass, and SG are effectively the same, since 1 L of fresh water (at 4°C) has a mass of 1 Kg.

Temperature and pressure effects on density: Fresh water density will vary with temperature. Fresh water has a relative density of 1 SG at 4°C.

Density will decrease with increasing temperature. At 100°C the density of fresh water is 0.95 SG. As pressure increases, density will increase. However, water has a very low compressibility, so any density increase for water-based fluid is negligible. If oil-based fluids are used, the density increase becomes more significant.

Note: Fluid density corrections for temperature and pressure are described in Chapter 5, Completion, Workover, and Intervention Fluids.

Force: Consider a mass of 1 lb suspended by a length of string. A force will keep the string in tension. The product of gravitational acceleration and the mass causes the force. Force can be expressed in unit pound-force, which can be defined as:

One pound-force is the force which will influence a body with 1 lb mass when subjected to gravitational acceleration of 9.80665 m/s^2 (32.147 ft/s^2). Gravitational acceleration varies between the equator and the poles. For example, gravitational acceleration at the North Pole is equal to 9.831 m/s^2, which gives a force influence on a mass of 1 lb according to the following:

$$G = 1 \times \frac{9.831}{9.80665} = 1.0025(\text{lbs})$$

The value 9.80665 expressed here is used as standard, and represents the acceleration of gravity at 45 degree latitude North—midpoint between the pole and the equator.

When using oilfield units, the variation in gravitational acceleration is ignored, and a 1-lb mass is considered to exert a 1 lb-force influence. Mass and weight become synonymous, since a 1 lb-force and 1 lb weight are one and the same. This may upset pure scientists (especially astrophysicists), but for practical terrestrial purposes there is no difference and it seems OK.

In SI units, one kilo is, strictly speaking, mass. Force (mass × gravitational acceleration) is measured in newtons. The newton is the SI unit for force, and is equal to the amount of net force required to accelerate a mass of 1 kg at a rate of 1 m/s^2.

$$1\ \text{kg} = 9.81\ \text{N. Hence,}\ \ 1\ \text{N} = 1/9.81\ \text{kg} = 0.102\ \text{kg}$$

A newton is approximately equivalent to 0.102 kilo (102 g).

Pressure: Pressure is defined in physics as force per unit area:

$$P = \frac{\text{Force}}{\text{Unit Area}} \tag{1.4}$$

This formula can be rearranged to calculate the force from a given pressure and a unit area:

$$\text{Force} = P \times A \tag{1.5}$$

When using oilfield units, pressure is expressed as the pounds of force applied against a one square inch area, i.e., pounds per square inch, and is commonly abbreviated as *psi*.

Pressure in SI units is expressed in pascals (Pa). A pascal is a pressure of 1 N (0.102 kg) per square meter. Since this unit is impractically small for most oil industry applications, the kilopascal (kPa) equal to 1000 N/m^2 is more commonly used (1 psi = 6.895 kPa).

In some locations, *bar*[a] is used for pressure measurement. One bar is equal to 100 kPa, and is slightly less than atmospheric pressure (0.987 atm).

1.1.6.1 Hydrostatic pressure

Hydrostatic pressure is the total fluid pressure created by the weight of a column of fluid (liquid or gas) acting at any given point in the well. The derivation of the word comes from "hydro," Greek for water and static, meaning not moving. To calculate hydrostatic pressure, well depth, and the density of the fluid in the well must be known.

Fluid density is converted to a pressure increase per unit depth, or *fluid gradient*.

When using oilfield units, fluid weight in ppg is converted to a gradient (psi/ft) by multiplying the fluid weight by a conversion factor of *0.052*.

It is worth understanding the derivation of this number, since it is used so frequently. It is also one worth committing to memory.

One cubic foot contains 7.48 US gallons. A cubic foot filled with a fluid having a density of 1 ppg would weigh 7.48 lbs, and the pressure at the bottom of the container will be:

$$\frac{\text{Weight(force)}}{\text{Area}} = \frac{7.48\text{lbs}}{1\text{ft}^2} = 7.48\text{lbs}/\text{ft}^2$$

[a] Although the bar is a metric unit and used for pressure measurement in some locations, it is not approved as part of the International (SI) System of Units, probably because it was developed in Norway and not France.

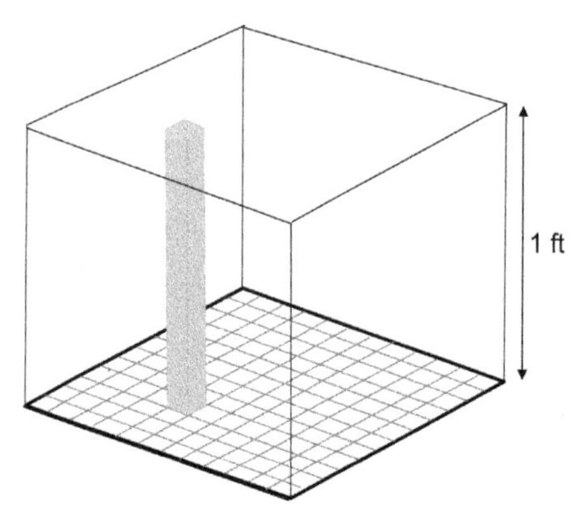

Figure 1.11 The pressure exerted by 1 ft of liquid over an area of 1 in.^{2}

Converting the area of the cube's base from ft^2 to in^2, the weight acting on each square inch is:

$$\frac{7.48}{144} = 0.051944 \rightarrow 0.052$$

The gradient of a 1 ppg fluid is therefore 0.052 psi/ft (Fig. 1.11).

It can also be derived by dimensional analyses. One US gallon contains 231 in.^{3}.

$$1\frac{\text{psi}}{\text{ft}} = \frac{1\text{ft}}{12\text{in}} \times \frac{1\frac{\text{lb}^2}{\text{in}}}{\text{psi}} \times \frac{231^3}{1 \text{ US gallons}} = 19.250000 \text{ lbs/gallon}$$

$$1/19.25 = 0.0519 \rightarrow 0.052.$$

Although it is more accurate to divide fluid weight (ppg) by 19.25 to obtain the gradient, this is rarely done. The factor 0.052 is widely used, and the conversion factor is listed in all well control manuals and text books. The magnitude of the error when using 0.052 is approximately 0.1%.

SI units: To convert density to a pressure gradient:

$$\text{Pressure gradient (kPa}/m) = \frac{\text{Fluid density in kg}/m^3}{102} \tag{1.6}$$

$$\text{Pressure gradient (kPa}/m) = \text{Kg}/m^3 \times 0.00981$$

1.1.6.1.1 Metric (bar/m)

$$\text{Pressure gradient (bar}/m) = \text{Fluid density kg}/L \times 0.0981$$

$$\text{Pressure gradient (bar}/m) = \text{Fluid density Kg}/m^3 \times 0.0000981$$

Example: What is the gradient of a 10 pgg (1.20 SG) brine.

$$\begin{array}{ll} \text{Oilfield}: & 10 \times 0.052 = 0.52 \text{ psi}/\text{ft} \\ \text{SI} & 1200/102 \ = 11.76 \text{ kPa}/m \ or \ 1200 \ x \ 0.00981 \ = \ 11.772 \end{array}$$

Once the fluid gradient is known, hydrostatic pressure at any point in the well can be determined.

$$P_{\text{hy}} = \text{Gradient} \times \text{TVD} \tag{1.7}$$

Where P_{hy} = hydrostatic pressure (psi or kPa or bar); TVD = True vertical depth (feet or meters).

Example (oilfield units): What is the hydrostatic pressure at 11,500 ft TVD in a well filled with 10.5 ppg brine?

$$10.5 \times 0.052 \times 11,500 = 6279 \text{ psi}.$$

Example (SI): What is the hydrostatic pressure at 3500 m TVD in a well filled with 1250 kg/m^3 brine?

$$(1250/102) \times 3500 = 42,892 \text{ kPa}$$

When calculating hydrostatic pressure, the vertical depth TVD must be used. Measured depth must be used for calculating volume and capacity.

1.1.6.1.2 Crude oil density

Crude oil density is commonly expressed as API gravity (°API). To convert API gravity to SG:

$$\text{SG} = \frac{141.5}{\text{Deg API} + 131.5} \tag{1.8}$$

To convert SG to oilfield units:

$$\text{SG} \times 8.33 = \text{ppg (8.33 is fresh water density in ppg)}$$

$$\text{SG} \times 0.433 = \text{psi/ft (0.433 is the gradient of fresh water)}$$

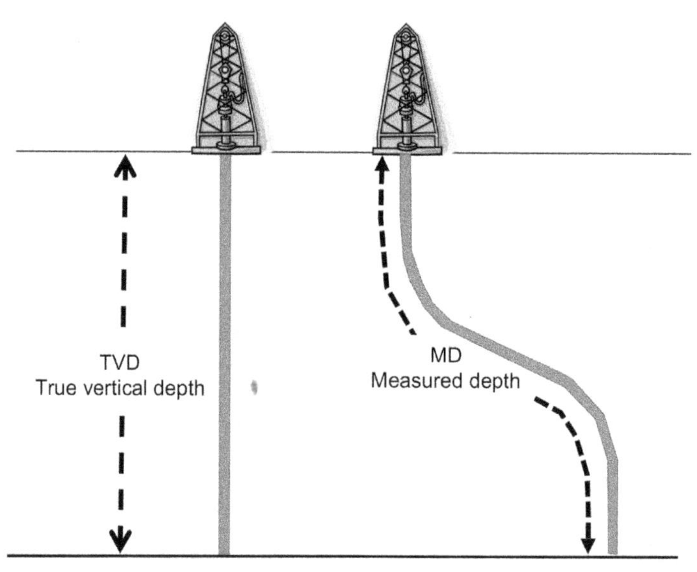

Figure 1.12 True vertical depth (TVD) and Measured Depth (MD).

Example: Determine the fluid density of 36° API gravity oil:

$$SG = \frac{141.5}{36 + 131.5} = 0.8447\ (SG)$$

Density in ppg = 0.8447 × 8.33 = 7.04 ppg.
Fluid gradient (psi/ft) = 0.8447 × 0.433 = 0.3657 psi/ft.
Fluid gradient (kPa/m) = 844.7/102 = 8.28 kPa/m

1.1.6.1.3 True vertical depth and measured depth

Unless a well is drilled with absolutely no deviation from the vertical, measured depth will be greater than the vertical depth (Fig. 1.12).

1.1.6.2 Calculating bottom-hole pressure

When a wellbore is filled with a single fluid of the same density, calculating the hydrostatic pressure is straightforward. Where a wellbore contains fluids of different densities, the hydrostatic pressure of each fluid must be calculated, and the result for all the fluids added. The depth of each fluid interface is required to accurately determine bottom-hole pressure.

Example: Oilfield units:

Consider the well shown here.

Full column of (mixed) fluids to surface

Pressure at surface 0 psi

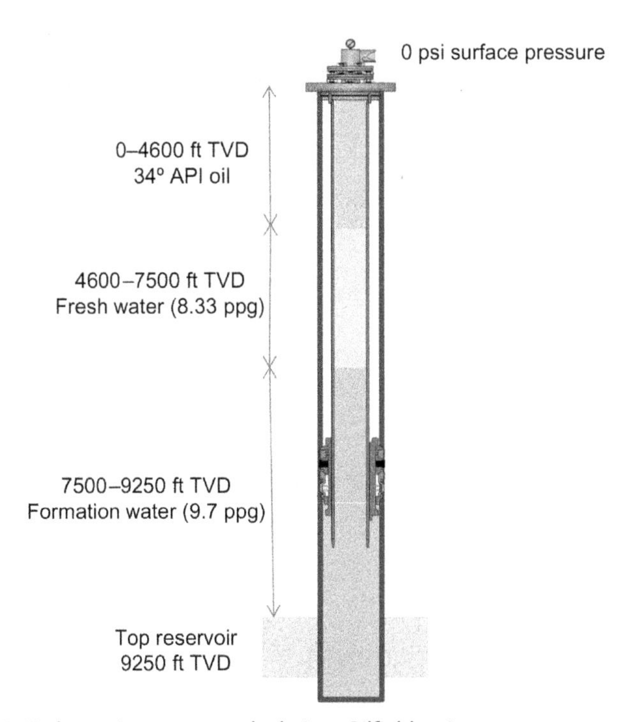

Figure 1.13 Hydrostatic pressure calculation. Oilfield units.

34° API oil from surface to 4600 ft TVD

Fresh water from 4600 to 7500 ft TVD (8.33 ppg)

Formation water from 7500 to 9250 ft TVD; top of reservoir (9.7 ppg) (Fig. 1.13).

Calculate the hydrostatic pressure at the top of the reservoir.

1. Calculate the hydrostatic pressure of the oil column:

$$SG = \frac{141.5}{34 + 131.5} = 0.8549$$

Density in ppg $= 0.8549 \times 8.33 = 7.12$ ppg $\times 0.052 \times 4600 =$ **1704 psi**

2. Calculate the hydrostatic pressure of the fresh water column.

$$(7500 - 4600) \times 8.33 \times 0.052 = \mathbf{1256\ psi}$$

3. Calculate the hydrostatic pressure of the formation water column.

$$(9250 - 7500) \times 9.7 \times 0.052 = \mathbf{883\ psi}$$

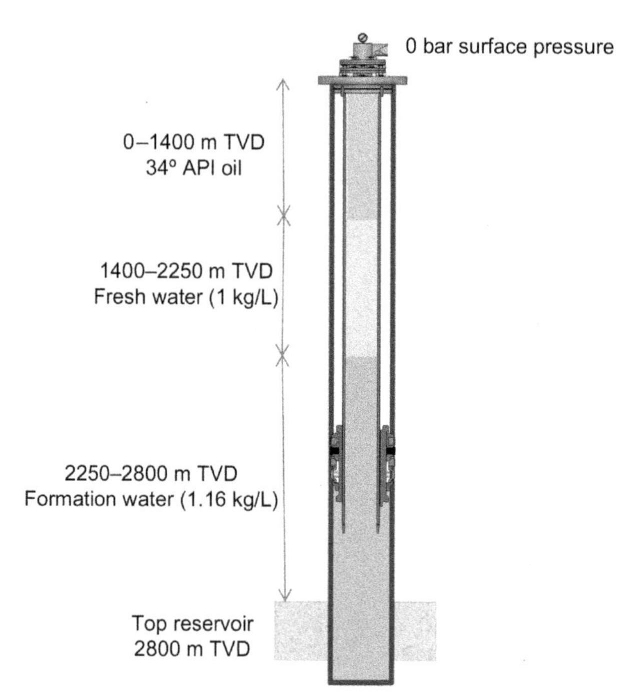

Figure 1.14 Hydrostatic pressure calculation. SI units.

4. Total hydrostatic pressure at the top of the reservoir.

$$1704 + 1256 + 883 = \mathbf{3843\ psi}.$$

Example SI units:
Full column of (mixed) fluids to surface
Pressure at surface 0 psi
34° API oil from surface to 1400 m TVD
Fresh water from 1400 m TVD to 2250 ft TVD 1000 kg/m^3
Formation water from 2250 to 2800 m TVD. 1160 kg/m^3 (Fig. 1.14).

1. Calculate the hydrostatic pressure of the oil column:

$$SG = \frac{141.5}{34 + 131.5} = 0.8549$$

2. Calculate the hydrostatic pressure of the oil column:

$$(854.9/102) \times 1400\ \text{m} = \mathbf{11{,}734\ kPa}$$

3. Calculate the hydrostatic pressure of the fresh water column.

$$(2250 - 1400) \times (1000/102) = \mathbf{8333\ kPa}$$

4. Calculate the hydrostatic pressure of the formation water column.

$$(2800 - 2250) \times (1160/102) = \mathbf{6254\ kPa}$$
$$11,734 + 8333 + 6254 = \mathbf{26,321\ kPa}$$

1.1.6.2.1 Surface pressure

Intervention work is frequently carried out on wells with pressure to surface. Surface pressure (Shut in Tubing Pressure, or SITP[b]) must be included when calculating bottom-hole pressure (BHP). Add SITP to the hydrostatic pressure.

1.1.6.3 Gas hydrostatic pressure

Gas hydrostatic pressure is calculated in one of three ways:

1. Use of a gas correction factor (from tables).
2. By calculation using formula.
3. If the gas gradient is known, by multiplying by the depth (TVD).

In dry gas wells, the gas column will reach from the surface to the reservoir. In liquid-producing wells, the depth of the gas/liquid contact is needed. Some of the pressure calculations require additional information that may or may not be available at the well site; average wellbore temperature and the compressibility (Z factor) of the gas.

Method 1: Using gas correction tables (Tables 1.5 and 1.6)

To calculate gas hydrostatic pressure:

- Find the appropriate gas gravity in the row along the top of the table.
- Find the well depth (TVD) in left hand column.
- Use the correction factor where the depth row intersects the gas gravity column.
- Correction factor × SITP = gas pressure at the required depth. *The result includes both surface pressure and gas hydrostatic pressure.*

The gas correction factors in the tables were calculated using the formula below:

$$\text{Correction factor} = e^{\left(\frac{SG \times D}{28812.47}\right)} \tag{1.9}$$

[b] The abbreviation CITP' (closed in tubing pressure) is sometimes used instead of SITP. Also used are the abbreviations SITHP (shut in tubing head pressure) and CITHP, closed in tubing head pressure.

Table 1.5 Gas correction factors (oilfield units)

Depth (ft)	Gas gravity—Relative to air										
	0.5	0.55	0.6	0.65	0.7	0.75	0.8	0.85	0.9	0.95	1
1000	1.018	1.019	1.021	1.023	1.025	1.026	1.028	1.030	1.032	1.034	1.035
2000	1.035	1.039	1.043	1.046	1.050	1.053	1.057	1.061	1.064	1.068	1.072
3000	1.053	1.059	1.064	1.070	1.076	1.081	1.087	1.093	1.098	1.104	1.110
4000	1.072	1.079	1.087	1.094	1.102	1.110	1.117	1.125	1.133	1.141	1.149
5000	1.091	1.100	1.110	1.119	1.129	1.139	1.149	1.159	1.169	1.179	1.190
6000	1.110	1.121	1.133	1.145	1.157	1.169	1.181	1.194	1.206	1.219	1.232
7000	1.129	1.143	1.157	1.171	1.185	1.200	1.215	1.229	1.244	1.260	1.275
8000	1.149	1.165	1.181	1.198	1.215	1.232	1.249	1.266	1.284	1.302	1.320
9000	1.169	1.187	1.206	1.225	1.244	1.264	1.284	1.304	1.325	1.345	1.367
10,000	1.190	1.210	1.232	1.253	1.275	1.297	1.320	1.343	1.367	1.391	1.415
11,000	1.210	1.234	1.257	1.282	1.306	1.332	1.357	1.383	1.410	1.437	1.465
12,000	1.232	1.257	1.284	1.311	1.338	1.367	1.395	1.425	1.455	1.485	1.517
13,000	1.253	1.282	1.311	1.341	1.371	1.403	1.435	1.467	1.501	1.535	1.570
14,000	1.275	1.306	1.338	1.371	1.405	1.440	1.475	1.511	1.549	1.587	1.626
15,000	1.297	1.332	1.367	1.403	1.440	1.478	1.517	1.557	1.598	1.640	1.683
16,000	1.320	1.357	1.395	1.435	1.475	1.517	1.559	1.603	1.648	1.695	1.742
17,000	1.343	1.383	1.425	1.467	1.511	1.557	1.603	1.651	1.701	1.752	1.804
18,000	1.367	1.410	1.455	1.501	1.549	1.598	1.648	1.701	1.755	1.810	1.868
19,000	1.391	1.437	1.485	1.535	1.587	1.640	1.695	1.752	1.810	1.871	1.934
20,000	1.415	1.465	1.517	1.570	1.626	1.683	1.742	1.804	1.868	1.934	2.002

Table 1.6 Gas correction factors (SI units)

Depth (m)	Gas gravity—relative to air										
	0.5	**0.55**	**0.6**	**0.65**	**0.7**	**0.75**	**0.8**	**0.85**	**0.9**	**0.95**	**1**
500	1.029	1.032	1.035	1.038	1.041	1.044	1.047	1.050	1.053	1.056	1.059
1000	1.059	1.065	1.071	1.077	1.083	1.089	1.096	1.102	1.108	1.114	1.121
1500	1.089	1.099	1.108	1.118	1.127	1.137	1.147	1.157	1.167	1.177	1.187
2000	1.121	1.134	1.147	1.160	1.173	1.187	1.200	1.214	1.228	1.242	1.256
2500	1.153	1.170	1.187	1.204	1.221	1.239	1.256	1.274	1.293	1.311	1.330
3000	1.187	1.207	1.228	1.249	1.271	1.293	1.315	1.338	1.361	1.384	1.408
3500	1.221	1.246	1.271	1.296	1.322	1.349	1.376	1.404	1.432	1.461	1.491
4000	1.256	1.285	1.315	1.345	1.376	1.408	1.441	1.474	1.508	1.543	1.578
4500	1.293	1.326	1.361	1.396	1.432	1.470	1.508	1.547	1.587	1.629	1.671
5000	1.330	1.369	1.408	1.449	1.491	1.534	1.578	1.624	1.671	1.719	1.769
5500	1.369	1.412	1.457	1.504	1.552	1.601	1.652	1.705	1.759	1.815	1.873
6000	1.408	1.457	1.508	1.560	1.615	1.671	1.729	1.789	1.852	1.916	1.983
6500	1.449	1.504	1.560	1.619	1.681	1.744	1.810	1.878	1.949	2.023	2.099
7000	1.491	1.552	1.615	1.681	1.749	1.820	1.894	1.972	2.052	2.135	2.222

Where SG = density of gas; D = depth (feet); 28812.47 is a product of multiplying together; 53.34 (constant); 558.6 degree Rankin; 0.967–Z factor (compressibility).

For SI units the following values were used in the formula:

$$\text{Correction factor} = e^{\frac{SG \times D}{8765.275}} \tag{1.10}$$

For SI units the following values were used in the formula:

Where SG = density of gas; D = depth (m); 8765.275 is a product of multiplying together; 29.24 (constant); 310°Kelvin; 0.967–Z factor (compressibility).

Alternative formulas for gas pressure at depth are:

$$\text{BHP} = \text{SITP} + (2.5)\left(\frac{\text{SITHP}}{100}\right)\left(\frac{\text{Depth}}{1000}\right) \tag{1.11}$$

This is a "drillers" estimation that includes an approximate 30 psi overbalance. It underestimates pressure in deep wells with high density gas.

An alternative (more accurate) equation is:

$$\frac{\text{SITP}}{\text{BHP}} = \frac{1}{2.718^{0.000034 \times \text{SG} \times \text{Depth}}} \tag{1.12}$$

If the gas SG, the gas compressibility (Z) factor, and the average wellbore temperature are known, a more accurate determination of bottom-hole pressure can be calculated from:

$$\text{SITP} \times e^{\frac{0.01877 \times SG \times D}{Z \times T}} \tag{1.13}$$

Where SG = gas specific gravity (dimensionless); D = depth in feet; Z = gas correction factor (compressibility); T = temp in °Rankin (°F + 460); e = exponential (approximately 2.718).

Example: Calculate the bottom-hole pressure in a gas well using the following data:

SITHP:	2500 psi
Well depth:	9000 ft TVD
Gas gravity:	0.80
Average temperature	190°F
Gas Z factor	0.96

Method 1: Using Table 1.5:

Select the correction factor at the intersection of depth (left hand column) and gas SG (top row). Multiply the chosen correction factor by the SITP:

$$1.284 \times 2500 = 3210 \text{ psi.}$$

Method 2: Using the equation $\text{BHP} = \text{SITP} + (2.5)\left(\frac{\text{SITP}}{100}\right)\left(\frac{\text{Depth}}{1000}\right)$

$$\text{BHP} = 2500 + (2.5)\left(\frac{2500}{100}\right)\left(\frac{9000}{1000}\right)$$

$$\text{BHP} = 2500 + (2.5) \times (25) \times (9) = 3062.5 \text{ psi}$$

Using the equation $\text{Ratio}\ \frac{\text{SITP}}{\text{BHP}} = \frac{1}{2.718^{0.000034 \times \text{SG} \times \text{TVD}}}$

$$\frac{1}{2.718^{0.000034 \times 0.8 \times 9000}}$$

$$\frac{1}{2.718^{0.2448}} = 0.7829$$

$$2500/0.7829 = 3193.33 \text{ psi.}$$

Using the equation $\text{SITP} \times e^{\left(\frac{0.01877 \times \text{SG} \times \text{TVD}}{Z \times T}\right)}$

$$\left(\frac{0.01877 \times 0.8 \times 9000}{0.96 \times (190 + 460)}\right) = 0.21657$$

$$2500 \times e^{0.21657} = 3104.5 \text{ psi.}$$

1.1.7 Underbalance and overbalance pressure

If hydrostatic pressure in a well is *higher* than the reservoir pressure, the difference is called *overbalance* pressure, or simply overbalance. Conversely, if reservoir pressure is more than hydrostatic pressure, the difference is called *underbalance*.

Monitoring and controlling overbalance and underbalance pressure is a critical element of well control. Most completion and workover operations are carried out with overbalanced (kill) fluid in the well. Kill weight fluid is normally designed to provide 200–300 psi (1379 kPa) overbalance. Insufficient overbalance will result in a kick. On the other hand, too much overbalance can result in fluid loss to the formation, resulting in a loss of overbalance and a kick.

Most interventions are carried out on underbalanced (live) wells, where surface equipment, such as a wireline lubricator, is used to contain the pressure.

1.1.8 Tubing and casing volume and capacity

Tubing, casing, and annular volume can be looked up in tables or calculated. To calculate pipe capacity all that is needed is the ID. To calculate the annulus capacity, the pipe OD and casing ID is needed. Capacity × depth gives the volume of a tubing string or annulus.

1.1.8.1 Using tables

A simple way of calculating tubing, casing, or annular volume is to use one of the many readily available paper or electronic tables, such as the Baker "Tech Facts" book, Halliburton's "Red Book",[c] and the Schlumberger digital i-handbook.

Tables provide the capacity of industry standard sizes for open hole, drill-pipe, tubing, casing, and annulus. For example, the widely used "Baker Tech Facts" book displays casing data using four different measures of capacity. When working with oilfield units, the most frequently used measure of capacity is barrels per linear foot (bbl/ft). Linear feet per barrel is the reciprocal $\frac{1}{x}$ of barrels per linear foot, for example for 5 ½″ 17 lb/ft tubing: 1/0.02324 = 43.01 bbls.

Example: Using Table 1.7, find the volume of fluid needed to fill 5000 ft of 5 ½″ 17 lb/ft tubing.

$$5000 \text{ ft} \times 0.9764 = 4882 \text{ US gallons}$$
$$5000 \text{ ft} \times 0.1305 = 652.5 \text{ ft}^3$$
$$5000 \text{ ft} \times 0.02324 = 116.2 \text{ bbls.}$$

[c] Halliburton's Red Book is available as an App. Download from their website or from iTunes.

Table 1.7 Casing capacity and dimensions

OD (in.)	Weight (lb/ft)	ID (in.)	Capacity					Weight (kg/m)	ID (mm)	Wall Thickness *(mm)*	OD *(mm)*
			Gallons per Lineal Foot	Cubic Feet per Lineal Foot	Barrels per Lineal Foot	Lineal Feet per Barrel	Liters per Meter				
	15.00	4.974	1.0094	0.1349	0.02403	41.609	*12,538*	*22,320*	*126.3*	*6.68*	*139.7*
	15.50	4.950	0.9997	0.1336	0.02380	42.013	*12,417*	*23,064*	*125.7*	*6.98*	*139.7*
5½	17.00	4.892	0.9764	0.1305	0.02324	43.015	*12,128*	*25,296*	*124.2*	*7.72*	*139.7*
	20.00	4.778	0.9314	0.1245	0.02217	45.093	*11,569*	*29,760*	*121.3*	*9.16*	*139.7*
	23.00	4.670	0.8898	0.1189	0.02118	47.202	*11,052*	*34,224*	*118.6*	*10.54*	*139.7*
	26.00	4.548	0.8439	0.1128	0.02009	49.769	*10,482*	*38,688*	*115.5*	*12.09*	*139.7*
	14.00	5.290	1.1417	0.1526	0.02718	36.786	*14,182*	*20,832*	*134.3*	*5.84*	*146. 0*
	17.00	5.190	1.0989	0.1469	0.02616	38.217	*13,651*	*25,296*	*131.8*	*7.11*	*146. 0*
5¾	19.50	5.090	1.0570	0.1413	0.02516	39.734	*13,130*	*29,016*	*129.2*	*8.38*	*146. 0*
	22.50	4.990	1.0159	0.1358	0.02418	41.342	*12,619*	*33,480*	*126. 7*	*9.65*	*146. 0*
	25.20	4.890	0.9756	0.1304	0.02322	43.051	*12,118*	*37,497*	*124. 2*	*10.92*	*146. 0*
	15.00	5.524	1.2449	0.1664	0.02964	33.736	*15,464*	*22,320*	*140. 3*	*6.04*	*152. 4*
	16.00	5.500	1.2342	0.1649	0.02938	34.031	*15,330*	*23,808*	*139.7*	*6.35*	*152.4*
	17.00	5.450	1.2118	0.1619	0.02885	34.658	*15,053*	*25,296*	*138.4*	*6.98*	*152.4*
6	18.00	5.424	1.2003	0.1604	0.02857	34.991	*14,909*	*26,784*	*137.7*	*7.31*	*152.4*
	20.00	5.352	1.1686	0.1562	0.02782	35.939	*14,516*	*29,760*	*135.9*	*8.22*	*152.4*
	23.00	5.240	1.1202	0.1497	0.02667	37.492	*13,915*	*34,224*	*133.0*	*9.65*	*152.4*
	26.00	5.140	1.0779	0.1440	0.02566	38.965	*13,389*	*38,688*	*130.5*	*10.92*	*152.4*

Source: From Baker Tech Facts.

When calculating capacity remember to use the measured depth of the well

Capacity from formula:

If tables are not available, pipe and annular capacity can be obtained using formula providing the pipe dimensions are known (ID and OD). For oilfield units, tubing capacity in cubic inches is calculated from:

$$1\ \text{ft}^3 = 1728\ \text{in.}^3$$
$$1\ \text{bbl} = 5.6146\ \text{ft}^3$$
$$1\ \text{bbl} = 9702\ \text{in.}^3$$

Tubing capacity can therefore be expressed as: $\frac{\pi \times \text{ID}^2 \times 12}{4 \times 9702} = \frac{\text{ID}^2}{1029.4}\ \text{bbls/ft}$

$$\text{Tubing or casing capacity}(\text{bbls/ft}) = \frac{\mathbf{ID^2}}{\mathbf{1029.4}} \tag{1.14}$$

$$\text{Tubing or casing capacity}(\text{gallons/ft}) = \frac{\mathbf{ID^2}}{\mathbf{24.51}} \tag{1.15}$$

$$\text{Tubing or casing capacity}(\text{ft}^3/\text{ft}) = \frac{\mathbf{ID^2}}{\mathbf{183.35}} \tag{1.16}$$

Where ID is in inches.

Annular capacity between casing (or open hole) and tubing (or drill-pipe).

$$\text{Annular capacity in bbls/ft} = \frac{\text{Casing ID}^2 - \text{Tubing OD}^2}{1029.4} \tag{1.17}$$

$$\text{Annular capacity in ft/bbl} = \frac{1029.4}{\text{Casing ID}^2 - \text{Tubing OD}^2} \tag{1.18}$$

$$\text{Annular capacity in gallons/ft} = \frac{\text{Casing ID}^2 - \text{Tubing OD}^2}{24.51} \tag{1.19}$$

$$\text{Annular capacity in ft/gallons} = \frac{24.51}{\text{Casing ID}^2 - \text{Tubing OD}^2} \tag{1.20}$$

Annular capacity between casing and multiple strings of tubing, i.e., dual or triple string completions.

$$\text{Annular capacity in bbls/ft} = \frac{\text{Casing ID}^2 - [\text{Tubing1 OD}^2] - [\text{Tubing2 OD}^2]}{1029.4} \tag{1.21}$$

$$\text{Annular capacity in ft/bbl} = \frac{1029.4}{\text{Casing ID}^2 - [\text{Tubing1 OD}^2] - [\text{Tubing2 OD}^2]} \quad (1.22)$$

Note: Since the denominator for calculating capacity in bbl/ft (1029.4) is close in value to 1000, it follows that pipe ID^2 *gives an approximate value for bbls per 1000 ft. This is useful as a rough approximation for tubing capacity if a calculator is not available.*

For example, 5½" 17 lb/ft tubing has an ID of 4.892. Round up to 5" to simplify the calculation: 5 × 5 = 25 bbls per 1000 ft of tubing.

Calculating the figure from tables, the answer is 1000 × 0.02324 = 23.24 bbls.

Mixed units: (diameter in inches—capacity in L/m).

$$\text{Tubing or casing capacity liters per meter}, L/\text{m} = \frac{\text{ID}^2(\text{in.})}{1.974} \quad (1.23)$$

$$\text{Annulus Capacity: L/m} \frac{\text{Casing ID}^2 - [\text{Tubing1 OD}^2] - [\text{Tubing2 OD}^2]}{1.974} \quad (1.24)$$

SI units: where the ID is in mm.

$$\text{Tubing or casing capacity: m}^3/\text{m} = \frac{\text{ID}^2}{1273000} \quad (1.25)$$

$$\text{Tubing or casing capacity: L/m} = \frac{\text{ID}^2}{1273} \quad (1.26)$$

$$\text{Annular capacity in m}^3/\text{m} = \frac{\text{Casing ID}^2 - \text{Tubing OD}^2}{1,273,000} \quad (1.27)$$

$$\text{Annular capacity in L/m} = \frac{\text{Casing ID}^2 - \text{Tubing OD}^2}{1273} \quad (1.28)$$

Annular capacity between casing and multiple strings of tubing, i.e., dual or triple string completions.

$$\text{Annular capacity in m}^3/\text{m} = \frac{\text{Casing ID}^2 - [\text{Tubing1 OD}^2] - [\text{Tubing2 OD}^2]}{1273000} \quad (1.29)$$

$$\text{Annular capacity in L/m} = \frac{\text{Casing ID}^2 - [\text{Tubing1 OD}^2] - [\text{Tubing2 OD}^2]}{1273} \tag{1.30}$$

Example volume calculation

How much liquid is needed to fill up 5000 ft of 5 ½″ 17 lb/ft tubing, ID 4.892″.

$$\frac{4.892^2}{1029.4} = 0.02324 \quad 0.02324 \times 5000 = 116.24 \text{ bbls}$$

Calculate the annulus volume. Casing size = 9 5/8″ 47 lb/ft. ID = 8.681″. Dual string completion 3½″ OD long string tubing.

2 7/8″ OD short string tubing.

Dual packer at 7500 ft MD.

$$\frac{8.681^2 - (3.5^2 + 2.875^2)}{1029.4} = 0.05327 \text{ bbls/ft}$$

$$7500 \times 0.05327 = 399 \text{ bbls}$$

1.1.9 Gas hydrates

A live well perforating operation was being conducted on a North Sea platform. Wireline conveyed guns were used to open a new producing zone in an existing gas well. After successfully perforating the new zone, the spent gun was being pulled back towards surface. At the approximate depth of the seabed, the gun became stuck. It was suspected that a hydrate was preventing further progress. The wireline crew responded by pumping methanol into the tubing to disperse the hydrate. However, a supervisor became impatient. To speed up the removal of the hydrate, the platform production technicians were instructed to close the downhole safety valve (located below the gun), and vent the tubing into the production facilities. It was thought, by the supervisor, that reducingwell pressure would speed up the removal of the hydrate. Soon after the tubing pressure had been lowered, the wireline cable was observed to jump, followed by a complete loss of tension. The loss of weight coincided with a rapid increase in wellhead pressure. The cable was blown out of the well, and gas escaped through the grease head until the valves on the Christmas tree were closed.

The subsequent investigation concluded that this sequence of events had occurred.

After reducing pressure in the production tubing, high pressure remained trapped below the hydrate (above the closed safety valve). The high differential pressure across the hydrate caused it to release. It was propelled up the tubing, probably taking the perforating gun with it. As soon as pressure equalized, the gun and hydrate dropped back down the well. The falling gun parted the wireline, allowing the gun to drop onto the closed safety valve where the impact sheared the hinge pin in the flapper valve, damaging it beyond repair.

Following a fishing operation to recover the lost toolstring, the well had to be recompleted to replace the damaged safety valve.

Gas hydrates (or clathrates) are a solid structure of water and gas that closely resemble dirty ice or snow. Under the right conditions they can form rapidly, blocking production systems. Hydrates form when low molecular weight gases mix with water at relatively low temperature and high pressure. Hydrates are characterized by a rigid network of water molecules that cage in gas molecules of another substance. Gas stabilizes the crystalline structure of water (Fig. 1.15).

Since methane (CH_4) is the most abundant gas in hydrocarbon producing wells, nearly all hydrates are methane hydrates. Although having similarities with ice, if pressure is high enough, hydrates will remain stable well above the freezing point of water. For example, at 2000 psi, a typical natural gas hydrate can remain solid at up to 70°F.[4] Approximately 85%–90% (by weight) of a hydrate is water, with the remainder being gas. When a gas hydrate dissociates, a large volume of gas is released. One cubic foot of hydrate contains between 160–180 ft^3 of gas, and 0.8 ft^3 of water.

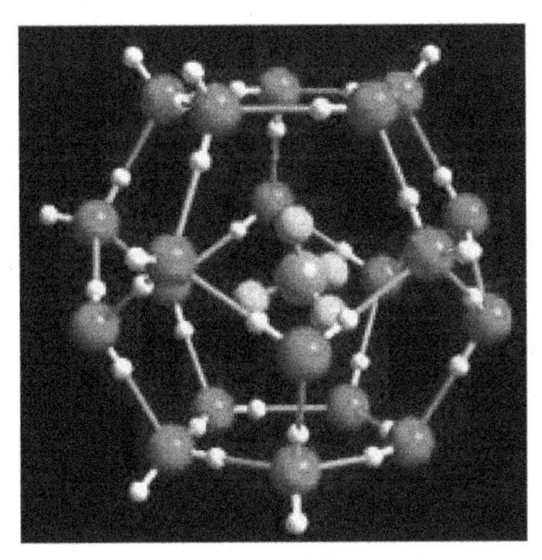

Figure 1.15 A methane hydrate. A methane molecule (in the center of the structure) is trapped inside a latticework of water molecules.

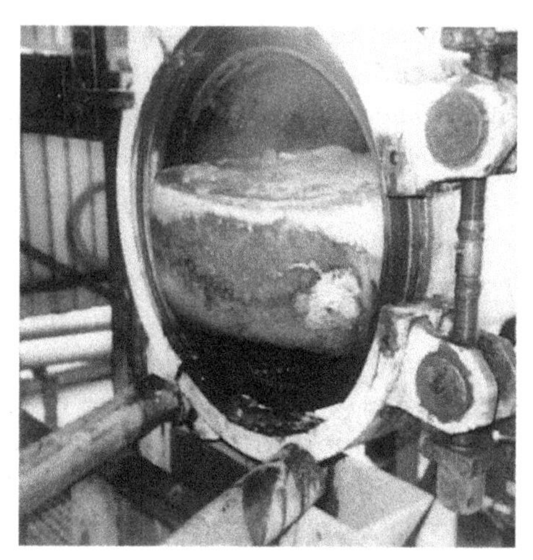

Figure 1.16 A hydrate in a "pig receiver" on an offshore platform.

Under specific conditions, hydrates form rapidly, blocking tubing, and flowlines, and trapping pressure. They can be extremely difficult to remove, and are a potential hazard unless handled correctly (Fig. 1.16).

Knowing where and when there is the potential for a hydrate to form is essential. If temperature is low, pressure high, and water and gas are present in the system, preventative measures will be needed to stop hydrates forming. If hydrates do form, they will need to be safely removed.

Hydrates form and exist for indefinite periods inside the hydrate stability zone. This can be plotted on a hydrate disassociation curve. Outside the zone hydrates do not form. However, if the pressure/temperature combination is above the stability curve (Fig. 1.17), the potential exists for hydrates to form if gas and water mix.

Predicting closed-in pressure will normally be based on the pressure record for a recent shut-in. Predicting temperature against depth can be more problematic, particularly if an intervention requires intermittent periods of flow, shut in, and possibly injection. Temperature modeling using proprietary software such as WellCAT is useful in this regard.

1.1.9.1 Hydrate risk during well interventions

During production, a combination of high temperature and relatively low pressure will normally prevent hydrates from forming. However, most interventions require the well to be closed in for some or all of the time. After shut-in, pressure increases and temperature decreases. If the post shut-in combination of temperature and pressure fall in the hydrate formation zone, the potential for water/gas contact must be evaluated.

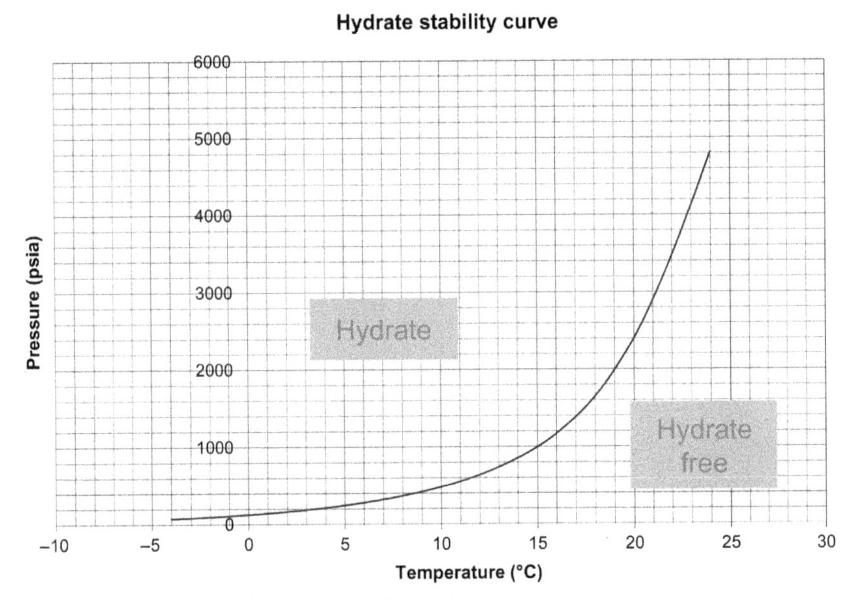

Figure 1.17 Hydrate stability curve—Example.

During some interventions, there are circumstances where the co-mingling of water and gas occurs. These include:

- *Integrity testing of pressure control equipment using water.* When performing interventions on a live well, it is standard practice to pressure test any surface pressure control equipment, such as a wireline lubricator or coiled tubing stripper. These tests are normally conducted using water. On completion of the test, the valves in the Christmas tree are opened to give access to the well. In most wells, including those producing liquids, there will be a gas cap at the surface. Water falling from the surface as the tree valves are opened will mix with the gas. A hydrate can form if the pressure and temperature are in the hydrate zone.
- *Production start-up of cold wells.* Wells that produce water are particularly vulnerable. Consider a situation where a well has been closed-in and water remains in the low-lying sections of the flowline. During the shut-in, gas accumulates below the closed tree valves. When the well is opened, gas comes into immediate contact with water in the flowline. Similarly, a newly perforated well can be vulnerable if it was filled with water when perforated. As production starts, gas migrating up through the water can lead to the formation of hydrates at the surface.

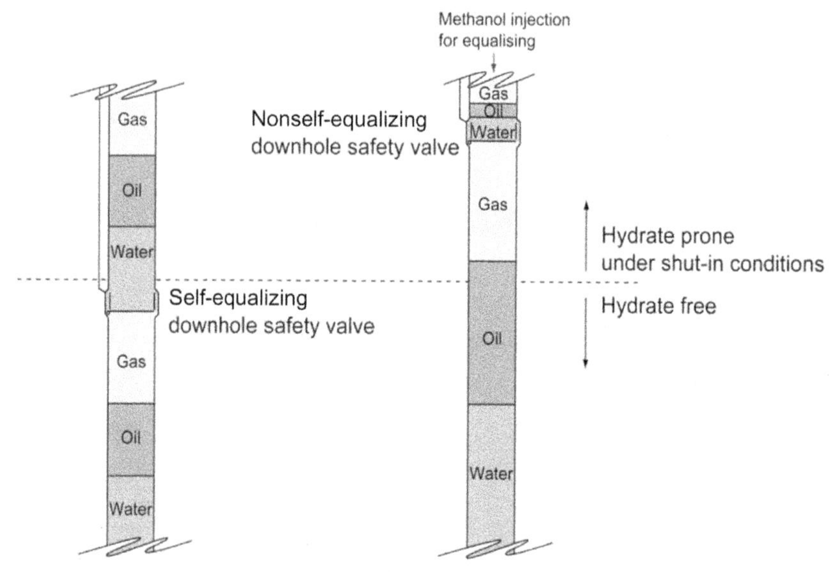

Figure 1.18 Hydrates and safety valves (Illustration courtesy of Jonathan Bellarby).

- *Equalization of surface controlled subsurface safety valves.* There is a risk of hydrate blockage when opening a SCSSSV, self-equalizing valves are particularly vulnerable. If pressure is bled off above a closed valve, phase separation will take place both above and below the closed valve. (Fig. 1.18). Applying control line pressure to a self-equalizing valve, opens the equalizing port. Gas from below the valve will immediately mix with water above, allowing a hydrate to form.

  To open a nonequalizing valve, fluid is pumped into the well above the closed valve until it equalizes with the pressure below. If water is used, the potential for a hydrate exists.
- *Completion and workover operations on dead wells.* When brine or seawater is used to maintain a hydrostatic overbalance, there is a risk of a hydrate if gas enters the wellbore. Gas, migrating through a water-based fluid, could result in the formation of a hydrate, particularly close to the surface, where temperature is low.

1.1.9.2 Hydrate prevention

A cliché perhaps, but the expression "prevention is better than cure" is particularly apt when applied to hydrates. Once formed, they can be difficult to remove. Hydrates can be prevented by keeping temperature high and pressure low. For example, nitrogen in the annulus, or even vacuum insulated tubing, can be used to insulate the tubing and retain more heat in the produced fluids. However, for many phases of a completion or intervention, the combination of high pressure and low temperature is unavoidable. If gas and water cannot be separated, then chemical inhibition is the only option.

Several chemicals are used to inhibit (and remove) hydrates. Those that remove or prevent the accumulation of ice also work well with hydrates. The general class of chemicals that are effective for de-icing and hydrate control are alcohols. Methanol (CH_3OH) is widely used and easily available. Its relatively low molecular weight enables it to permeate the hydrate, making it an effective inhibitor. However, there are some safety concerns:

- Toxicity.
- Low flash-point.
- It burns with a near invisible flame.
- It is a contaminant in oil as it interferes with catalysts in refineries, and adversely affects produced water treatment and discharge.
- It is a cause of stress corrosion cracking in titanium components such as heat exchangers.

Monoethylene glycol (C_2H_3OH) is also widely used. Although not as effective as methanol, since it has a higher molecular weight, it is still a useful inhibitor and is much less toxic than methanol. It is routinely used to prevent the formation of hydrates when pressure testing intervention equipment (Fig. 1.19).

Methanol or glycol mixed with water at varying concentrations lowers the hydrate free temperature at a given pressure (Fig.1.19). For example, at 3000 psi, hydrates can form in seawater at below 72°F. Mixing the seawater 50/50 with glycol lowers the temperature at which hydrates can form from 72°F to 35°F. Mixed with methanol, the temperature is reduced further still, to well below the freezing point of fresh water (32°F).

Glycol is routinely used to inhibit when testing well intervention pressure control equipment. There is anecdotal evidence to suggest that, if left for several hours, the heavier glycol (1.113 SG) separates out from the water and hydrate problems can occur. Some operators advocate the use of neat glycol for pressure testing. However, methanol and glycol are soluble in water, so providing they are properly mixed and agitated a problem should not occur.

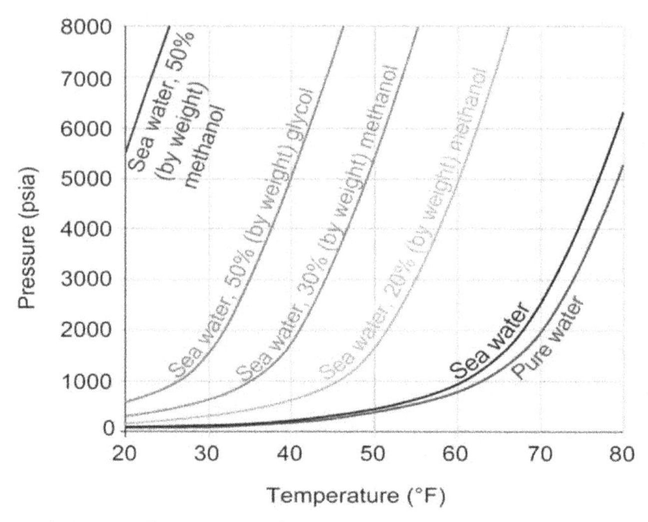

Figure 1.19 Inhibition effectiveness of methanol and glycol.

Some wells, subsea wells in particular, have chemical injection lines at the tree and/or downhole, specifically intended for methanol injection. Similarly, during well testing operations methanol is normally injected into the produced fluids upstream of the choke. This is particularly important immediately after start-up, when temperature is likely to be low and water-based fluid is being produced back to clean up the well.

1.1.9.3 Hydrate removal

Hydrate removal is normally accomplished by three methods. Lowering pressure, raising temperature, or chemical dispersal.

1.1.9.3.1 Chemical disassociation

If a hydrate forms in the well, the biggest problem is often finding a way to get chemical inhibitor to the hydrate. Methanol, although an excellent inhibitor, has a relatively low density at 0.79 SG (6.58 ppg). If pumped in at the surface, it is unlikely to reach the hydrate unless the liquids in the well are of a significantly lower density. Glycol, 1.113 SG (9.3 ppg), might be more successful in reaching the hydrate, but will take longer to work. It can take a long time to remove hydrates by pumping inhibitors from the surface, as the contact area is small (cross-sectional area of the tubing) and a hydrate can be large.

If it is not possible to reach the hydrate by pumping from the surface, coiled tubing can be used.

1.1.9.3.2 Pressure reduction

Where possible, pressure should be reduced on both sides of a hydrate, although this is not normally possible if hydrates have formed downhole. Bleeding off on one side of a hydrate is hazardous. If pressure differential across the hydrate is high, the hydrate can partially dissolve, detach from the tubing wall, and become a projectile that can travel at very high velocity. The hard, heavy projectile, propelled by gas from below can cause severe damage to surface equipment.

1.1.9.3.3 Elevate the temperature

If a hydrate is not completely blocking the tubing, flow the well to raise the temperature and remove the hydrate. Similarly, in dual string completions where one string is blocked, producing the adjacent string can raise the system temperature sufficiently to disperse any hydrates present.

Where a hydrate has formed at the surface, and ambient temperature is high, simply waiting for the system to warm up should be enough to disperse the hydrate. Heating the area where the hydrate has formed can be hazardous. Trapped pressure can increase dramatically as the hydrate dissociates; pipelines have been ruptured (Fig. 1.20).

Hydrates can form single or multiple plugs, and high pressure can be trapped between plugs.

1.1.10 Hydrogen sulfide (H_2S)

Hydrogen sulfide, sometimes called sour gas, occurs naturally in many hydrocarbon reservoirs. In some circumstances, "sweet" reservoirs become "sour," because of bacteriological action. The primary mechanism for reservoir souring is the reduction of the sulfates present in seawater and other completion brines to H_2S through the action of sulfate-reducing bacteria (SRB) under anaerobic (oxygen-free) conditions.

Hydrogen sulfide is both corrosive and highly toxic. It kills at even low concentrations. Hydrogen sulfide data is normally recorded in ppm

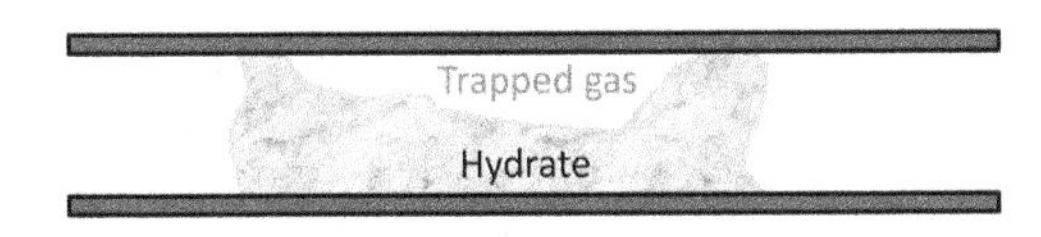

Figure 1.20 Pressure trapped by hydrates can increase as temperature rises.

(parts per million) rather than percent. Working on wells with hydrogen sulfide (H_2S) requires additional precautions to protect personnel and equipment.

The danger from H_2S to personnel depends on its concentration and the duration of exposure. Factors like wind strength, direction, and ventilation must also be taken into account. As H_2S is denser than air, it will accumulate in low-lying areas such as the wellhead cellar.

Human beings can tolerate very low concentrations of the gas, but with increasing concentration the risk increases (Table 1.8).

There are several recommendations regarding exposure limits for people working in the presence of H_2S.

The API base their recommendations on those provided by the Atlanta-based National Institute for Occupational Safety and Health (NIOSH), and state that personal protection should be provided if the work area concentration of H_2S exceeds 10 ppm 8-hour time weighted average (TWA), or 15 ppm as a Short Term Exposure Limit (STEL), averaged over 15 minutes. Personal protection is not required when the atmospheric concentration of hydrogen sulfide could not exceed 10 ppm in the breathing zone (API RP 49).[5]

In the United Kingdom, the Health and Safety Executive are rather more conservative, setting the workplace exposure limits (WELs) as 5 ppm for an 8-hour time weighted average (TWA), and 10 ppm for a 15-minute TWA.[6]

1.1.10.1 Hydrogen sulfide safety precautions

As H_2S is toxic, it is crucial that all necessary precautions are in place and implemented:

- Where possible, position personnel upwind of the H_2S source.
- Blowers can be used dissipate H_2S in sheltered locations, or on days when there is little or no wind.
- If H_2S well concentrations are above the prescribed threshold for protective equipment, self-contained breathing apparatus (SCBA) must be available at the worksite, and all personnel trained in its use. Normally this equipment is supplied by specialist vendors who will train well site personnel and ensure the equipment is properly maintained.
- Position H_2S detectors at any location where personnel would be at risk in the event of a gas escape. The positioning, monitoring, and maintenance of the H_2S detection system is often carried out by the vendor who supplies the breathing apparatus.

Table 1.8 Physiological effect of hydrogen sulfide (H_2S)

Concentration in air	Effect
<1 ppm	Odor of rotten eggs can be clearly detected
10 ppm	Unpleasant odor. Possible eye irritation
20 ppm	Burning sensation in eyes and irritation of the respiratory tract after 1 h or more exposure
50 ppm	Loss of sense of smell after about 15 or more minute's exposure. Exposure over one hour may lead to headache, dizziness, and/or staggering. Pulmonary edema reported following extended exposure to greater than 50 ppm. Exposure at 50 ppm or greater can cause serious eye irritation or damage
100 ppm	Coughing, eye irritation, loss of sense of smell after 3–15 min. Altered respiration, pain in eyes, and drowsiness after 15–20 min, followed by throat irritation after 1 h. Prolonged exposure results in a gradual increase in the severity of these symptoms
200 ppm	The sense of smell will be lost rapidly, and it will irritate the eyes and throat. Prolonged exposure (>20–30 min) may cause irreversible pulmonary edema, i.e., accumulation of fluid in the lungs
300 ppm	Marked conjunctivitis and respiratory tract irritation. Concentration considered immediately dangerous to life or health
500 ppm	Unconsciousness after short exposure, breathing will stop if not treated quickly. Dizziness, loss of sense of reasoning and balance. Victims need prompt artificial ventilation and/or cardiopulmonary resuscitation (CPR) techniques
700 ppm	Unconscious quickly. Breathing will stop and death will result if not rescued promptly. Artificial ventilation and/or CPR is needed immediately.
>1000 ppm	Unconsciousness at once. Permanent brain damage or death may result. Rescue promptly and apply artificial ventilation and/or CPR

- Train all personnel in cardiopulmonary resuscitation (CPR), and any other first aid techniques relevant to H_2S poisoning.

1.1.10.2 Hydrogen sulfide and equipment corrosion

There are two forms of H_2S corrosion. A pitting type corrosion and sulfide stress cracking (SSC). Stress cracking is a form of hydrogen embrittlement. Steels react with hydrogen sulfide, forming metal sulfides and atomic hydrogen as corrosion by-products. Atomic hydrogen either

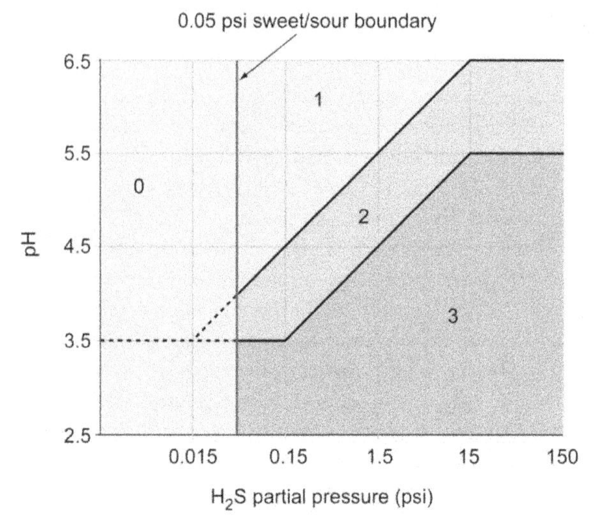

Figure 1.21 National Association of Corrosion Engineers (NACE) standard MR0175/ ISO 15156 limits for sour corrosion.

combines to form H_2 at the metal surface, or diffuses into the metal matrix. As the sulfides on the metal surface inhibit hydrogen recombination, the amount of atomic hydrogen that recombines to form H_2 on the surface is reduced, and more migrates into the metal. Once in the metal, and away from the sulfides, the H_2 recombines to form hydrogen molecules at dislocations and grain boundaries in the steel structure. Under conditions of low stress, blistering can occur below the surface of the steel. If the steel is stressed, it can crack and fail. High yield and inherently hard materials are more prone to stress cracking.

The most widely used definition of sour service is provided by the National Association of Corrosion Engineers (NACE) in standard MR0175. In the United States, this standard is legally enforceable, and equipment intended for sour service must comply with the service limitations (Fig. 1.21).

Sulfide stress cracking can occur at a very low concentration of H_2S, especially at high pressure. Completion engineers and intervention supervisors should be aware of the partial pressure of H_2S, and ensure that any pressure retaining equipment used in the completion, or for interventions is suitable for the intended service.

The boundary between "sweet" and "sour" service is defined by NACE as 0.05 psi (0.3 kPa) H_2S partial pressure.[d] Partial pressure can be defined as the pressure exerted by a single gas in a mixture of gasses, and is the pressure that would be exerted by a single gas if it alone occupied the volume originally occupied by the mixture. To establish the partial pressure of H_2S from mole % in the gas phase:

(Mole % H_2S ÷ 100) × bubble point pressure.

To convert ppm to percent:

$$1\% = 10,000 \text{ ppm therefore } \%H_2S = \frac{\text{ppm } H_2S}{10,000}$$

Alternatively, to establish partial pressure directly from ppm

Partial pressure of H_2S = (PPM H_2S ÷ 1,000,000) × bubble point pressure.

1.1.10.3 Carbon dioxide (sweet) corrosion

Carbon dioxide (CO_2) or "sweet" corrosion is far more common than H_2S corrosion in most producing wells. Carbon dioxide attacks metal due to the acidic nature of dissolved carbon dioxide (carbonic acid). The acidity (pH) is largely dependent on the partial pressure of CO_2 in the produced fluids. To calculate the partial pressure of CO_2 (Mole % CO_2 ÷ 100) x bubble point pressure.

Chromium is added to steel to reduce CO_2 corrosion, with 13% chrome tubing being widely used for completions. Pressure control equipment that is only exposed to static well fluids is not normally constructed using corrosion resistant alloys, since CO_2 corrosion rates are negligible when the produced fluid is static.

1.1.11 Roles and responsibilities

Completion and Well Intervention Operations often involve many individuals from a range of disciplines, all coming together to perform a task that is complex and potentially hazardous. Each individual must have the necessary training and skills to enable them to perform their assigned task safely and efficiently. Crucially, they must have the competence, experience, and confidence to know what to do in the event of an emergency.

[d] Petroleum and natural gas industries. Materials for use in H_2S-containing Environments in oil and gas production. Part 1: General principles for selection of cracking-resistant Materials. NACE MR0175/ISO 15156-1:2001 ISBN 1-57590-1765, 2001.

1.1.11.1 Person in charge of well control

Before work starts, the person in charge of well control must be clearly identified. Where a drilling rig is used to carry out completion or workover operation, or where it is being used in support of an intervention, the person in charge of well control is normally the operating company representative at the well site. Where well intervention operations are carried out without rig support, the person in charge of well control is normally an operating company supervisor. In some circumstances, the senior specialist from the intervention service provider will be the person in charge of well control. For example, slickline crews working in a remote location are often expected to work without the direct on-site supervision of a client company representative. The slickline operator is the person in charge of well control at the worksite. The person in charge is not necessarily the person who performs the vital actions needed to make a well safe when a well control incident occurs. For example, if a well begins to flow whilst completion tubing is being run, it would normally be up to the driller to shut it and make it safe.

Whoever is in charge of well control must be clear about their responsibilities.

1.1.11.1.1 Responsibilities during rig supported completion and workover operations

Well Control Prevention.

- Seek assurance that all personnel are competent to perform their assigned tasks and have the requisite well control training/certification.
- Examine the work program and any accompanying risk assessment documentation. Identify any well control risks and concerns. Ensure that any identified risks and the appropriate mitigations are communicated to the crew.
- Ensure members of the crew know how to respond to well control emergencies.
- Ensure members of the crew have knowledge of location specific emergency plans, i.e., the correct response to fire, explosions, spillage, etc.
- Ensure all the necessary documentation is easily accessible, easy to use, complete, accurate, and up-to-date. This can include information relating to well control equipment status, well status, kick sheets, shut in instructions, and check lists. Consider compiling vital information into a work-pack.[7]

- Ensure that the unit system (oilfield or SI) used during the operation is the same as that used in the completion or intervention program.
- Ensure regular scenario-based well control drills are carried out. Where appropriate report on crew performance.
- Ensure that there is an adequate supply of lost circulation material and kill weight brine on location.
- For work on wells where H_2S is present or expected, confirm that there is sufficient breathing apparatus and monitoring equipment at the well site and all personnel have been adequately trained in its use.

1.1.11.1.2 Responding to a well control incident

In the event of a well control incident on a well where the primary barrier is a column of kill weight fluid, and that barrier has been compromised, the person in charge of well control will:

- Confirm that the crew on the drill floor have made the well safe.
- Gather together as much information as possible.
- Convene a meeting with all of the key personnel who will be involved with the well control (kill) operation.
- Be present at the drill floor at the beginning of the well control (kill) operation.
  - Most operating companies will insist on having their own representative, or the senior representative of the drilling contractor (the Toolpusher) present on the drill-floor until the well control situation has been resolved.
- Maintain communications with the operating company base office.
- Assign responsibility for keeping a record of events.

Produce reports as necessary in accordance with company and jurisdictional requirements. In addition to these responsibilities, the person in charge of the well control incident will ensure that the other key members of the team are carrying out their assigned duties.

Drilling contractor's Toolpusher (if not the person in charge):

- Overall responsibility for implementing the well control operations.
- Ensure that the driller and the drill crew are properly deployed.
- Should be present on the rig floor when the well control operations begin.
- Ensure that handover between shifts is accomplished with the minimum disruption to the operation—ensure continuity.

Driller:

- Responsible for detecting kicks or losses and making the well safe.

- Must immediately notify the client company representative and contact drilling company Toolpusher.
- Implement the well kill or contingency procedure.
- Direct supervision of the drill crew during the well control operations.

Mud engineer (if present) or assistant driller:

- Responsible for the fluid system—preparing LCM, weighting up or cutting brine weight as required.

Mud logger:

It is unlikely that mud loggers would be on location during a completion or workover operation. However, if they were present they can assist with:

- Share responsibility with driller for early kick detection.
- Monitor and record parameters during well control operation including:
  - Times
  - Shut in pressure
  - Fluid loss rate—cumulative losses
  - Pump strokes (cumulative and SPM)
  - Pump pressure
  - Casing pressure
  - Gas
  - H_2S
  - Pit volume.

If no mud logger is present, the person in charge of the well control operation will have to allocate these tasks to other responsible personnel.

Drilling contractor Subsea Engineer (where relevant):

- Available for consultation at all times during the well control operation.
- Responsible for the supervision of the subsea BOP system operation.

1.1.11.2 Responsibilities during Interventions on live wells (independent of rig support)

During intervention operations on live wells, the person in charge of well control will normally be a well intervention (well service) supervisor. Some operations, especially in remote locations, are supervised by the senior member of the intervention vendor crew.

1.1.11.2.1 Prejob checks

The supervisor with responsibility for well control will:

- Ensure that any pressure control equipment is in good condition, of the correct pressure rating and service type, and that all necessary test certificates are in date.
- Ensure that any diesel operated power packs are in conformance with zoning requirements.
- Ensure that the equipment is rated for the anticipated service loads.
- Ensure that the contractor's crew are competent to perform interventions on a live well.
- Liaise between process operators and the intervention crew during handover of the well, and at any time the program requires the well to be flowed/shut in.
- Confirm all necessary plant isolations are in place (double valve isolation where appropriate) and documented.
- Ensure that all necessary work permits are complete, and that the crews are familiar with any permit conditions.
- Where relevant, confirm that the remote (single) well control panel is properly configured and all members of the crew are trained in its use.
- Confirm that the crew know what to do in the event of an emergency at the host facility—i.e., muster alarm or abandonment alarm on an offshore platform.
- Ensure the crew carry out well control drill, and where necessary report on their performance.

1.1.11.2.2 During a well control incident

The supervisor with responsibility for well control will:

- Confirm the intervention crew have made the well safe.
- Where relevant, inform production staff (control room technicians and production supervisors). This is of particular importance if the well control incident is affecting, or likely to affect, the operation of the production facilities.
- Gather the relevant data and inform the base office.
- Coordinate any necessary remedial action.

The intervention technician at the well site will:

- Take whatever immediate action is necessary to contain a hydrocarbon escape from the pressure control equipment.

- Alert the operating company supervisor.
- Where a hydrocarbon release could trigger gas alarms at the facility, the control room operator will need to be informed immediately.

1.1.12 Human and organizational factors

Human rather than technical failure now represents the greatest threat to complex and potentially hazardous systems[8]

Between the 1960s and the 1990s, it has been estimated that the contribution of human error to accidents in hazardous technologies underwent a fourfold increase.[9] This is not because people are becoming more accident prone, but more a reflection of the way accidents are analyzed and the improved understanding of human factors. By the 1990s, the oil industry was beginning to realize that human and organizational factors were important considerations in accident cause and analysis. Following the Piper Alpha disaster, and the publishing of the Cullen report (November 1990), Robert Gordon University were commissioned by the UK's Health and Safety Executive to investigate and report on "Human and Organisational Factors in Offshore Safety". The findings, published in 1997, made several observations and recommendations. These included a recommendation that "*Crew Resource Management training should be used to teach 'human factor skills such as leadership, team working, decision making, assertiveness and communication, with the aim of reducing human error*".[10] The review also noted that most of the industry used a coding system for accident reporting that was limited in terms of human and organizational factors. They recommended using a wider range of human factor codes, more in line with those used by the marine and aviation industries. Nevertheless, oil industry progress towards Crew Resource Management (CRM) training remained limited, and human factors barely featured in formal well control training, where the emphasis remained on kill sheets and equipment. The Macondo disaster (April 2010) dramatically changed attitudes to the importance of human factors in well control training.

The Macondo blowout killed 11 workers. It caused enormous environmental damage along the Gulf coast, and has had a serious adverse effect on the livelihood of the local people. As of July 2016, Macondo is estimated to have cost BP almost $62,000,000,000.[11] A report by the Chief Counsel to the National Commission on the BP Deepwater Horizon Oil Spill and Offshore Drilling concluded that "*all of the technical failures at Macondo can be traced back to management errors by the companies involved in the incident.*"[12] Professor Andrew Hopkins, writing in his book "Disastrous Decision" reached a similar conclusion, noting that the causes of the accident were "*mundane, involving a series of human and organisational factors similar to those identified in other high-profile accidents.*"[13]

Professor Hopkins lists several contributory factors that led to the blowout, starting with an observation that the concept of "defense-in-depth" failed. One element of the critical well control defenses, was the 9 5/8" × 7" production casing. Once cemented, the casing would form part of the primary well control envelope. It is now known that the cement failed to isolate the reservoir. Crucially, the crew misinterpreted the results of the inflow test performed to confirm the integrity of the cement. They concluded that the barrier was intact. It was not. During the inflow test, drill pipe pressure built to over 1400 psi (had the barrier been intact there would have been no pressure increase). The crew bled off the pressure, only for it to build up again. These clear warning signs, indicating that reservoir fluids were entering and pressurizing the casing, were ignored by the crew; or rather, they were explained away. Although there was justified criticism of BP for having inadequate inflow test procedures, it is still hard to understand how those at the well site could rationalize such obvious anomalies. The explanation provided by Professor Hopkins (and others) is that the group was "*subject to powerful confirmation bias because the cement job had already been declared a success.*"

The Oxford Dictionary of Psychology defines confirmation bias as, "*the tendency to test one's beliefs or conjectures by seeking evidence that might confirm or verify them and to ignore evidence that might disconfirm or refute them.*"[14] The concept is not new, in 1620 Francis Bacon noted that, *once a human intellect has adopted an opinion (either as something it likes or as something generally accepted), it draws everything else in to confirm and support it. Even if there are more and stronger instances against it than there are in its favour, the intellect either overlooks these or treats them as negligible.*[15]

For the crew on the Macondo well, the cementing operation had been declared a success. The belief that the cement was good was reinforced by a successful positive pressure test on the casing (2700 psi for 30 minutes). They expected the inflow test to confirm the cement integrity; and they believed it had despite evidence to the contrary.

Thinking that the inflow test had confirmed the integrity of the cement, the crew subsequently failed to properly monitor fluid returns during the displacement to underbalanced fluid. As the heavy mud was displaced by the much lighter seawater, underbalance was lost, allowing reservoir fluid to flow into the well. Sadly, by the time the crew realized there was a problem, it was too late. Mud oil and gas were blowing out onto the rig floor. Although the crew were able to activate the BOPs annular preventer, the high velocity of flow eroded and washed out both the elastomer seal and the drill pipe; the well continued to blow out. Closing the variable bore pipe rams (VBR) momentarily stopped the flow, but pressure building in the drill pipe caused a rupture where it had eroded (at the annular preventer). Oil and gas continued to flow onto the rig where it ignited. The shear rams closed automatically, but failed to seal the well.

Macondo dramatically changed attitudes to training. It was clear from the more measured and thoughtful reports on the accident that human and organizational failings had been largely to blame. It was also clear that well control training had failed. Some operating companies, recognizing limitations in the standard International Well Control Forum and International Association of Drilling Contractors (IWCF/IADC) well control syllabus, began to develop their own in-house training courses. Shell, e.g., developed advanced well intervention well control courses that placed a greater emphasis on human factors and scenario-based exercises. Using a new well intervention simulator,[e] supervisors are subjected to a range of well control emergencies. From the service sector, Maersk Drilling have recognized the importance of human factors, and have integrated it into their well control training. Other industry bodies were also addressing human factors in well control training. In Europe, the "North Sea Offshore Authorities Forum"[f] performed a multinational audit that examined "Human and Organisational Factors in Well Control."[16] This audit investigated:

- The engineering system.

  How well control equipment controls are configured, and how critical well control information is displayed for the user.
- Human factors.

  Competency, situational awareness, availability of procedures, and team working.

[e] The well intervention simulator was developed by Drilling Systems Ltd. with support from Shell. It includes wireline, coiled tubing, hydraulic workover, and rig-based workover simulations.

[f] The NSOAF has representatives from the regulatory authorities of the following nations: The Netherlands, Denmark, Germany, Norway, The United Kingdom, The Republic of Ireland, and the Faeroe Islands.

- Organizational factors.
  Management of competence and training. Safety auditing and safety leadership.
- Company interface.
  Safety management interface between operating companies and vendors.

The audit supported the view that the industry was supplying equipment that provided key personnel with comprehensive and readable data, enabling them to make the correct decision in the event of an emergency. The audit also found that key personnel were generally empowered to make decisions on well control matters, but concluded that scenario-based training would improve matters.

Areas of concern included a lack of experienced personnel, the fact that well control was not one of the key performance indicators used by many operating companies, and poor implementation of a lessons-learned system. The auditors were also concerned about the quality of interface arrangements between the client company and drilling contractors.

Also responding to events on the Macondo well, the International Association of Oil and Gas Producers (IOGP) issued a report (476) that made several recommendations for enhancements to well control training, examination, and certification.[g] The report recommends CRM be included in well operations training. It expresses an expectation that this would be implemented by the industry and hopes it would "*evolve to be embedded in the well control curriculum as well as industry practices and procedures.*"

Crew Resource Management,[h] was developed in the late 1970s (by NASA) to help prevent accidents in the airline industry. Now, almost universally applied, it undoubtedly saves lives. In the widely reported ditching of US Airways flight 1549 in the Hudson River on the January, 15, 2009, the National Transportation Safety Board (NTSB) concluded that "*the professionalism of the flight crewmembers and their excellent CRM during the accident sequence contributed to their ability to maintain control of the airplane, configure it to the extent possible under the circumstances, and fly an approach that increased the survivability of the impact.*"[17]

Before recommending the inclusion of CRM into well operations training, the IOGP had commissioned a study to investigate how such

[g] IOGP. Report 476. Recommendations for enhancements to well control training, examination, and certification.

[h] Originally called Cockpit Resource Management.

training could be implemented and the content of the syllabus. The report findings were released in April of 2014.[18] Subsequently the findings were summarized in the IOGP report "Guidelines for Implementing Well Operations Crew Resource Training." This report lists six distinct competencies for inclusion in a training syllabus. They are reproduced here, along with the main learning objectives[19]:

WOCRM Competencies are:

- situation awareness (SA)
- decision-making
- communication
- teamwork
- leadership
- factors that impact human behavior.

1.1.12.1 Situation awareness

Developing and maintaining a dynamic awareness of the situation and the risks present during a Wells Operation, based on gathering information from multiple sources from the task environment, understanding what the information means and using it to think ahead about what may happen next.

Learning objectives

1. Describe or explain common causes and symptoms of SA problems, e.g., inattention, distraction, cognitive bias, and tunnel vision.
2. Develop SA skills relevant to Well Operations environments:
   a. actively seeking relevant information
   b. correctly interpreting and understanding information
   c. being able to foresee what is likely to happen next or the effect of current events on future states
   d. recognizing mismatches between your own SA and that held by others.

1.1.12.2 Decision-making

The ability to reach a judgment or choose an appropriate option to meet the needs of an assessed or anticipated situation.

Learning objectives

1. Recognize different approaches to decision-making, including strengths and weaknesses, e.g., following procedures, use of expert judgment, intuition.
2. Explain how problems to be solved need to be correctly defined and that difficulties can be caused in Wells Operations with decision

errors, e.g., confirmation bias and task fixation, often due to inadequate comprehension of the problem.

3. Understand workplace, personal, and interpersonal factors affecting decision-making:
   a. develop decision-making skills relevant to Well Operations environments
   b. recognize the situations where a decision is needed
   c. recognize where different approaches to decision-making are appropriate
   d. identify personal role and contribution in making decisions
   e. recognize where bias, such as group think, and other factors may result in poor decisions.

1.1.12.3 Communication

The exchange (transmission and reception) of information, ideas and beliefs, by verbal and nonverbal methods.

Learning objectives

1. Describe the characteristics of effective communication.
2. Recognize the personal, interpersonal, and workplace factors that can impair effective communication.
3. Explain the difference between an instruction/order and a dialogue, and recognize where each is appropriate.
4. Practice the importance of timely and effective feedback appropriate to the operational situation, e.g., in ensuring an instruction has been understood.
5. Recognize situations where different types of communication are appropriate, e.g., radio communication, briefing and de-briefing, and shift handovers.
6. Develop communication skills relevant to Well Operations environments:
   a. creating a clear message (how, what, where, why, when, who)
   b. delivering a clear message
   c. effective listening skills and seeking clarification
   d. tuning in to nonverbal responses
   e. being appropriately assertive for the situation (delivering and receiving communication)
   f. seeking and providing feedback and confirmation of understanding
   g. avoiding jumping to conclusions in critical time-pressured situations.

1.1.12.4 Teamwork

The ability to work effectively and interdependently in groups of two or more to achieve a shared goal.

Learning objectives

1. Explain the critical importance of being an effective team member to the safety of Well Operations.
2. Describe the characteristics of effective teamwork.
3. Provide examples of how different roles and responsibilities contribute to effective team performance.
4. Recognize how factors such as personal, interpersonal, workplace, cultural, contractual, and dispersed location can impair effective teamwork.
5. Recognize how individual behavior influences team dynamics and team performance.
6. Develop team working skills relevant to Well Operations environments:
   a. effective team coordination
   b. cooperation and collaboration
   c. recognize when team members do not have a common understanding of a shared situation or goal
   d. avoid creating situations of unnecessary conflict within a team
   e. detect and resolve disagreements and differences within a team
   f. show courage and ability to challenge when necessary.

1.1.12.5 Leadership

The ability to successfully influence others to achieve a shared goal by providing guidance, direction, coordination, and support.

Learning objectives

1. Recognize the critical importance of effective leadership to the management of safety.
2. Recognize the characteristics of effective leadership and be aware of how a leader's personal behavior affects others.
3. Describe how to motivate a team and what techniques and behaviors can work effectively.
4. Explain the importance of setting and maintaining high standards.
5. Develop leadership skills relevant to Well Operations environments:
   a. provide feedback, motivate and support the team and individual team members
   b. set and communicate expectations appropriate to the situation

c. convey the importance of leadership decisions and the reasons for them
d. adopt leadership styles and practices suitable to the situation.

1.1.12.6 Factors that impact human performance

Many factors affect the ability of people to perform reliably. These include stress, fatigue, health, distractions, and environmental stressors. They can arise from sources personal to the individual or can be imposed by external factors such as organizational and task design, team structure and work schedule, and the design and layout of plant and equipment, as well as cultural and environmental factors.

Learning objectives

1. Recognize that an individual's ability to remain alert and perform to a high standard is influenced by a wide range of factors: organizational, personal, psychological, physiological, and environmental.
2. Explain the importance of non-technical skills (SA, decision-making, communication, teamwork, and leadership) to operational safety.
3. Show awareness of major accidents within the industry where limitations in human performance have been significant contributory factors.
4. Provide examples of how the loss of alertness, distraction can increase risk to operations.
5. Describe the key types of human failure, e.g., slips, lapses, mistakes, violations, and how these represent risks to safe operations.
6. Recognize the type of operational situations where the risk of human error can be significantly increased.
7. Recognize strategies and actions that can be taken to minimize the potential for human failure on critical activities.
8. Explain the importance of sleep, work schedules, and shift patterns to effective performance, and the effect of time of day on alertness.
9. Recognize cultural differences, potential impacts, and mitigations.
10. Identify factors in the design and layout of plant and equipment that can impact on human performance, and how these can be mitigated.

Where possible, real industry incidents should be used to illustrate the potential for how a wide range of factors can impact on human performance. Where these are not available, invented case scenarios can be used for practical exercises. Specific examples of where CRM skills could be practiced are performing handovers, tool box talks, and task risk assessments.

1.1.13 Well control certificates

Almost without exception, operating companies and regulatory authorities mandate that everyone working in drilling, completions, workovers, and interventions must have a valid well control certificate. Currently, the industry recognizes and accepts well control certificates administered by two organizations. The European-based, IWCF, formed in 1992, and the American-based IADC, formed in the 1940s.

Following Macondo, well control training came under a great deal of scrutiny. Many in the industry were critical of the course content. For example, several of the equipment questions were criticized for being little more than a memory exercise and not a test of well control competence. There was also criticism of the way the training schools were operating. Although the IWCF and IADC are responsible for course content, the actual training is carried out by commercial organizations who must be approved of, and receive accreditation from, the IWCF or IADC (some schools have accreditation from both).

There is a great deal of pressure on these schools to get people through the exam. Understandable, since a candidate's continued employment may be conditional on having a valid well control certificate. There was a tendency to "teach to the exam," at the expense, many argued, of proper well control training.

There have been significant changes to the syllabus of both organizations since Macondo, well control training is now under continued scrutiny and review.

At the time of writing (August, 2017), the IWCF have four certification levels, and have plans to introduce a level five course aimed at experienced candidates "*who play a critical role in well design and the approval of well designs.*"[i]

1.1.13.1 IWCF well control training

Level one is a free, on-line, introduction to well control with seven modules covering:

- An overview of oil and gas
- The life cycle of a well
- Drilling rigs
- Well control during drilling operations
- Well interventions and workovers

[i] Quote from IWCF website.

- Pressure control during well interventions
- Conclusion.

At level two and above, candidates can choose to be examined in Drilling Well Control or Well Intervention Well Control.

1.1.13.2 Drilling well control

Level 2: Basic training for drilling and wells personnel

Level 3: Available for anyone expected to shut in a well.

Level 4: Well Site Supervisor Training.

Candidates are examined on:

- Surface equipment
- Surface principles and procedures
- Combined surface and subsea equipment
- Combined surface and subsea principles and procedures.

Candidates for levels 3 and 4 are given a practical assessment on a drilling simulator.

1.1.13.3 Well intervention well control

Level 2: Basic training for drilling and wells personnel.

Level 3: Anyone who might be expected to close in the well.

Level 4: Well site supervisor training.

Candidates are examined on:

- Completion Operations (compulsory)
- Completion Equipment (compulsory)
- Coiled Tubing Operations (optional)
- Wireline Operations (optional)
- Snubbing Operations (optional).

In response to the IAOP reports on Human Factors and CRM, the IWCF now have a CRM course and are able to combine it with "Enhanced and Specialist Well Control Courses."

1.1.13.4 IADC well control

The IADC support several well control courses under the name WellSharp, which updates and replaces the WellCAP program.

WellSharp introductory level

WellSharp Drilling. Driller level course

WellSharp Drilling—Supervisor

WellSharp Subsea—Supervisor

WellSharp Workover—Supervisor

WellSharp Workover Well Servicing—Supervisor.

REFERENCES

1. King G. *An introduction to the basics of well completions, stimulations and workovers*. 2nd ed Oklahoma: University of Tulsa; 1998.
2. Workover well control and blowout prevention guide. *Chevron drilling reference series*, vol. 15. 1994.
3. Aberdeen Drilling Schools: Well intervention pressure control. 2001.
4. Ellison B.T., Gallagher C.T., Frostman L.M., Lorimer S.E. The physical chemistry of wax, hydrates, and asphaltene. Offshore technology conference paper OTC 11963; 2000.
5. Recommended practices for drilling and well servicing operations involving hydrogen sulfide. API Recommended Practices. 3rd ed. May 2001.
6. HSE offshore COSHH essentials. http://www.hse.gov.uk/pubns/guidance/oce6.pdf.
7. Odgaard J., Morton T. Integrating human factors into well control training. SPE/IADC—184648-MS; 2017.
8. Reason J. Understanding adverse events: human factors. *Quality in healthcare* 1995;**4**:80−9 Downloaded from http://qualitysafety.bmj.com.
9. Hollnagel E. *Reliability of cognition: foundations of human reliability analysis*. London: Academic Press; 1993.
10. Health and Safety Executive Mearns K., Flin R., Fleming M., Gordon R. *Human and organizational factors in offshore safety*. OTH 543; 1997.
11. BP's total costs for deadly oil spill hit $62 billion. *Environ Leader J*, July 14, 2016.
12. Macondo: the Gulf oil disaster. Chief counsel's report. National Commission on the BP Deepwater Horizon Oil Spill and Offshore Drilling; 2011.
13. Hopkins A. *Disastrous decisions: the human and organisational causes of the Gulf of Mexico blowout*. CCH Australia Ltd; 2012.
14. Colman AM. *Oxford dictionary of psychology*. 3rd ed Oxford University Press; 2009.
15. From "Novum Organum" (1620). Francis Bacon. Lord Chancellor of England (1561−1626).
16. North Sea Offshore Authorities Forum. Multi-national audit. Human and organisational factors in well control, 2012−13.
17. Loss of thrust in both engines after encountering a flock of birds and subsequent ditching on the Hudson River. US Airways Flight 1549 Airbus A320-214, N106US Weehawken, New Jersey January 15, 2009 Accident report NTSB/AAR- 10/03 PB2010-910403 National Transportation Safety Board. May 4, 2010.
18. Crew resource management for well operations teams. International Association of Oil and Gas Producers. Report Number 501. April 2014.
19. Guidelines for implementing well operations crew resource management training. International Association of Oil and Gas Producers. Report No 502. December 2014.

CHAPTER TWO

Well Construction and Completion Design

Well control during completions, workover, and interventions requires a broad understanding of how the well is constructed, familiarity with different completion designs, and a good understanding of wellhead and downhole completion equipment. Armed with this knowledge, those planning and executing completion and intervention activities will be able to identify potential leak paths, and maintain an increased level of vigilance during operations where the risk of loss of containment is increased. A good understanding of the well construction process and completion architecture will also enable rig site personnel to respond more effectively in the event of a well control incident.

This chapter of the book describes the well construction process (drilling and casing the well) before going on to discuss completions and completion design.

2.1 WELL CONSTRUCTION

Nearly all wells are constructed using a series of casing strings, where each string reduces (tapers down) in diameter as the well gets deeper. Each casing is cemented in place, bonding it to the borehole. A properly cemented casing string will prevent formation pressure migrating to the surface, provide isolation between layers in the formation, support the borehole, and strengthen the casing. A series of tapered and cemented pipes is a construction concept that will be instantly recognized by anyone who has worked in the upstream side of the oil industry for any time.

Casing design is normally the remit of the drilling engineer, who will strive to find a design that allows the well to be drilled safely and economically, whilst at the same time resisting the many forces that are a necessary part of the drilling process. Casing should be designed to maintain integrity throughout the life of the well.

Well Control for Completions and Interventions.
DOI: https://doi.org/10.1016/B978-0-08-100196-7.00002-6

Drilling the well normally starts with the placement of the surface conductor. For land wells, a cellar is usually constructed. The cellar accommodates the wellhead, and allows the tree to be positioned with the base at ground level. The first casing string, conductor, is piled in to the depth of refusal. Alternatively, a post hole may be drilled and the casing run and cemented in place. On offshore platforms, the surface casing is normally run through guide slots that are part of the jacket structure and piled in to the seabed. In subsea wells, the casing is normally drilled and cemented into place. If the seabed is soft, it may be jetted into place.

With the conductor in place, the well is drilled to a depth where the surface casing can be run. Whilst the surface casing hole is being drilled, a diverter is in place on the surface to deal with shallow gas blowout. The surface casing is normally the first pipe that will support the drilling BOP, and the hole should be deep enough to ensure that formation fracture pressure is high enough to withstand kick pressure as the next hole size is being drilled. In most cases, the surface casing is cemented to the surface (or seabed). Once the surface casing has been run, successfully cemented, and tested, the drilling BOP will be installed.

Intermediate casing is used to deepen the wells where kick tolerance or unstable formation mean that it is not possible, or desirable, to drill from the surface casing to the reservoir in one continuous hole section. In very deep or difficult-to-drill wells, more than one intermediate casing string may be required. Cementing of intermediate strings will depend on a number of factors. For example, government regulation may require all or some formations to be isolated, particularly before abandonment. This may mean having to put the top of the cement above the shoe of the previous (surface) casing string. In other cases, a deliberate decision might be made not to cement above the shoe. Placing the top of the cement below a shoe might allow the casing to be cut and pulled, should a side track be needed. It may also allow annulus pressure to leak off into the formation at the exposed shoe. This is deliberate, and can help reduce annulus pressure build-up because of rising temperature as a well begins to produce (Fig. 2.1).

Some wells are drilled from the surface (or intermediate) casing shoe all the way through the reservoir to total depth (TD). Production casing is run across the producing formation and cemented in place. The casing is then perforated to allow hydrocarbons to enter the wellbore, where they will be produced through the completion.

In many wells, the production casing is set (and cemented) above the producing interval. Once the casing has been tested, the shoe of the

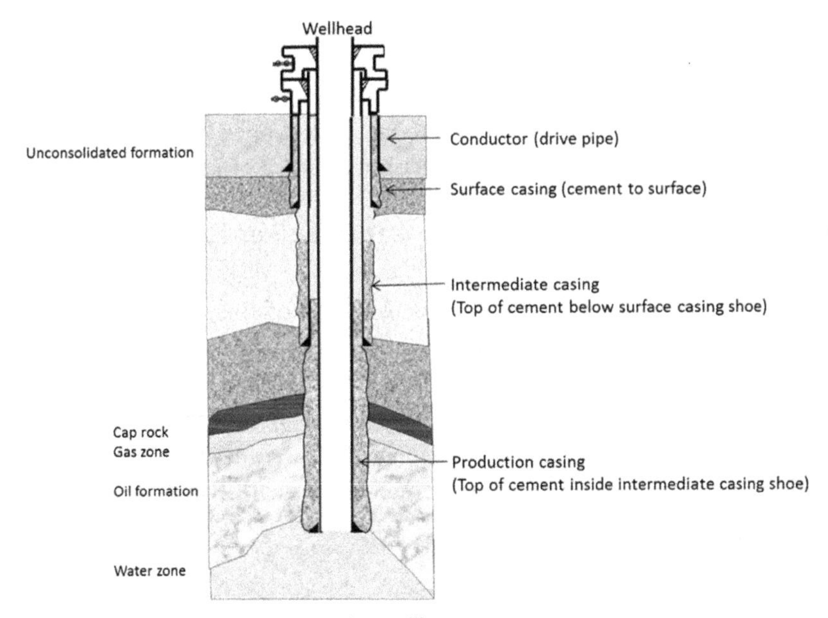

Figure 2.1 Casing and cementing the wellbore.

production casing is drilled out, the reservoir drilled, and the well completed. The type of completion will depend on the nature of the reservoir, and the different types of reservoir completion are discussed next.

2.2 TYPES OF COMPLETION

A completion is the final phase of the well construction process, and forms the interface between the reservoir and surface production facilities.

Wells can be producers or injectors, and completions may have to be able to produce (or inject) hydrocarbon liquids and gas, water, steam, and a range of stimulation chemicals. In addition, completions may have to withstand the corrosive effects of nonhydrocarbon products such as CO_2 and H_2S. Some completions will combine functions, e.g., annulus water injection and through tubing production. Over the life of the well, conditions can change and some wells may start life as production wells, and be converted to injection as conditions in the reservoir change.

Completion engineers often differentiate between reservoir completions (alternatively called the lower or sand-face completion) and the upper completion. The reservoir completion forms the connection between the reservoir and the wellbore. The upper completion is the conduit between the reservoir completion and the surface facilities.

2.2.1 The reservoir completion (sand-face or lower completion)

A good completion engineer will focus much of their attention on the lower completion design, since it is the efficiency of the lower completion that has the biggest influence on well productivity. Lower completion design will fall into one of only five categories. Within those five options there are a huge number of variables. Getting it wrong will be costly.

The main design decisions affecting the choice of reservoir completion are:

- Well trajectory and inclination
- Open hole vs cased hole
- Sand control requirements
- Stimulation requirements (propped frac or acid stimulation)
- Reservoir conformance – single or multi-layered reservoir and isolation between zones.

The five categories of reservoir completion are:

1. Open hole barefoot. This type of completion is simply a hole drilled through the hydrocarbon producing zone.
2. Open hole with slotted or predrilled liner.
3. Cased and perforated. A cemented liner or casing string is run across the reservoir and cemented in place. The casing is then perforated to provide a communication path between the reservoir and the wellbore.
4. Sand control completions in open hole. Sand control completions are required where sand production is anticipated. Sand production has detrimental effects on completions if not controlled. Open hole sand control completions include stand-alone screens and gravel packed screens.
5. Sand control screens in cased hole. Sand control screens can be run in to existing cased and perforated wells, or newly completed cased wells, and include screen-only completions, gravel pack, or frac pack. Expandable sand screens can and have been run in both open hole and cased hole (Fig. 2.2).

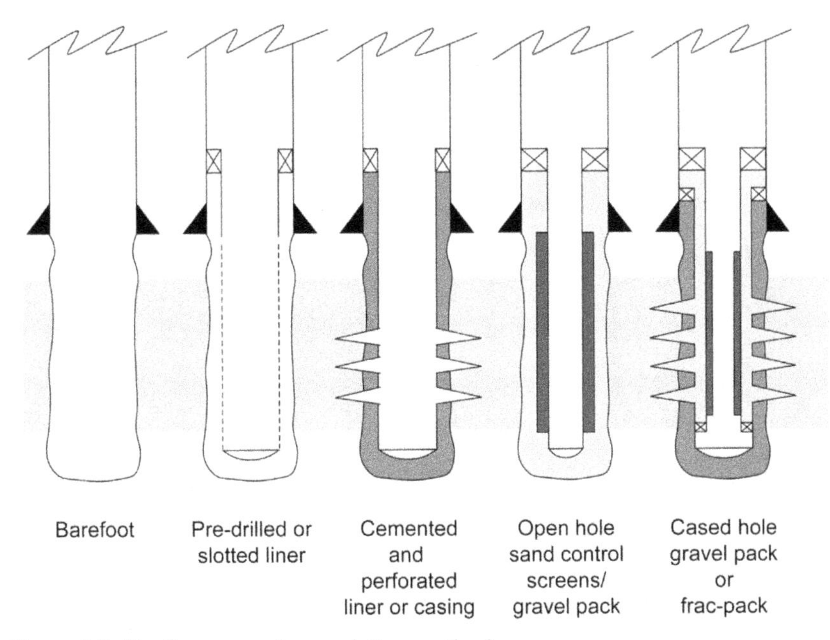

Figure 2.2 The five reservoir completion methods.

There are advantages and disadvantages associated with each of the five main completion types, and these are worth exploring further.

2.2.2 Open hole (barefoot) completions

Open hole completions are fairly common in homogeneous carbonate (limestone and dolomite) reservoirs, where rock strength is generally high and drilling induced formation damage can easily be removed with acid (Fig. 2.3). They are less common in sandstone formations. The main advantages and disadvantages of open hole (barefoot) completions are summarized in Table 2.1.

2.2.3 Open hole completions with predrilled and slotted liner

Predrilled and slotted liners are generally used where there is a risk of gross hole collapse. Using a predrilled or slotted liner also allows isolation between production zones if the liner is run in combination with External Casing Packers (ECP), or more commonly nowadays, swell packers. With a liner in place the deployment of intervention tools becomes simpler. For example, bridge plugs can be set in the blank pipe inside a Swell packer to isolate water production from a deeper zone.

Figure 2.3 Open hole barefoot completion.

Table 2.1 Open hole completions summary

Advantages	**Disadvantages**
No casing and perforating costs	Water or gas entry difficult to control
Saves rig time	Not easy to selectively stimulate different parts of the producing interval
Easy to deepen or side-track if desired	Open hole section subject to collapse—may require frequent clean-out
Can be converted to screen or liner completion at a later date	Only applicable to well consolidated formation
Large inflow area—high rate production	Drilling damage can be difficult to remove and can lead to reduced inflow

Predrilled and slotted liners are not normally used for sand control; it is difficult to cut slots fine enough to hold back the sand pack. Where very small slots are used (laser cut) the flow area is small, and they become susceptible to plugging. Predrilled liners are generally preferred because of better inflow performance; they are also cheaper to manufacture.

Predrilled and slotted liners can be deployed with washpipe if there are concerns about reaching bottom. In many cases, the liners are run with mud still in the hole. This reduces the risk of surge and swab induced kicks or losses (Figs. 2.4 and 2.5; Table 2.2).

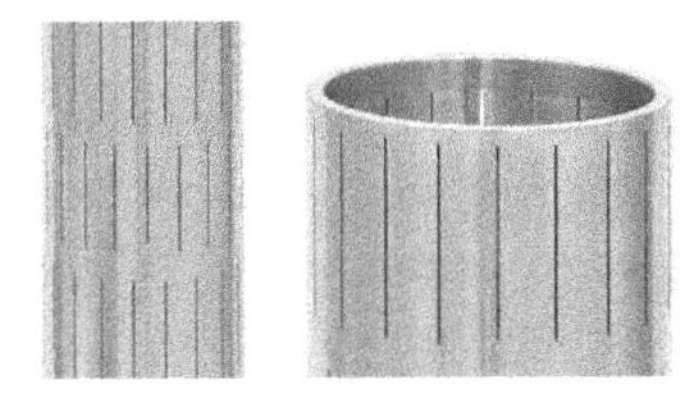

Figure 2.4 Slotted liner. Source: *Image courtesy of Euroslot.*

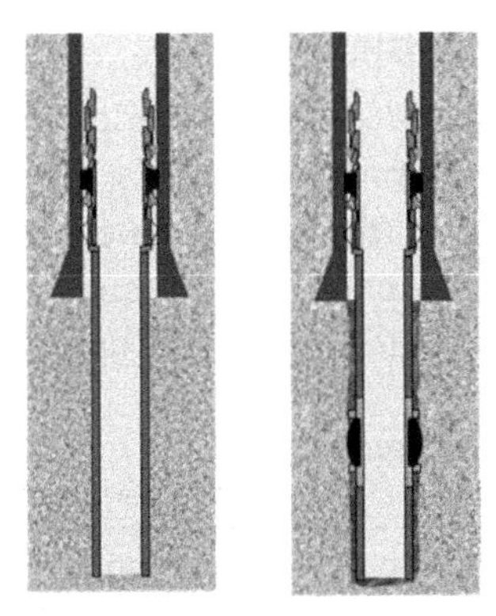

Figure 2.5 Predrilled or slotted liner completion. Left: Single zone. Right: Multiple zone with ECP or swell packer zonal isolation.

Table 2.2 Predrilled and slotted liner completions summary

Advantages	Disadvantages
Relatively cheap—elimination of cementing costs (and time)	Water or gas entry difficult to control unless external casing packers or "Swell" packers are used
Elimination of perforating costs	Not easy to selectively stimulate different parts of the producing interval unless ECP's or swell packers are used
Slots can be sized to give a degree of sand control in course sand formations	Mud filter cake very difficult to remove and can leave residual damage
Some degree of borehole stability achieved	Reduction in diameter over producing interval
	Some production logs difficult to interpret—at worst they become meaningless

2.2.4 Cased and perforated completion

Cased and perforated completions are used extensively in a variety of reservoir types. Wells can be completed with either a cemented production liner or cemented production casing across the reservoir. Cased and perforated completions have many advantages over open hole completions. These are summarized in Table 2.3: Cased and perforated completions (Fig. 2.6).

2.2.5 Sand control completions

Although approximately 60% of the world's hydrocarbon production comes from carbonate reservoirs, almost 90% of the wells are in sandstone formations.[1] About 30% of these sandstone formations are unconsolidated, and weak enough to be at risk of sand production.[2] The type of sand control completion used is dependent upon the consolidation of the

Table 2.3 Cased and perforated completions

Advantages	Disadvantages
Selectivity in production/injection	Perforating costs can be significant—particularly for long horizontal wells
Ability to shut off water, gas, or sand production through relatively straightforward interventions—plugs, straddles, cement squeeze	Some risk of productivity impairment from "completion skins" depending on how the well has been drilled and completed
Good productivity—drilling related formation damage can be by-passed	Adequate cement bond may be difficult to obtain in wells with long high angle liners
Zones can be added at a later date. Poorly producing zones can be re-perforated to by-pass damage	
Suitable for fracturing/stimulation, particularly where fracture containment or multiple fracturing is required	
Reduced sanding through careful perforating design	
Ease of application of chemical treatments—diversion	
Easily adapted for "smart" completions—single string multiple zone with multiple packers and inflow valves	

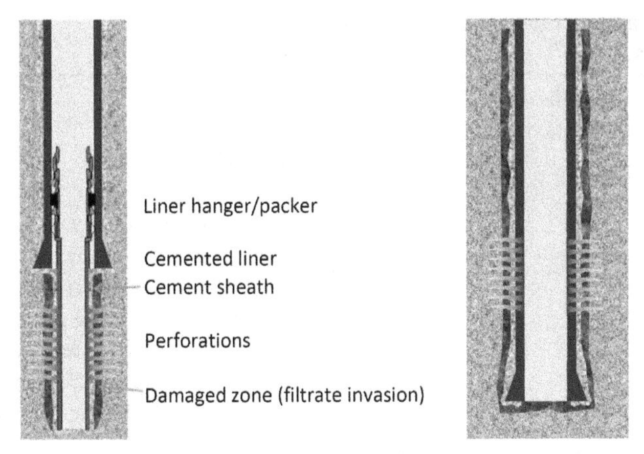

Figure 2.6 Cased and perforated completions. Left: Perforated production liner completion. Right: Perforated production casing completion.

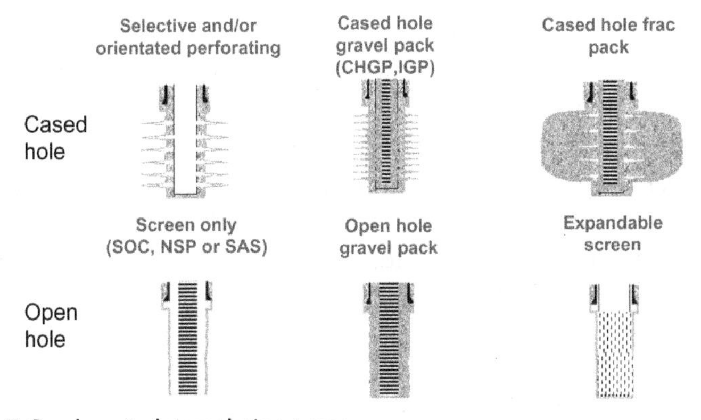

Figure 2.7 Sand control completion types.

formation (rock strength), homogeneity within the reservoir, and the quality of the sand pack (grain size distribution). Good quality well sorted coarse grained sandstone with a small percentage of fines can be completed with screen-only completions. As the quality of the sand deteriorates, more complex solutions are needed; open hole gravel pack or cased hole frac pack. The main sand control completion options are shown in Fig. 2.7.

A screen type sand control completion is run in two parts. Firstly, the screens are run into the open hole and hung off using a gravel-pack packer, an operation analogous to running a liner. If the well is to be gravel packed, it would take place at this point—before the running string is pulled from the well. With the lower completion in place, the upper

Table 2.4 Sand control completions summary

Advantages	**Disadvantages**
Sand production eliminated or significantly reduced	Costly
	Difficult to install
	Risk of high skin and increasing skin over time with many gravel pack/frac pack completions
	Poor record of reliability—mean time to failure for screen-only completions is 7 years
	Zonal isolation, water and gas shut-off difficult in open hole sand control completions
	Screen failure normally means extensive length and costly workover. Side track normally needed

completion (production tubing) can be installed. Screens can be run with (highly conditioned) mud still on the well, and the mud flowed back through the screens and gravel (if used). Some operators displace the mud from the well, replacing it with clear filtered brine, before the screens are run. There is a risk from losses unless a properly formulated lost circulation material is used. The LCM must be sized to prevent losses, but still enable flow-back through the screens (and gravel). There are well control problems particular to the running of screens, and these are described fully in Chapter 7, Well Kill, (Table 2.4).

2.2.6 The upper completion

The upper completion is the conduit taking the hydrocarbons from the lower (reservoir) completion to the surface. A number factors will influence the configuration of the upper completion. These will include:

- Pressure isolation requirements between reservoir and surface—is a packer required?
- Integrity requirements—Surface Controlled Sub Surface Safety Valve (SC-SSSV) and Annulus Safety Valve (ASV).
- Reservoir management requirements—single string multizone completion or dual string completion.
- Tubing size (driven by production potential and expected decline).
- Artificial lift requirement.
- Life of field concerns.
- Intervention requirements.

Some of the more common upper completion configurations are illustrated and discussed in the next few pages.

2.2.7 No tubing: Flow through production casing

Tubingless completions, if completion is indeed the right word, are still in use in many places. They are popular first and foremost because they are cheap and simple. They are primarily used on land locations where conditions are benign. There are clear disadvantages to this type of "completion."

- Well integrity—The production casing and wellhead is exposed to reservoir pressure. There are no barriers in the event of a loss of wellhead integrity. Quite simply, they are not safe.
- The casing is exposed to wellbore products, and is at risk of erosion and corrosion leading to well integrity problems.
- Flow stability concerns in larger casing sizes.
- Well kill (when necessary) is complicated by the absence of a tubing string. Bullheading is the only option, unless coil tubing or a snubbing unit is used to create a circulation path.

From a well integrity and well control standpoint, these wells fall far short of the ideal and are not allowed by some regulatory authorities—particularly if the well is capable of flow to surface without reliance on artificial lift (Fig. 2.8).

2.2.8 Tubing only completion: no production packer

Running a string of completion tubing gives number of distinct advantages over the tubingless configuration.

- Although the casing is still exposed to hydrocarbons and other wellbore products, the annulus between the tubing and casing is static. The risk of erosion is virtually eliminated, and whilst corrosion is still possible it will be at a much reduced rate since the fluids in the annulus are static (not being replenished by flow), unless the well is deliberately produced through the annulus.
- Worn or corroded tubing can be recovered and replaced—whereas the repair of corroded casing would be more complex and costly, perhaps requiring a side track or even abandonment.
- Tubing can be correctly sized to stabilize flow.
- Well kill is simplified—reverse circulation or conventional circulation now becomes possible.

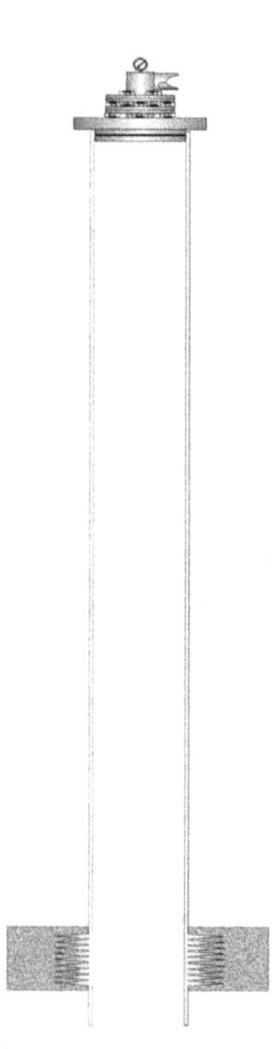

Figure 2.8 A tubingless completion.

The tubing only (packerless) completion is the most common method of upper completion in use today, since it includes almost all rod pump wells (Beam Pump and Progressive Cavity Pump) as well as many wells equipped with Electrical Submersible Pumps (ESP). However, if the well is capable of flowing to surface naturally, i.e., with the artificial lift pump switched off, then integrity remains a concern. The wellhead and casing will be exposed to reservoir pressure unless a packer is installed (Fig. 2.9).

2.2.9 Single production tubing string with production packer (reverse taper configuration)

Adding a production packer to the string of tubing removes many of the integrity concerns associated with the packerless completion.

- The casing above the packer is isolated from wellbore fluids, therefore corrosion is much less likely to occur.
- The wellhead and casing (above the packer) are no longer exposed to reservoir pressure.

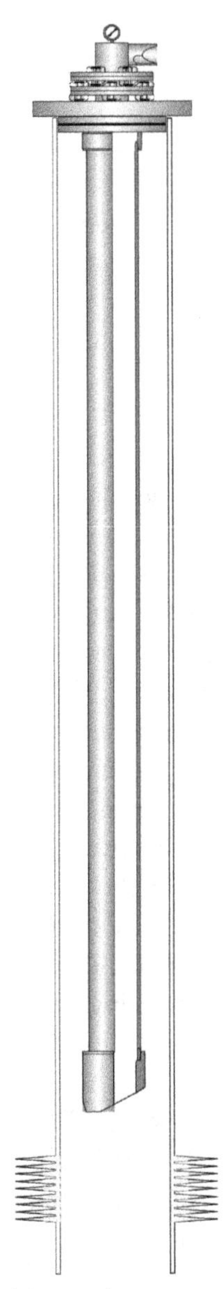

Figure 2.9 Packerless (tubing only) completion.

- On the annulus side there are two mechanical barriers between the reservoir and surface—packer and tubing hanger. These will be integrity tested at the time of installation, and form part of a well integrity barrier system.

- On the tubing side, integrity is provided by closure of the Christmas tree valves. An additional barrier can be provided in the form of a SC-SSSV.

Installing a packer in the well brings with it additional complications.

- Annulus pressure build-up (APB) caused by fluid expansion must be anticipated and controlled by regular monitoring and bleed off. Failure to do so could result in tubing collapse or casing burst.
- The annulus usually becomes an integral part of the well integrity envelope (the secondary barrier), and therefore annulus integrity must be monitored and maintained. Loss of integrity will almost certainly require a well intervention.
- With a packer in place, a circulation path is no longer available. Circulation, for well kill or any other purpose, can only take place if a circulation path is created. This must be close to the packer (for maximum effect). Some completions will include a circulation device, such as a sliding sleeve or side pocket mandrel. In the absence of circulation equipment, the only method of creating a circulation path is to perforate (punch) the production tubing above the packer.
- Workover (re-completion) becomes more complex. If the tubing is connected (anchored) to the packer it must be released before it can be recovered. If a dynamic (moving) seal assembly is stabbed into the packer, then the completion below the packer remains in place unless the packer is recovered separately. Packer replacement can be problematic, if permanent packers have to be removed.
- The tubing will be subjected to much higher loads (axial, burst collapse, and tri-axial). Detailed tubing stress analysis will need to be carried out to confirm the design (Fig. 2.10).

2.2.10 Single production tubing string with production packer (mono-bore configuration)

A mono-bore completion is generally defined as one where the internal diameter of the production tubing is the same or larger than the internal diameter of the production liner.

A completion configured in this way has some distinct advantages.

- Water and gas shut-off interventions can be easily accomplished using full bore (liner ID) mechanical bridge plugs, packers, and straddles.
- There is no need to kill the well and recover the production tubing to gain access to the reservoir.
- Prior to a workover (re-completion), full size mechanical bridge plugs can be run through the upper completion and set in the liner, above

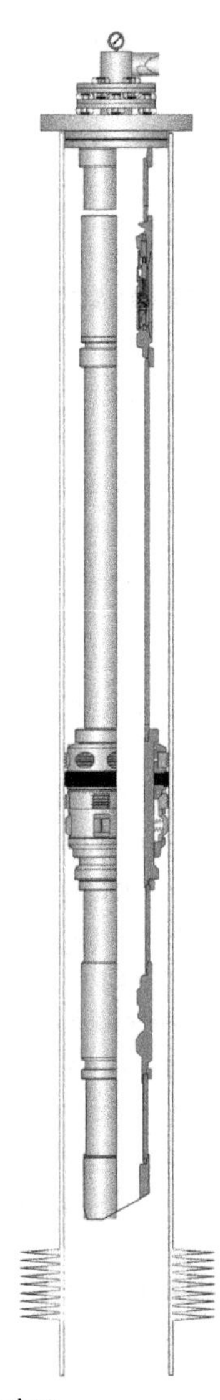

Figure 2.10 Single string with packer.

the reservoir. Having a mechanical barrier in the well during a workover reduces the risk of fluid loss or a kick, as well as preventing potentially damaging kill fluid from contacting the formation.

- Interventions to remove oilfield scale from the liner can be performed using a full size scale mill. There is no requirement to use more troublesome underreamers.

There are some disadvantages with the mono-bore design. Frictional pressure drop, particularly in long small ID liners, can be significant. In addition, obtaining a good cement bond in high angle wells with long liners can be problematic (Fig. 2.11).

2.2.11 Single string production tubing completed across multiple reservoir zones

Many fields have a number of separate reservoir compartments stacked vertically one above another. Reservoirs of this type can be exploited using a single string, multiple zone, completion. Production zones are isolated from one another using production packers, and there will normally be an additional (top) packer isolating the production annulus from produced fluids.

Fig. 2.12 illustrates a single string multizone completion. In the example illustrated, a well is completed across three separate zones. A configuration like this allows simultaneous production from any combination of zones, or would allow a single zone to be produced on its own. In the example shown, production from the lower zone enters the tubing through the tail-pipe below the lower packer. Flow from this zone could be isolated by installing a wireline set plug (mechanical barrier) in a nipple profile below the packer. Production from the middle and upper zone would be controlled by opening or closing a communication port (sliding sleeve) located between the packers. In most wells the communication port is manipulated using a slickline deployed mechanical shifting tool. In wells where access is difficult or expensive, for example subsea wells, communications ports are hydraulically operated from the surface ("intelligent" or "smart" well technology).

One advantage of this design is the number of different reservoir layers that can be completed using a single wellbore and a single completion string—in theory there is no limit. Fields are in production where eight or nine zones are accessed through a single completion string. There are some obvious disadvantages to completing in this way. Intervention access is very limited at the zones that are behind pipe. Pressure contrast between zones can lead to problems with cross-flow. This can compromise well productivity, create problems during intervention, and make well kill very problematic. There will be additional well control concerns when running the completion, since the production casing will have to be perforated before the completion is run.

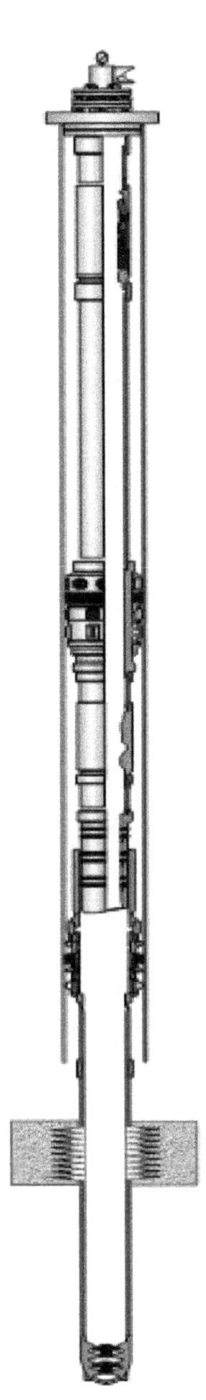

Figure 2.11 Mono-bore completion.

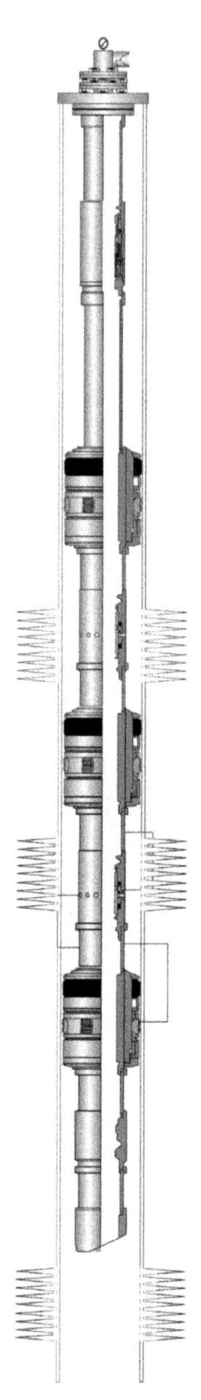

Figure 2.12 Single string multizone completion.

2.2.12 Multilateral Wells

A multilateral well is a well with one or more branches coming off the main bore. Because of the difficultly in drilling and completing multilateral wells, they are still relatively rare. The main reason for drilling a multilateral well is to increase reservoir exposure. For example, in the Troll Field, Statoil have drilled a well that uses five branches to expose 14,200 m (46,590 ft) of producing formation. Whilst there are clear benefits to having such extensive reservoir exposure from a single wellbore, the drawbacks are equally as obvious. Multilateral wells are complex; difficult to drill and complete. Any requirement to control production from a lateral means complex completions. The ability to carry out mechanical interventions is normally limited to the main bore, unless very complex junctions are installed. There are also additional well control problems associated with some multilateral wells—for example, a bullhead may be the only viable option in some wells, since it is usually not possible to circulate below the junction (Fig. 2.13).

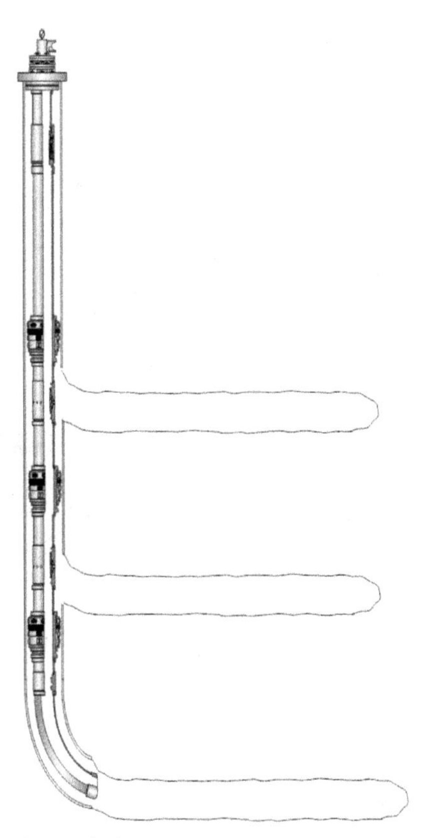

Figure 2.13 Multilateral completion.

In 1997 a joint industry taskforce, the Technical Advancement for Multilaterals (TAML) established a six tier classification for isolation at the junction. The six junction categories are illustrated, each with a brief description (Fig. 2.14).

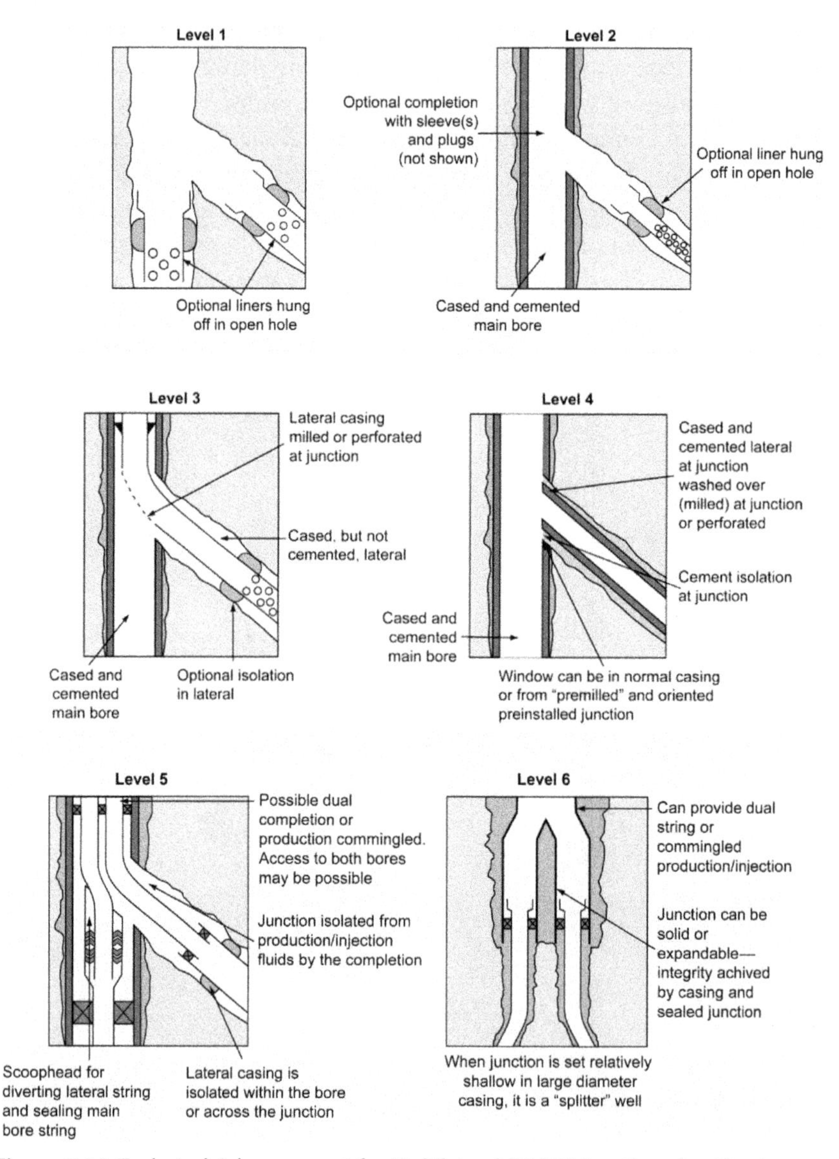

Figure 2.14 Technical Advancement for Multilateral (TAML) junction classification.

2.3 SUMMARY

Anyone with responsibility for well control or well integrity during completion, workover, or intervention operations must have at least a basic understanding of the well construction process, as well as an in-depth knowledge of completion types and completion architecture. Knowing how the well works as an integrated system will aid identification of potential leak paths, and how best to respond to a well control incident should one occur.

REFERENCES

1. Bellarby J. Well Completion design. Elsevier; 2009.
2. Walton IC, Attwood DC, Halleck PM, et al. Perforating unconsolidated sands: an experimental and theoretical investigation. SPE 71458; 2001.

CHAPTER THREE

Completion Equipment

Completions vary hugely in their complexity, and the range of equipment components is correspondingly large. Some items of completion equipment are safety critical and in many instances the failure of a critical component equates to a barrier failure, perhaps even a loss of integrity. Completion engineers, well site supervisors, and those responsible for producing the well need to have a broad understanding of what each component is used for, where it is located in the well, and how it is installed and maintained.

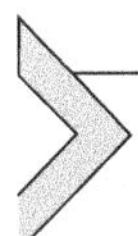

3.1 THE WELLHEAD, TUBING HANGER, AND CHRISTMAS TREE

Although the wellhead, tubing hanger, and Christmas tree are separate components they form part of an integrated assembly. As such they must be compatible, and are usually supplied by the same manufacturer. In a well equipped with a conventional (vertical) tree, the tubing hanger is designed to sit within the wellhead. In some cases a separate tubing hanger spool is attached to the top of the wellhead to accommodate the tubing hanger. The Christmas tree is attached to the top of the wellhead (or tubing hanger spool), and seals around the neck of the tubing hanger.

Fig. 3.1 shows a wellhead with the tubing hanger in place. The tree is above.

3.1.1 The wellhead

The wellhead is an integral and important part of the well integrity envelope and is designed to:

- Support each casing string.
- Support the production tubing (the completion).
- Provide a base (support) for the drilling blowout preventer (BOP) while the well is being drilled.

Well Control for Completions and Interventions.
DOI: https://doi.org/10.1016/B978-0-08-100196-7.00003-8

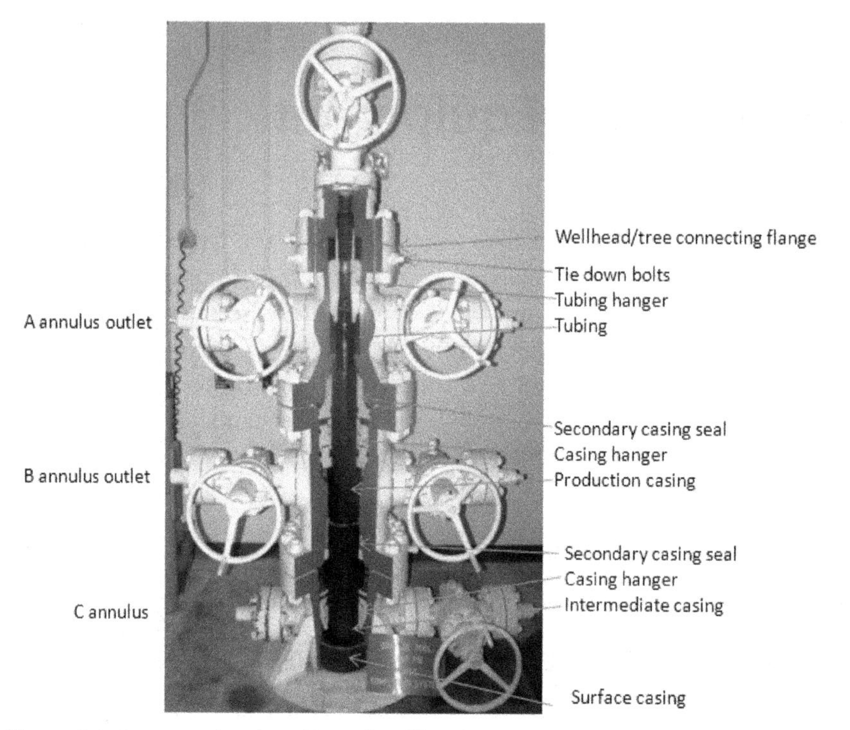

Figure 3.1 A conventional multispool wellhead.

- Support the Christmas tree (if fitted) once the well is completed, and while it is in service.
- Provide pressure barriers between each annulus and atmosphere.
- Provide annulus access through side mounted gate valves. On subsea wells only the inner (A[a]) annulus can be accessed—all other annulus are inaccessible and therefore effectively trapped.
- Where relevant, provide pressure tight exit spool for surface controlled subsurface safety valve (SC-SSSV) control lines, instrument cables, ESP cables, and other remotely operated downhole tools.

Many wells are still equipped with traditional, multiple spool wellheads where each casing string is suspended from a separate and appropriately sized casing spool. This means that each time a casing string is run

[a] In this context "A" refers to the innermost (production) annulus. Most operating companies, by convention, label each annulus starting with the innermost (A) and work out. B is the intermediate and C the outer annulus. There are exceptions, however. Some operators, particularly in Canada, label the outermost annulus A, since it is the first to be installed and progress inwards. This can lead to confusion.

(and cemented) the BOP must be removed, a new casing spool fitted, and the BOP rigged back up. This process is both time consuming and costly. Moreover, a multispool construction means multiple flanges and gasket seals, and therefore many potential leak paths. Conventional, multiple spool wellheads are relatively tall and need a deep cellar (land wells). On offshore production platforms, conventional trees need more elevation between the cellar (wellhead) deck and the deck giving access to the trees.

In 1961, Gulf Oil introduced the first compact (Unihead) spool wellhead. Originally developed for subsea wells, the technology now finds widespread applications on land and offshore platform wells. In all respects the compact spool is a technically superior design.

- Compact design—less space needed.
- Uninterrupted BOP cover—the BOP remains in place for the whole drilling operation.
- Reduced wellhead related flat time—no need to remove the BOP at each casing point.
- Improved pressure retention—fewer flanges and gaskets (Fig. 3.2).

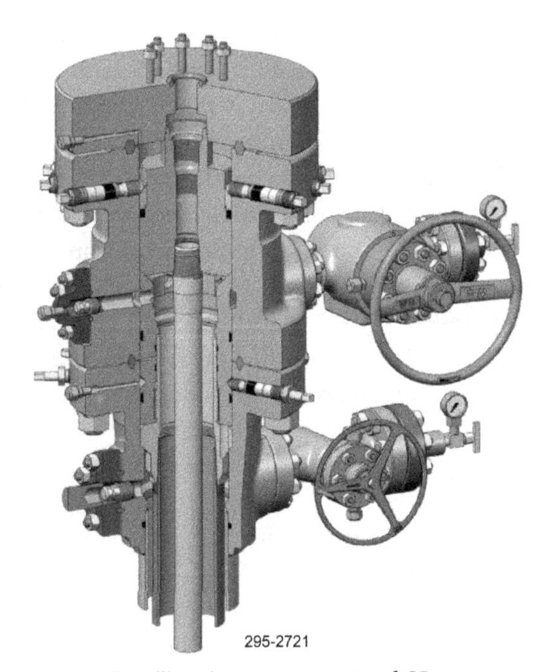

Figure 3.2 Compact spool wellhead. *Image courtesy of GE.*

Completion and intervention personnel are unlikely to be directly responsible for the installation of the wellhead and its associated components. However, many completion, workover, and intervention operations involve working on the wellhead. For example, wellhead maintenance and integrity assurance work is often carried out by well intervention crews. Some intervention operations will require the inner (production) annulus to be accessed and monitored, and there will be occasions where there is a requirement to pump fluid into, or take returns from, the inner annulus.

Ideally each annulus will have a pressure monitoring facility and a sample/bleed point.

- *In subsea wells only the "A" annulus can be monitored and bled off. The outer (B and C) annulus cannot be accessed. Subsea wellheads and trees are covered in the subsea section of this book.*

The "A" annulus is treated a little differently to the B and C annulus. This is because there is a greater potential for it to become pressurized, either because of leaks from the production tubing and packer, or because of annulus fluid expansion.

Although all of the annulus have the potential to undergo pressurization when production begins, the effect is normally greater in the innermost annulus. Hot fluids, flowing from the reservoir, begin to heat up the tubing, casing, and surrounding annulus. As fluid in the annulus heats up it begins to expand. Since the annular volume is finite, fluid expansion will be limited and where fluid cannot expand, pressure will increase. If measures are not taken to mitigate against this pressure increase by venting the annulus, there is a very real risk of tubing collapse or casing burst.

The other major cause of annulus pressure build-up is a leak from the production tubing or past the packer. If this occurs the annulus will be exposed to tubing pressure. The worst case would be maximum wellhead closed-in pressure. For this reason, the production annulus is normally tested to the same pressure rating as the completion string (production tubing).

Well integrity requirements for the inner (production) annulus will normally be:

- Two flanged gate valves, each with the same pressure rating as the Christmas tree on each outlet.
- Valves should be sized to allow adequate pump rate for well kill/circulation of gas injection. Most valves are 2″ nominal diameter.

Intermediate (B) annulus: One outlet should have a single gate valve with a sampling/bleed down arrangement. The other side may be terminated with a flange, needle valve, and pressure gauge.

Outer (C) annulus: One outlet should have a single gate valve with a sampling/bleed down arrangement. The other side may be terminated with a flange, needle valve, and pressure gauge. When there are more than three annuli on the wellhead, these should be treated in the same way as the "C" annulus.

3.1.2 Tubing hangers

The tubing hanger rests in a prepared profile (tapered step) in the wellhead. It must be able to support the hanging weight of the tubing, and any additional tension or compression force over and above the tubing string weight. It also seals the production annulus, forming a mechanical barrier between the reservoir and surface. Since the hanger seals the top of the annulus, any pressure will generate a significant upward piston force. For this reason, the hanger must be locked into place using tie-down (hold-down) bolts, or some form of locking mechanism (Fig. 3.3).

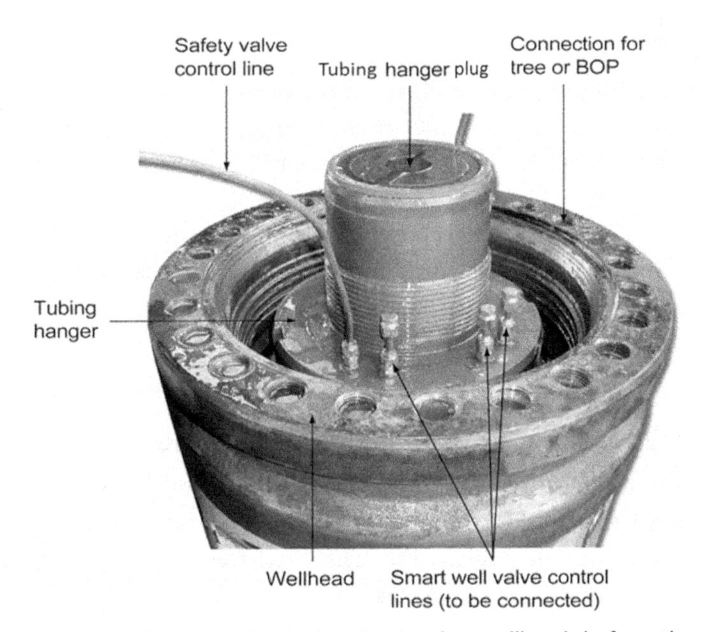

Figure 3.3 Tubing hanger—shown in situ in the wellhead before the tree is installed.

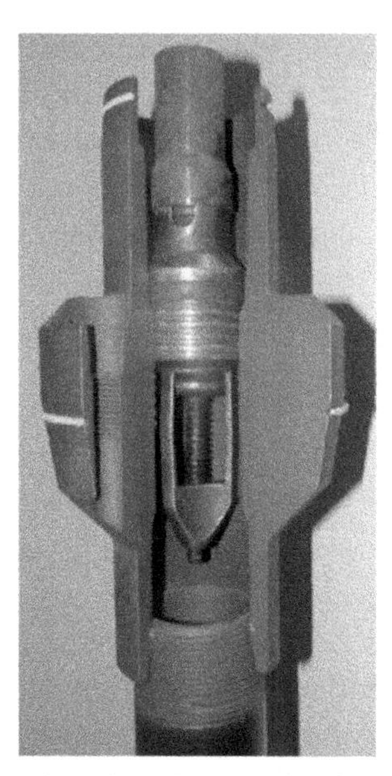

Figure 3.4 A Cameron type tubing hanger two-way check.

Nearly all tubing hangers have a profile machined into the bore designed to accept a plug, back pressure valve or check valve. The most common use for the tubing hanger plug is to provide a well control barrier when removing the drilling BOP and installing and testing the Christmas tree (or vice versa). A tubing hanger can also be used to provide a well control barrier when repairing Christmas tree valves (Fig. 3.4).

Many tubing hangers have one or more specially prepared penetrations that enable hydraulic control lines, instrument cables or electric submersible pump (ESP) power cables to be fed past the hanger and out through an exit block in the wellhead.

3.1.3 Wellhead and Christmas tree service tools

Because the wellhead is required to last for the life of the well there will, from time to time, be a requirement to repair or replace the side outlet valves, or remove and replace the Christmas tree. A range of tools has been developed to allow this to take place in a safe and controlled manner. These are summarized in Table 3.1 (Figs. 3.5 and 3.6).

Table 3.1 Wellhead and Christmas tree service tools

Valve removal plug	On many wellheads, the outlet port for each annulus is machined with an internal thread profile. A valve removal plug can be screwed into the profile sealing off the annulus. With a plug set, the annulus gate valve can be safely removed. To comply with standard barrier policy (two barriers), annulus integrity would need to be confirmed.
Valve removal tool	The valve removal tool enables a valve removal plug to be placed in the side outlet profile. Most removal tools are designed to enable plugs to be installed and recovered when there is annulus pressure present.
Back pressure valve/ two-way check valve	Most tree/wellhead designs have a profile machined into the tubing hanger. A wireline set plug or threaded back pressure valve placed in the hanger profile allows the tree to be removed or tested, and the drilling BOP be installed and tested.
Polished rod lubricator	Used to install and remove threaded back pressure valves in live wells.

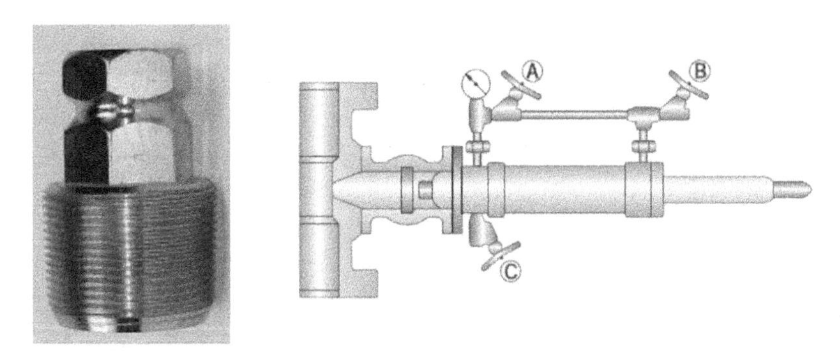

Figure 3.5 Valve removal plug (left) and valve plug setting tool (right).

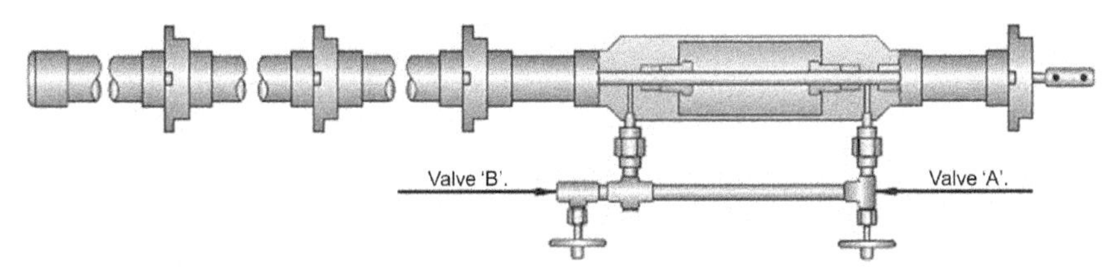

Figure 3.6 Polished rod lubricator—for setting/pulling tubing hanger plugs and back pressure valves.

3.1.4 The Christmas Tree

Not every well is equipped with a Christmas tree. Beam pump wells, for example, only have a stuffing box. Where a tree is installed its main functions are:

- Enable production (or injection) from the well to be controlled through opening and closing of pressure containing valves. When closed and tested, these valves form a mechanical barrier between the reservoir and surface.
- Provide access for well intervention work.
- Provide a connection point for the flowline (or injection line).
- Provide a connection and seal at the tubing hanger and the wellhead.
- Isolate the well from other adjacent wells.

The configuration of the Christmas tree will vary depending on well location, flow control requirements, pressure and temperature. Christmas trees fall into one of two categories: conventional (vertical) trees and horizontal (or spool) trees.

3.1.4.1 Conventional (vertical) Christmas trees

Most trees used on land or offshore platforms are the conventional (vertical) design. On a vertical tree, the valves used to control flow are arranged vertically through the center-line of the tree body. The bottom flange on the tree is connected to, and will have the same pressure rating as, the tubing head flange on the wellhead. The internal profile in the base of the tree (or in the adapter flange) will mate over the tubing hanger neck and create an internal seal. Conventional trees will either be composite (made up from individual valves flanged together), or solid block (sometimes called mono-block) where the valves are housed in a solid casting (Fig. 3.7). Each valve on a vertical Christmas tree will have a specific function (Table 3.2).

3.1.4.2 The horizontal or spool tree

Horizontal trees were first used in the early 1990s, and were developed primarily for subsea use. Unlike the vertical tree, the horizontal tree does not need a bespoke riser system for running and interventions.[b] More recently, "dry" horizontal trees have been developed for land and platform applications. It is important to understand the distinct differences between the two tree types, as well intervention pressure control

[b] Riser and intervention system for both vertical subsea trees and horizontal subsea trees are described fully in Chapter 13, Subsea BOP and Marine Riser Systems.

Figure 3.7 Conventional mono-block Christmas tree on an offshore platform.

Table 3.2 Christmas tree valve function

Valve name	Function
Lower master valve (LMV)	The lower master is not normally used for day-to-day operations. It is left open and only used to provide a barrier if other valves in the tree have failed.
Upper master valve (UMV)	This is the valve that is normally used to open and close the wellbore. On many trees the UMV is operated by a hydraulic (or pneumatic) actuator. The actuator is configured for "fail-safe closed" operation, and connected to the facilities automated shut-down panel.
Swab valve (SV)	The swab valve (sometimes called the crown valve) is normally closed while the well is producing (injecting). It gives access to the wellbore during interventions, where it used as one of two mechanical barriers (the UMV being the other).
Flow wing valve (FWV)	Used, along with the UMV, to control flow from the well. Like the UMV, many wing valves are fitted with fail-safe actuators and are tied into the facility ESD system.
Kill wing valve (KWV)	This valve enables fluid to be pumped into the tubing. As the name suggests, it allows the well to be killed while the flowline is still attached. It is also used to pump stimulation and inhibition chemicals down the well, probably a more common application.
The tree cap	Sometimes called the swab cap, the tree cap is a high pressure cap with the same working pressure as the tree body. It is removed to allow intervention pressure control equipment to be rigged upon the well. Most tree caps have "quick union" type thread, compatible with those used on intervention pressure control equipment.

requirements differ significantly. In addition, the sequence of events during a completion or workover is very different when using a horizontal tree.

As described, a conventional (vertical) Christmas tree has the valves controlling flow arranged vertically through the bore of the tree. A concept that will be very familiar with anyone working on wells (Fig. 3.8).

During the construction of a well with a vertical tree a typical sequence of events might be:

- Drill to TD.
- Run and cement liner.
- Clean out wellbore.
- Run and land completion. When the completion is landed the tubing hanger rests inside the uppermost spool in the wellhead—the tubing hanger spool.
- Install mechanical barriers in the well.
- Nipple down (remove) drilling BOP.
- Install and test Christmas tree.
- Remove mechanical barriers.

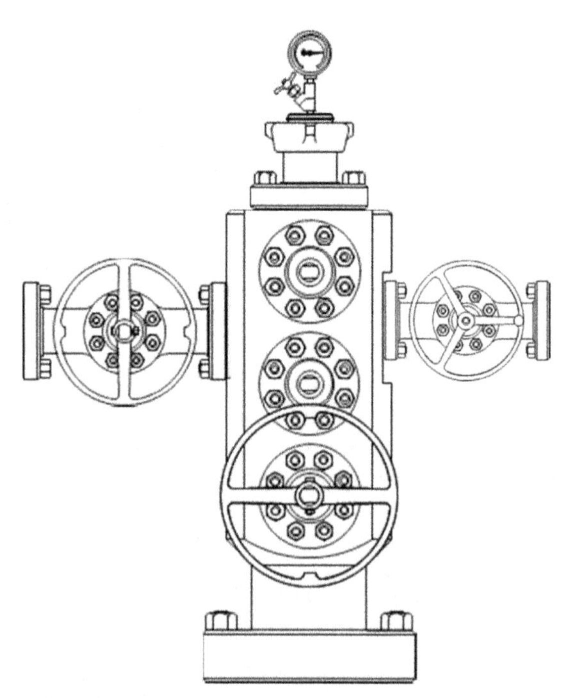

Figure 3.8 Simplified schematic of a conventional Christmas tree with the valves arranged vertically.

A horizontal tree is configured very differently from a vertical tree. Valves are positioned on the outside of the tree block, not through the vertical bore of the tree. The other fundamental difference from the vertical tree is that the tubing hanger rests inside the tree spool, not the wellhead (Fig. 3.9).

Because the hanger sits inside the tree, it follows that the completion cannot be run until after the tree block has been installed. A drilling BOP still needs to be in place during the running of the completion, but now has to be connected to a hub on the top of the tree. A typical sequence of events for installing a completion on a well equipped with a horizontal tree is:

- Drill to top reservoir.
- Run bridge plug or mechanical set, retrievable packer.
- Nipple down BOP.
- Install horizontal tree.
- Nipple up BOP (on top of horizontal tree) and install wear bushing in tubing hanger profile inside tree body.
- Drill to TD.
- Run and cement liner (or run lower completion).
- Recover wear bushing.

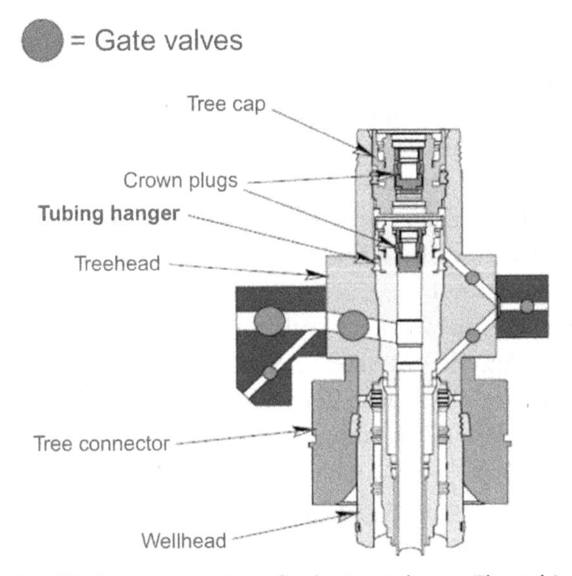

Figure 3.9 A simplified representation of a horizontal tree. The tubing hanger the yellow block in the center of the diagram.

- Run completion.
- Install wireline plug in tubing hanger profile (tubing hanger plug).
- Install high pressure internal cap. The cap is usually run with the wireline retrievable "crown plug" preinstalled.
- Recover drilling BOP and install debris cap.

Live well intervention work on a well equipped with horizontal tree is more time consuming and complex than the equivalent operation with a conventional tree. With a conventional tree, the swab cap is removed and pressure control equipment connected directly to the tree block. Double valve isolation is provided by the closed and tested upper master valve and swab valve. Once the pressure control equipment is tested both tree valves are opened, giving access to the wellbore. To gain access to the wellbore through a horizontal tree, both wireline plugs (tubing hanger plug and crown plug) must be removed. Removal of these plugs requires the rigging up of gate valves to provide well control barriers. Access to the wellbore requires the following actions:

- Remove the tree cap.
- Rig up dual valve isolation. This is normally two gate valves. Ideally at least one of the valves will be actuated and connected to a well control panel.
- Pull the crown plug.
- Pull the tubing hanger plug.
- Carry out intervention work using the two temporary valves for double barrier isolation.
- Replace and test the tubing hanger plug.
- Replace and test the crown plug.
- Replace and test the tree cap.

Since the tubing hanger plug and crown plug are permanent barriers and an integral part of the Christmas tree, they are normally equipped with metal-to-metal seals (Fig. 3.10).

Energizing wireline set metal seals can be problematic. To add to the difficulty, these plugs do not have an equalizing device. Without this critical feature, pressure differential (force) from above will make pulling the plug impossible. Potentially more hazardous, pressure differential from below risks blowing the plug (and the wireline recovery tool string) up the hole, with the attendant risk of breaking the wire and damaging the surface pressure control equipment. Pressure must be balanced across the plugs before recovery can be safely accomplished, and effectively balancing pressure can be tricky. For these reasons, horizontal trees are not ideal where frequent mechanical interventions (coiled tubing, wireline,

Figure 3.10 Horizontal tree plug. *Image courtesy of NOV Elmar.*

snubbing) are needed. From the well intervention perspective, horizontal trees have few advantages. They are, however, advantageous on wells that are likely to require frequent recompletion (replacement of the tubing string), the best example being wells with ESPs. ESPs fail with depressing regularity, requiring frequent replacement. Use of a horizontal tree can result in considerable time savings, as the tree does not have to be removed to allow the completion to be recovered. Horizontal trees are increasingly used on modern offshore platforms and satellite towers. The compact nature of the tree enables designers to reduce the overall size of the platform structure.

3.2 TUBULARS

Rig site supervisors need to have a working knowledge of the properties of oilfield tubular goods. Completion tubing is described using terminology that is in widespread use amongst industry professionals. Tubing is defined with regard to:

- size (outer diameter, OD),
- weight in pounds per linear foot (lb/ft),
- yield (usually in psi),
- grade (metallurgy).

For example, a tubing string selected for a completion might be described as: 5½″ 17 lb/ft L-80 13% chrome.

3.2.1 Tubing size

Tubing size is defined by the nominal OD of the base pipe and excludes connections. Integral connections and threaded couplings have a larger diameter than the base pipe (except where flush joint connections are used). Tubing is supplied in a range of industry standard sizes. The American Petroleum Institute (API) defines pipe with a diameter of 4½″ and smaller as tubing,[1] whereas pipe having a diameter of 4½″ and larger is defined as casing. Somewhat confusingly, 4½″ can be either tubing or casing. The definition is further confused, since pipe with a diameter larger than 4½″ is often used as production tubing. In fact, pipe as large as 9⅝″ has been used in some fields.[2]

3.2.2 Tubing weight

Tubing weight is recorded in pounds per linear foot of the base pipe (it does not include the connections). Since the API defines tubing by OD, increasing tubing weight per foot for a given OD must logically (and obviously) come about through an increased wall thickness (reduced ID). Tubing is specified by OD and weight of pipe per linear foot. If OD and weight are known, the internal diameter (ID) can be found using tables, such as the Halliburton Red Book, Baker Tach Facts or the Schlumberger iHandbook. For example, 5½″ 17 lbs/ft base pipe has an ID of 4.892″.

3.2.3 Tubing grade and tubing yield

The grade of the tubing refers to the metallurgical properties. API designated grades are prefixed with a single letter, for example, L, N, P, and Q. The properties associated with each letter are laid out in API 5CT . For example, L grade material is suitable for sour service conditions, while N grade is not. In addition to the API defined grades, most manufacturers have their own proprietary grade designations. For example, Sumitomo prefix many of their grades with the letters SM. Information about the properties of a proprietary grade must be obtained from the manufacturer. Most manufacturers have comprehensive websites that list the properties of their grades. Grade designations are often stenciled on the pipe body. In addition, most manufacturers will color code the couplings and pipe to aid grade identification (Fig. 3.11).

Generally, the letter denoting tubing grade is followed by a number; for example, a tubing joint might be stenciled L-80. The number 80 denotes yield, and is given in thousands of psi, 80 equating to 80,000 psi, 110 equates to 110,000, and so on. The axial strength of a piece of tubing

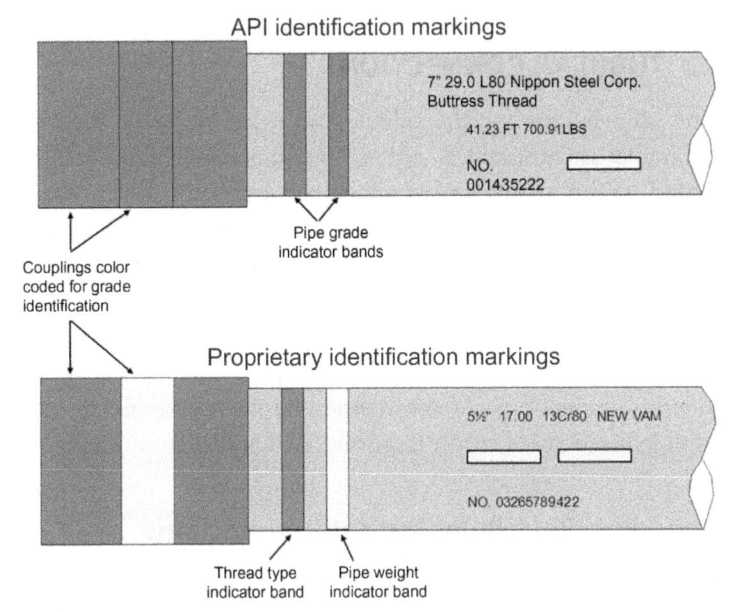

Figure 3.11 Pipe markings are used to indicate grade and yield.

can easily be calculated by working out the cross-sectional area of the tubing wall (in in^2) and multiplying this value by the tubing yield (in psi).

For example, to calculate the yield of 4½" 12.6 lb/ft tubing:

The ID must be obtained from tables. The ID that corresponds to 4½" 12.6 lb/ft tubing is 3.958".

The cross-section of the tubing is therefore $\pi/4 \times (OD^2 - ID^2)$, becomes $\pi/4 \times (4.5^2 - 3.958^2) = 3.6$ in^2.

The axial strength is 80,000 psi $\times$ 3.6 in^2 = 288,000 lbs.

3.2.4 Tubing length range

The API specification 5CT defines tubing and casing range, and refers to the length of casing and tubing joints. Tubing joints will fall within one of three ranges.

API tubing ranges:

- Range 1:20–24 ft,
- Range 2:28–32 ft,
- Range 3:32–48 ft.

Casing ranges are different:

- Range 1:16–25 ft,
- Range 2:25–34 ft,
- Range 3:34–48 ft.

3.3 TUBULAR CONNECTIONS

Pipe connections for the oilfield use many different thread forms, but can broadly categorized as either "integral connector" or "threaded and coupled." With an integral connector thread, both box and pin connections are cut into the body of the base pipe. In coupled connections, the pipe body is cut with a pin connection at each end of the joint. The joints are then connected using a separate box/box coupling. Threaded and coupled is by far the most common type of connection, as they are simpler and cheaper to produce, especially where corrosion resistant alloys are required. Connections are also defined by the sealing mechanism, and can be divided into two broad categories, API and premium (Fig. 3.12).

3.3.1 American Petroleum Institute connections

API connections are widely used. They are cheap, simple, and readily available. However, they are not suitable for all wells, having a number of limitations.

- They are weaker than the tubing when the connection is in tension.
- Axial loads are taken through the threads.
- Because of the thread flank angle, they are prone to "jump-out."
- The sealing mechanism is thread interference, and sealant (pipe dope) is needed to prevent liquids escaping through the spiral leak path of the thread. Time and differential pressure can extrude the pipe dope, and leaks will develop. The API connections are not suitable for use in gas wells.

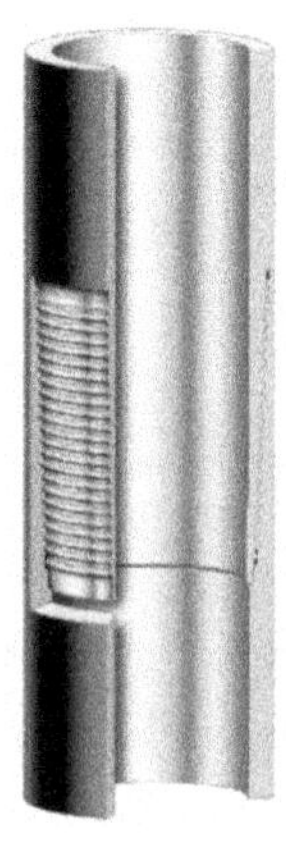

Figure 3.12 Thread and coupled connection (left) and integral connection (right).

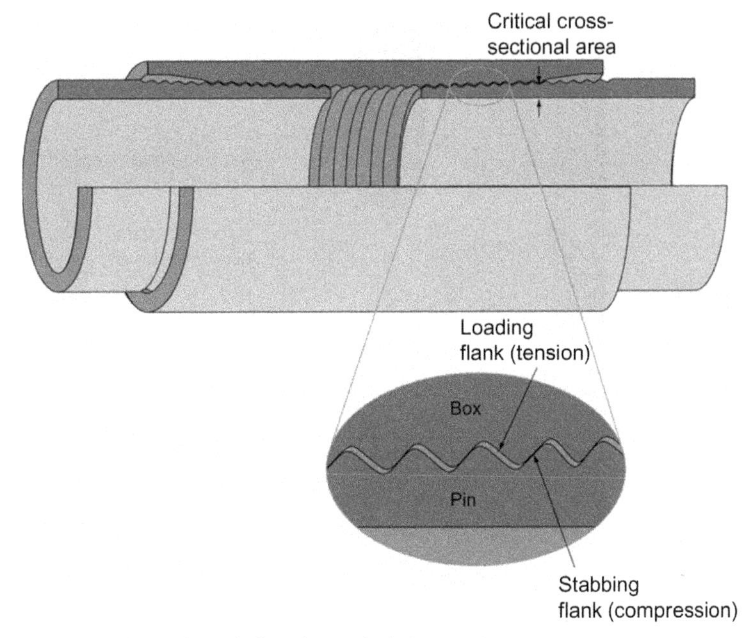

Figure 3.13 API long threaded and coupled connection.

- Part of the thread is exposed on the internal face of the connection, and can induce turbulence (Fig. 3.13).

3.3.2 Premium connections

Premium connections use a metal-to-metal seal face to ensure a gas-tight seal. The thread provides mechanical strength, and thread geometry is designed to allow easy make up and break out. Many connections incorporate a torque shoulder to reduce compression loading on the threads. Most premium connections are matched to the pipe body yield in tension, as well as burst and collapse limits. Although most premium connections are matched to pipe body yield when under compression loads, a significant minority are weaker in compression than the base pipe. Completion engineers need to be aware of these limitations, and ensure they are included in stress analysis calculations. Failure to do so could compromise the integrity of the production tubing. ISO standard 13679 details testing protocols for tubing connections (Fig. 3.14).[3]

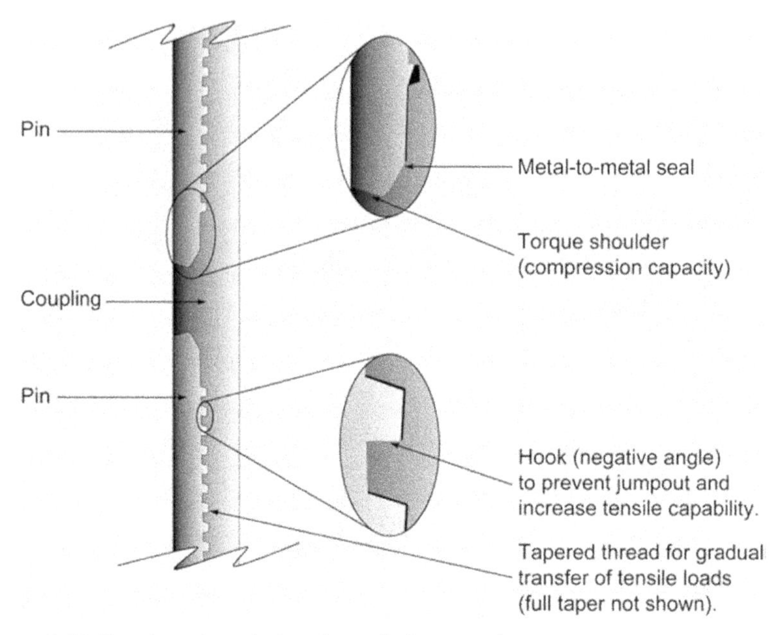

Figure 3.14 Premium threaded and coupled connection.

3.4 PRODUCTION LINERS

A liner can be defined as a casing string that does not extend back to the wellhead. Instead it is set (hung off) above the shoe of the previous casing string. Liners are used for drilling purposes and as production liners. Drilling liners are used to extend intermediate casing through troublesome zones where further drilling is planned. Many subsea wells use drilling liners, because of the limited number of casing strings that can be installed in a subsea wellhead. Production liners are set across the reservoir, and are an integral part of the lower (sand-face) completion. Since they are exposed to flow, they must be compatible with produced (and injected) fluids and robust enough to withstand any additional stress from production enhancement operations, for example, reservoir stimulation. Liners are normally cemented in place, usually along their entire length. Slotted and predrilled liners are not cemented, although increasingly swell packers and open hole packers are used for conformance (zonal isolation) purposes.

3.4.1 The liner hanger

A liner hanger, as the name suggests, secures and supports the liner. It uses mechanical slips to grip the inside of the casing a pre-determined distance above the casing shoe. The space between the liner hanger and the casing shoe is called the liner lap. Liner hangers can be set hydraulically, mechanically, or a mixture of the two. Most liners are cemented back to the liner hanger. Some systems are designed to allow liner rotation after the hanger is set as the ability to rotate has been shown to improve cement bond, particularly in long high angle wells.[4] The price of having improved cement integrity is a mechanically more complex hanger system.

Some liner hanger systems have a polished bore receptacle (PBR) above the liner hanger. This enables the base of the production tubing to be stabbed into the liner top, providing both a seal and a continuous conduit for produced fluids (Fig. 3.15).

Figure 3.15 Weatherford rotatable liner hanger with integral liner top packer.

3.4.2 Liner top packers

Liner top packers can be run as an integral part of the liner system, and are normally positioned above the hanger slips. They create a seal between the liner and the supporting casing. There are several reasons why this may be necessary:

- Isolation of formation pressure below the liner top and from the casing above.
- Isolation of pressure during stimulation treatment.
- Isolation of formation fluids while cement is setting—prevention of gas migration.
- Isolation of lost circulation zones.
- Isolation of the production zone in uncemented liner completions.

It is possible to run a liner top packer retrospectively. This would generally be done where an existing packer failed to set, or had failed an integrity test.

3.4.3 External casing packers

External casing packers (ECPs) are run with casing or liner. They may also be called annulus casing packers, or casing annulus packers. ECPs are run to provide an annular seal between the running string (liner) and the borehole. They enable multiple zone reservoirs to be completed using open hole completion techniques (uncemented liner). Communication devices can be run between ECPs to enable each zone to be produced. ECPs are run to depth as part of the liner string. Once on depth, an elastomer bag is inflated to provide a seal against the open hole. Inflation is usually carried out using mud, and/or water. More recently they have been inflated using cement.[5]

The ECP is made up of a metal mandrel and collars, an inflation element, an inflation valve system, and sealing mechanisms for parts that move during installation. They are normally identified by the casing OD, the uninflated seal element OD, and the seal length.

3.4.4 Swell packers

Swell packers use an elastomer seal element bonded to the outside of a length of pipe. Swell technology uses the swelling properties of different elastomers that expand to create a seal. Elastomers can be water sensitive, oil sensitive, or a hybrid that combines both water and oil sensitivity.

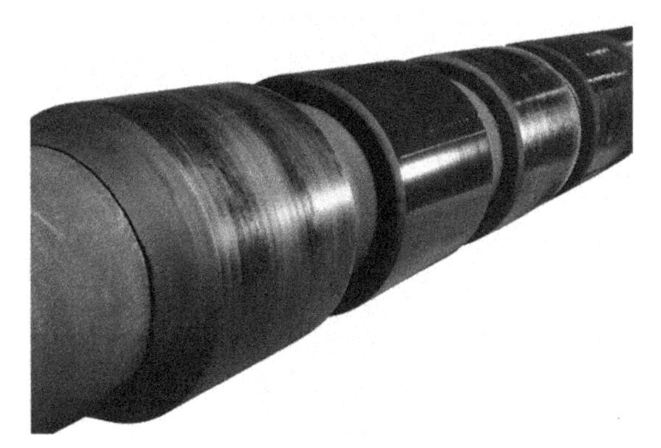

Figure 3.16 Swell packer.

They are have largely replaced ECPs as the method of choice for open hole multiple zone completions, where they are used to provide isolation between zones. They are also used as an alternative to cemented and perforated liners,[6] and have been used at the liner top as a cement assurance tool (Fig. 3.16).

3.5 WIRELINE ENTRY GUIDES

The wireline entry guide (WEG) or mule shoe is commonly used at the bottom of a tubing string. It has two functions. To guide the tubing into a liner top PBR or packer seal bore, and to guide through tubing intervention tools back into the tubing. To perform these functions, a properly designed WEG should have internal and external tapers. Some entry guides are "self-indexing." These self-indexing mule shoes make it easier for seals to engage in a PBR in deviated wells.

3.6 LINER TOP SEAL ASSEMBLY

A liner top seal is constructed from a series of chevron (V) packing stacks or bonded seals mounted on a flow mandrel (Fig. 3.17).

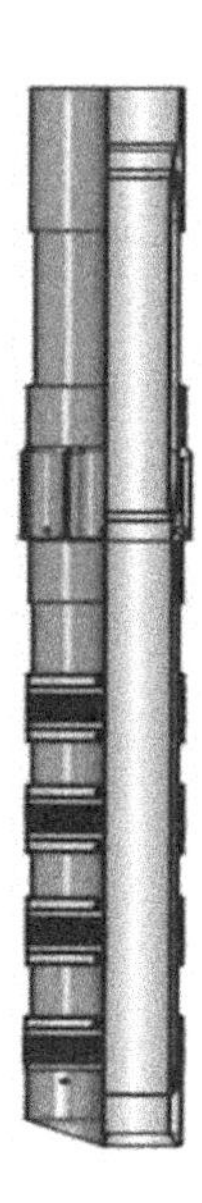

Figure 3.17 Seal assembly.

When a seal assembly is stabbed into a liner top PBR, a barrier is created, isolating the annulus. However, if the tubing above the seal assembly is not anchored the seal will be dynamic (moving), and less reliable than a static seal. Most completion engineers will not rely solely on a liner top seal for well integrity. More usually a production packer (see Section 3.11) will be set a short distance above the liner top.

Fig. 3.18 shows a WEG, seal assembly, and no-go locator stabbed into a polished seal bore located above a liner hanger/packer. This configuration provides a continuous conduit from the upper completion and down into the liner. In the example shown, the production packer is close to the liner top. This is representative of many completions. While this type of design is common, it has the disadvantage of creating a trapped annulus between the packer and the liner top. To overcome this, many completion engineers require the seal assembly to be nonsealing—a contradiction in terms. Seals are replaced by wiper rings, or the mandrel is machined with a spiral groove below the elastomer seals, allowing trapped pressure to escape. A simple alternative is to drill a hole in the tubing between the packer and liner top.

The rig site supervisor should be aware of the planned status—sealing or nonsealing—since it defines the location of critical well barriers. An unplanned trapped annulus could lead to a tubing collapse.

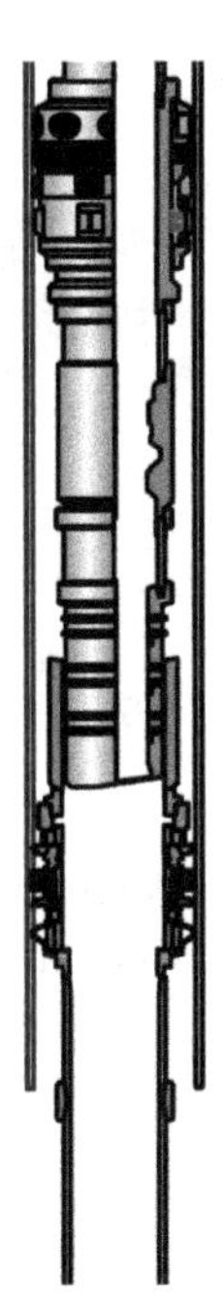

Figure 3.18 Seal assembly and liner top run in conjunction with a production packer.

3.7 FLUID LOSS CONTROL VALVES

Fluid loss is a problem during many open hole completions. A range of downhole valves have been developed that can help control losses while the upper completion is being run. Different vendors call the loss control valves different names, but they all have the same function; to provide a downhole mechanical barrier.

- Halliburton—formation saver valve (FSV)
- Schlumberger—formation isolation valve (FIV)
- Baker—fluid loss control valve (FLCV)
- Weatherford—fluid loss valve (FLV).

Loss control valves are usually run with the lower (sand-face) completion. For example, in gravel pack wells the valve is run below the gravel – pack packer and above the sand control screen. After gravel packing the well, the service tool and washpipe are recovered to the surface. A shifting tool on the end of the washpipe closes the FLCV, isolating the reservoir and forming a well control mechanical barrier. Once the upper completion

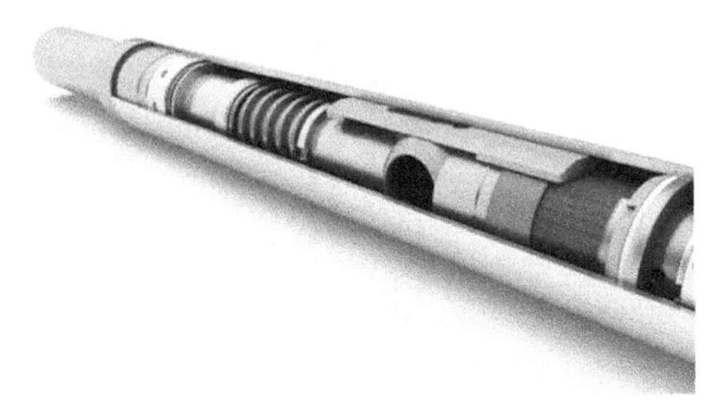

Figure 3.19 Formation saver valve (FSV). Source: *Photo courtesy of Precision Oiltools Ltd.*

has been installed, the barrier is opened allowing the well to produce. Most valves open following a predetermined number of pressure cycles. This eliminates the need for mechanical intervention. Having a mechanical barrier undoubtedly reduces the possibility of losses (and therefore a kick) during completion operations. However, there are some issues potential problems associated with these valves.

- Debris can accumulate on top of the valve, preventing it from opening. Keeping the wellbore clean and performing a dedicated clean out trip is essential.
- There have been incidents where the valves have been opened inadvertently. On one occasion a valve was mechanically opened after the tail-pipe of a wellbore cleaning work string caught on the mechanical shifting profile in the body of the loss control valve. The fluid in the casing was underbalanced and the well kicked. On other occasions, poor record keeping on pressure cycles has resulted in the valves opening unexpectedly.

In spite of these problems, having a barrier in the well is a good thing and one to be recommended, mainly because of well control considerations, but also because of reduced formation damage (Fig. 3.19).

3.8 LANDING NIPPLE

A locating nipple or landing nipple allows a range of wireline (or coiled tubing) conveyed tools to locate and, where necessary, seal inside the tubing string. A nipple is a short length of heavy wall tubing machined with an internal profile that will receive and secure a lock mandrel. Most nipples have a honed bore that enables seals mounted on the lock mandrel to form a pressure barrier at the nipple. Most completions have one or more nipple profiles. The common locations and uses of these nipple are:

- Tubing hanger. This profile is most commonly used to plug the well. The plug would be one of the two mechanical barriers normally required to allow the BOP to be removes and the Christmas tree installed. This profile can also be used to test the tree and allow tree valve repairs to be carried out.
- A set distance below surface (or mudline) and used for the setting of a downhole safety valve. If a tubing retrievable valve is used, it will have an integral nipple profile used to locate an insert valve.
- Below the production packer: A plug placed in this nipple would be used for hydraulic packer setting and tubing tests.
- Installation of memory gauges for reservoir monitoring purposes.
- Deployment of downhole chokes.
- Landing of syphon or velocity strings.
- Installation of pack-off and straddle tools.

Although there are many different nipple systems available, they all fall into one of two broad categories; no-go and selective. No-go nipples have a small internal step (no-go) that is sized to allow a lock mandrel to locate inside the profile. The position and size of the no-go, the profile of the locking dog recess, and the length and diameter of the seal bore are all design specific. Lock mandrels are not interchangeable between different nipple systems (Fig. 3.20).

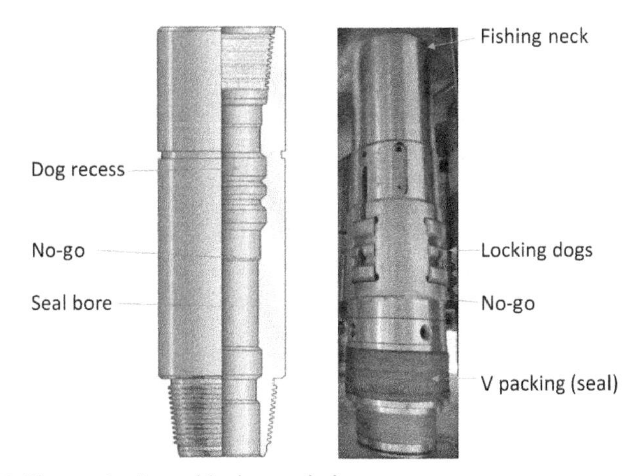

Figure 3.20 No-go nipple and lock mandrel.

If more than one no-go nipple is needed for the same size tubing string, then the no-go diameter must be progressively reduced with depth. If a large number of nipples are required, the cumulative reduction can be significant, seriously reducing the through bore of the completion. This cumulative restriction is compounded if the nipple has a no-go that is designed to take load. In some older systems, force (load) from above, for example, during a pressure test against a plug, acts on the no-go shoulder. To withstand this force, a substantial no-go is needed, resulting in a significant reduction in ID for each nipple size. Modern nipple systems have a much smaller step change, since the no-go is designed for location only.

If a no-go system is limiting size too severely, a selective nipple system might be used. If selective nipples are used, any number of the same size seal bores can be run in a completion; in theory at least. Since selective nipples do not have a no-go shoulder, a different method of locating the lock in the nipple profile is needed. Most systems require careful manipulation of the setting tool. Some utilize different shaped locking dogs. There are limitations with selective systems. To reach the lowest nipple in the string, the lock will have to pass through one or more nipples of the same size. Because the packing is designed with an interference fit, a degree of mechanical manipulation or jarring will be required to push the lock past each nipple. Premature shearing of running tool setting pins is not uncommon.

3.9 FLOW COUPLINGS AND BLAST JOINTS

Blast joint and flow couplings are short joints of heavyweight tubing. They are usually made from coupling stock, so it is common for the OD to match the tubing coupling OD, and ID to match the tubing ID. They are run as part of the completion, and located in places where a higher than normal rate of erosion is expected.

3.9.1 Flow coupling

Flow couplings are run where there is a large change in diameter, and where turbulent flow might lead to accelerated erosion. For example, at the safety valve, cross-overs, side pocket mandrels, and nipple profiles.

3.9.2 Blast joint

Blast joints are used in multiple zone wells where the completion runs across open perforations. Blast joints are used to counteract the erosion effects caused by jetting from perforation tunnels.

3.10 PRODUCTION PACKERS

Production packers create a seal between the outside of the production tubing and the casing or liner. They protect the casing from the potentially corrosive effects of produced fluids, and form a well control barrier. Packers are used in multiple layer reservoirs to provide isolation between each different producing zone. They are also used as service tools during well tests, stimulation operations, and during remedial cementing.

Not all wells require packers, they are rarely used in rod pump wells, and many ESP completions are run without packers.

Although packers are used for a wide variety of applications, most share some common characteristics. They all have a mandrel to enable fluid flow through the packer, slips to anchor the packer to the casing wall, and an elastomer element that extrudes to seal against the casing wall. There are three classifications of packer: permanent, retrievable, and permanent/retrievable. They are set using mechanical manipulation, hydraulic or hydrostatic pressure, or wireline conveyed explosive setting kits.

3.10.1 Permanent packers

Permanent packers are usually simple, cheap, and robust. In general they are able to withstand much higher tension, compression, and differential pressure than retrievable packers. They generally have a larger through bore than the retrievable equivalent, and are ideal for the following applications:

- high pressure completions,
- high volume completions,
- high pressure/high volume frac jobs with anchored or floating tubing, and
- isolation of a lower zone by using a permanent packer as a bridge plug.

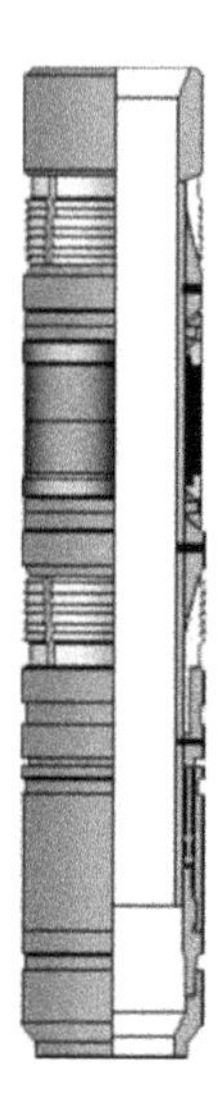

Figure 3.21 A permanent packer.

The only notable disadvantage of using a permanent packer is its permanency, removal requires a milling operation. In a well where the packers are constructed from high grade stainless steel alloys, milling will be more difficult; casing damage becomes a risk (Fig. 3.21).

Permanent packers can be run ahead of the completion. Packers that have a long, heavy tail-pipe would be run on drill-pipe and set hydraulically or mechanically. More usually, the packer is run on e-line, enabling very accurate depth determination. With the packer in place, the upper completion is run. Seals on the tail pipe of the production tubing are stabbed into the packer seal bore, isolating the annulus. Most permanent packers only have short seal bores. If a high degree of seal movement is expected, a seal bore extension can be run below the packer to protect the seals during production.

3.10.2 Retrievable packers

Many retrievable packers are service packers of the type used during well tests, interventions, cementing operations, and for temporary well abandonment. When run on a work string, they are normally set mechanically by rotating the string and setting down-weight. Some are designed to be set, released, and re-set a number of times in a single trip.

Retrievable packers used for completions are usually run with the completion tubing and set hydraulically. In most cases they are retrieved

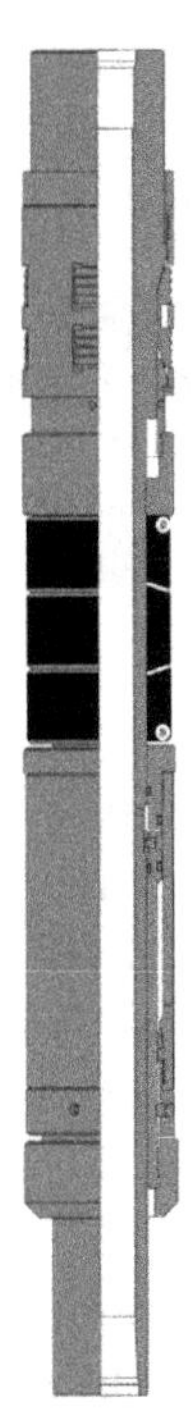

Figure 3.22 Hydraulic set, pull to release retrievable production packer.

with a straight pull (tension) on the tubing. This simple recovery mechanism makes them unsuitable for some completions. For instance, pumping cold fluid down the tubing during a scale inhibitor squeeze would risk unseating the packer, because thermal contraction and ballooning would significantly increase string tension. Similarly, a bullhead well kill could also result in a premature release of the packer. The consequences of such an event need to be properly considered, although in the case of a well kill, it might actually be advantageous. Potential difficulties aside, the ease with which retrievable packers can be pulled means that workover operations are very much simplified. They are a popular choice for wells where frequent workovers are required; ESP completions are a good example (Fig. 3.22).

3.10.3 Permanent retrievable packers

With the introduction of intelligent completions, the need to feed hydraulic, electrical, and fiber optic lines through packers has become

Figure 3.23 Permanent releasable packers ready for deployment in an intelligent completion. Note the various feed-through conduits serving hydraulic and instrument cables.

commonplace. Intelligent systems are often used in multiple zone, single string completions. This has led to the development of the "Permanent Retrievable" packer—sometimes called "Perma-Trieve."[c] A permanent retrievable packer is usually run with the completion tubing, then set hydraulically or hydrostatically. Once set, these packers have high load and pressure retention capability, on a par with many permanent packers (Fig. 3.23).

Permanent releasable packers are pulled with the completion tubing, but first need to be released, usually by some form of mechanical manipulation; normally cutting or punching the inner mandrel. Well-designed permanent/retrievable packers have many features in common:

- Moderate (variable) setting pressure to reduce tubing stretch while setting. This also reduces residual tension in the tubing once the packer is set.
- No mandrel movement during setting, allowing the stacking of a number of packers in a multiple zone completion.
- Multiple feed through for control line by-pass in intelligent completions.
- Simple and reliable release mechanism.

3.10.4 Multistring packers

Dual and triple string completions require single, dual, and triple string packers to maintain separation between zones (Fig. 3.24).

Dual and triple string packers are usually hydraulically set retrievable packers. They can be configured to set with the application of pressure to either string.

[c] Perma-Trieve is a registered trademark of Halliburton.

Figure 3.24 Hydraulic set pull release dual string production packer.

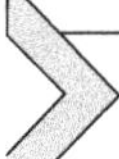

3.11 PACKER SETTING

Packers will be set in one of the following ways:

- Hydraulic set. The packer is run on the completion string. When it is on depth, the tubing must be plugged and pressure applied. The packer will set with the application of pressure at a predetermined value.
- Hydrostatic set. The packer is run on the completion tubing. Once on depth, the wellbore is pressurized to a predetermined value. The pressure needed to set the packer is hydrostatic pressure at the setting depth plus a predetermined margin than must be applied at the surface. A hydrostatic pressure can only be used if there is something in the wellbore to pressure against, for example, unperforated casing or a FLCV. They cannot be used where the wellbore is open to the reservoir. Where multiple hydraulic or hydrostatic packers are run, they can be set sequentially or simultaneously.
- Mechanical setting. These packers can be run on a drill-pipe and set before the completion is run. Alternatively, they can be run with the completion tubing. Once on depth the packer is set by mechanical manipulation.
- Wireline setting. The packer and tail-pipe is run on e-line, and set using a slow burn explosive charge that drives a setting piston, compressing and setting the packer. With the packer and tail-pipe in place, it is possible to install a wireline retrievable plug before the completion is run, thus providing a well control barrier. Alternatively, the plug can be preset in the tail-pipe before the packer is run.

3.12 PACKER-TO-TUBING CONNECTION

There are a number of methods of connecting the completion tubing to a production packer. Of the most commonly applied, three provide a static connection, the fourth uses a dynamic seal, allowing tubing movement.

3.12.1 Premium threads

The simplest, and the most reliable, method is to screw the tubing directly into the top of the packer with a premium thread connection. When tubing is connected to a packer using a thread, pressure integrity comes from the seal face on the thread profile. All of the alternative methods of connection rely on elastomeric seals, and are not as reliable in terms of pressure integrity.

3.12.2 Ratch latch or anchor latch

Tubing can be connected to the packer using a stab-in ratchet mechanism commonly referred to as a "Ratch Latch". This has a coarse thread (usually left hand) that can be snapped into a profile in the top of the packer by the application of down-weight. Elastomer seals, positioned below the latch mechanism, seal in a honed bore in the top of the packer. Most latches are removed by clockwise rotation (Fig. 3.25).

3.12.3 Locator seal assembly

A locator seal assembly is a series of elastomer seals arranged along the length of a mandrel that are designed to locate in the packer seal bore. They are free to move (dynamic) when the well is in service. Having a

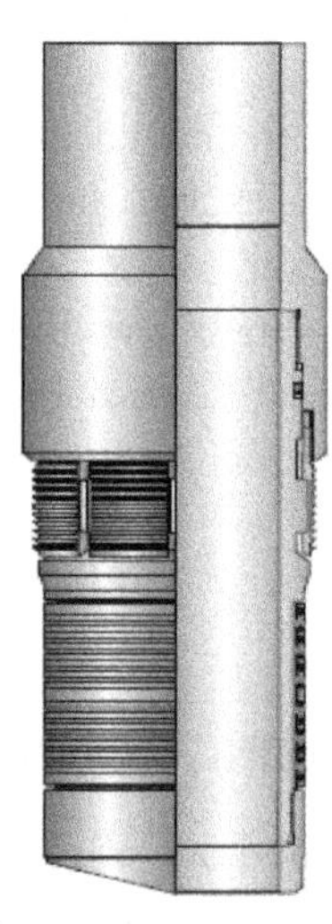

Figure 3.25 Ratch latch assembly.

dynamic seal helps reduce axial loads caused by pressure and temperature changes in the tubing. Most dynamic seals will no-go at the packer, limiting downward movement. Upward movement is not normally restricted, and so extremes of temperature (cooling) and pressure (ballooning and piston forces) could cause the seals to stroke out the top of the packer. Seal length can be varied, and will depend on how much movement is expected when the well is in service.

3.12.4 Seal bore extension

Some seal assemblies can be quite long, in excess of 25 ft. The seal bore section of the packer, on the other hand, is usually short. Two or three feet is typical. If seals are continually moving in and out of the seal bore they will be prone to damage. To prevent this, an extension of the seal bore can be run on the bottom of the packer.

3.12.5 J-Latch connector

Some seal stacks will be anchored to the packer using a j-slot or j-latch device above the seals. This allows the seals to be un-jayed from the packer, and the upper completion removed for workover.

3.12.6 Seal assembly and polished bore receptacle (ELTSR)[d]

A seal assembly/PBR is subtly different from the locator seal assembly, but fulfills the same purpose as the locator seal. Whereas the locator seal stabs into a seal bore within the body of the packer, the PBR is positioned above the packer. PBRs generally have a larger through bore than a locator, usually matched to the tubing ID. They are used where a lot of seal movement is anticipated (Fig. 3.26).

3.12.7 Telescoping joint

The telescoping joint can be used instead of the seal assembly. It performs a similar function—expansion and contraction of the tubing string. The only advantage of the telescoping (expansion) joint is that it can be positioned anywhere in the string. However, they are rarely used in modern completion as they are not normally as reliable or robust as a seal assembly. They are still widely used to allow string movement in well test DST strings.

[d] ELTSR-Extra long tubing seal receptacle.

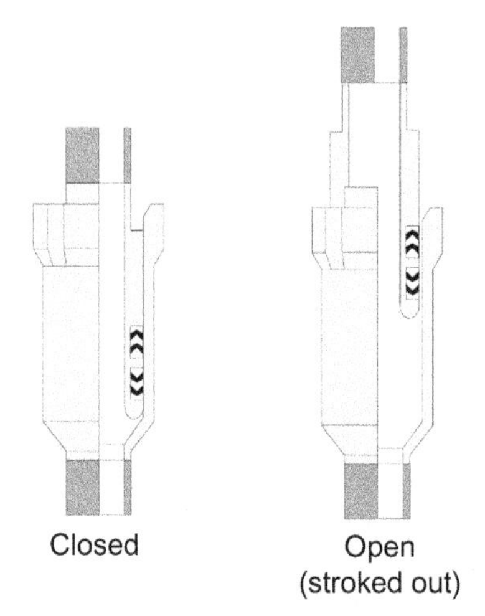

Figure 3.26 Seal assembly and PBR.

3.13 CHEMICAL INJECTION MANDRELS

Chemical injection mandrels enables chemicals to be injected into the produced fluid within the confines of the tubing. They are mainly used to inject scale, wax, and corrosion inhibitors, and methanol for the prevention and dispersal of hydrates. Chemical injection mandrels are run with the completion tubing. A stainless steel, Monel, or alloy 825 chemical injection capillary tube runs from the mandrel back to the surface. Chemical injection lines can be ¼″, but are more generally ⅜″ or even ½″ diameter. Most will have a plastic encapsulation to reduce vibration-induced fatigue.

Mandrels should be fitted with check valves (preferably dual checks) to prevent ingress of hydrocarbons into the annulus if the injection line fails. Many mandrels will have a rupture disk fitted. This enables the control line to be run with pressure applied. Once the completion has been landed, pressure is increased to burst the disk and allow injection through the line.

Chemical injection systems that protect against calcite scales will normally be run as close to the reservoir as possible. The mandrel should be positioned below pressure gauge mandrels and gas lift mandrels (if run), as this protects them against scale build-up

Where possible, injection mandrels for the dispersal of hydrates and wax will be run to below the hydrate or wax formation depth (Fig. 3.27).

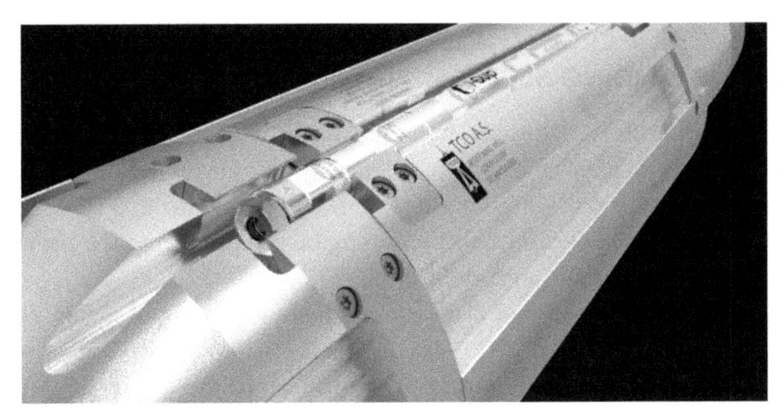

Figure 3.27 Chemical injection mandrel. *Illustration courtesy of TCO™ group.*

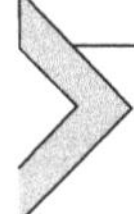

3.14 DOWNHOLE PRESSURE AND TEMPERATURE GAUGES

Downhole pressure gauges can be a valuable source of data for reservoir engineers, and are widely used in offshore wells, especially subsea wells. Having accurate and up-to-date reservoir pressure data can be invaluable if preparations need to be made to kill a well in advance of a workover. Accurate pressure data can also reduce the risk of overpressure during interventions where fluid is pumped into the well, for example, scale inhibitor squeezing or stimulation treatments.

A basic quartz gauge will provide continuous surface monitoring of pressure and temperature at gauge depth. The gauge is mounted in a mandrel that is run with the completion. A ¼″ hollow steel control line containing an insulated electrical conductor is attached to the gauge and clamped to the outside of the tubing as the completion is run. After the completion has been landed, the cable is fed through the tubing hanger and exits the wellhead using electrical penetrators (Fig. 3.28).

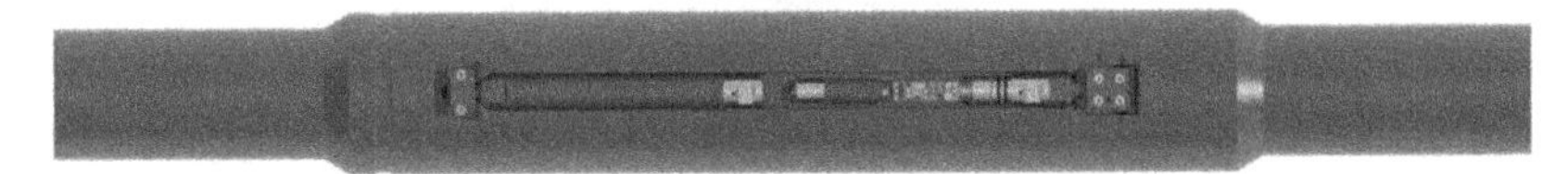

Figure 3.28 Gauge mandrel. *Image courtesy of Omega.*

3.15 SLIDING SLEEVES

A sliding sleeve allows communication between tubing and annulus, and is usually opened and closed mechanically, using wireline or coiled tubing deployed tools. Common uses of sliding sleeves are:

- To circulate a light fluid into the tubing before producing the well.
- To circulate kill fluid into the well.
- To allow production or isolation of each zone in a single string multi-zone completion.

It is common to have a nipple profile built into the top of the sliding sleeve. This can be used to install a straddle if a sleeve is leaking. It can also be used to locate plugs, hang gauges, or install a choke to control flow. In some completions, a sleeve is run immediately above the packer and used to circulate fluids. This practice has been discontinued by most completion engineers, and for a very good reason. Over time the sleeve can, and often does, begin to leak. The result is a live annulus. Added to this obvious problem is a tendency for the sleeve to stick in the closed position after a prolonged period in the well. Gaining a circulation path before a workover is now more commonly achieved by punching a hole in the tubing (Fig. 3.29).

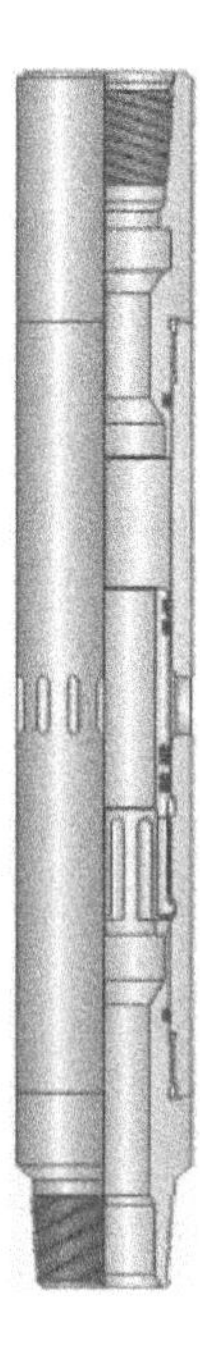

Figure 3.29 Sliding sleeve.

3.16 PORTED NIPPLES

Ported nipples can be used as an alternative to sliding sleeves. Honed bores above and below ports allow straddles to be located in the nipple isolating flow from behind pipe. Production from a zone can be reinstated by pulling the straddle.

3.17 INFLOW VALVES (INTELLIGENT COMPLETIONS)

In multiple zone single string intelligent completions, production from each zone is controlled by opening and closing a hydraulically operated communication device. These are variously called:

- Inflow control valve (ICV)—Halliburton
- Flow control valve (FCV)—Schlumberger
- Remotely operated sliding sleeve (ROSS)—Weatherford.

Inflow valves for intelligent completions vary in design and complexity, from a simple open/close valve through valves with incremental choke (multiple position), and valves with an infinitely variable choke and position sensor (Fig. 3.30).

3.18 SIDE POCKET MANDRELS

Side pocket mandrels are elliptically shaped with a valve pocket located to one side of the main bore. This enables different types of flow control devices to be installed and retrieved using standard slickline tools (Fig. 3.31).

Figure 3.30 Hydraulically operated Inflow Control Valve.

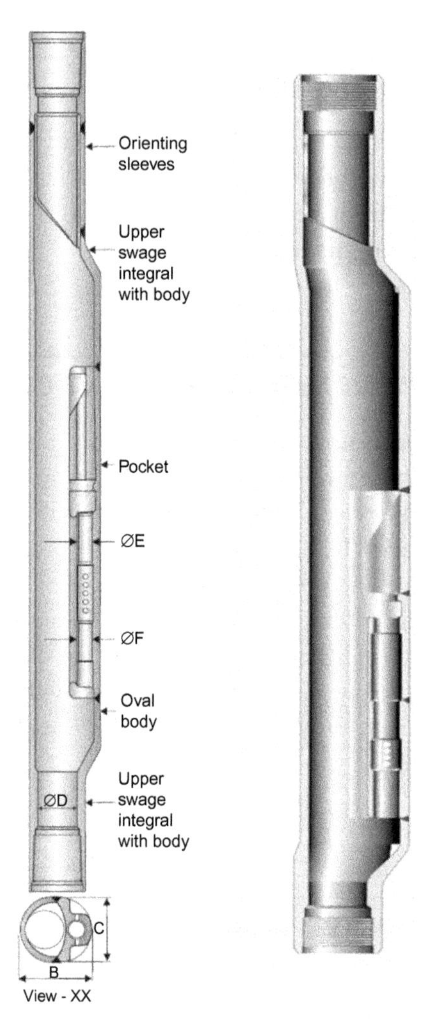

Figure 3.31 Side pocket mandrel.

Common applications for side pocket mandrels are:

- Circulation device. The mandrel can be run with a dummy valve straddling the ports in the side pocket. This can be pulled to provide a circulation path between the annulus and tubing. The blanking valve can be replaced once the circulation is completed.
- Chemical injection. A chemical injection valve can be located in the side pocket. Chemical inhibitors are pumped through a control line to the valve.
- Gas lift. By far the most common use for a side pocket mandrel. Gas injected into the annulus passes through a sized orifice in a gas lift

valve located in the side pocket. The gas mixes with the produced fluids, enhancing production. In some systems a series of side pocket mounted unloading valves are needed to unload fluid from the well before flow can begin.

3.19 SUBSURFACE SAFETY VALVES

The API defines a subsurface safety valve as a "device whose design function is to prevent uncontrolled well flow when closed." It can protect against collision, fire, equipment failure, unforeseen catastrophic events, and sabotage. The need to run a safety valve will in part be determined by local regulations, in part by company policy. Nowadays many wells are equipped with some form of downhole protection.

Although subsurface safety valves are intended for emergency use only, they have been used to provide an additional barrier when working on the Christmas tree and wellhead. Some operating companies now accept the subsurface valve as a barrier for well intervention work, but this is by no means universal.

Modern tubing retrievable SC-SSSVs have evolved from simple poppet type valves developed in the 1930s. Those valves were subsurface controlled. They relied upon an increase in flow velocity to close a tapered valve against a valve seat. A catastrophic failure of the wellhead would cause an increase in flow. Subsurface controlled valves were often placed in Gulf Coast wells to afford protection during the hurricane season, so became known as storm chokes. This term is still used today by some of the older generation!

Early valves had a number of disadvantages. Flow area was restricted and tortuous, they could not withstand high differential pressure, and calibration (flow velocity to close) was problematic. Nevertheless, they were used extensively, and led ultimately to the modern surface controlled high pressure ball and flapper valve systems in use today.

The modern SC-SSSV is fail-safe. A ball, or more usually a flapper, is held in the open position for as long as pressure is maintained in a hydraulic control line running from the surface to the valve operating piston. If, for any reason, pressure is lost, the valve will close and can only be reopened by once more applying pressure to the line. Clearly, the integrity and function of the valve and control line is of great importance to the well. Careful installation and management of subsurface valves can help improve reliability.

3.19.1 Subsurface controlled valves

Subsurface controlled subsurface safety valves (SSC-SSSV) are, as the name suggests, operated by flow (or injection) parameters below the surface. In simple terms, they close if flow velocity exceeds a predetermined value. These types of valves are run postcompletion, and are usually set in a conventional nipple profile. There are two types of valve for production wells.

3.19.1.1 Pressure differential safety valves

Pressure differential valves are normally an open valve. If flow velocity across an internal choke exceeds a predetermined velocity, the pressure drop across the choke will be sufficient to force piston travel against the operating spring. Piston movement will close the valve. Opening the valve requires pressure to be equalized from above (Fig. 3.32).

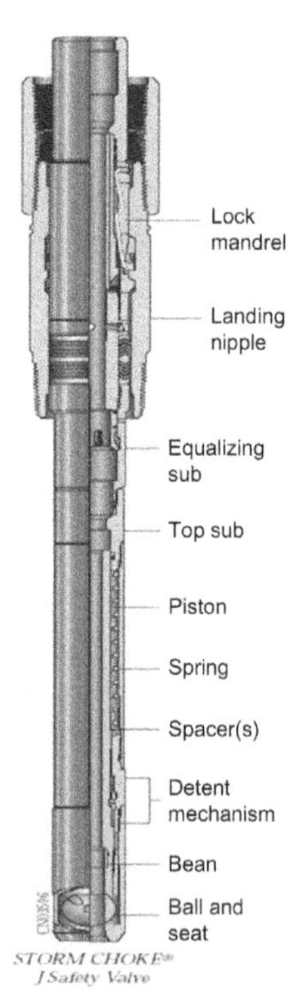

Figure 3.32 Flow-operated safety valve (storm choke). *Image courtesy of Halliburton.*

3.19.1.2 *Ambient pressure operated valves*

This type of valve uses a nitrogen charge to hold the valve open. If ambient pressure at the valve drops below a predetermined value, as may occur during a large volume leak at the surface, the valve will close. This type of valve has the advantage of not needing a choke to control its operation, and is therefore more suited to high velocity wells. However, it may require regular replacement in locations where there is a rapid decline in flowing pressure.

3.19.1.3 *Injection valves*

Not all injection wells will be fitted with a safety valve. However, where injection is into a hydrocarbon bearing formation, and where reservoir pressure is high enough for the well to flow back to the surface, a safety valve will often be used.

These are usually a simple "normally closed" valve that allows the injection of fluids or gas, but closes if the direction of flow reverses. Ball and flapper closure mechanisms are both used. However, they are to some degree flow sensitive. At low injection rates the valves can be prone to throttling. This will damage the valve if it goes on for too long. For this reason, the "poppet" type valve is now becoming increasingly common. This is a simple dart type check valve, and is not flow sensitive (Fig. 3.33).

3.19.2 Surface controlled valves

Modern SC-SSSVs operate on the fail-safe principle, and form an integral part of a production facilities emergency shut-down system. Where a number of wells are clustered together, for instance on an offshore platform, individual well shut-down systems will be interlinked.

During normal operating conditions, a surface panel supplies hydraulic pressure to the downhole safety valve via a hydraulic control line. The pressure drives down a piston, compressing a powerful return spring. At the same time the piston also moves the inner flow mandrel, or flow tube. Flow tube movement opens the valve; either by rotating a ball, or by pushing a flapper off seat and in to a recess away from the flow path (the flow tube will fully cover and protect the flapper). As long as control line pressure is maintained, the valve remains open. If control line pressure is lost, the spring pushes the flow tube up, allowing the flapper (or ball) to close. Closure can be triggered by pressure sensing pilots attached to the flowline, flame and heat detection equipment, gas detection equipment, or by manual activation (Fig. 3.34).

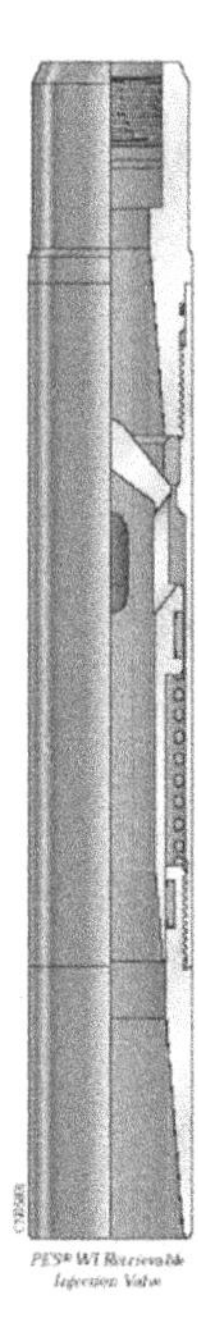

Figure 3.33 Injection valve. (Image courtesy of Halliburton).

3.19.3 Equalizing or nonequalizing valve?

Early equalizing valves were less reliable than their nonequalizing equivalent. In recent years, the reliability of equalizing valves has improved to the point where there is little to choose between them. Most, although not all, equalizing valves are equipped with a simple check seat recessed into the flapper, when control line pressure is applied to the valve the flow tube moves down until it contacts the flapper. If there is differential pressure across the valve it cannot open. However, the bottom of the flow tube pushes the equalizing port open. This allows pressure to equalize across the valve. Once pressure across the flapper is equalized, continued application of hydraulic pressure moves the flow tube down, fully opening the valve. Older types of equalizing valves had more complicated, hence less reliable, equalizing systems.

3.19.4 Tubing retrievable or wireline retrievable valves?

Surface controlled safety valves can either be run as an integral part of the completion tubing (tubing retrievable), or run through the tubing and set in a ported safety valve landing nipple (wireline retrievable).

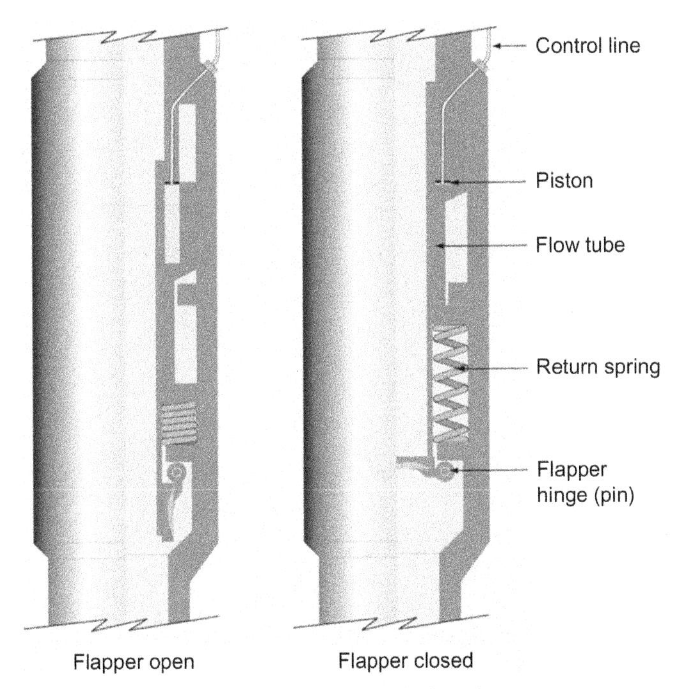

Figure 3.34 A SC-SSSV shown in the open and closed position. *Image courtesy of Jonathan Bellarby*

The tubing retrievable valve (TR SC-SSSV) has four significant advantages over the wireline retrievable (WR SC-SSSV) valve.

- Improved reliability.
- Improved integrity—the valve is an integral part of the completion string.
- Large through bore. The pressure drop and turbulent flow cased by the restricted ID of wireline set valves can lead to the formation of scale, compromising the valves integrity.
- No requirement to pull the valve for through tubing interventions.

Tubing retrievable valves have become the valve of choice for subsea completions, intelligent completions, and completions in locations where interventions are complex, time consuming, or expensive. A wireline retrievable valve functions in the same way as a tubing retrievable valve. Internally, the valve configuration of piston, flow tube, spring, and closure mechanism is the same. The only real difference is the method of delivering control fluid to the piston. In the tubing retrievable valve, the control line is connected directly to the hydraulic chamber. In wireline

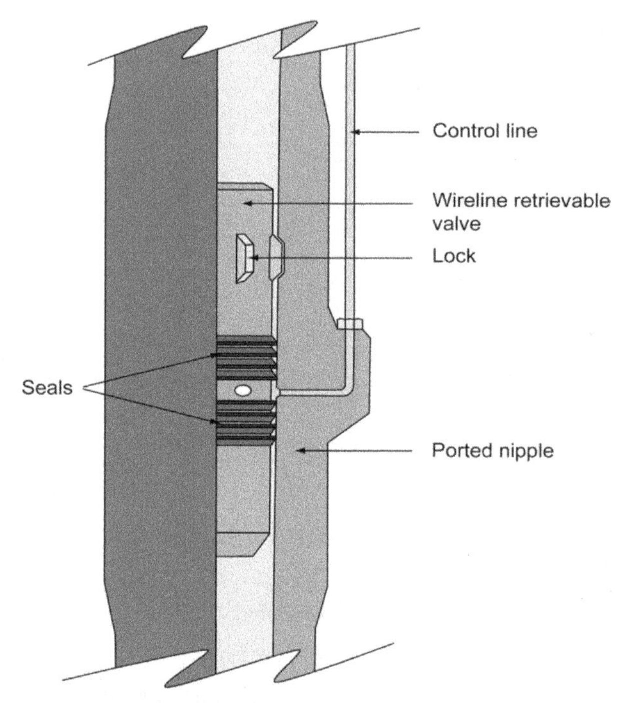

Figure 3.35 A wireline retrievable safety valve. (Image courtesy of Jonathan Bellarby).

retrievable valves, the control line is connected to a ported wireline nipple that is run as part of the completion tubing. Hydraulic fluid passes through a port into the tubing where it is trapped between two sets of seals on the body of the wireline retrievable valve (Fig. 3.35).

It then passes through a port in the valve body to reach the piston.

Wireline set safety valves can be set inside the body of a tubing retrievable valve, should the tubing retrievable valve fail an integrity test. A wireline intervention is required to prepare the tubing retrievable valve before the wireline set valve can be installed.

3.19.5 Installation of surface controlled downhole safety valves

A surface controlled downhole safety valve is a safety critical component and as such, particular care needs to be taken when installing the valve. Points to note are:

- Control line fluid and the fluid in pump used to pressurise the control line must conform to a high standard of cleanliness. Some companies use NAS (National Aerospace Standard) 6 or 8 specification[e]:
- The valve must be function tested at the rotary table before it is run. A record should be made of the fluid volume required to fully open the valve, and the fluid volume returned from the control line when the valve is closed.
- The control line should be run with pressure applied (valve open). Loss of pressure will give an early indication of line damage. An open valve also allows the well to be monitored.
- If the control line is damaged, the valve must be pulled back to the surface, and the line replaced. It is not normally permissible to splice a safety valve control line.
- The valve should be run as close as possible to the programme depth. Do not run valves deeper than the programme depth, they may fail to close.
- Monitor control line pressure at production start-up. Thermal expansion of the fluid in the control line caused by increasing flowing temperature can cause the hydraulic piston operating pressure to be exceeded.

3.19.6 Annulus safety valves

Annulus safety valves (ASVs) were developed primarily for use in gas lift wells. They are designed to retain high pressure hydrocarbon (lift) gas in the annulus in the event of a loss of integrity at the wellhead. An annular safety valve is basically a modified permanent releasable production packer. The packer is equipped with a port that allows annulus fluids to flow past the packer. Within the port is a fail-safe valve held open using hydraulic pressure. Like the conventional safety valve, the pressure to keep the ASV open is supplied from the surface through a control line. Loss of hydraulic pressure causes the valve to close (Fig. 3.36).

ASVs are nearly always run just below the tubing safety valve. If they were positioned above, the control line for the tubing safety valve would have to be cut and fed through the ASV. Splicing the control line to feed it through the ASV would have a detrimental effect on the reliability of the tubing valve.

[e] NAS Grade 1638 is a widely adopted standard to measure the contamination degree of hydraulic oil.

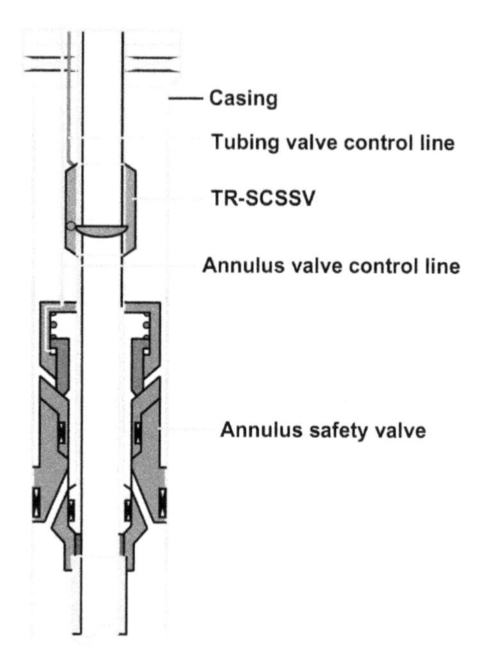

Figure 3.36 Annulus safety valve.

ASVs are mandatory in some European jurisdictions where gas lifted wells are located on manned offshore platforms. They are not a requirement for subsea wells, unmanned platforms, or land wells.

3.20 LUBRICATOR VALVES

A lubricator valve is a surface controlled downhole ball valve designed to hold pressure from above and below. Lubricator valves are used to allow the deployment of a long intervention tool string without the need to kill the well. For instance, running long perforating guns allows a long interval perforating in a single run. Lubricator valves are commonly installed in completions where rig up of intervention pressure control equipment is height limited. Unlike the downhole safety valve, the lubricator valve is not fail-safe. The Valve is opened and closed by pressuring a hydraulic control line(s) from the surface. If a lubricator valve fails, it will fail "as is," be that open or closed.

A lubricator valve is run as part of the completion string, and is located above the tubing or wireline retrievable downhole safety valves. This gives the safety valve protection while long tool strings are being loaded into the well. In addition, the safety valve can be closed and inflow tested to create a second barrier.

Running the lubricator valve should be handled in a manner similar to that used for running a tubing retrievable safety valve, that is, care of the control line is paramount. It is standard practice to run the valve in the open position with pressure applied to the control line (both lines in a dual line design valve). The valve will be pressure tested and function tested once the completion has been installed. Well control considerations when using a lubricator valve are covered in detail in Chapter 9, Wireline Operations.

3.21 CONTROL LINES

Complex modern completions can have several control lines, each one serving a different function. Control lines are made from 316 stainless steel or high grade stainless steel alloys (Alloy 825 or 925).

Common applications are:

- Hydraulic control lines.
  - For functioning downhole safety valves and flow control devices (sliding sleeves and inflow valves). These are usually ¼″ in diameter with a 0.049″ wall.
- Chemical Injection lines. Size varies, but are normally 3/8″ diameter.
- Instrument cables—electric and fiber optic.
  - Delicate electrical and fiber optic cables are run inside ¼″ stainless steel lines. This affords protection from annulus fluid and pressure.

Most control lines are protected with a hard plastic encapsulation. Encapsulation reduces vibration in the line and improves fatigue life. Encapsulation can be color coded to ease identification if multiple lines are being run. It is common practice to bundle the lines together into "flat packs" (Fig. 3.37). Steel cables are sometimes embedded in the

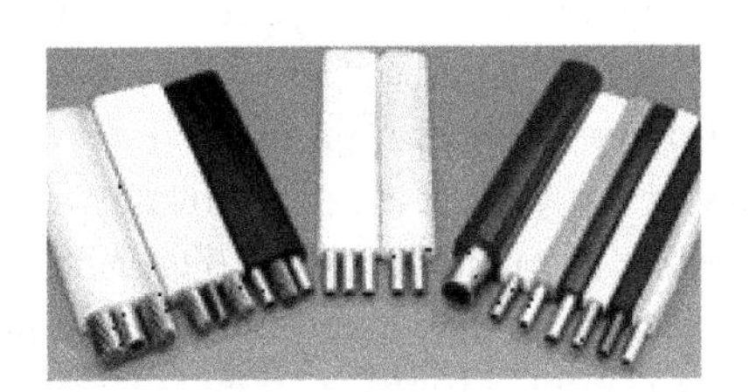

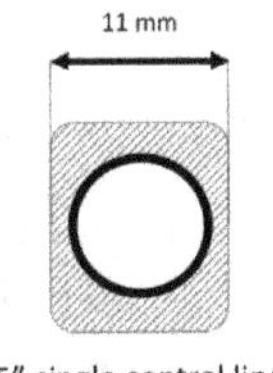

0.25″ single control line with
11 mm x 11 mm encapsulation

Figure 3.37 Flat packs—multiple control lines bundled in plastic encapsulation. *Photograph courtesy of TubeTec Ltd*

plastic encapsulation to help protect against crush damage. Bundling the control lines also makes handling simpler when the completion is being run. However, control lines complicate matters if there is a well control incident as they prevent pipe rams and annulus rams from obtaining an effective seal around the tubing. Additional measures need to be in place when control lines are being run (or pulled) during a completion or workover. Shut-in procedures for control line running are covered in Chapter 7, Well Kill.

3.22 CONTROL LINE CLAMPS

Control lines and flat packs need to be adequately supported and protected while the completion is being run, and throughout the life of the well. Control line protection in early completions was rudimentary. It was not uncommon to find control lines held in place with duck tape or bits of rope. Steel "band-it" strapping was common, and is still used in some very basic completions to the present day. The problem with these "cheap and cheerful" techniques is that they often fail to do what they are supposed to. Poorly supported control lines are susceptible to fatigue failure, because of vibration when the well is on line. In addition, band-it straps, duck tape, and rope are all a source of junk. A piece of duck tape can easily block the port in a gas lift valve. Snapped band-it tape can find its way into a PBR during a workover.

The first attempts to make dedicated control line clamps did not improve matters a great deal. Clamps were made from hard rubber and held in place with a steel pin. They were prone to damage, and were often missing when tubing was pulled back. These early problems more or less disappeared with the introduction of dedicated cast steel control line clamps.

Modern cast steel clamps are designed to sit across a tubing coupling. Hinged fastenings are held in place by threaded or toggle connections. A well-designed cross-coupling clamp both protects and supports all the control lines. They are manufactured with a variety of internal profiles, cut to match the size and number of control lines or flat packs being run in the well. Cross-coupling clamps are normally run one clamp per coupling (Fig. 3.38).

"Mid-joint" clamps are also available. These can be run one per joint in addition to the cross-coupling clamps. However, they have a tendency

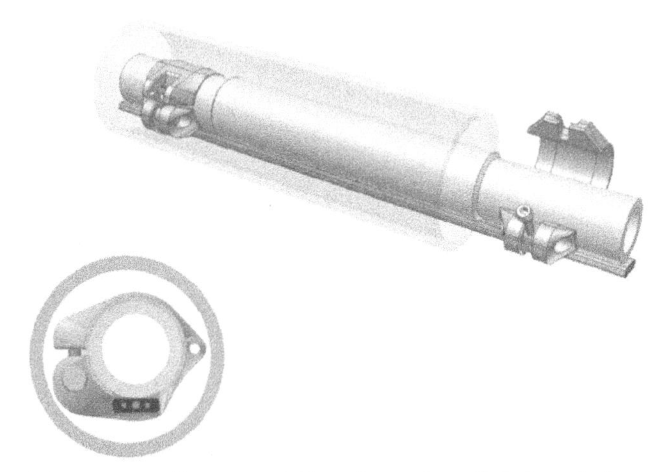

Figure 3.38 A cross-coupling control line clamp. A well-designed clamp will both protect and support all the control lines.

Figure 3.39 Fitting special clamps to blast joints in a multizone, single string, intelligent completion.

to catch on BOP ram cavities and other upsets in the well. When this happens, the clamp can slide along the tubing and damage the control line. In recent years the practice of running mid-joint clamps has been largely discontinued. However, it is still common to send a small number of mid-joint clamps to the rig site. They can be used to tidy the lines immediately below the tubing hanger.

The standard cross-coupling or mid-joint clamp is not suitable for all applications. For instance, in multiple zone single string intelligent completions, control lines may have to be run adjacent to open perforations, and will therefore have to be properly protected from jetting action. In the upper completion, special clamps will be needed at some of the completion components, such as gas lift mandrels and tubing retrievable safety valves (Fig. 3.39).

REFERENCES

1. American Petroleum Institute (API) Specification 5CT, 9th ed., June 2011. The Specification for Casing and Tubing.
2. Hartmann et al. Big bore high flowrate deep water gas wells for Oman Lange. OTC 16554.
3. ISO 13679: 2002. Petroleum and natural gas industries—procedures for testing casing and tubing connections.
4. M.A. Arceneaux, R.L. Smith 1986. Liner rotation while cementing: an operator's experience in South Texas. SPE 13448.
5. He, Yurong, Yu, Jinling, Liu, Qingyou, Hu, Qin, 2004. The cement slurry inflating external casing packer technology and its applications. SPE 88019
6. Kennedy et al. 2005. The use of Swell Packer's as a replacement and alternative to cementing. SPE 95713.

CHAPTER FOUR

Well Control Surface Equipment

4.1 INTRODUCTION

During most completion and workover operations an overbalanced fluid column forms the primary well control barrier; overbalance pressure prevents an influx of reservoir fluid. If the primary barrier fails and reservoir fluids enter the wellbore, surface well control equipment is used to secure the well. Since surface equipment is classified as the secondary well control barrier, it must be able to close in and secure the well at all the different stages of a completion or workover operation. It must also be configured in a way that allows the primary barrier to be restored. During drilling operations, restoration of the primary fluid barrier is normally achieved by circulating heavy (overbalanced) fluid into the well and circulating out the formation fluid. During completion and workover operations the primary barrier might be restored by circulation of kill fluid, but it is equally as likely to be a bullhead or lubricate and bleed kill.

Properly configured, the well control equipment will allow shut-in with and without pipe in the well. It will also allow drill pipe and completion tubulars to be cut, allowing the drill string or completion to be hung off in the rams and stripping of pipe into the well if a kick occurs off the bottom.

In the event of a kick with pipe in the well, it is secured using the following barriers:

Inside pipe barriers:

- Full opening safety valve (FOSV).
- One-way valve (internal blow out preventer (BOP) or dart sub).
- Check valves (drill pipe float valve).

Annulus (outside pipe) barriers. BOP equipped with:

- Annular preventer.
- Pipe ram (fixed diameter or variable bore).

Well Control for Completions and Interventions.
DOI: https://doi.org/10.1016/B978-0-08-100196-7.00004-X

Having secured the well, additional equipment is needed to restore the primary well control barrier. This includes:
- Fluid pumps.
- Fluid storage tanks.
- Choke manifold.
- Mud gas separator (poor-boy).
- BOP control system.

Most of the components that make-up the well kill systems are permanent and necessary parts of the drilling equipment, and have functions unrelated to well control. For example, the mud pits and fluid pumps that are used daily during the drilling and completion of the well become a necessary part of well kill equipment following a kick.

This chapter provides an overview of well control surface equipment. It concentrates on equipment requirements and configuration, rather than providing detailed nuts and bolts descriptions of every component.

4.2 THE BLOW OUT PREVENTER STACK

A BOP is a large valve designed to shut-off flow from the well. To enable flow to be shut-off under all circumstances during completion of a well or during a range of workover operations, several BOP valves, each with a different function, are assembled to make the BOP stack.

Configuring the stack will depend on the maximum anticipated surface pressure, and the type of operation to be performed. Most completions are installed immediately after drilling the well. Since the stack will have been configured for the drilling operation some modifications may be required before the completion can commence. For example, pipe ram inserts may have to be changed to match the diameter of the production tubing.

Stack configuration has a direct impact on operability. Optimum configuration of the rams and outlet spools allows more operability during a well control incident. When configuring the stack, the following should be considered:
- Positioning the pipe ram above blind rams allows the well to be isolated and the ram inserts changed out.
- There should be a drilling spool (inlet/outlet) below the blind ram to allow circulation and pressure monitoring.
- To enable tubing to hang off in pipe rams before disconnection or shearing above the rams, the stack should be configured to allow the

blind ram to close above the suspended pipe. An inlet is required between the closed blind ram and the pipe ram where the pipe is suspended to allow pressure to be monitored and circulation to occur through the suspended pipe.

- Variable bore rams (VBRs) have reduced load-bearing capability, and can only be used to suspend a limited length of pipe. Stripping through VBRs is not recommended, since they wear easily. They also have lower temperature limitations (225°F) than fixed diameter pipe rams. Nevertheless, they are often used during completion and workover operations, since tapered productions strings are common.
- Inlets/outlets below the lowermost ram should only be used to monitor pressure. A leak in the stack or associated lines below the bottom ram would result in loss of control and the release of fluids. In exceptional circumstances, it may be necessary to kill the well through the line at the base of the stack, i.e., where the well is shut-in on the lowermost ram and the kick is migrating.

Whilst having more rams in a BOP stack will improve safety and operability, it also adds to the weight of the stack. More rams also mean more expense. A well-designed stack is one that is both functional, safe, and cost-effective.

4.2.1 Blow out preventer classification

BOPs are classified and configured based on well pressure. The American Petroleum Institute (API) defines six working pressure classifications[1] with higher pressure systems requiring more complex configurations to meet the API minimum requirement. These are:

1. 2000 psi (13,800 kPa)
2. 3000 psi (20,700 kPa)
3. 5000 psi (34,500 kPa)
4. 10,000 psi (69,000 kPa)
5. 15,000 psi (103,500 kPa)
6. 20,000 psi (138,000 kPa)

In addition to these classifications, NORSOK[a] also include a 25,000 psi (172,369 kPa) rating.

These API pressure classifications are not the same as the working pressure rating of the BOP components. For example, a stack might be configured for a 2000 psi maximum surface pressure but be constructed

[a] NORSOK

from 5000 psi components. Similarly, the working pressure of surface BOP equipment will sometimes be higher than the wellhead to which it is attached. Working pressure limits must be defined by the rating of the weakest component in the system.

In addition to the pressure classifications, the API list component codes that are used when describing a BOP stack configuration. These are:

A = annular preventer.

G = rotating head.

R = single ram type BOP with one set of rams, blind or pipe depending on operator preference.

R_d = double ram type preventer with two sets of rams, blind or pipe depending on operator preference.

R_t = triple ram type preventer with three sets of rams, blind or pipe depending on operator preference.

CH = high remotely operated connector attaching wellhead or preventers.

CL = low pressure remotely operated connector attaching; the marine riser to the BOP.

S = spool with side outlet for choke and kill lines.

M = 1000 psi.

These codes can be used to describe how a BOP is, or needs to be, configured. By convention the code should be listed from the bottom of the stack upwards. For example, 15 M 13⅝ in. RSRRA refers to a stack that has a 15,000 psi working pressure with a through bore of 13⅝ in. (13.625). It is equipped (bottom to top) with a ram preventer, drilling spool, ram preventer, ram preventer, and annular preventer. The abbreviated description does not provide any information about the type of ram preventers used.

Figs. 4.1–4.3 show example API configurations for each pressure classification. It should be noted that these configurations represent the minimum requirement for each working pressure classification. Different operators and drilling contractors often have more stringent requirements. Similarly, more stringent requirements are called for by some regulatory authorities. For example, the NORSOK minimum requirement for BOPs is one annular preventer, one shear/seal ram, and two pipe rams.[2]

1. *Up to 2000 psi (13,800 kPa) working pressure*

   For pressure below 2000 psi the minimum requirements are:

   - Annular preventer and a ram preventer or two-ram preventers.

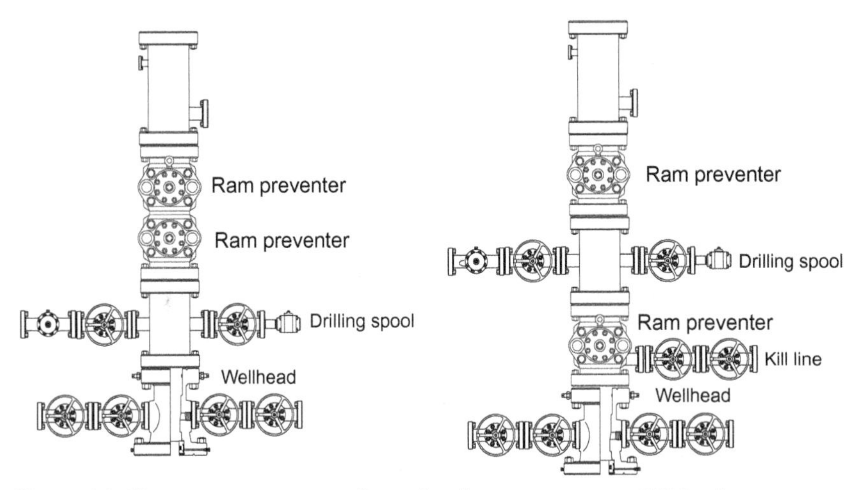

Figure 4.1 Blow out preventer configuration for pressure up to 2000 psi.

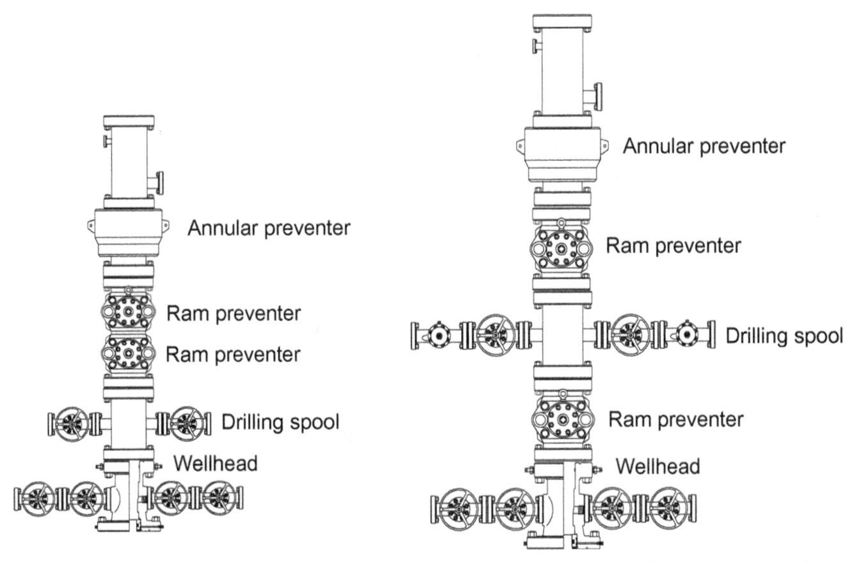

Figure 4.2 Blow out preventer configuration for 3000 and 5000 psi working pressure.

- The kill and choke lines can be connected to the inlet of the ram BOP or to a drilling spool.
- The kill and choke line must be at least 2 in. nominal diameter.
- If used, the drilling spool can be positioned between the BOP rams or below the lower ram. It must have two side outlets of at least 2 in. nominal diameter.

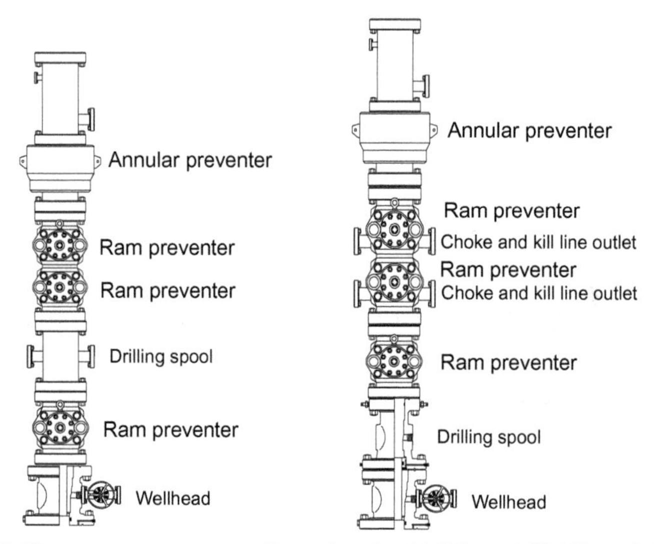

Figure 4.3 Blow out preventer configuration for 10,000 and 15,000 and 20,000 psi working pressure.

For a two-ram configuration, most operators and drilling contractors recommend configuring the stack with a shearing blind ram (SBR) below a pipe ram (or annular preventer). This allows a joint of pipe to be placed on top of a leaking blind ram and the closure of the pipe ram above a tool joint. The pipe in the BOP must have a FOSV installed and the space between the bottom of the pipe ram and the top of the blind ram must be long enough to accommodate the tool joint.

Having the blind ram below the pipe ram (or annular) allows the pipe ram seals (or annular seal) to be changed out when the blind ram is shut.

2. *Configuration for 3000 and 5000 psi working pressure BOP stacks*

   The minimum requirements for 3000 and 5000 psi BOP stacks are fundamentally the same. The stack should have:
   - Annular preventer.
   - Two pipe ram preventers (either a double ram preventer or two singles).
   - Kill and choke lines can be connected to the inlet of the ram BOP or to a drilling spool.
   - Kill and choke line must be at least 2 in. nominal diameter.

- If used, the drilling spool can be positioned below or between the ram preventers. It must have two side outlets of at least 2 in. nominal diameter.

There are several possible configurations where an annular preventer and two pipe rams are used. One option is to place the blind ram below the pipe ram with the drilling spool between the two-ram preventers. Positioning the blind ram at the bottom allows pipe to be placed in the pipe ram to back-up a leaking blind ram. In addition, the pipe rams can be changed or the annular seal replaced and the blind ram closed.

An alternative configuration places the drilling spool between the annular preventer and the ram preventers. When configured this way, it is common to place the bind rams above the pipe rams. This allows the tubing or drill string to be hung off in the pipe rams and the blind rams closed.

3. *Configuration for 10,000 15,000, and 20,000 psi working pressure BOP stacks*

The minimum requirements for 10,000 15,000, and 20,000 psi BOP stacks is to have:

- Annular preventer.
- Three ram type preventers (usually a double ram and a single ram preventer).
- Kill and choke lines can be connected to the BOP ram outlets providing they meet the required dimensional requirements—nominally 3 in. diameter.
- If a drilling spool is used it should have two outlets, one of 3 in. nominal diameter, and a second with a 2 in. nominal diameter.

The stack can be configured with a double ram preventer block below the annular preventer and immediately above a drilling spool. A single ram preventer is positioned below the drilling spool. Placing the blind (or blind shear) ram between the two pipe rams (above the drilling spool) allows pump in and bleed off with the blind ram closed. Some operators recommend having a second pressure monitoring and emergency kill line below the lowest ram. In a situation where the bottom pipe ram needs to be closed it would still be possible to monitor well pressure. It could be used to perform a bullhead or lubricate and bleed well kill.

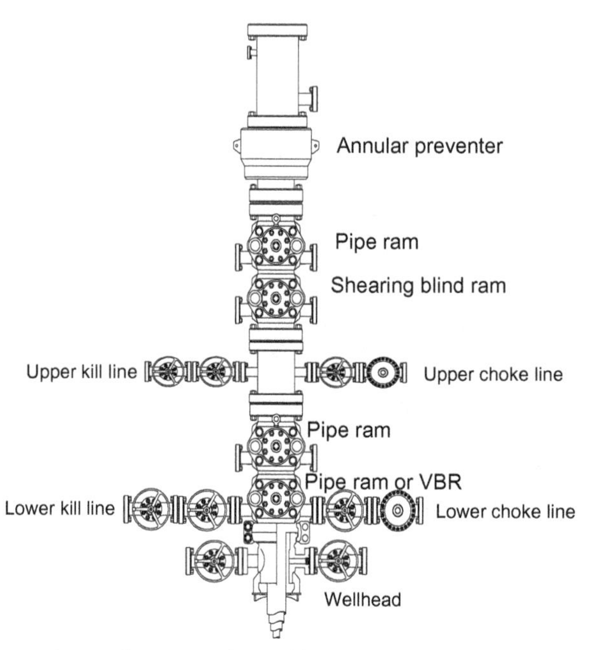

Figure 4.4 Example configuration for a blow out preventer stack with four ram preventers.

4. *Annular preventer and four ram preventers*

Many operating companies have a policy of using four ram preventers on high-pressure wells (<10,000 psi) and hazardous wells (H_2S). Having one more rams than the API minimum recommendation has several advantages:

- Two pipe rams allow ram-to-ram stripping with the third ram used as a safety ram.
- Allows two-ram sizes when running a tapered completions string. A VBR in the position provides back-up for both pipe rams.

A typical configuration is to position a drilling spool between two double ram blocks with the rams equipped (bottom up) with pipe (or VBR), pipe, shear/blind, and pipe (Fig. 4.4).

4.3 ROUTINE TESTING OF BLOW OUT PREVENTION EQUIPMENT

Blow out prevention equipment is emergency equipment and must be maintained in proper working order. A strict regime of routine maintenance, function testing, integrity testing, and crew drill must be practiced. The minimum requirement is to follow the maintenance and testing requirements recommended in API RP 53. Several types of test are defined by the API. These are:

- Inspection test: Defined by the API as "procedural examinations of flaws that may influence equipment performance."
- Function test: Operating an item of equipment to verify it functions as planned.
- Pressure test: Verification of pressure containment.
- Hydraulic operator test: Pressure testing of hydraulic operating components.
- Crew drills.

4.3.1 Function test

Function tests are normally carried out on a weekly basis. They should be performed from the driller's console, and from any remote panels that are in use at the location.

4.3.2 Pressure tests

Pressure tests must be carried out before operations begin. For drilling operations this normally means prior to spudding the well. For interventions and workovers, testing should take place before the equipment is exposed to well pressure, or has the potential to be exposed to well pressure. Once in use, equipment is normally tested every 21 days. Equipment should also be tested after a disconnect or repair.

Tables 4.1 and 4.2 detail the test requirements as recommended in API RP 53.

4.3.3 Crew drill

The API RP 53 has no specific recommendations with respect to crew drills; instead it refers to API RP 59 (Recommended Practice for Well Control Operations). That document outlines the requirements for crew drill frequency and content.[3] The drills described and recommended focus mainly on drilling operations. During completion, workover, and intervention operations, additional drills and briefings will be required to familiarize the crew with the actions to take when running non-standard equipment, or equipment that can compromise the ability of the BOP to properly shut-in the well.

Table 4.1 Initial test requirements

Component to be tested	Recommended pressure test	Recommended pressure test
	Low pressure (psi)	**High pressure (psi)**
Annular preventer		
• Operating	200–300	Minimum of 70% of annular BOP working pressure
chambers	N/A	Minimum of 1500
Ram preventer		
• Fixed pipe	200–300	Working pressure of ram BOPs
• Variable bore	200–300	Working pressure of ram BOPs
• Blind/blind shear	200–300	Working pressure of ram BOPs
• Operating chamber	N/A	Maximum operating pressure recommended by ram BOP manufacturer
Choke line and valves	200–300	Working pressure of ram BOPs
Kill line and valves	200–300	Working pressure of ram BOPs
Choke manifold		
• Upstream of last high-pressure valve	200–300	Working pressure of ram BOPs
• Downstream of last high-pressure valve	200–300	Optional
Blow out prevention control system		
• Manifold and BOP	N/A	Minimum of 3000
lines	Verify precharge	N/A
• Accumulator	Function test	N/A
pressure	Function test	N/A
• Close time	Function test	N/A
• Pump capability		
• Control stations		
Safety valves		
• Kelly, kelly valves, and floor safety valves	200–300	Working pressure of component
Auxiliary equipment		
• Mud/gas separator	Flow test	N/A
• Trip tank, flow show	Flow test	N/A

As per API RP 53 (Table 1).

Table 4.2 Initial test requirements

Component to be tested	Recommended pressure test	Recommended pressure test
	Low pressure (psi)	**High pressure (psi)**
Annular preventer		
• Operating chambers	200–300 N/A	Minimum of 70% of annular BOP working pressure N/A
Ram preventer		
• Fixed pipe • Variable bore • Blind/blind shear • Operating chamber	200–300 200–300 200–300 N/A	Greater than the maximum anticipated surface shut-in pressure Greater than the maximum anticipated surface shut-in pressure Greater than the maximum anticipated surface shut-in pressure
Diverter flowlines	Flow test	N/A
Choke line and valves	200–300	Greater than the maximum anticipated surface shut-in pressure
Kill line and valves	200–300	Greater than the maximum anticipated surface shut-in pressure
Choke manifold		
• Upstream of last high-pressure valve • Downstream of last high-pressure valve	200–300 Optional	Greater than the maximum anticipated surface shut-in pressure Optional
Blow out prevention control system		
• Manifold and BOP lines • Accumulator pressure • Close time • Pump capability • Control stations	N/A Verify precharge Function test Function test Function test	Optional N/A N/A N/A N/A
Safety valves		
• Kelly, kelly valves, and floor safety valves	200–300	Greater than the maximum anticipated surface shut-in pressure

(Continued)

Table 4.2 (Continued)

Component to be tested	Recommended pressure test	Recommended pressure test
	Low pressure (psi)	High pressure (psi)
Auxiliary equipment		
• Mud/gas separator	Optional flow test	N/A
• Trip tank, flow show	Flow test	N/A

As per API RP 53 (Table 2).

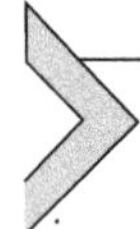

4.4 KILL AND CHOKE LINES AND THE CHOKE MANIFOLD

The choke manifold is an arrangement of high-pressure lines and valves that allow a kick to be removed from the well in a controlled manner using either manual or remotely operated chokes. Downstream of the choke, valving allows the fluids to be directed to the poor-boy degasser, the pits, or flare. The choke manifold can also be used to stop flow from the well at any time.

In many cases the choke manifold is supplied by the drilling contractor, and is part of the fixed rig equipment. When a mobile rig is brought to a location it is the responsibility of senior operating company personnel to approve (or otherwise) the choke manifold for use during the planned operation. Operating company wellsite supervisors arriving at a rig site for the first time should make themselves familiar with the layout and configuration of the manifold. The best way to do this is to physically "walk the line" noting the location of the main components and how the hook up is arranged. Consideration must be given to the routing of kill fluids where forward circulation is not used, since it is common during completion and workover operations to bullhead or reverse circulate. Reverse circulation normally means pumping into the well through a kill line. If the tubing hanger is still in place, the kill line will have to be connected to the wellhead, not the BOP.

Normally however, kill lines are connected to the BOP stack, either at an inlet on one of the ram preventers, or on the drilling spool. The positioning of the kill line will depend on the stack configuration, but ideally it will be placed below the ram preventer that is closed first.

The minimum requirements for choke manifolds and the associated kill and choke lines are set down in API 53. The main points are:

- All choke manifold components that may be exposed to well pressure must have a working pressure rating equal to or greater than that of the preventer stack in use.
- Choke manifolds, choke lines, and kill lines having a working pressure of 3000 psi and above must use flanged, welded, clamped connections on components subjected to well pressure.
- Where possible, the choke manifold should be situated in an easily accessible location, preferably outside the rig substructure.
- A minimum of one remotely operated choke should be installed on 10,000, 15,000, and 20,000 psi rated working pressure manifolds.
- Choke manifolds should be configured to allow re-routing of flow (in the event of eroded, plugged, or malfunctioning parts) without having to interrupt flow control.
- The choke line must have a working pressure that is equal to, or greater than the stack in use.

  The choke line (which connects the BOP stack to the choke manifold) and lines downstream of the choke should:
  - be as straight as practicable; turns, if required, should be targeted.
  - be firmly anchored to prevent excessive whip or vibration.
  - have a bore of sufficient size to prevent excessive erosion or fluid friction.
- The minimum recommended size of choke line for pressure up to (and including) 5000 psi is 2 in. nominal diameter. Some operating companies recommend using 3 in. lines for all pressure ratings.
- The minimum recommended size of choke line for 10,000, 15,000, and 20,000 psi equipment is 3 in. nominal diameter.
- For air or gas drilling operations the minimum recommended size for the choke line is 4 in. nominal diameter.
- The kill line must have a working pressure equal to or greater than the ram BOP in use.
- Two full bore manual valves plus a check valve, or a manual valve and a hydraulically actuated valve between the BOP stack and the kill line are recommended for systems with a rated working pressure more than 5000 psi.
- The minimum recommended size for choke lines is 2 in. (Fig. 4.5).

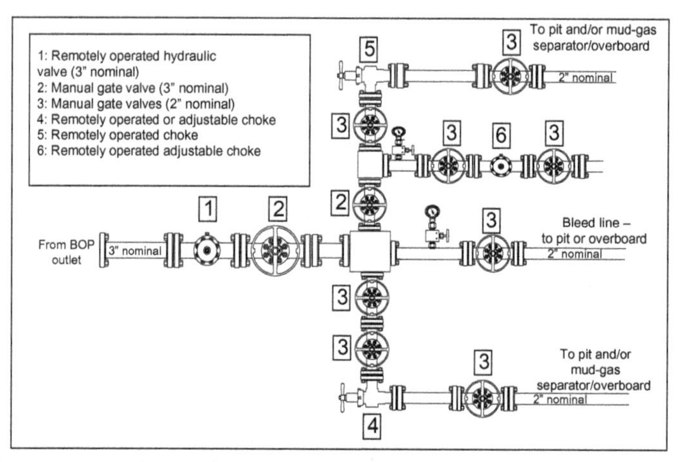

Figure 4.5 Example of a choke manifold suitable for up to 20,000 psi working pressure. Configuration in accordance with API RP 53.

4.5 CHOKES

During a well kill a choke is used to hold back pressure on the well and prevent any further influx of fluids from the formation. Chokes are either positive (fixed diameter) or adjustable. Adjustable chokes can better regulate pressure than fixed diameter chokes. Hydraulic chokes are more easily adjusted and allow accurate remote regulation of choke pressure.

Although the API minimum requirement is to have one remotely operated choke where pressure is 10,000 psi or more, most rig manifolds are equipped with two remotely operated chokes. During completion and workover operations, the production choke and chokes in a well test choke manifold might also be used. Chokes used in oilfield applications should be manufactured in accordance with API specification 6A.[4]

4.5.1 Positive (fixed diameter) chokes

Positive chokes use an insert with a fixed diameter to regulate flow and pressure. The choke insert is changed to suit well conditions and operational requirements. Choke inserts (or beans) are available in a range of sizes. Choke diameter is normally stated in 64ths of an inch; for example, a "32 bean" is equivalent to a ½″ choke. Fixed chokes are commonly used during well tests, but are not suitable for well kill operations where

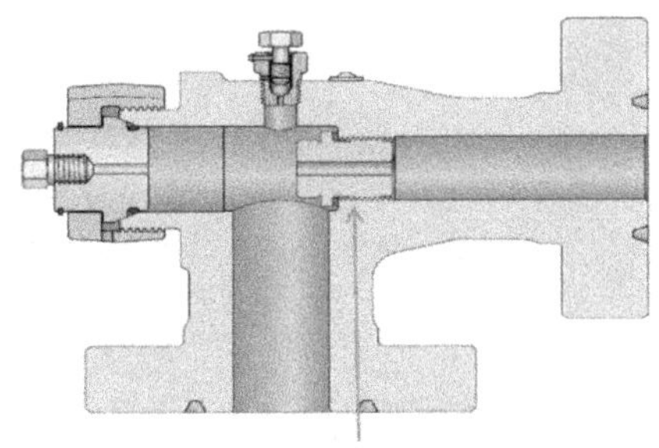

Figure 4.6 Fixed diameter choke.

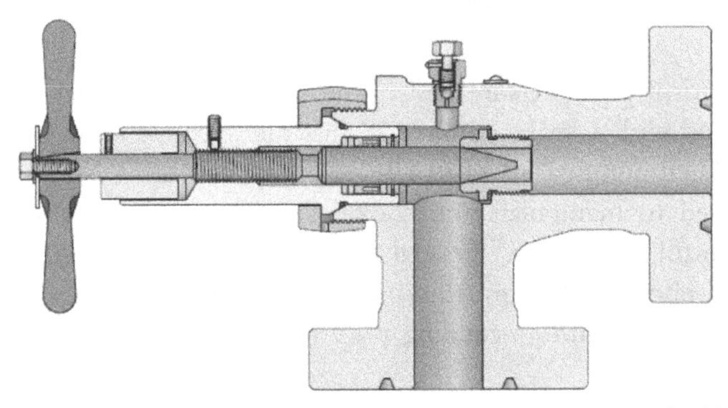

Figure 4.7 Adjustable needle and seat choke. Source: *Image courtesy of Schlumberger.*

regular adjustments are required to maintain a constant bottom hole pressure (Fig. 4.6).

4.5.2 Manually adjustable choke

The manually adjustable chokes most commonly used on rig choke and well test choke manifolds use a needle and seat to regulate flow. The tapered stem and valve seat are easily replaced and made from hardened material such as tungsten carbide (Fig. 4.7).

4.5.3 Remotely operated choke

If the choke manifold is equipped with a remotely operated choke, it will be worked from a control panel. The panel draws the hydraulic pressure needed to operate the choke from the BOP hydraulic supply. In addition to the control leaver for the choke, the panel should have pressure gauges

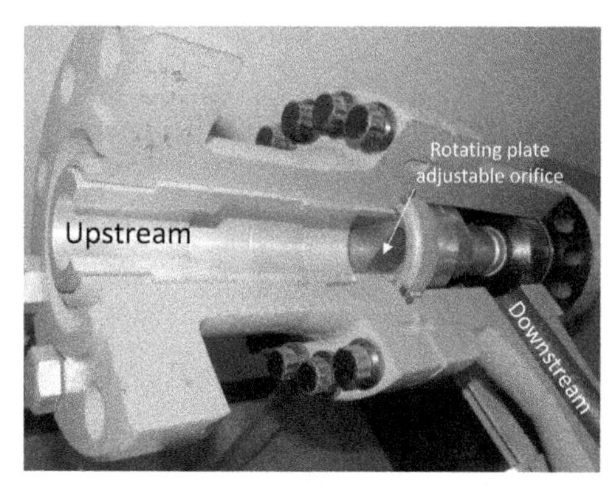

Figure 4.8 MI type remotely operated adjustable choke.

displaying drill pipe pressure and casing pressure. The panel should also be equipped with a stroke counter, so that the operator knows how much fluid has been pumped. Most panels are equipped with a hand pump that can be used to manipulate the choke if the hydraulic supply from the BOP control unit fails. Several types of mechanism are used in adjustable chokes. Perhaps the most widely used is the MI Swaco "Superchoke" that uses a rotating plate to adjust the choke (orifice) size (Fig. 4.8).

4.5.4 Choke operation

Chokes are prone to erosion. If there is a large pressure differential across the choke, the pressure drop causes bubbles to form as liquid vaporizes. A little further downstream (the recovery point) these bubbles collapse; cavitation takes place. Prolonged flow under these conditions can be very damaging, and the bigger the difference between inlet and outlet pressures, the greater the potential for damage. A rule of thumb to determine if damage is likely is to divide the pressure drop across the choke (Delta *P*) over the upstream pressure. If the result is 0.6 or more, damage is likely (Fig. 4.9).

$$\frac{P_u - P_d}{P_u} \tag{4.1}$$

where P_u is pressure upstream of the choke; P_d is pressure downstream of the choke.

Example:

Upstream pressure = 6500 psi.

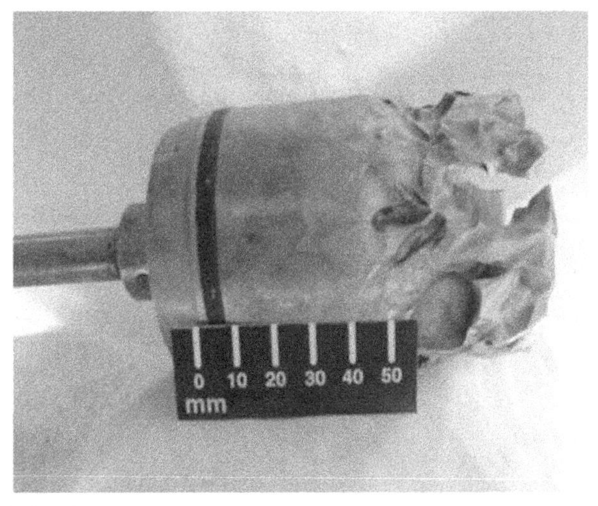

Figure 4.9 Choke damage.

Downstream pressure = 1500 psi.

$$\frac{6500 - 1500}{6500} = 0.77.$$

In this example, damage is likely to occur.

4.6 GATE VALVES

Gate valves are designed to create a seal, shutting off flow through a line by means of a gate moving within the body of the valve. They are the preferred type of valve for surface well control installations, since they can hold high differential pressure, and are generally reliable, easy to maintain, and simple to use. Gate valves are on/off valves and should never be used to choke or regulate flow; doing so risks damage to the gate, seat, and valve body and can lead to a loss of containment.

The many gate valves that are used with well control surface equipment are mostly manually operated. For surface installations, the API 5C does not make any recommendations concerning the use of actuated valves in the choke line or choke manifold. It does, however, recommend the use of an actuated valve in the kill line for pressure above 5000 psi. Many drilling contractors and operating companies will use more that the

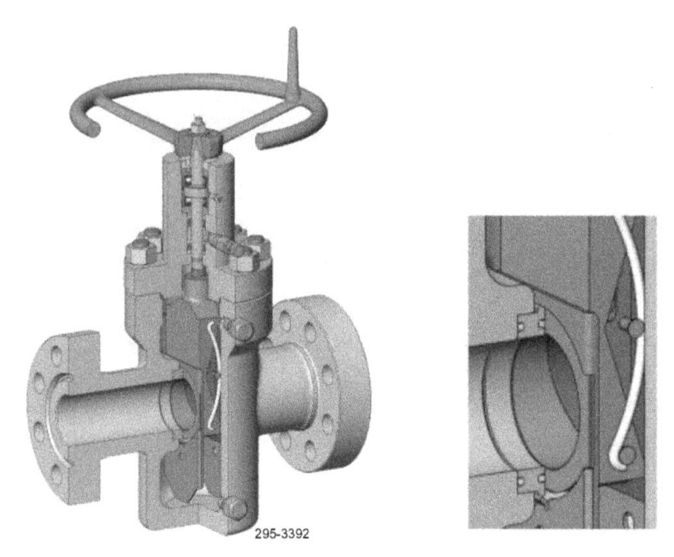

Figure 4.10 Gate valve with split gate. Source: *Image courtesy of GE.*

recommended minimum number of hydraulically actuated valves. Hydraulically actuated valves are a necessity on subsea BOP stacks. Gate valves are supplied with either solid slab gates or split gates. The closure mechanism falls into to two broad categories; rising stem and fixed stem.

4.6.1 Split gate valves

The gate mechanism in a split gate valve, as the name suggests, is constructed from two separate plates in the form of a wedge. When the gate is closed, the wedge forces the gate against the seats on each side. Split gate valves are normally bi-directional, although most have a preferred sealing side. The wedge design of the gate means that it does not rely on well pressure to initiate a seal and has good sealing properties at low pressure (Fig. 4.10).

4.6.2 Slab (floating) gate

A slab seat moves across the face of two (one each side) floating seals. Springs between the seat pocket and the seat keep the seat in constant contact with the gate, and prevent well fluids and contaminants from leaking into the valve cavity. The valve is designed to hold pressure in either direction and work best when well pressure assists with seating the

Figure 4.11 Rising stem (left) and non-rising stem (right). Note: these illustrations came from a plumbing catalogue. However, they clearly illustrate the difference between the two closure mechanisms. *Photo courtesy of Michels Plumbing, Aurora, IL.*

gate against the seat. Slab gates are generally robust, reliable, and easily maintained.

4.6.3 Rising and non-rising stem valves

Both slab and split gate valves can be configured with a rising or non-rising stem to move the gate. With a rising stem type valve, the gate is fixed to the stem. The stem travels pulling or pushing the valve gate across the seat to open or close the valve. Most actuated valves are of this design, with hydraulic or pneumatic piston force moving the valve stem. In a valve with non-rising stem, the stem is attached to the valve gate with a threaded connection. As the stem is rotated the gate moves up and down the threaded section of the stem (depending on the direction of rotation), opening or closing the valve (Fig. 4.11).

4.6.4 High closing ratio valves

High closing ratio (HCR) valves are hydraulically actuated gate valves. HCR relates to the design of the closing piston area. The HCR valve is held open against a power spring and well pressure using hydraulic pressure from the BOP control unit. If hydraulic pressure is bled off, the power spring and well pressure combine to close the valve. Most HCR valves operate with between 1500 and 3000 psi piston pressure.

4.6.5 Check valve or non-return valve

A check valve should be placed in the kill line to prevent backflow from the well and protect the fluid pump. The operating principal of a check valve is simple. Fluid can be pumped through the valve (left-to-right in

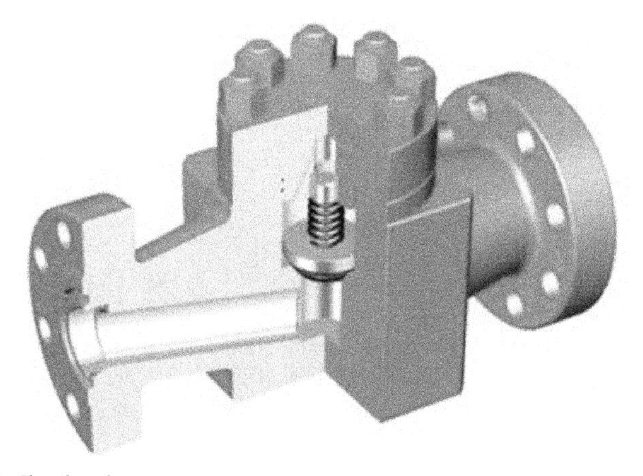

Figure 4.12 Check valve.

the figure shown). Flow towards the well lifts the valve off seat. When pumping stops, or if the direction of flow is reversed, the valve seats, stopping backflow from the well. Modern valves normally have metal-to-metal seats (Fig. 4.12).

4.7 ANNULAR PREVENTERS

The annular preventer, also called the bag preventer, or simply "the bag," is normally the first device to be closed in the event of a kick. The principal advantage of the annular preventer is that it is designed to close around any size of pipe, and can create a seal around some irregular and odd shaped completion component. However, annular preventers have their limits and will not seal around any of these items of completion equipment:

- Spent hollow carrier perforating guns.
- Slotted or pre-drilled liners.
- Sand control screens of any type.
- Completion tubing with external control lines or electric submersible pump (ESP) cable.
- Severely corroded (holed) pipes.

Most annular preventers will seal even if no pipe is in the well, although in some cases, this is not recommended by the manufacturers. Typically, an annular preventer will hold about 50% of the working pressure when closed on open hole. Annular preventers also allow pipe to be stripped into the well under pressure.

Most annular preventers are designed for a maximum recommended closing pressure of 1500 psi (10,342 kPa). Some annular preventers operate with a higher chamber working pressure of 3000 psi (20,684 kPa). The minimum pressure necessary to obtain a seal is dependent on several factors such as bore size, outer diameter (OD) of the pipe, and the wellbore pressure. In general, the greater the difference between annular BOP bore size and pipe OD, the more closing pressure is required to ensure a seal.

Annular preventers have a doughnut shaped multi-segment seal element with integral bonded steel reinforcement. Closure of the seal occurs when a hydraulically actuated piston is driven upwards. This extrudes the seal into the wellbore and compresses it around any pipe in the well. Closure pressure is hydraulically regulated and most systems are configured to allow control fluid to flow in both directions, so closure pressure remains constant as pipe (and tool joints) are stripped through the seal. For surface installations, the API recommend a maximum closure time of 30 s for annular preventers smaller than 18¾ in. and 45 s for equipment of 18¾ in. and larger. For subsea installations closure time should not exceed 60 s.

Most of the annular preventers currently in use are made by three manufacturers, or will be copies of these basic types, made in the emerging markets of India and the Far East.

- Cameron cooper:
    - Type D and DL.
- Hydril (a subsidiary of GE):
    - Model GK.
    - Model GL.
    - Model GX.
- Shaffer:
    - Shaffer spherical.
    - Shaffer "Spherical" annular preventer.

Since the annular preventer is part of the BOP stack, and therefore normally provided by the drilling contractor, operating company supervisors will not always have adequate knowledge of the workings of the annular preventer. A brief description of the most commonly used annular preventers follows.

4.7.1 Cameron (Sclumberger) D and DL annular preventers

The Cameron Model "D" preventer is available in sizes between 7(1/16) in. and 21¼ in., with working pressures between 2000 and 10,000 psi. It uses a

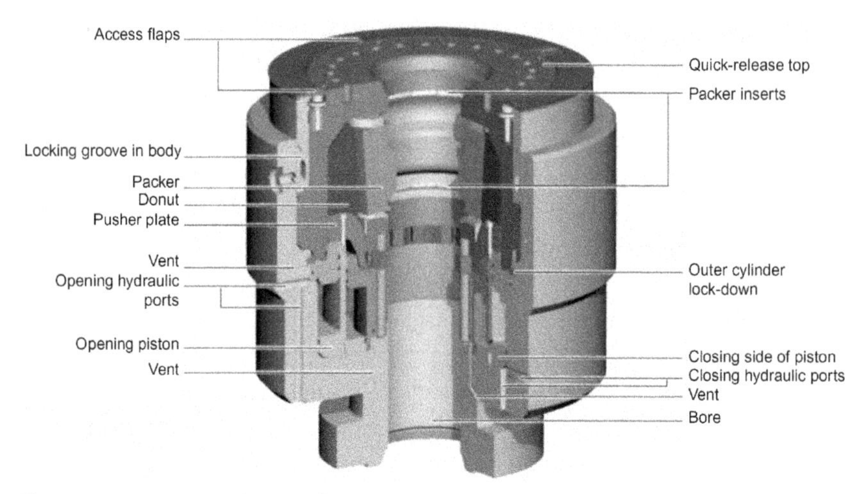

Figure 4.13 Cameron DL annular preventer.

two-piece (donut and packer) seal. When closing pressure is applied, the main piston moves up, forcing the pusher plate against the elastomer donut and extruding the packer radially inwards to create the seal. Interlocking steel reinforced insets rotate inwards to form a continuous support ring above and below the seal.

The main features, as described by the manufacturers, are[5]:

- Quick release top that allows easy change out of the packing element.
- Components subject to wear are field-replaceable, and the entire operating system may be removed in the field for immediate change out, without removing the BOP from the stack.
- Packers for DL BOPs have the capacity to strip pipe, as well as close and seal on any size object that fits in the wellbore. These packers will also close and seal on open hole at half working pressure.
- Shorter height and lighter weight than comparable BOPs.
- Most sizes of the DL preventer use less control fluid than the Hydril or Shaffer preventers of an equivalent size (Fig. 4.13).

4.7.2 Hydril GK preventer

The "GK" annular preventer is designed primarily for surface installations. It comes in a range of sizes from 7(1/16) in. to 16¾ in. with working pressure ranging from 3000 to 15,000 psi. The packing unit is designed to hold full-rated working pressure. It is tested to 50% of

Figure 4.14 Hydril GK annular preventer.

working pressure when closed on open hole. The manufacturers list the following features:

- Single packing unit that closes on any size pipe or open hole and handles stripping.
- Only two moving parts: the piston and packing unit, for less wear and maintenance
- Optional latched head for fast, easy access to the packing unit and wear seals: the majority of GK annulars have a screwed head design.
- Bolted-in inner sleeve that is field-replaceable.
- Replaceable wear plate that eliminates metal-to-metal contact between the packing unit inserts and the BOP head—extending time between major overhaul and repair (Fig. 4.14).

4.7.3 Hydril GL preventer

The "GK" annular preventer is designed primarily for sub surface use but can be used for platform and land operations. Three sizes are available; 13⅝, 18¾ and 21¼ in. The working pressure is 5000 psi.

The manufacturers list the following features:

- Packing unit specially molded with two different styles of inserts to reduce closing volume and increase closing cycles.
- Replaceable wear plate that eliminates metal-to-metal contact between the packing unit inserts and the BOP head.
- Piston design that provides complete balance, reliable operation, and ease of assembly.
- Latched head for fast, easy access to the packing unit and wear seals.
- Opening chamber head that prevents debris from falling into the chambers when replacing the packing unit.

4.7.4 Hydril GX annular preventer

The GX annular preventer can be used for surface and subsea operations. It is available with 11, 13⅝ and 18¾ in. through bore, and 5000 or 10,000 psi working pressure. The manufacturers state that the "*GX annular BOP can hold full-rated working pressure after more closing cycles and stripping than any other unit available.*"[6] They also list the following features:

- Single packing unit that closes on any size pipe or an open hole—and handles stripping.
- Only two moving parts, the piston and packing unit, for less wear.
- Latched head for fast, easy access to the packing unit and wear seals.
- Replaceable wear plate that eliminates metal-to-metal contact between the packing unit inserts and BOP head.
- Pressure balanced piston design that allows use in ultra-deep water.
- Opening chamber head that prevents debris from falling into the chambers when replacing the packing unit.

4.7.4.1 The Shaffer's (NOV) spherical annular preventer

The Shaffer's spherical BOP is so-called because of the shape of the seal element and the internal chamfer on the top cover (either wedge or bolted). It is available in a range of sizes from 4(1/6) in. to 21¾ in.; NOV also makes a 30 in. version of the spherical BOP, but it is not API classified.

Working pressure ranges from 1000 psi for the 30 in. preventer, through to 5000 psi for the 7(1/16), 9, 11, and 13⅝ in. models, and 10,000 psi for 4(1/6) in. and 7(1/16) in. models.

The Shaffer BOP is relatively simple to operate compared with other annular BOPs. Maximum closure pressure is 1500 psi. When hydraulic pressure is applied, the piton pushes up against the seal element. The seal

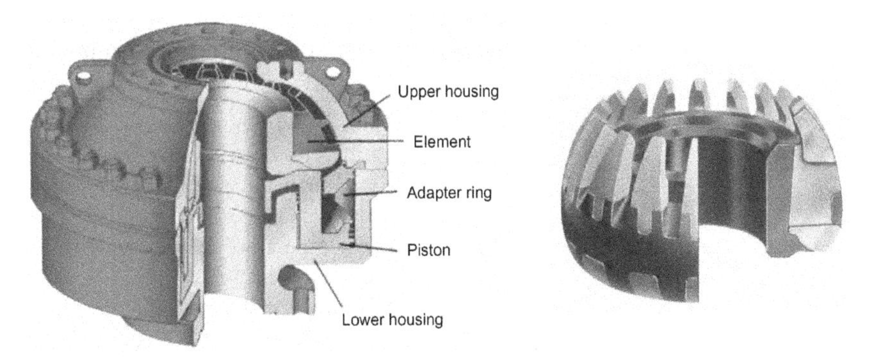

Figure 4.15 Shaffer (NOV) spherical annular preventer.

elements spherical shape causes it to close as it moves upwards. The Shaffer BOP can close on open hole (Fig. 4.15).

4.7.5 Care and use of annular preventers

Although the drilling contractor is normally responsible for the installation, use, maintenance, and testing of the annular preventer, supervisors should have a good understanding of the basic principles of operation. This will enable them to ensure that the system is being properly maintained and operated. The following guidance is relevant to most operations:

- The lifting eyes on the annular body are for lifting the annular only, and not the entire BOP stack.
- If visible, check the color code on the annular element. It should correspond to the elastomeric compound that is compatible with the completion or workover fluid being used. For a description of the various packer compounds and color code (see Table 4.3).
- Shut-off valves are often placed in the open and close lines (for line testing and isolation purposes). Make sure these valves are fully open whenever the annular is in use, so it can be closed if necessary.
- H_2S exposure causes a slow hardening of materials and loss of elasticity in most annular elements.
- Ensure that opening hydraulic pressure is applied to the annular when tripping pipe. This pressure keeps the element packing fully retracted and avoids mechanical damage.
- Keep the manufacturer's operating manual on location. It contains valuable information about preventer weight and dimensions, closing

Table 4.3 Annular preventer seal selection table

Manufacturer	Elastomer	Color	Supplier code	Manufacturers recommended use
Hydril	Natural rubber	Black	R	Water-based fluids with less than 5% oil and operating temperature greater than −30F. Suitable for H_2S service
Shaffer	Natural rubber	Red	1 or 2	Low temperature operations with water-based fluids
Hydril	Nitrile	Red	S	Oil-based muds with aniline points between 165 and 245F; suitable for H_2S service and operating temperatures greater than 20F
Shaffer	Nitrile	Blue	5 or 6	Oil- and water-based muds; suitable for H_2S service
Cameron	Nitrile	Black	n/a	Oil and water-based fluid; suitable for H_2S service Temperature range from −30 to 250°F
Hydril	Neoprene	Green	N	Oil-based muds with operating temperatures between 20 and −30°F; suitable for H_2S service

fluid requirements, replacement part numbers, internal seal locations, testing, etc.

- Check the annular manifold pressure gauge for the correct reading according to the equipment operating manual.

4.7.6 Stripping with annular preventers

Annular preventers allow drill pipe to be stripped into a well, since they can maintain a seal as a drill pipe connection passes. They also have better abrasion resistance than ram preventers. However, stripping is unlikely to be possible during most completion or workover operations, as threaded and coupled connections rarely have enough taper to prevent seal damage. Where stripping is possible, attention should be given to the annular preventer. Accumulator pressure should be adjusted to maintain constant closing pressure. Since most respond slowly to the pressure change caused by a tool joint passing the seal, tool joints should be moved through the preventer slowly.

It is possible to change the seal element with pipe in the hole. To change the seal element under live well conditions, the pipe rams (or VBR) below the annular preventer must be closed, locked, and integrity tested. To conform with a two-barrier policy, two pipe rams would have to be available. With the well secure, the top of the annular preventer can be removed and the worn rubber sealing element extracted. The element must be cut to enable it to be removed from around the pipe. The new rubber sealing element is cut (never sawed) between the metal ribs and the new element is installed in a reversal of the removal sequence.[b]

4.8 RAM PREVENTERS

Ram preventers are used to shut-in the well in the event of a kick, and are used in addition to, or instead of, the annular preventer. They can be equipped with several different types of ram block, each one performing a different function. These are:

- Pipe rams: Seal around a specific size of production tubing or drill pipe. Most rams are designed for a single string of pipe, but dual or offset rams are available for dual string completions.
- VBRs: Designed to seal around a range of pipe sizes.
- Shear rams: To cut tubing and drill pipe.
- Blind rams: To seal the wellbore when there is no pipe across the BOP, i.e., no pipe in the well, or after the shear ram has been used to cut and drop the string.
- SBRs: Combine the cutting action of the shear ram and the sealing action of the blind ram.
- Production rams: To seal around rod pump sucker rods.

Although ram preventers are designed to only hold pressure from below, most can hold a small differential from above. Shorter in profile (height) than annular preventers, they are sometimes the only preventers installed where space is limited (Fig. 4.16).

Early ram preventers were manually operated. Rotation of the threaded stem moved the ram blocks back and forth between the open and closed position. It soon became apparent that a faster method of

[b] It is not possible to cut Cameron sealing elements as described above. There is also a documented case of a Shaffer element coming out of the annular body after being split.

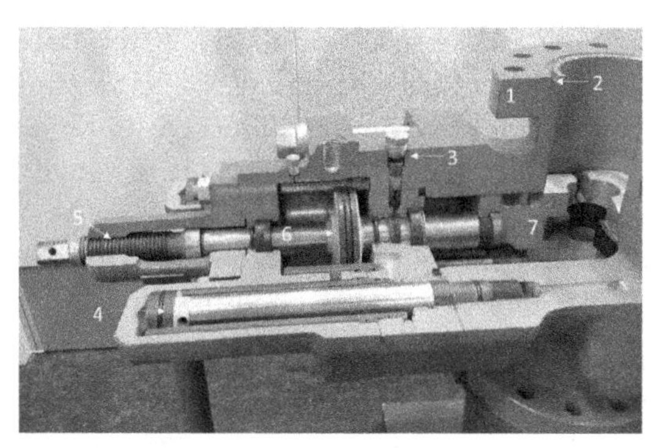

1: Top flange connection
2: Gasket ring groove
3: Plastic injection port
4: Ram change piston
5: Locking screw
6: Operating piston
7: Ram block (pipe ram)

Figure 4.16 Cutaway view of a Cameron "U" ram preventer. This is the most widely used of all the ram preventer types available.

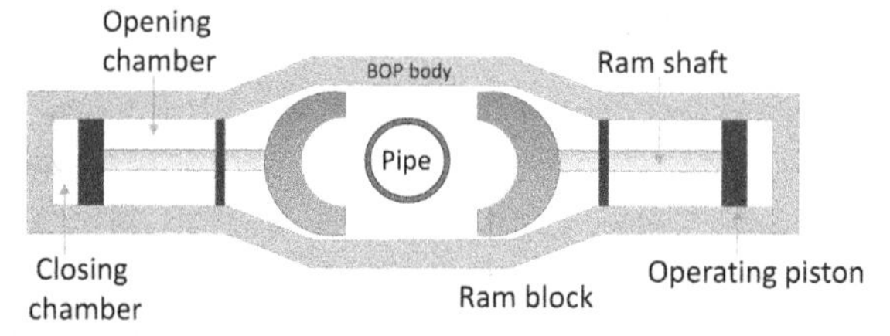

Figure 4.17 Simplified diagram of a ram preventer.

shutting in the well was needed. This led to the development of hydraulically operated ram preventers. Modern ram preventers normally close more quickly than annular preventers as they require less hydraulic fluid. The API RP 53 stipulates a closure time of 30 seconds for ram BOP closure (land and platform) and 45 seconds for subsea applications. Most will easily beat this. As well as being able to shut the well in much more quickly than a manually operated system, hydraulics have the obvious benefit of remote operation. There is no need to expose the drill crew to the risks associated with closing in the well manually.

Fig. 4.17 is a simplified illustration showing the normal configuration of hydraulic chamber and hydraulic piston used in most ram type BOPs. Pressure applied to the opening or closing chamber is used to move the piston, opening or closing the preventer.

When rams are closed, hydraulic pressure acts on the surface area of the operating piston. Well pressure acts on the surface area of the ram shaft in opposition to the opening force. Since the piston area is larger than the shaft area, closure force provided by the hydraulic pressure will be proportionally higher than the opening force from well pressure. The

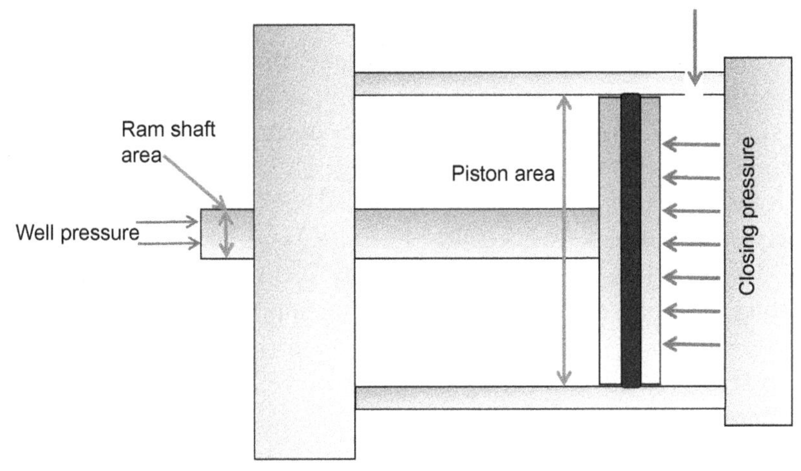

Figure 4.18 Closing ratio.

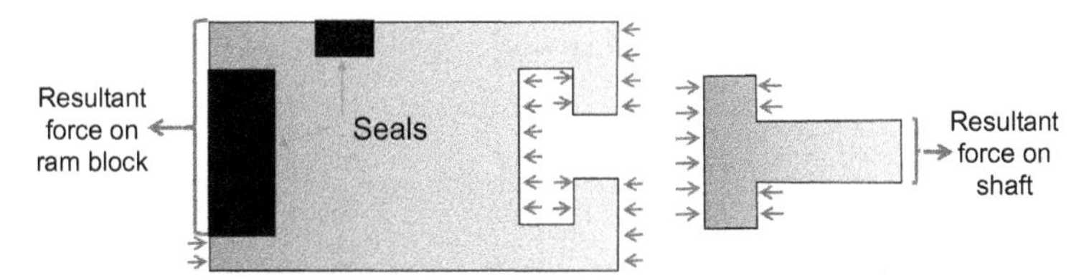

Figure 4.19 Opening forces.

ratio of piston size to shaft size will give the closing ratio, which can be defined as "*a dimensionless factor equal to the wellbore pressure divided by the operating pressure necessary to close the BOP ram against wellbore pressure*" (Fig. 4.18).

Closing ratios are generally in the range from 6:1 to 9:1, meaning 1 unit of force of closing pressure is needed for every 6–9 wellbore pressure units of force to close the preventer. For example, if a preventer has a closing ratio of 6:1, and wellbore pressure is 3000 psi, then 500 psi hydraulic pressure will close the preventer (ignoring friction). Maximum ram operating pressure is the working pressure rating of the BOP divided by the closing ratio.

It is highly unlikely there will ever be a need to open rams when pressurized, and opening rams against well pressure is certainly not recommended. If an attempt is made to open the rams, hydraulic pressure needs to overcome the wellbore pressure that acts against the large surface area at the back of the ram (Fig. 4.19).

Wellbore pressure is, in fact, helping keep the ram in the closed position. Since the area of the ram block is large, the opening ratio is much lower than the closing ratio. Whilst there are preventers that have an opening ratio of less than 1:1, requiring an opening pressure greater than well pressure, most are in the 1:1–1:4 range.

Rams should not be closed on pipe tool joints or tubing upset areas, otherwise seal damage will result. Most ram preventers can be locked in the closed position, desirable during well control operations or when shutting in for the night (daylight-only operations).

Nearly all ram preventers are based on designs from three leading equipment providers:

1. Cameron (Schlumberger).
2. Hydril (GE).
3. Shaffer (NOV).

Many smaller independent manufacturers make copies of the original Cameron, Hydril, and Shaffer BOP designs, as well as a range of spares. Not all of them conform to the exacting standards of the mainstream manufacturers.

4.8.1 Pipe rams

Pipe ram are constructed with a semi-circular cut-out matched to a specific pipe size. Most have a front packer that seals around the pipe and an upper seal that prevents pressure by-passing the ram block. Pipe rams should never be closed on an open hole, as this risks damaging the elastomer.

During a completion or workover, it is normal to have at least one pipe ram for each size of tubing that is being run. However, if completion complexity or space prevents this, VBRs can be used. If control lines, chemical injection lines, instrument, or ESP cables are being run, closing the pipe ram is ineffective. In these circumstances, the control lines would have to be cut and then lowered below the stack. This is normally done using a kick stand (drill pipe), and the rams should be sized accordingly; i.e., matched to the kick stand diameter.

Most pipe rams can support the weight of the string when a tool joint is lowered onto the top of a closed pipe ram (Fig. 4.20).

4.8.2 Variable bore rams

VBR can close on a range of pipe sizes, and are particularly useful when running a tapered completion string. All VBRs use a similar method of

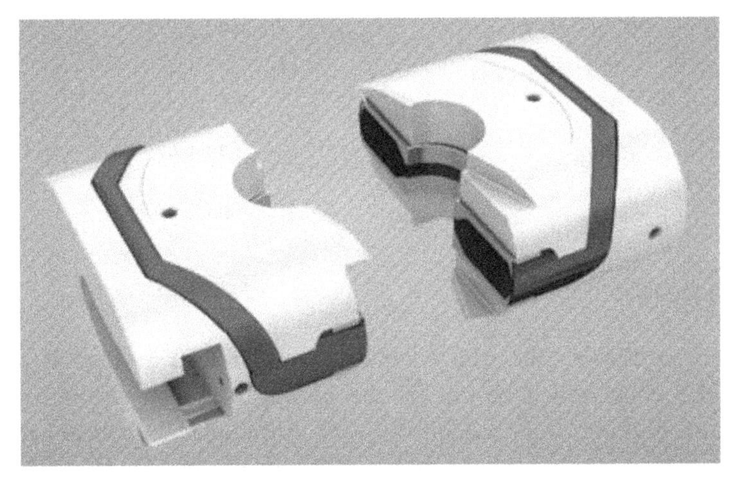

Figure 4.20 Pipe ram for a Cameron type "U" preventer. Source: *Image courtesy of Schlumberger.*

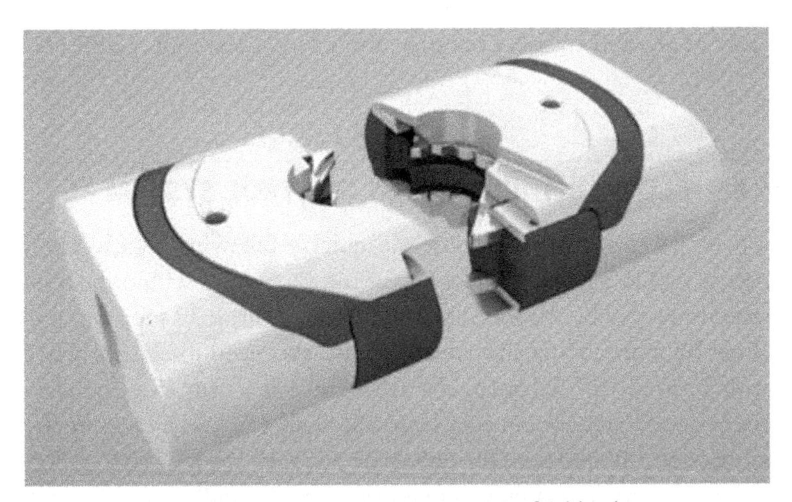

Figure 4.21 Variable bore ram. Source: *Image courtesy of Schlumberger.*

construction, using a feedable rubber packing element. However, the design means the ability to hang the tubing string from the ram is severely limited (Fig. 4.21).

4.8.3 Blind rams/shearing blind rams

Blind rams have a flat elastomer face and are designed to shut-in the well when there is no pipe across the BOP stack. Blind/shear rams are

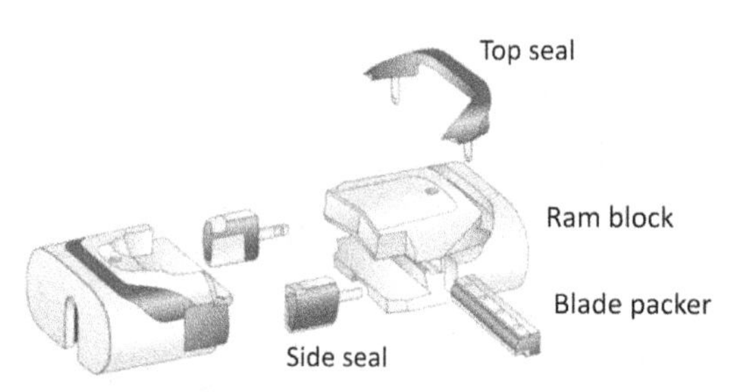

Figure 4.22 Blind/shear ram.

designed to close in the well whether or not pipe is present. If pipe is across the stack it will be cut by the closing blades in the ram block. Elastomer packings then seal off the well. Although blind/shear rams are mainly used in subsea applications, they should also be considered for surface applications on high-pressure wells, wells where H_2S is present, or wells in urban and environmentally sensitive areas.

Depending on pipe size, wall thickness, and material yield, high pressure may be needed to shear the pipe and effect a seal (Fig. 4.22).

4.9 BLOW OUT PREVENTER CONTROL SYSTEM

It is vital that the well is closed in quickly following a kick. The control system should be able to close any of the BOP rams within the response time recommended by the API. As already stated, this is 30 s for ram preventers. Annular preventers smaller than 18¾ in. also need to close within 30 s, whilst those sized 18¾ in. or larger must close in less than 45 s. Annular preventers require a large volume of hydraulic fluid to close. For example, the Hydril 13⅝ in. 10,000 psi GK preventer requires over 37 US gallons to close. A pump that can supply enough pressurized fluid within the required closure time would be impractical in terms of size and cost. Moreover, pumps cannot operate if there is a loss of power at the wellsite. For these reasons, accumulator bottles are used to store pressurized hydraulic fluid for quick release. The volume of fluid stored in the accumulator is set out in API RP 53, specification 16D,[7] and recommended practices 16E.[8] They require a BOP system to have enough usable hydraulic fluid volume (with pumps inoperative) to satisfy the greater of the two following requirements:

- Close from a full open position at zero wellbore pressure all of the BOPs in the BOP stack plus 50% reserve.

- The pressure of the remaining stored accumulator volume after closing all of the BOPs should exceed the minimum calculated (using the BOP closing ratio) operating pressure required to close any ram BOP (excluding the shear ram) at the maximum rated wellbore pressure of the stack.

Some jurisdictions and many operating companies have more stringent requirements. For example, NORSOK, the Norwegian regulatory authority, stipulate that the accumulator capacity of a BOP stack used for snubbing operations "*shall as a minimum have sufficient volumetric capacity to close, open and close all the installed BOP functions plus 25% of the volume for one closing operation for each one of the said BOP rams.*"[9]

4.9.1 Operating principles and main components

When the well needs to be closed in, the relevant valve on the accumulator unit manifold is moved to the close position. Pressurized fluid from the accumulator bottles flows through steel and coflex lines to reach the opening chamber of the desired BOP ram or HCR valve. Similarly, placing the valve in the open position allows fluid to flow into the closing chamber. Hydraulic fluid returning from the BOP stack flows back to the accumulator reservoir, so that the entire system is a "closed loop." The accumulator unit, sometimes called the Koomey unit, consists of four basic components:

- Accumulator bottles.
- Pumping system (air and electric).
- Manifold.
- Fluid reservoir (Fig. 4.23).

1: Air pump
2: Electric pump
3: Accumulator bottles
4: Four way valves
5: Manifold
6: Fluid reservoir

Figure 4.23 Blow out preventer control unit or "Koomey" unit.

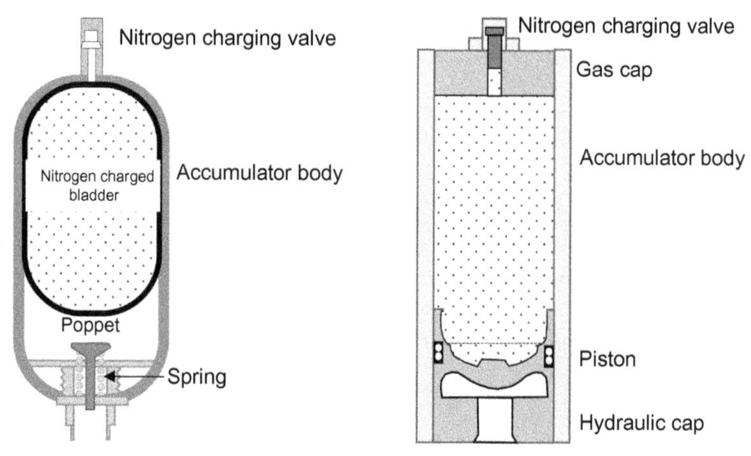

Figure 4.24 Bladder accumulator (left) and piston accumulator (right).

4.9.2 Accumulator bottles

Accumulator bottles are available in several sizes and types. By far the most common is the 11 or 34 US gallon cylindrical bottle containing a flexible rubber bladder. The bladder is attached to the top of the cylinder using a threaded connection, allowing the bladder to be filled with nitrogen. Since nitrogen is inert it is safe for use at the wellsite. The quantity of nitrogen used depends on the size of the bottle, but in most cases the bladder will be pressurized to 1000 psi. This is termed the precharge pressure. Hydraulic fluid is pumped into the bottom of the bottle using the accumulator unit pumps. The nitrogen charged rubbed bladder is compressed as the bottle fills. When the bottle pressure reaches the operating pressure for the system, the pump stops, leaving the hydraulic fluid stored under pressure. When released, the pressurized fluid is forced out of the bottle by the nitrogen in the bladder expanding.

Other cylindrical bottles use a buoyant float instead of a bladder. In these bottles, the nitrogen is pumped into the top of bottle above the float. Fluid pumped into the bottom of the bottle below the float forces the float up, compressing the nitrogen.

Accumulator bottles should be protected by a pressure relief valve, normally set to 3300 psi (Fig. 4.24).

4.9.3 Accumulator system pumps

The pumping system supplies hydraulic fluid to the accumulator bottles. Most accumulator units have two independent pumps: a triplex pump driven by an electric motor, and a separate air driven plunger pump. The API specification 16D requires that:

- With the accumulator isolated and with one pump or one power system out of service, the remaining pump system shall have the capacity within 2 min to:
  - Close one annular BOP on open hole;
  - Open the hydraulically operated choke valve;
  - Provide final pressure at least equal to the grater of the minimum operating pressure recommended by the manufacturer of both the annular BOP and choke valves.
- The cumulative output capacity of the pump systems shall be sufficient to charge the entire accumulator system from precharge pressure to the system rated working pressure within 15 min.

4.9.4 The manifold system

The manifold directs and regulates the flow of pressurized fluid from the accumulator bottles to the BOP, and directs hydraulic fluid back to the reservoir. The accumulator bottles are connected to the high-pressure section of the manifold. For most API approved units, this section of the manifold has a working pressure of 3000 psi. In a 3000 psi operating pressure system, the manifold regulator takes fluid from the accumulator bottles and regulates it down to 1500 psi for the ram type preventers and HCR valves. Similarly, a separate regulator reduces the pressure to that required to operate the annular preventer. In most, system annular pressure can be adjusted at the accumulator or the remote station. Gauges on the manifold should read accumulator bottle, ram preventer, and annular preventer pressure. The setting on most 3000 psi systems are 3000 psi for the accumulator, 1500 psi for control the manifold, and 800–1200 psi for the annular preventer. The manifold returns fluid from the BOP to the fluid tank.

Manifold valves are used to direct accumulator pressure to a BOP preventer or HCR valve, with each valve controlling a single function. These valves are normally three position four-way valves. The three positions are:

- Open: Connects the manifold with the opening hydraulic chamber on the BOP.
- Closed: Connects the manifold to the closing chamber on the BOP.
- Neutral: Connects the BOP to the fluid reservoir (Fig. 4.25).

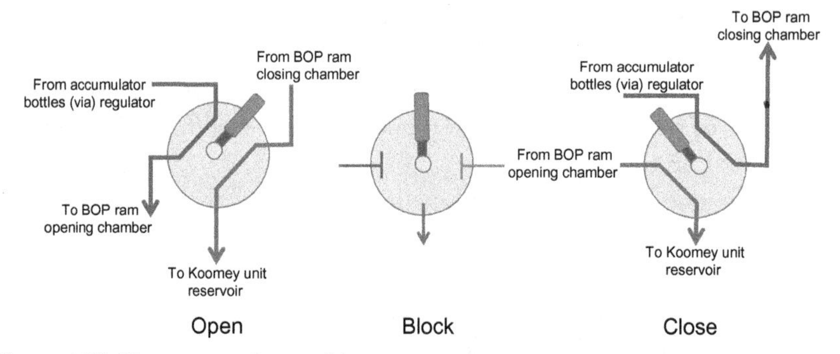

Figure 4.25 Three-way valve positions.

4.9.5 Fluid reservoir

The reservoir feeds the pumps and takes return fluid from the BOP stack, storing the spare hydraulic fluid that is not under pressure. The reservoir is normally equipped with a level indicator and filters to prevent contaminated fluid entering the pumps.

4.9.6 Remote blow out preventer control panel

Most systems will have at least one remote control panel in line with API recommendations. This enables the well to be closed in from at least two locations. Any of the BOP rams can be operated from the accumulator unit or from the remote BOP panel. If only one remote panel is used, it should be located by the drill floor. For snubbing operations, the remote panel would be in the work basket. At many locations, more than one remote panel will be used. In addition to the panel by the drillers console it is common to have a second panel in the toolpusher's (or company man's) office. Additional panels are sometimes located by escape routes on offshore installations.

Three types of remote panel are available:

- *Air*: Not recommended for a distance of more than 150 ft or freezing temperatures.
- *Electro–Pneumatic*: Not recommended for freezing temperatures.
- *Electro–Hydraulic*: Suitable for freezing temperatures.

Remote panels should have clear, unambiguous labeling of the ram functions. A simple method of achieving clarity is to mount the controls on a graphic representation of the BOP stack. Most control panels will have simple hinged plate covers over critical controls (shear rams) to prevent accidental closure (Fig. 4.26).

Figure 4.26 Remote control blow out preventer panel.

4.9.7 Accumulator volume calculations

Accumulators must be sized in accordance with regulatory requirements. The first step in calculating the volume of fluid available in a system is to determine how much usable fluid each accumulator bottle can deliver.

Example: Using an 11 gallon, bladder-type accumulator bottle with a 1000 psi nitrogen, the precharge volume of usable fluid can be calculated using Boyle's law:

$$P_1 V_1 = P_2 V_2 \text{ also } P_1 V_1 = P_3 V_3$$

where P_1 is the nitrogen precharge pressure of 1000 psi; P_2 is the minimum operating pressure of 1200 psi; P_3 is the maximum operating pressure of 3000 psi; V_1 bladder internal volume at precharge pressure (10 gallons); V_2 bladder internal volume at the minimum operating pressure, P_2 (gallons); V_3 bladder internal volume at the maximum operating pressure, P_3 (gallons).

Volume at minimum operating pressure, $V_2 = \frac{1000 \times 10}{1200} = 8.33$ galls.

Volume at maximum operating pressure, $V_3 = \frac{1000 \times 10}{3000} = 3.33$ galls.

The usable volume of hydraulic fluid in each bottle is $8.33 - 3.33 = 5$ gallons.

The compressibility (Z factor) of nitrogen at (80F) 1000 psi is 1.01, at 1200 psi it is 1.02, and at 3000 psi it is 1.06. Since these values are relatively small, temperature and compressibility are normally ignored, as is the difference between gauge pressure and absolute pressure (14.7 psi). Including nitrogen compressibility into the calculation gives a slightly smaller result, 4.91 gals.

In subsea wells the precharge pressure in subsea accumulator bottles needs to be increased to account for the hydrostatic pressure of the hydraulic fluid in the power fluid supply umbilical. A conservative approach is to use the gradient of seawater, 0.445 psi/ft, rather than the gradient of the power fluid.

Having calculated the useable volume in each accumulator bottle, the number of accumulator bottles needed to supply the BOP can be determined. For example, a surface BOP stack has been configured with:

Blow out preventer stack component	Volume to close (gallons)	Volume to open
1 × annular BOP	24.1	24.1
3 × ram BOP	11.8	11.8
2 × HCR valves	0.46	0.46

The total fluid volume required to close from a full open position at zero wellbore pressure all of the BOPs in the BOP stack plus 50% reserve is 90.63 gallons. The number of 11-gallon accumulator bottles is therefore:

$$90.63/5 = 18.126 \rightarrow 19 \text{ bottles.}$$

4.10 IN PIPE SHUT-OFF DEVICES

If it becomes necessary to shut-in a well, flow through the annulus is stopped by closing either the annular preventer or the pipe rams. It is also necessary to prevent flow to the surface through the pipe in the well. There are several "in-pipe" shut-off devices that can be used.

4.10.1 Kelly valves

On rigs where a kelly is still in use, an upper kelly valve (sometimes called a kelly cock) is positioned between the swivel and the kelly. A second, lower kelly valve, is placed on the bottom of the kelly. Kelly valves are normally manually operated full opening ball valves. They are closed by rotating the ball through a quarter turn using a hexagonal wrench. Most drilling contractors and operating companies recommend using the upper kelly when working with a BOP stack that has a rated working pressure of 5000 psi or more. The upper valve isolates the kelly hose, swivel, and surface equipment from the high well pressure.

A lower kelly valve is used as a back-up to the upper valve, and is also often used as a mud saver.

4.10.2 Top drive valves

Rigs equipped with a top drive have two ball valves located on the top drive equipment, in an arrangement very similar to that used with a kelly. Indeed, they are sometimes referred to as kelly valves, kelly cocks, or inside BOPs. In most installations the lower valve is a standard manually operated ball kelly valve. The upper valve is hydraulic or pneumatically operated from the driller's console, and is normally the first valve to be closed.

4.10.3 Full opening safety valve (FOSV)

When running or pulling of completion tubulars, the pipe in the rotary table is not connected to the top drive (or kelly). If there is a kick, flow through the pipe is normally shut-in by stabbing a FOSV into the tubing in the rotary table, and then closing it. Most valves are uni-directional, only holding pressure from below, and are closed by rotating a ball through a quarter turn using a hexagonal wrench. Both valve and wrench must be kept close to the rotary table for immediate access. The valve must be left in the open position to enable it to be made up even if the well has begun to flow. Since FOSVs are manufactured with drill pipe connections (box up, pin down), a range of cross-overs from the FOSV thread to each of the thread forms used during a completion must be available. The relevant cross-over should be kept with the FOSV ready for immediate use. The reason for having the box up connection on top of the FOSV is so drill pipe can be made up above the valve, and the tubing stripped back into the well below an inside BOP (Fig. 4.27).

In some circumstances, the response to a kick is to shut-in the annulus by closing the annular BOP, whilst at the same time making up the FOSV to the tubing in the rotary table and then shutting the ball. There are, however, circumstances where additional measures are needed to properly secure the well. These are fully described in Chapter 7, Well

Figure 4.27 Full opening safety valve (FOSV).

Kill, Kick Detection and Well Shut In, but in summary the main concerns are:

- If control lines, instrument cables, or ESP cables are being run with the tubing, it will not be possible to close in the annulus, as the cables will prevent both annular and pipe preventers from sealing around the tubing. To enable the well to be closed in when cables are across the BOP stack, a "kick stand" should be racked back in the derrick. The stand should be made up from joints of drill pipe with a cross-over to whatever tubing is in the rotary table at the bottom, and a FOSV at the top.

  In the event of a kick, the control lines must be cut and the kill-stand made up to the tubing in the well. Once made up, the FOSV is closed and the stand lowered across the BOP stack, only then can the annular or pipe rams be closed.
- There will be circumstances during the tripping of completion tubulars and equipment when closing the annular preventer or the pipe rams will not prevent flow. This would include tripping badly corroded or holed pipe and the running of sand control screes or slotted

liner. The well should be secured by using the kill-stand, or by cutting the pipe and closing the blind rams.

- Installing and closing a FOSV in response to a kick results in an upwards piston force equal to well pressure times the cross-sectional area of the pipe OD. If there is not much tubing in the well, piston forces will force the pipe up through which ever preventer has been closed. Tubing couplings do not usually have enough taper to prevent them damaging the annular preventer if they are forced through the closed seal. Closing the pipe ram preventer may be enough to hold the tubing, but not if it is weakened by corrosion. The safest option is to use the kill-stand, the couplings on drill pipe having enough upset to be retained by closed pipe rams.

It is good practice to stab a FOSV any time the tubing needs to be left in the slips, for example whilst carrying out repairs or during a shift change.

4.10.4 Inside blow out preventer (gray valve)

An inside BOP, sometimes called a gray valve, is a non-return valve that is used when drill pipe or tubing needs to be stripped into the well. It prevents upwards flow, but allows fluid to be pumped down the pipe. The most commonly used internal BOP features a rod that is used to hold the valve off-seat until it has been made up to the tubing. With the valve in place, the rod is released, allowing the valve to seat and shutting off flow. The upper sub can then be removed and pipe can be made up above the internal BOP and stripped into the well (Fig. 4.28).

4.10.5 Drill-string float valve

Non-return valves prevent an influx into the drill string during a kick. They also prevent backflow of annular cuttings plugging bit nozzles. Non-return valves are required by some drilling contractors and operating companies when drilling and opening the hole before setting the surface casing, or at any time the well control plan is to divert. They are often used for drilling deeper sections of the hole.

There are two types of valve in widespread use, the flapper and the spring-loaded dart (or ball).

These types of valves are not normally used during completion or workover operations, since they prevent the tubing from filling as it is run. When completion tubing is run with a valve in place, the tubing needs to be filled from the top. The tubing displacement volume is much

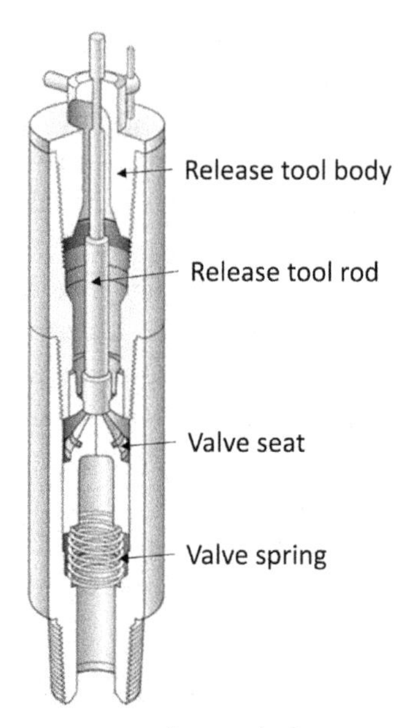

Figure 4.28 Inside blow out preventer (gray valve).

more than for open ended pipe, so small pit gains or losses become more difficult to detect, moreover, reverse circulation is not possible.

4.11 MUD GAS SEPARATOR

Gas handling equipment is vital for the safe removal of the large volume of gas that is often present following a kick. The mud gas separator, also called the gas buster or poor-boy degasser, is the first device downstream of the choke during a well control operation. It is used following a kick to separate kick gas from the brine or mud as it is circulated out of the well, allowing mud or brine that was in the well to be recovered and reused; an important consideration when pit space is limited. The mud gas separator is not part of the normal circulation system, and is only normally used for well control.

A mud gas separator is basically a gravity separation tank. Mud or brine flowing from the choke enters the vessel and drops down through several baffle plates. This causes the light gas to break out from the liquids and flow up into the vent line, where it is vented to atmosphere through the derrick line (offshore), or directed away from the rig, usually to the

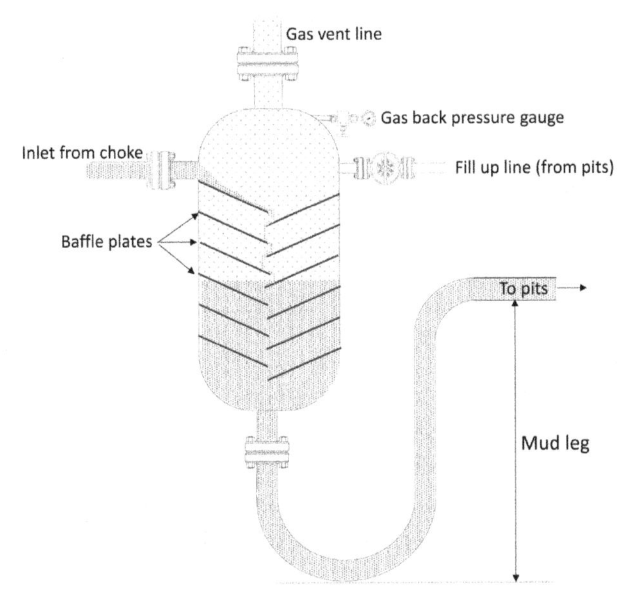

Figure 4.29 Mud gas separator.

flare pit (land operations). Fluids are taken from the separator vessel at the lowest point, exiting through the mud leg (a liquid seal U tube). The U tube is necessary to prevent blowdown of the vessel. During completion and workover operations, it is good practice to have the separator vessel filled with the kill fluid in use (mud or brine) (Fig. 4.29).

There are several design features which affect the volume of gas and fluid that the separator can safely handle. Production and well test separators are sized and internally designed to efficiently separate gas from the liquid. This is possible because the fluid and gas properties are known, and design flow rates can be readily established. Gas separators for drilling rigs need to be able to cover a wide range of conditions. Gas volume, as well as drilling and completion fluid rheology, varies considerably. For practical (and cost) reasons, rig mud gas separators are not designed for maximum possible gas release rates. However, they should be able to handle most kicks. Should gas flow rates exceed the separator capacity, flow will have to by-pass the separator and go directly to the flare line. This prevents the liquid being blown out of the mud leg (bottom of separator) and discharging gas into the mud system, which could cause a very hazardous situation. The maximum pressure that can be applied to the separator vessel is equal to the hydrostatic pressure of the fluid in the mud leg. If this pressure is exceeded, the blow through will occur. An accurate, easy-to-read, and properly calibrated low pressure gauge must be fitted to the separator vessel. Gauges are typically in the 15–20 psi range, and ideally should be visible from the choke console. Many gas separators are fitted with audible alarms to warn of overpressure.

4.12 FLUID STORAGE

During drilling and completion operations, a large volume of fluid needs to be held at the wellsite. On land rigs, most of the fluid is stored in partitioned rectangular steel tanks, with each compartment holding approximately 200 bbls. In addition to the bulk storage, several smaller tanks are used for specific purposes, for example the storage of a kill pill. At some land locations, large volume pits are made by forming earth into raised banks then lining the construction with impermeable sheeting (normally plastic). In recent years, earth pits have become less common, principally because of environmental concerns. Where they are used, it is usually for the storage of waste mud and cuttings prior to disposal.

On offshore drilling rigs and platforms, fluid is stored in several large tanks each holding as much as 1000 bbls. Although individual tanks are generally larger, the cumulative volume stored is often less than at a land location, since space is limited. Table 4.4 is representative of the pit capacity of a medium sized jack-up drilling rig (Fig. 4.30).

Insufficient fluid storage can create logistical problems, especially when transitioning from the drilling to the completion phase of the well construction process. For some completion operations, for example open-hole gravel packing, a large volume of completion brine is required. Moreover, the entire fluid system needs to be thoroughly cleaned before the fluid can be stored. In practical terms, this normally means that when the drilling mud is displaced from the well and replaced by completion brine, it must be sent to a support vessel rather than returned to the pits. The need to transfer the drilling mud to a vessel rather than back to the pits has obvious implications for well control. It becomes more difficult to monitor fluid returns, and an influx can go unnoticed for longer. In some circumstances, the overbalanced drill fluids are replaced with underbalanced completion fluids, adding to the risk.

Table 4.4 Jack-up rig mud pit capacity

Pit number	Working volume (bbl)	Max volume (bbl)
Active 1	450	460
Active 2	490	500
Active 3	440	450
Reserve 1	540	550
Reserve 2	510	520
Reserve 3	575	580
Reserve 4	250	260
Pill pit	80	90
Slug pit	80	90
Water pit	180	190
Degasser 1	40	50
Degasser 2	40	50
Desilter 1	40	50
Desilter 2	40	50
Centrifuge	40	50

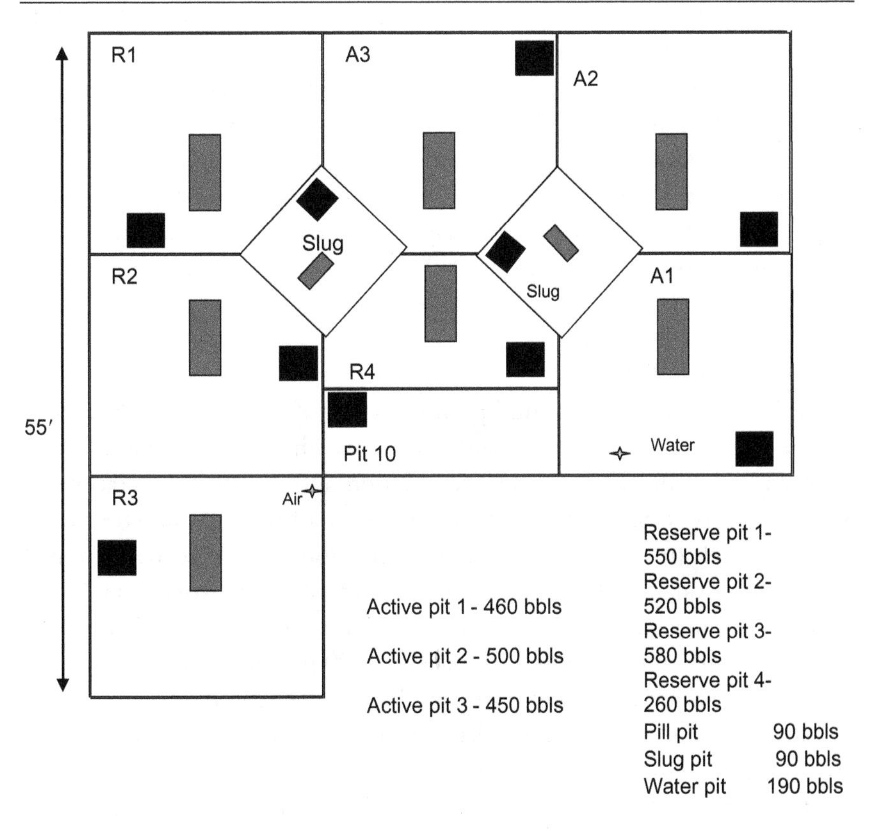

Figure 4.30 Mud pit layout on a jack-up rig—corresponding to Table 4.4.

For both land and offshore fluid systems, the pits are normally categorized as "active" and "reserve."

Active pits: During most operations that require fluid circulation, fluid from the active pit is returned to the active pit in a closed loop system. Consequently, if the active pit capacity is too large, it becomes more difficult to spot the pit gains or losses that are an early indication of a well control problem. The active pits should ideally be equipped with a pit volume totalizer (PVT) that accurately monitors the level (volume) in the pit. Modern PVT systems normally feature visible and audible alarms set to warn the rig crew if there are pit gains.

Reserve pits: Where possible, the reserve pits should be large enough to store all surface volumes required for the completion of the well. Ideally there should, as an absolute minimum, be enough excess to deal with losses equivalent to 1−1½ times the hole volume.

Trip tank: A trip tank is a low-volume (100 barrels or less), calibrated tank that can be isolated from the remainder of the surface fluid system. It is used to accurately monitor the amount of fluid going into or coming out of the well whilst pipe is being tripped. Properly configured, maintained, and monitored, small volume fluid gains or losses can be detected. Trip tanks are also used to monitor fluid level during static conditions, for example when logging operations are taking place. If pipe is being stripped into the well, a second, stripping tank is used to measure fluid transferred from the trip tank (Fig. 4.31).

4.12.1 Fluid pumps

Most rigs have at least two mud pumps in case of breakdown. They must be able to pump the maximum anticipated weight of kill fluid at the rate required to kill the well. During completion and workover operations, a bullhead kill is common and normally requires a higher pump rate than a circulation kill (to overcome gas migration). Pumps need to be equipped with a functional stroke counter and the number of strokes per barrel recorded on any kill sheets. The maximum discharge pressure that can be obtained from mud pumps is typically 7500 psi, if higher pressure is required to initiate a well kill normally the cement unit would be used.

Cement pump: Most offshore drilling units and offshore platforms with drilling derricks have a cement pump as part of the permanent facilities. Most are diesel driven, and are therefore independent of the rig power supply. On land locations, a trailer-mounted cement unit is brought to the location as and when required. Cement pumps provide a back-up to

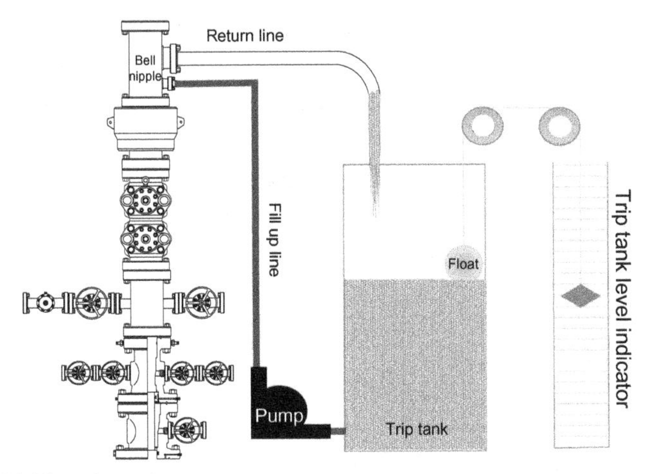

Figure 4.31 The trip tank.

the mud pumps, and typically operate at much higher discharge pressure (up to 15,000 psi), but at lower rates (6–8 bbls/min).

Positive displacement fluid pumps are described in more detail in Chapter 8, Pumping and Stimulation.

4.13 FLANGES, RING GASKETS, AND SEALS (API 6A)

Flange connections using metal ring gaskets, sometimes called ring type joints, are widely used for the assembly of the BOP stack, choke manifold, wellhead and Christmas tree. Since the flange and the gasket ring are both metal, a correctly assembled ring joint forms a robust and resilient metal-to-metal seal. All flanges, ring gaskets, and seals used on well control equipment, including the wellhead, kill and choke lines, choke manifold, and the BOP stack should conform to the recommendations laid out in API Specification 6A and API Specification 16A.[10]

4.13.1 Flanges

There are two main categories of flange designated by the API in specification 6A.

- 6B flange. Connected with R or RX ring gaskets.
- 6BX flanges. Connected with BX ring gaskets.

The API also recognizes segmented flanges used with dual string wellheads and Christmas trees.

Table 4.5 American Petroleum Institute specification flange working pressure

Rated working pressure	Flange size range	
	Type 6B (in.)	Type 6BX (in.)
2000	2(1/16)−21¼	26¾−30
3000	2(1/16)−21¾	26¾−30
5000	2(1/16)−11	13⅝−21¼
10,000		1(13/16)−21¼
15,000		1(13/16)−18¾
20,000		1(13/16)−13⅝

Type 6B flanges have a maximum working pressure of 5000 psi, and 6BX flanges are rated to 20,000 psi (Table 4.5).

Some API 6B flanges are manufactured with a raised face. A raised face on a 6B flange has no functional purpose, since the R or RX gasket seal means there will always be a stand-off between to two flange faces. At present, most manufacturers are making 6B flanges with a flat (sometimes called full) face, i.e., without a raised face.

For a 6BX flange to conform to the API 6A specification it must have a raised face of at least ⅛ in. height, unless it is a studded face BX flange. API Specification 6A gives the following reason for having a raised face, "*Depending on tolerances, the connection make-up bolting force may react on the raised face of the flange when the gasket has been properly seated. This support prevents damage to the flange or gasket from excessive bolt torque.*"

Normally, 6BX flanges with raised faces contact when fully made up. This improves bending resistance at the connection, and limits further deformation of the ring gasket.

The API 6A requires that flanges integral with "drill through" equipment be marked as follows:

Manufacturer's name or mark, API monogram (when authorized by API), flange or hub size, pressure rating, ring joint type and number, minimum vertical bore, model and serial number and date of manufacture, and clamp number where applicable.

4.13.2 Ring gaskets

Ring gaskets used with API flanges are identified by type and number:

- The types are R, RX, and BX.
- A number denotes the dimension or size.

Ring gaskets are available in a range of materials, and must meet the API specification for maximum hardness to ensure that the gasket

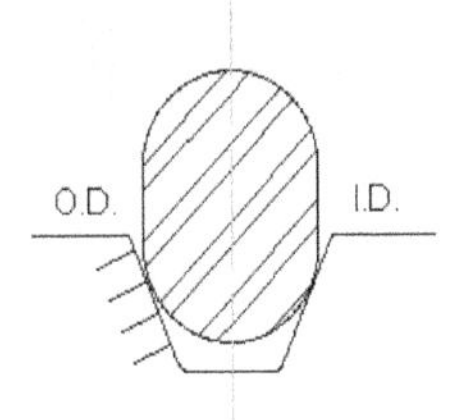

Figure 4.32 "R" type ring gasket.

deforms to form a seal (coining) without deforming the ring groove in the flange face. A ring gasket should never be reused.

4.13.2.1 "R" ring gasket

The type "R" ring gasket seals along a small band of contact on both the OD and ID of the gasket. It is not energized by internal pressure. These gaskets are either oval or octagonal in cross-section. Since the "R" gasket does not allow face-to-face contact between the flange face, any external loads are transferred through the sealing surface to the ring. Vibration and external loads can cause small bands of contact between the ring and the groove to deform plastically, so that the joint may develop a leak unless the flange bolting is periodically tightened. Standard procedure with type "R" joints in the BOP stack is to tighten the flange bolting weekly (Fig. 4.32).

4.13.2.2 "RX" pressure energized ring gasket

Originally a Cameron development, the "RX" gasket is now recognized by the API. Octagonal in cross-section, it is energized by well pressure along a band of contact between the groove in the flange and the OD of the ring. Since the gasket is slightly larger in diameter than the groove, tightening the flange bolts compresses the ring and creates an initial seal that is fully energized by internal pressure. The "RX" gasket does not allow face-to-face contact between the flange face, so any external load is transferred through the sealing surface to the ring. However, as the gasket has a large load-bearing surface on its inside diameter, loads are transmitted without excessive plastic deformation of the sealing surface (Fig. 4.33).

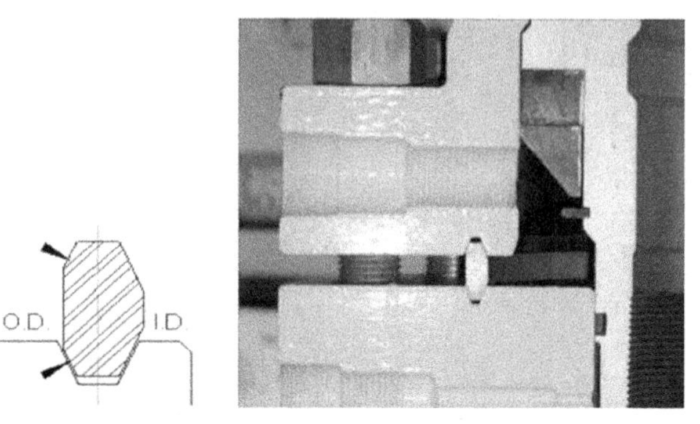

Figure 4.33 Type "RX" ring gasket.

4.13.2.3 "BX" pressure energized ring gasket

The "BX" ring gasket is used with 6BX flanges where flange face-to-face contact is allowed. Contact between the flange faces enables gasket compression and plastic deformation to be controlled. As both the ID and OD of the gasket may contact the grooves, holes are normally drilled through the gasket to ensure pressure either side of the joint is balanced.

Sealing takes place along small bands of contact between the groove in the flange and the OD of the gasket. The gasket is slightly larger in diameter than the grooves, and so is compressed slightly to obtain the initial seal as the joint is tightened. Variances in groove and ring gasket tolerance mean that face-to-face contact is not always achieved. When this occurs, vibration and external loads can cause deformation of the ring, leading to a leak.

REFERENCES

1. *Recommended practices for blowout prevention equipment systems for drilling wells.* API Recommended Practice 53. 3rd ed.; 1997.
2. NORSOK standard D001. Drilling facilities. 3rd ed.; 2012.
3. *Recommended practice for well control operations.* API RP 59. 2nd ed.; 2006.
4. API specification 6A. Specification for Wellhead and Christmas Tree Equipment. 20th ed.; 2010.
5. Information from Cameron Schlumberger web page: <https://cameron.slb.com/products-and-services/drilling/pressure-control-equipment/annular-bops/dl-annular-bop>.
6. Information from GE/Hydrill web page: <https://www.geoilandgas.com/sites/geog/files/gea31279_gx_annular_blowout_r2.pdf>.

7. *Specification for control systems for drilling well control equipment and control system diverter equipment*. API Specification 16D (Spec 16D). 2nd ed.; 2005.
8. *Recommended practice for design of control systems for drilling well control equipment*. API RP 16E. 2nd ed.; 2004.
9. NORSOK. *System requirements well intervention equipment*. D-002. Rev 1; 2000.
10. *API specification 16A*. *Specification for drill through equipment*. 4th ed.; 2017.

CHAPTER FIVE

Completion, Workover, and Intervention Fluids

5.1 INTRODUCTION

Nearly all completions are run with a clear solids free, brine in the well. During workover operations, brine is used to kill the well and remains in the wellbore until the new completion has been installed. Brine is also used as a packer fluid; the fluid left in the annulus at the end of the completion or workover. Finally, brine is used for perforating, gravel packing, and fracturing operations.

For many completion and workover operations, a brine column is the primary well control barrier. Where brine has a well control function, it must be dense enough to overbalance the formation pressure by a margin sufficient to prevent an influx. Since brine is nominally solids free, any overbalance pressure acting against a permeable formation will result in fluid loss to the formation. Uncontrolled or unmonitored losses would mean a loss of hydrostatic overbalance, causing the well to kick. Fluid loss can also damage the formation, and consequently reduce productivity.

5.2 BRINE SELECTION

Where brine forms the primary well control barrier, the first step in selecting a brine is to calculate the density needed to overbalance formation pressure. Brine density is calculated with respect to well depth (true vertical depth, TVD) and reservoir pressure (bottom hole pressure, BHP). Once brine density has been calculated, it must be adjusted to correct for wellbore average temperature and pressure. With this done, a suitable brine can be selected.

A wide range of brine densities can be achieved by dissolving different salts in freshwater (Table 5.1).

Well Control for Completions and Interventions.
DOI: https://doi.org/10.1016/B978-0-08-100196-7.00005-1

Table 5.1 Brine density

Brine type	Formula	Density range (ppg)	Density range (SG)
Sodium chloride	NaCl	8.4–10.0	1–1.20
Potassium chloride	KCl	8.4–9.7	1–1.164
Ammonium chloride	NH_4Cl	8.4–8.9	1–1.06
Sodium bromide	NaBr	8.4–12.7	1–1.52
Sodium chloride/bromide	NaCl/NaBr	8.4–12.5	1–1.50
Sodium formate	NaCOOH	8.4–11.1	1–1.33
Potassium formate	KCOOH	8.4–13.3	1–1.60
Cesium formate	CsCOOH	13.0–20.0	1.56–2.40
Potassium/cesium formate	KCOOH/CsCOOH	13.0–20.0	1.56–2.40
Sodium/potassium formate	NaCOOH/KCOOH	8.4–13.1	1–1.57
Calcium chloride	$CaCl_2$	8.4–11.3	1–1.36
Calcium bromide	$CaBr_2$	8.4–15.3	1–1.84
Calcium chloride/bromide	$CaCl_2/CaBr_2$	8.4–15.1	1–1.81
Zinc bromide	$ZnBr_2$	12.0–21.0	1.44–2.52
Zinc/calcium bromide	$ZnBr_2/CaBr_2$	12.0–19.2	1.44–2.30

The densities listed in Table 5.1 can only ever be approximate, since density varies with temperature, and to a lesser extent, pressure. Increasing temperature causes fluid to expand, with a consequent decrease in density. Increasing pressure causes a density increase.

In addition to density requirements, brine selection must also take account of the true crystallization temperature (TCT). Brine should be compatible with both the formation and formation fluids, as well as with any exposed completion components; all elastomers and metal. Selecting a compatible brine that has both a suitable crystallization temperature, and is compatible with the formation and the completion, requires an iterative approach and can require extensive testing.

5.3 BRINE DENSITY

Where brine forms the primary well control barrier, calculating the correct density is crucial. Density calculations will always be the first step in the brine selection process. Brine must have sufficient density to exceed formation pressure by the required margin. That margin (overbalance) is typically between 200 and 300 psi (14–20 bar).

To enable brine density to be calculated, both the vertical depth (TVD) to the reservoir and the reservoir pressure must be known. Eq. (5.1) is used to find the brine density necessary to overbalance reservoir pressure by the desired amount.

$$\text{Density (ppg)} = \frac{\text{Reservoir pressure (psi)} + \text{required overbalance (psi)}}{0.052 \times \text{vertical depth (feet)}}$$

$$\text{Density (SG)} = 10.2 \times \frac{\text{Reservoir pressure (bar)} + \text{required overbalance (bar)}}{\text{vertical depth (m)}} \tag{5.1}$$

5.3.1 Adjusting brine density for wellbore temperature and pressure

Having calculated a surface brine density, it must be adjusted to correct for temperature and pressure in the wellbore. Brine expands when heated, lowering the density. Since the wellbore temperature is, almost without exception, higher than ambient surface temperature, this density decrease must be calculated and surface brine weight increased to compensate. Pressure has the opposite effect; increased pressure in the wellbore compresses the brine, increasing density. Since the density decrease due to increased temperature is significantly more than any density increase due to pressure, the net result is always to increase brine weight to adjust for wellbore conditions. To enable density loss downhole to be calculated, it is necessary to know the wellbore average temperature and pressure, as well as the brine expansion and compression coefficients. Expansion and compression coefficients vary considerably and are a function of brine composition and weight, as well as temperature and pressure. Brine vendors are usually able to supply the user with accurate expansion and compression coefficients for any combination of brine weight and brine composition across a wide range of temperatures and pressures. These vendor supplied coefficients are based on empirical tests and theoretical models.

In addition to vendor supplied data, expansion and compression coefficients are widely available from society of petroleum engineers (SPE) papers and technical publications. Two of the most widely quoted examples are listed in Tables 5.2 and 5.3. These list expansion and compression

Table 5.2 Brine expansion coefficients (Krook and Boyce)

Brine type	Density (ppg)	Expansion coefficient (vol/vol/°F)	Density decrease in lb/gal/100°F
NaCl	9.49	2.54×10^{-4}	0.24
$CaCl_2$	11.46	2.39×10^{-4}	0.27
NaBr	12.48	2.67×10^{-4}	0.33
$CaBr_2$	14.3	2.33×10^{-4}	0.33
$ZnBr_2/CaBr_2/CaCl_2$	16.0	2.27×10^{-4}	0.36
$ZnBr_2/CaBr_2$	19.27	2.54×10^{-4}	0.48
		At 12,000 psi between 76 and 345°F	

Table 5.3 Brine compressibility coefficients (Krook and Boyce)

Brine type	Density (ppg)	Compression coefficient (vol/vol/°F)	Density increase in lb/gal/1000 psi
NaCl	9.49	1.98×10^{-6}	0.019
$CaCl_2$	11.46	1.50×10^{-6}	0.017
NaBr	12.48	1.67×10^{-6}	0.021
$CaBr_2$	14.3	1.53×10^{-6}	0.022
$ZnBr_2/CaBr_2/CaCl_2$	16.0	1.39×10^{-6}	0.022
$ZnBr_2/CaBr_2$	19.27	1.64×10^{-6}	0.031
		At 198°F from 2000 to 12,000 psi	

coefficients were derived experimentally by Krook and Boyce in 1984.[1] They are the coefficients listed in American Petroleum Institute (API) Recommended Practices 13J; "Testing of Heavy Brines."[2]

Although the Krook and Boyce coefficients are widely used, they represent expansion coefficients at a given pressure; 12,000 psi. At lower pressure coefficients are significantly different, resulting in higher expansion ratios.

Tetra in their "Engineered Solutions Guide for Clear Brine Fluids"[3] use expansion and compression coefficients compiled by Bridges.[4] The values used are calculated at atmospheric pressure and 77°F, and are reproduced in Table 5.4.

Comparing Tables 5.2 and 5.3 (Krook and Boyce) with Table 5.4 (Bridges) reveals significant differences between the expansion coefficients of a brine under pressure and one at atmospheric pressure.

Using the relevant data, calculating density loss and an adjusted (increased) surface brine density is relatively straightforward. Two different equally valid methods are described:

Table 5.4 Brine compressibility coefficients (Bridges)

Brine type	Density (ppg)	Density decrease in ppg/100°F	Fluid compression lb/gall/1000 psi
NaCl	9.0	0.314	0.0189
NaCl	9.5	0.386	0.0188
NaBr	12.0	0.336	0.0190
$CaCl_2$	9.5	0.285	0.0188
$CaCl_2$	10.0	0.289	0.0187
$CaCl_2$	10.5	0.273	0.0186
$CaCl_2$	11.0	0.264	0.0187
$CaCl_2/CaBr_2$	12.0	0.325	0.0190
$CaCl_2/CaBr_2$	12.5	0.330	0.0193
$CaCl_2/CaBr_2$	13.5	0.343	0.0201
$CaCl_2/CaBr_2$	14.5	0.362	0.0212
$CaCl_2/Zn\text{-}CaBr_2$	15.5	0.387	0.0226
$CaCl_2/Zn\text{-}CaBr_2$	16.5	0.416	0.0224
$CaCl_2/Zn\text{-}CaBr_2$	17.5	0.453	0.0264
$CaCl_2/Zn\text{-}CaBr_2$	18.0	0.475	0.0276
$CaCl_2/Zn\text{-}CaBr_2$	18.5	0.501	0.0288
$CaCl_2/Zn\text{-}CaBr_2$	19.0	0.528	0.0301

In "Well Completion Design,"[5] the following equations are used to calculate brine weight adjusted for temperature and pressure.

Temperature correction:

$$\rho_T = \rho_{70}\left(2 - \left(\frac{1}{1 - (\overline{T} - 70)\alpha}\right)\right) \tag{5.2}$$

where:

ρ_T is the adjusted brine density (at average wellbore temperature),
ρ_{70} is the brine density at the API reference temperature of 70°F,
$\overline{T}$ is the average wellbore temperature, and
α is the expansion coefficient in vol/vol/°F (Table 5.2).

Pressure correction:

$$\rho_p = \rho_T\left(2 - \frac{1}{1 + (\overline{P})\beta}\right) \tag{5.3}$$

where:

ρ_p is the pressure corrected density,
$\overline{P}$ is the average pressure of the fluid in the wellbore, and
β is the compressibility coefficient in vol/vol/psi (Table 5.3).

Example calculation:
Well depth: 10,000 ft TVD.
Bottom hole temperature: 260°F.
Ambient temperature (surface): 80°F.
Reservoir pressure: 4690 psi.
Required overbalance: 250 psi.
Using Eq. (5.1), calculate the required density under surface conditions.

Density (ppg) $= (4690 + 250)/(0.052 \times 10,000) = 9.5$ ppg. The expansion coefficient for a 9.49 NaCl brine will be used: 2.54×10^{-4} (Table 5.2).

Next, calculate the average temperature in the wellbore.

$$\frac{\text{Surface temp} + \text{bottom hole temperature}}{2}$$

$$\frac{60 + 260}{2} = 160°F$$

The temperature adjusted brine weight becomes:

$$\rho_T = 9.5\left(2 - \frac{1}{1 - (160 - 70)0.000254}\right) = 9.277 \text{ ppg.}$$

The temperature adjusted brine weight is adjusted for pressure using Eq. (5.2).

BHP using the corrected brine density of 9.277 ppg is:

$$10,000 \times 0.052 \times 9.277 = 4824 \text{ psi}$$

Average wellbore pressure (between TD and surface) is therefore $4824/2 = 2412$ psi.

The pressure correction is:

$$\rho_p = 9.227\left(2 - \frac{1}{1 + (2412)0.00000198}\right) = 9.27 \text{ ppg.}$$

This would, by convention, be rounded up to 9.3 ppg.

A slightly different method is used by Tetra and illustrated in their "Engineered Solution Guide for Clear Brines." First, a density correction is calculated.

$$C_t = \frac{\alpha(\text{BHT} - \text{surf})}{200} \tag{5.4}$$

where:

C_t = average temperature correction in ppg,

α = expansion factor in lb/gal/100°F (from Table 5.4),

surf = ambient (surface) temperature °F, and

BHT = bottom hole temperature (°F).

A pressure correction is obtained from the equation:

$$C_p = \frac{\beta(\text{BHP})}{2000} \tag{5.5}$$

where:

C_p = average pressure correction in ppg,

β = compression factor in lb/gal/1000 psi (Table 5.4), and

BHP = bottom hole pressure in psi.

Brine density corrected for both pressure and temperature is then obtained:

$$d_c = d_u + C_t - C_p$$

where:

d_c = corrected density (temperature and pressure) in ppg,

d_u = uncorrected density in ppg,

C_t = average temperature correction in ppg (Eq. (5.4)), and

C_p = average pressure correction in ppg (Eq. (5.5)).

Example calculation (using the same data as the previous example):

Well depth 10,000 ft TVD.

Bottom hole temperature 260°F

Ambient temperature (surface) 80°F

Reservoir pressure 4690 psi.

Required overbalance 250 psi.

The brine density will be as before, 9.5 ppg. The expansion and compression factors for a 9.5 ppg NaCl will be used (Table 5.4).

Density loss (ppg) is:

$$C_t = \frac{0.386\,(260 - 80)}{200} = 0.3474 \text{ ppg}$$

To calculate the density increase, bottom hole pressure using an uncorrected brine weight must be calculated, $9.5 \times 0.052 \times 10{,}000 = 4940$ psi; the same as reservoir pressure + overbalance:

$$C_p = \frac{0.0188(4940)}{2000} = 0.0463 \text{ ppg}$$

Adjusted brine density is therefore:

$$d_c = 9.5 + .0.3474 - 0.0463 = 9.8$$

The difference of 0.5 ppg between the two example calculations is a result of using different expansion and compression coefficients. In a 10,000 foot well (as per the example), a 0.5 ppg difference in brine weight is equivalent to 260 psi; in many cases more than the required overbalance. These differing results only go to emphasize the importance of obtaining accurate expansion and compression data directly from the fluids vendors.

Deepwater wells require the calculations to be performed twice.

- Fluid density in the marine riser (from surface to seabed) where cooling will increase density.
- Fluid density in the wellbore (from mudline to TD).

Having made the density corrections for both sections of the well, the values are combined to give an overall density for the well.

5.4 CRYSTALLIZATION TEMPERATURE

Brine crystallizes at low temperature. The crystallization temperature of a brine is the temperature at which solids start to form. Those solids will either be ice crystals formed by water freezing, or salt crystals forming as brine salts come out of solution. Allowing brine temperature to fall to the crystallization temperature is to be avoided. Salt crystals can plug lines and interfere with pumping operations. They can also plug filtration units. Of greater concern is the potential for crystallization to result in a reduction of brine hydrostatic pressure, leading to a well control incident. Since it is the salts dissolved in water that gives brine the required density, crystallization of those salts cause the density to be reduced. If reduced density fluid is pumped into the well, the hydrostatic overbalance is reduced or lost altogether, and a kick can occur. Conversely, if ice crystals form, the fluid being pumped into the well will be denser than expected, increasing the potential for fluid loss, perhaps even fracturing the formation. Knowing the brine crystallization temperature is therefore essential.

Freshwater freezes at 32°F (0°C). Adding salt to freshwater lowers the freezing point. For example, dissolved sodium chloride (NaCl) reduces the freezing point of seawater to 28.4°F or −1.8°C. Similarly, salt dissolved in completion brine reduces the temperature at which freezing occurs. If freezing does occur, water ice crystals begin to form in the brine. Dissolving more salt both increases brine density and reduces the freezing point down to the eutectic point. If the density is increased further still, the crystallization temperature increases, but here, at higher densities, cooling causes salt crystals to form.

Fig. 5.1 shows the phase diagram for a single salt $CaCl_2$ brine. Most single salt brines have similar phase diagrams, since the relationship between temperature and density is predictable. With multisalt brines, the crystallization temperature can be adjusted by varying the concentration of the different salts in solution, meaning a multisalt brine can be formulated with a wide range of crystallization temperatures.

In oilfield application, brines are often mixed at relatively high degrees of saturation. It is therefore common to operate toward the right hand side of the phase envelope—to the right of the eutectic point. In locations where ambient temperatures are low, brine selection is often determined primarily by the cryztallisation point.

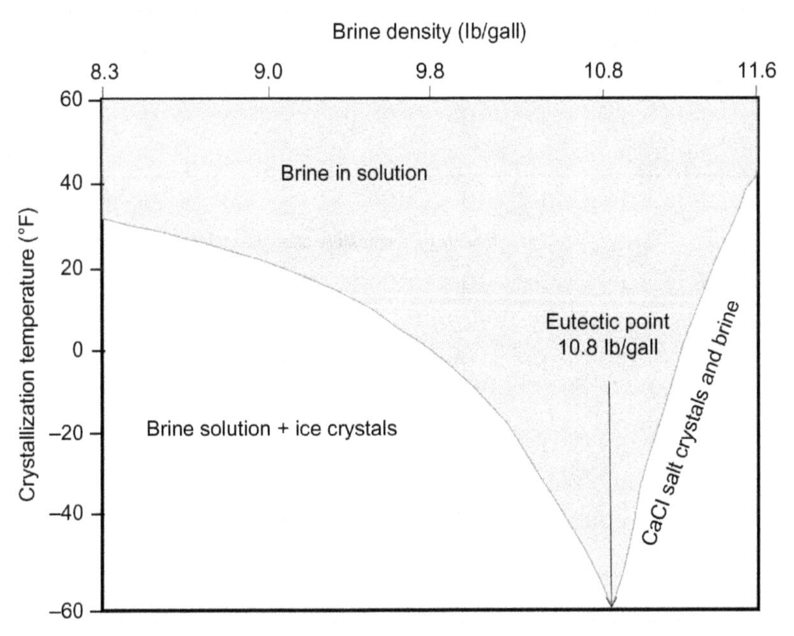

Figure 5.1 Crystallization temperature for a $CaCl_2$ brine.

5.4.1 Determining crystallization temperature

Brine crystallization temperature is determined experimentally by repeatedly cooling and heating a brine sample.

If a brine solution is cooled, at some point crystals will appear. The temperature at which this occurs is called "First Crystal to Appear," and is commonly abbreviated to FCTA. The FCTA temperature can include some super-cooling effect; cooling below the actual FCTA temperature but with no crystal formation. This can be minimized by slowing the rate at which the brine is cooled.

Crystal formation creates a small amount of heat that raises the temperature of the brine slightly. The slightly higher temperature is termed the "True Crystallization Temperature" or TCT. To remove crystals from a brine it must be heated until all of the crystals have gone back into solution. The temperature at which the "Last Crystal to Dissolve" occurs is recorded (Fig. 5.2).

The API 13J calls for the addition of a fourth temperature measurement. This is the "Maximum Temperature after Last Crystal," and is "the maximum temperature at which the second and subsequent cooling cycles begin."[6]

The reported brine crystallization temperature is normally the TCT, as this best represents the temperature at which crystal precipitation will occur.

Where multisalt brines are used, the least soluble salt will be the first to crystallize as temperature reduces (Table 5.5).

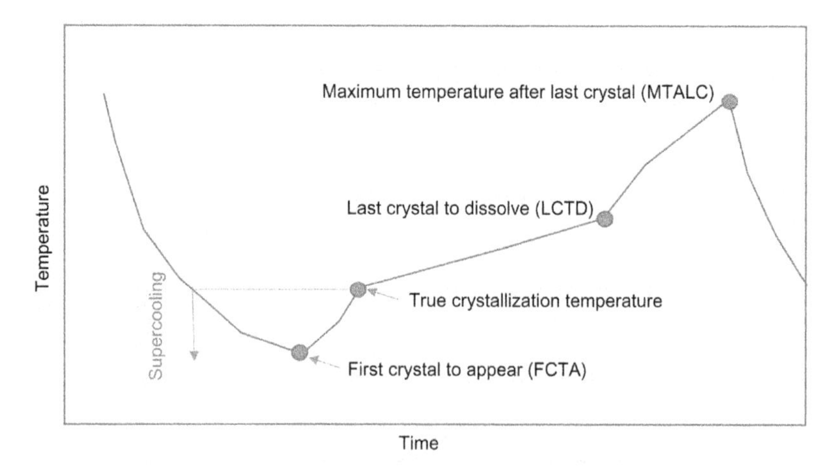

Figure 5.2 Determining the crystallization temperature of a brine.

Table 5.5 Multisalt brine: least soluble component

Brine	Least soluble salt
$NaCl/CaCl_2$	$NaCl$
$CaCl_2/CaBr_2$	$CaCl_2$
$CaBr_2/ZnBr_2$	$CaBr_2$
$CaCl_2/CaBr_2/ZnBr_2$	$CaCl_2$

Cross-contamination of a heavy completion brine with less soluble NaCl or KCl brine from the formation can raise the TCT temperature significantly.

5.4.2 Pressure effect on crystallization temperature

Brine crystallization temperature is measured at atmospheric pressure. At high pressure the crystallization temperature (TCT) increases. This is not normally a problem, since the highest pressure in the well is normally encountered where temperature is highest, that is, at the bottom of the well. However, as the industry began to develop deepwater offshore fields, the combination of high pressure and low temperature became more common. High pressure testing of surface equipment can raise the temperature at which crystallization will occur. When ambient temperature is low, a pressure test can lead to crystallization. In one case, a brine at 5000 psi had a crystallization temperature 15°F higher than that recorded at ambient temperature and pressure.[7] Where a combination of high pressure and low temperature is expected, additional testing is recommended.

5.5 SAFETY AND THE ENVIRONMENT

Completion brines are highly concentrated solutions of inorganic salts, mostly chlorides and bromides, and are potentially harmful to people and the environment. Brines are hygroscopic, meaning they will absorb moisture from their surroundings, including drawing moisture from the air and, of more concern, the skin of anyone who comes into direct contact with the brine. Brine exposure can cause skin irritation and eye damage. Generally, brine becomes more harmful as density increases. While potassium, sodium chloride, and sodium bromide brines are mildly irritating, calcium chloride, calcium bromide, and zinc bromide are extremely hazardous, with the potential to blind and cause painful reddening and blistering of the skin similar to a burn.

Material Safety Data Sheets should be available for any brine, as well as brine additives that will be used at the wellsite. Moreover, personnel should be properly briefed on any risks associated with mixing, storing, and pumping brine. Where appropriate, personal protective equipment should be used to prevent exposure. Should exposure to harmful brine occur, it must be dealt with immediately. Skin contact should be treated with prolonged washing with freshwater, and eye contact with eye wash solution, or if none is available, prolonged washing with freshwater.

Brines are harmful to the environment. While some of the brine salts can be diluted to render them harmless, brine containing zinc bromide cannot. Indeed, in some jurisdictions the use of brine containing zinc bromide is discouraged.

5.6 BRINE COMPATIBILITY

The well control function of brine determines brine density, and crystallization temperature is related to brine density. The next step is to consider brine compatibility with the formation, formation fluids, and the completion materials. While brine incompatibility with the formation can have a detrimental effect on productivity, an incompatibility with the completion materials (metals and elastomers) can have a direct effect on well integrity. Poor brine selection can lead to accelerated corrosion problems in metal components and failure of elastomers. Compatibility can only be established through testing. Some of the tests can take many weeks, even months, to complete so changes to proven brine formulations should not be undertaken lightly.

5.6.1 Fluid compatibility with completion materials (metals)

Dissolving salt such as potassium chloride and sodium chloride in water will increase corrosion rates, since the brine becomes more ionic, relative to freshwater. However, some high density brines will become less corrosive, as the high concentration of salt reduces oxygen solubility. For example, corrosion in a sodium chloride solution peaks when the chloride concentration is approximately that of seawater, 20,000 mg/L.

Gases such as oxygen, carbon dioxide, and hydrogen sulfide are soluble in some brines. The combination of dissolved gas and salinity can lead to very high rates of corrosion. Since corrosion could ultimately cause the failure of completion tubulars or components, resulting in a possible loss of containment, selecting an appropriate brine, compatible with the completion tubulars, is essential. Costly corrosion resistant stainless steels with a high chrome and nickel content are still susceptible to corrosion in the form of chloride stress corrosion. This is a type of intergranular corrosion and occurs in austenitic stainless steel when they are under tensile stress in the presence of oxygen, chloride ions, and high temperature.

5.6.2 Fluid compatibility with completion materials (elastomers)

Many completion components that are critical to well integrity rely on elastomers seals; production packers, seal assemblies, and tubing hangers, for example. Some of the commonly used elastomers material degrade and can fail when exposed to particular types of brine, especially some of the heavier brines.[8] Some of the additives used with packer fluid can also lead to elastomer degradation. Elastomer degradation, and ultimately failure, is a function of the concentration of the aggressive species, pressure differential and absolute pressure, seal movement, and temperature. In general, large seal sections, for example, packer seals, have better chemical resistance than small components, such as T seals and O rings (Table 5.6).

Table 5.6 Fluid compatibility with completion elastomers

Elastomer material	Poor resistance to
NBR (nitrile)	Brines containing zinc bromide. Small section seals in sodium bromide (NaBr) H_2S. Amine based corrosion inhibitor.
HNBR (hydrogenated nitrile)	Brines containing zinc bromide, H_2S. Amine based corrosion inhibitor.
FKM fluoroelastomers (viton)	Some formate brines ($pH > 8$). Amine based corrosion inhibitor.
FEPM fluoroelastomers	Organic acids.
FFKM perfluoroelastomers (kalrez, chemraz)	Amine based corrosion inhibitor.

5.6.3 Brine compatibility with the formation

Although density and crystallization temperature are the main concerns when preparing a brine, it is also important to select a fluid system that keeps formation damage to a minimum. During drilling operations, fluid loss into the formation is controlled by the development of a thin, low permeability, filter cake on the wall of the wellbore. The solid material used to create the filter cake (and to add density to the mud) has the potential to damage the formation by bridging off in the formation pore spaces. Completion and workover brines are solids-free, meaning the usual problems associated with solids blocking pore throats are eliminated. However, the potential for damage remains. Indeed, in some poorly designed mud systems it might not be the solids content of the mud that causes the most damage, but the loss of base fluid (filtrate) into the formation.

Damage to the formation has serious economic consequences, and can have an indirect and negative influence on well control. The main fluid related formation damage mechanisms are:

- *Clay swelling and fines mobilization.*

  Many producing formations contain water sensitive clay such as smectite (montmorillonite). These swell when exposed to freshwater, low salinity, and (or) high pH brine. When clay expands it occupies more of the pore space in the formation. Permeability is therefore reduced. Freshwater or low salinity filtrate will almost always damage formation containing swelling clays. Damage is least likely to occur if the salinity of the brine is similar to the salinity of the formation water.

  Changes in salinity or the use of deflocculating chemicals can also lead to the mobilization of clay particles, allowing them to move through the formation until they become trapped in pores, again resulting in permeability reduction. Fines migration can also be caused by high velocity flow (production or injection) (Fig. 5.3).

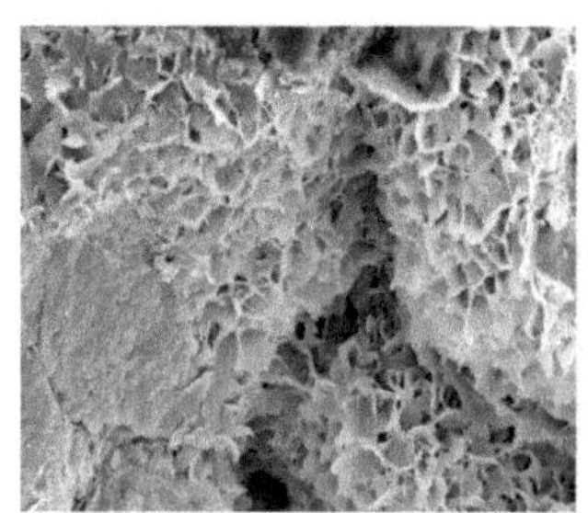
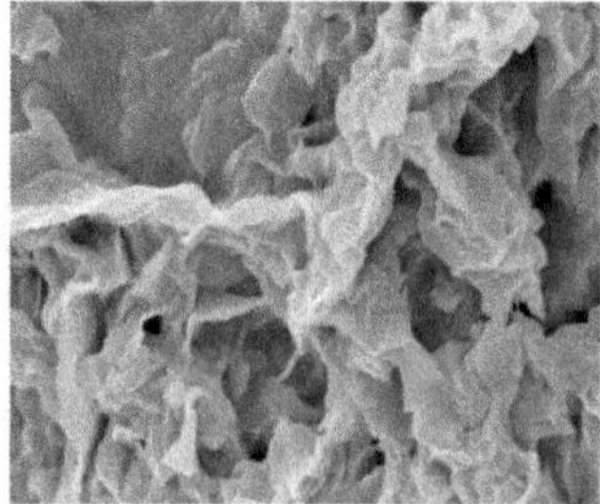

Figure 5.3 Smectite (swelling) clay.

- *High pH values.*

  High pH brines are known to release fines. Changing the salinity of the formation brine by the introduction of completion brines can cause increases in pH through a process of ion exchange.[9] Many heavier brines have low pH values, and pH is sometimes deliberately increased to reduce corrosion. This applies mainly to "packer fluids", that is, those fluids left in the annulus at the end of the completion.
- *Changes in water saturation.*

  If brine is lost to the formation, water saturation in the near wellbore region will increase, leading to a reduction in oil permeability, since water occupies more of the pore space. Producing the well will usually reverse the effects, but damage could become long lasting or permanent if brine additives are used that change the formation wettability. If the formation is depleted (low pressure) capillary forces can make clean-up of the lost fluid problematic, particularly in low permeability reservoirs.
- *Changing formation wettability.*

  Most sandstone formations are water wet. Sand grains in the formation are surrounded by a thin film (coating) of formation water held in place by surface tension. When the well is produced, oil moves through the pore, slipping past the immobile formation water. The use of oil wetting chemical additives (surfactants) can change formation wettability by breaking the surface tension, allowing the water to move into the pore spaces. Permeability to oil decreases, and the decrease can be severe. Wettability changes (water to oil wet) are usually permanent (Fig. 5.4).
- *Polymer invasion.*

  Polymers are added to brine for a number of reasons: as friction reducers in "slick-water," a viscosifying agent to control losses, to improve the carrying properties of fluids during hydraulic fracturing

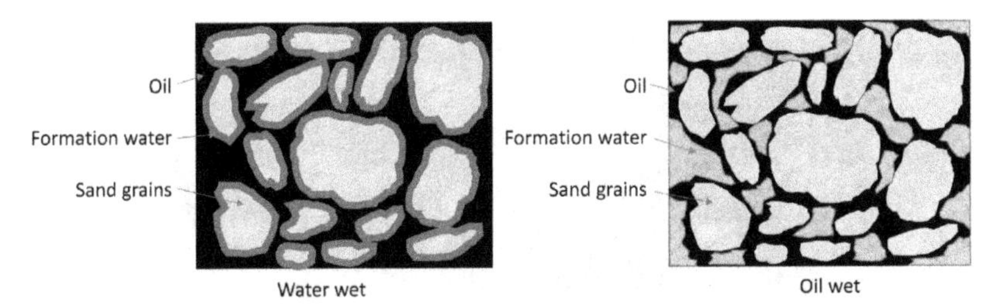

Figure 5.4 Water wet formation (left) and oil wet formation (right). Water trapped in the pore space of the oil wet formation reduces permeability to oil.

operations, and for wellbore cleaning operations. Polymers can be adsorbed onto the surface of a formation, leading to damage. Viscous polymers will impede flow, so must be designed to break.

Many polymers arrive at the wellsite as dry powder and need to be mixed (hydrated) before use. Poorly hydrated (fish eye) polymers are very damaging. If a polymer is not mixed properly, the unhydrated powder can enter pore spaces causing significant damage (Fig. 5.5).

- *Scale.*

  While oilfield scales are generally associated with production related precipitation, scale can and does form during interventions and workovers, and is the result of fluid incompatibility. The most common scales encountered during interventions are the carbonates or sulfates of the alkaline earth metals calcium, strontium, and barium. Scale has the potential to damage the formation, and of more importance in the well control context, scale can prevent the proper functioning of safety critical completion components, such as downhole safety valves and Christmas tree valves.

  Scale can form when incompatible fluids mix. The most common brine incompatibility occurs when carbonate ions in formation water and calcium ions (Ca^{++}) in brine are allowed to mix. The result can lead to calcium carbonate scale ($CaCO_3$). Similarly, barium ions in formation water, if allowed to mix with sulfate ions in seawater, will lead to the formation of a hard barium sulfate ($BaSO_4$) scale.

- *Emulsion blocking.*

  Water based completion fluid can form emulsions with oil from the formation. Similarly, if oil based fluids are introduced into the formation emulsions can form with formation water. Emulsions form more readily if surfactants are used with the brine. Since emulsions are more viscous than the constituent parts, they make it more difficult for oil to flow through the formation.

Figure 5.5 Partially hydrated polymer (fish eyes). Very damaging. Source: *Image courtesy of George King.*

5.6.4 Brine additives

A number of different chemicals are routinely added to completion and intervention brine. Adding these chemicals changes brine chemistry and can increase the potential for a brine to damage the formation. The most commonly used brine additives are listed and described in Table 5.7.

Table 5.7 Brine additives

Additive	Function
Clay inhibitors	Used mainly with low density (low salinity) brines, these organic compounds are used to inhibit the swelling of pore lining clays. They are mixed directly into a brine at the surface.
Caustic soda (sodium hydroxide) and magnesium oxide	Low pH brines are corrosive. Adding caustic soda or magnesium oxide increases the pH and reduces corrosion. Increasing pH can lead to fines mobilization.
Corrosion inhibitors	These are used to limit tubular corrosion. They are often oil wetting, and therefore damaging to the formation. Highly toxic in concentrated form, there are HSE risks associated with their use. *The use of thiocyanate corrosion inhibitor had been discontinued by some operating companies, as there is evidence to suggest that it breaks down at high temperature with corrosive H_2S as a by-product.*[a]
Surfactants	These are used to prevent secondary emulsions. Surfactants also improve the dispersion of other additives used in completion fluids.
Defoamer	As the name suggests, defoamer is used to reduce foaming when mixing surfactants and some types of corrosion inhibitors.
Friction reducers	Polymers (such as polyacrylamide) are used to reduce the coefficient of friction. This reduces tubing-to-casing friction when running coiled tubing and completion tubing, and also reduces frictional pressure drop when circulating fluids. They are widely used when fracturing unconventional (shale) formations.
Oxygen scavenger	Used to reduce/eliminate free oxygen from the fluid system.
Biocide	Used to control sulfate reducing bacteria and acid producing bacteria. They are not normally required in brine with a high salt content.

[a]Mingjie K, Qi Q. Thermal decomposition of thiocyanate corrosion inhibitors: a potential problem for successful well completions. SPE paper 98302, 2006.

5.7 BRINE CLARITY AND SOLIDS CONTENT

Minimizing formation damage relies on having a clean (solids free) brine. This is especially important if the brine is expected to come into contact with the reservoir. Brine clarity and solids content is measured at the wellsite. The three most common and useful measures of brine cleanliness are turbidity, volume percent of solids, and particle size distribution.

Turbidity meters work by shining a light (laser) through a fluid sample and observing the amount of scattering or reflection. Turbidity, a measure of the cloudiness of a fluid, is recorded in Nephelometric Turbidity Units (NTUs).

Five NTUs are just noticeable by the eye, while 50–100 NTUs are often used as an indication of brine clarity. More stringent NTU values are used for some completion and intervention fluids. For example, brines used for open hole gravel packing and frac-packs normally require very low limits, with some operating companies setting targets as low as 5NTU. However, an absolute reliance on NTUs as a measure of fluid cleanliness can be misleading. This is because the fluid sample may be discolored while being largely free of solids. Moreover, measuring NTU gives no indication of particle size or distribution (Fig. 5.6).

Figure 5.6 Low solids content, but NTU value?

A centrifuge can be used to separate solids from the liquid and a simple measurement of percentage solids by volume recorded. This is a coarse measurement, but quick and easy to obtain at the wellsite. As with the NTU value, there is no measurement of particle size or distribution. A maximum solids-by-volume content of 0.05% is widely used across the industry as an acceptable measure of fluid cleanliness.

Particle size and particle distribution (if required) can be measured using a Coulter counter or laser particle size analyzer.

5.8 BRINE FILTRATION

Removal of unwanted solids from brine requires the fluid to be filtered. Filtration is normally a two-stage process.

High volume coarse filtering is accomplished using large press filters. Press filters come in a range of size and throughput capacity. Large (1500 ft^2) units can filter up to 22 bbls/min, and will filter out any solids larger than 8–10 μm (nominal) using diatomaceous earth as the filter medium.

Where a very high degree of fluid cleanliness is required, for example, brines used with low permeability reservoirs, brines used to mix lost circulation polymers and brines used for gravel packing, cartridge filters are used. These can remove all particulate matter larger than 2 μm.

5.9 FLUID LOSS CONTROL

Where brine is used as a well control fluid barrier, overbalance pressure is typically 200–300 psi. This pressure, acting against a permeable formation, means fluid will continually seep away to the formation, unless preventative action is taken. In low permeability reservoirs, fluid loss is manageable and no additional action is necessary. In more permeable reservoirs, measures will have to be taken to slow or stop fluid loss. In some reservoirs controlling fluid loss can be difficult; for instance low pressure, naturally fractured, and highly permeable reservoirs. Losses can also be triggered, or made worse, through some mechanical actions such as surging when running packers and bridge plugs. In some cases, fluid can be lost to one layer in the formation while a gas influx is taking place at a different layer.

In addition to the clear and obvious need to maintain a well control overbalance, losses need to be controlled to prevent formation damage. Paradoxically, continual (but controllable) fluid loss can actually be an indication that the formation is not being damaged. If fluid losses were damaging the formation, losses would diminish, perhaps even stop. In addition to the well control and formation damage concerns, losses can be costly. Brine cost can be significant, some of the heavier formate brines cost in excess of $2000 bbl.

Reducing hydrostatic overbalance would reduce losses. However, this is unlikely to be a safe or practical option, since most completion and workover operations are conducted with a relatively low (200–300 psi) overbalance, and any reduction would risk allowing the well to kick. Fluid loss during completion and intervention operations is therefore almost always controlled by placing lost circulation material (LCM) in the form of a fluid loss control pill (FLCP) or "kill pill" across the zone of permeability. FLCPs are made using viscous fluid, bridging solids, or a combination of both.

5.9.1 Solids-free lost circulation material

Brine viscosity can be increased by mixing HEC (hydroxyethyl-cellulose) or XC (Xanthan gum) polymers. Concentration is normally in the range of 2–4 lb/bbl (5.5–11.5 kg/m^3).

Viscous fluid lost to the formation can cause formation damage, especially if they are slow to degrade. Viscous pills degrade fairly quickly if the temperature is above 220°F. While thermal degradation is an advantage, inasmuch as polymer related formation damage will be minimized, there is the obvious disadvantage of the resumption of fluid loss. In some cases FLCPs will need to be replenished. In low temperature wells, it may be necessary to use breakers to remove FLCPs. Hydrochloric acid is fast acting, but has the potential to cause formation damage. Enzyme breakers react more slowly, but are generally less damaging.

5.9.2 Example fluid loss calculations

It is possible to estimate the effectiveness of using viscous fluids to reduce fluid losses using Darcy's radial inflow equation:

$$Q = \frac{7.08 \times 10^{-3}\, kh(P_r - P_{wf})}{\mu B\left(\ln\left(r_e/r_w\right) - 0.75 + S\right)}$$

where:

Q is flow (or injection)rate,
7.08×10^{-3} is a correction factor for oilfield units,
k is permeability in millidarcy,
h is the net height of the formation in feet,
P_r is reservoir pressure,
P_{wf} is bottom hole flowing pressure (or kill weight hydrostatic pressure),
μ is fluid viscosity in cp,
B is formation volume factor,
r_e is drainage radius (ft),
r_w is wellbore radius (ft), and
S is skin.

Example calculation:

A 10,000 ft vertical well is filled with a 9.7 ppg brine with a viscosity of 1.2 cp and a formation volume factor of 1.

Reservoir pressure: 4750 psi.
Formation permeability: 120 mD.
Formation thickness (height): 135 ft.
Drainage radius: 2000 ft.
Wellbore diameter: 6 inches.
Skin: 0.

What loss rate could be expected if no loss circulation material is spotted across the reservoir?

$$Q = \frac{7.08 \times 10^{-3}\ 120 \times 135(4750 - 5044)}{1.2 \times 1 \times \left(\ln\left(\frac{2000}{0.25}\right) - 0.75 + S\right)} = 3411 \text{ bbls/day}$$

$$3411 \text{ bbls/day} = 142 \text{ bbls/h}$$

What loss rate could be expected if a 200 cp viscous pill is spotted across the reservoir?

$$Q = \frac{7.08 \times 10^{-3}\ 120 \times 135(4750 - 5044)}{200 \times 1 \times (\ln(2000/0.25) - 0.75 + S)} = 20 \text{ bbls/day}$$

$$20 \text{ bbls/day} = 0.83 \text{ bbls/h}$$

Viscous fluids make excellent loss control pills. However, viscosity decreases as temperature increases and will degrade over time.

5.9.3 Bridging solids

In high permeability or fractured formations, it may not be possible to control losses using viscosity alone, and it may be necessary to mix bridging solids into the viscous pill. The most commonly used bridging fluids are, in order of popularity, calcium carbonate ($CaCO_3$), sized common salt (NaCl), cellulose fibers, and oil soluble resin.

Bridging solids work most effectively when they are approximately one-third to one-fifth of the formation pore space diameter. Pore diameter can be measured from core, or if pore diameter is unknown it can be estimated. An often used short cut is to take the square root of permeability to represent the average pore diameter in microns (from the Kozeny equation). For example, a formation with 1 darcy (1000 md) permeability would have an average pore diameter of approximately 32 μm. Bridging particles would therefore need to be between 6 and 11 μm. In reality, bridging material for a 1000 md formation would have a wider range of particle size, for example, 1 μm up to 30 or 40 μm. A wide distribution provides particles large enough to bridge across a range of pore sizes (the calculation only provides an average), while containing enough smaller particles to create a low permeability cake.

Properly sized bridging solids should spontaneously lift off when the well is put into production. However, since the particles cover a wide range of sizes, it is possible than some particles will end up plugging the pores and will not be removed by flow.

5.9.3.1 Calcium carbonate

Calcium carbonate ($CaCO_3$) is the most common of all the bridging materials. Usually made from dolomite or marble, it is available in a wide range of sizes or can be made to a particular specification. Any residual damage caused by particulates not lifting off when the well is flowed can be removed using hydrochloric acid (Fig. 5.7).

5.9.3.2 Sized salt

Sodium chloride (NaCl) particles are frequently used as a bridging material. To prevent the solid salt particles from dissolving, the carrier fluid must be saturated with respect to sodium chloride. In theory, solution of the salt by formation brine or a low salinity wash should effectively remove the filter cake. However, in practice, clean-up effectiveness is reduced by the polymers used to keep the salt particles in suspension.

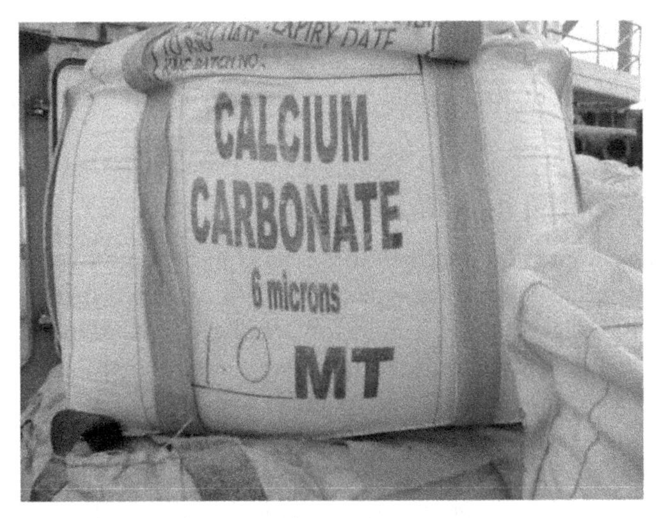

Figure 5.7 Bulk calcium carbonate at the rig site.

Polymer breaking enzymes are often run in conjunction with the low salinity wash, to aid in the removal of polymers in the filter cake.

Formation damage tests should be conducted to ensure that the high salinity fluid is compatible with the reservoir.

5.9.3.3 Cellulose fibers

Cellulose fibers are plant material, for example, peanut (groundnut) husks, or even sawdust. Cellulose fibers are more commonly used when circulation is lost during drilling operations, and are rarely used during completion or workover operations unless losses are severe. In most cases, $CaCO_3$ is equally as effective.

Cellulose fibers are removed by oxidizing with sodium hypochlorite (diluted to make household bleach), however, this should only be considered after careful evaluation of potential formation damage and downhole corrosion of tubulars.

5.9.3.4 Oil soluble resin

Oil soluble resins were developed in the 1970s as an alternative to calcium carbonate.[10] The aim was to develop a system that would develop a filter cake that was completely soluble in oil, therefore eliminating the need to use acid or low salinity water washing to remove residual particulates from the formation. Experience has shown that in some cases a treatment with an organic solvent has been required to remove soluble resin residue.

Moreover, there is some evidence to suggest some of the resin will liquefy when in contact with formation oil and invade the formation, causing damage. Because of these problems, oil soluble resins are not in widespread use.

5.9.4 Mechanical fluid loss control

The use of one or more mechanical barriers can be hugely beneficial during workover and completion operations. With a tested mechanical barrier, the risk of a kick, or of fluid loss leading to a kick, is significantly lower. In addition, a mechanical barrier can eliminate the need to place potentially damaging fluid in direct contact with the reservoir.

The use of mechanical barriers is largely dependent on the design of completion and the requirements of the workover.

The only disadvantage to using a mechanical fluid barrier is the potential for a difficult recovery operation, particularly if debris is allowed to settle on top of the plug. The use of debris catchers and proper cleaning of the wellbore will reduce the risk of a stuck plug.

5.10 HOW MUCH BRINE IS NEEDED?

Where brine has a well control function, it is important that enough brine is available at the wellsite to enable completion or workover operations to continue without interruption. The potential for brine to be lost must also be factored into any volume calculations. The volume of brine at the wellsite will be determined mainly by the wellbore capacity or circulating volume. The circulating volume must include any surface piping and throughput capacity of filtration equipment (if used). In addition to the circulating capacity, there should be enough brine at the wellsite to cover contingencies, principally losses.

5.10.1 Wellbore capacity

Wellbore or circulating volume are easily calculated. Casing or tubing volume can be looked up using widely available tables in publications such as "Baker Tech Facts" or the "Halliburton Red Book." Casing and tubing capacity data is also available on line, and there are now plenty of "Apps" that have tubing and casing capacity data. Alternatively, simple formula described in Chapter 1, Introduction and Well Control Fundamentals, can be used to calculate tubing and casing capacity.

During a completion or workover, there will be occasions when no tubing or drill-pipe is in the well. If losses were to occur at that point, the potential exits for the entire volume of the casing (and liner if run) from surface to the formation to be lost. Therefore, when calculating the wellbore circulating volume, the worst case assumption is based on casing capacity, that is, no allowance is made for drill-pipe or tubing displacement volume.

5.10.2 Surface piping

The volume required to fill the surface lines is normally small, unless the equipment needs to be positioned a distance from the well. In any case, the surface line volume should be included in the total circuiting volume.

5.10.3 Holding tanks

Where a brine is filtered, additional pits are used to store the brine. One tank is needed to store dirty brine coming back from the well, and a second tank for filtered brine ready for pumping into the well. This pit volume needs to be included in the total brine circulating volume. The volume should include any dead space in the tanks.

5.10.4 Filtration capacity

If brine is being filtered during circulation, then the capacity of the filtration system needs to be included in the total brine circulating volume. Large capacity (1500 ft^2) press filters hold approximately 34 bbls of fluid. Precoat and body feed tanks require an additional 20 bbls. Cartridge filter units hold about 4 bbls. A two-stage filtration system (press and cartridge) will therefore use almost 60 bbls of fluid.

5.10.5 Contingency

A volume of brine over and above the circulating volume needs to be stored at the wellsite, to enable the hydrostatic overbalance to be maintained in case of fluid loss to the formation. Keeping the hole full is crucial. If losses do occur, then it may be necessary to place LCM at the reservoir before they will stop. A worst case assumption is that losses begin when no drill-pipe or production tubing is in the well; there is no ability to circulate in a kill pill. In this case it may be necessary to control losses by pumping a kill pill into the casing at surface and waiting for it to reach the formation. The entire wellbore volume must be lost to the formation before the kill pill can work. For this reason, at least twice the circulating volume should be considered as a minimum requirement for brine quantity. In many instances, where losses are continual or difficult to control, considerably more fluid will be required.

5.10.6 Brine volume summary

Consider preparing a simple spreadsheet itemizing and quantifying brine volume requirement. The spreadsheet can incorporate simple tubing and casing capacity calculations. The total volume would include:

1. Wellbore capacity
2. Surface lines
3. Holding tanks
4. Filtration unit capacity
5. Contingency.

5.11 ALTERNATIVES TO BRINE

For fluid to be classified as a well control barrier, it must be monitored. In practical terms this means the fluid level must reach the surface, where it can be observed. If the reservoir pressure is too low to support the hydrostatic pressure that results from a full column of water based fluid, then a lower density alternative must be used if a fluid barrier is used. Nonwater based alternatives can be used across a range of lighter than water densities (Table 5.8).

Under exceptional circumstances, mud is used as a completion and workover fluid, normally where there are major safety concerns associated with well control. For example, in the Chevron operated Tengiz field

Table 5.8 Alternatives to brine (lighter than water)

Material	Density range (ppg)
Nitrogen gas	0.1–2.6
Foam	3.5–8.3
Kerosene, diesel, or base oil	6.7–7.1
20 API stock tank crude	7.8
30 API stock tank crude	7.3

(Kazakhstan), the produced gas has over 13% H_2S. A loss of containment is potentially fatal. The wells have barefoot lower completions in a thick, naturally fractured formation. The upper completion is run with mud still in the well. Post-completion the formation is acid washed using coil tubing. Well kill and workovers are similarly carried out with mud in the well. Mud is the correct fluid, given the circumstances, the hazardous nature of the produced gas, and the ability to easily remove mud damage with acid.

REFERENCES

1. Table 1 and 2 values from SPE paper 12490. G W Krook and T D Boyce. 1984.
2. API RP 13J/ISO 1353-3. Testing of heavy brines. 4th ed. American Petroleum Institute; 2006.
3. TETRA Technologies, Inc. The engineered solutions guide for clear brine fluids and filtration. 2nd ed. Chapter 2; 2007.
4. Bridges KL. *Completion and workover fluids. SPE monograph series.* Richardson, TX: SPE; 2000.
5. Bellarby J. *Well completion design.* New York: Elsevier; 2008.
6. API RP 13J/ISO 1353-3. Testing of heavy brines. 4th ed. American Petroleum Institute; 2006.
7. Murphey JR, Swartwout R, Caraway G, Weirich J. The effect of pressure and temperature on high density brines. IDAC/SPE paper 39318.
8. Buc Slay J, Ray, T.W. Halliburton energy services. Fluid compatibility of elastomers in oildfield completion brine. Corrosion 2003. Paper 03140.
9. Vaidya RN, Fogler HS. Formation damage due to colloidally induced fines migration. Department of Chemical Engineering. The University of Michigan, Ann Arbour, MI 48109-2136; March 1990.
10. Crowe CW, Cryar HB. Development of oil soluble resin mixture for control of fluid loss in water base workover and completion fluids. SPE paper 5662-MS.

CHAPTER SIX

Well Barriers

Well barriers are fundamental to well integrity and well control. An uncontrolled blowout can only occur if more than one barrier fails. Understanding the physical properties of a barrier, where the barriers are located, how they are tested, and the actions to take if a barrier fails, are critical skills for anyone working in completions or interventions. During interventions and workovers, barrier elements and the barrier envelope often change as the job progresses. Those working in completions and interventions must be able to clearly identify each change of barrier element, and how the barrier envelope changes throughout the operation. Many operating companies now incorporate barrier diagrams in work programmes. These diagrams clearly identify the primary and secondary barrier envelopes, all the individual barrier elements, and how they should be tested. Any change to the barrier configuration requires a new barrier diagram. This is an excellent system, and if properly implemented will mean those tasked with carrying out the work are always clear about where barriers are, how they change as work progresses, and crucially, how to respond to a barrier failure.

6.1 DEFINING WELL BARRIERS AND WELL BARRIER ELEMENTS

The Norwegian regulatory authority NORSOK define a well barrier as an "envelope of one or several dependent barrier elements preventing fluids[a] or gases from flowing unintentionally from the formation, into another formation or to surface." They further define a well barrier element as an "object that alone cannot prevent flow from one side to the other side of itself" (*sic*).[1] This is an important distinction, and not always made in some well control manuals and text books.

[a] A pedantic point. This should probably read *liquids or gases*. Gas is a fluid.

Well Control for Completions and Interventions.
DOI: https://doi.org/10.1016/B978-0-08-100196-7.00006-3

For example, a slickline stuffing box is often described as the primary well control barrier when wireline is in the hole. Although this is not incorrect, it is not a comprehensive answer. During slickline operations on live wells the well barrier, as defined by NORSOK, would include not just the stuffing box, but all the other "barrier elements" that, when combined, prevent the release of well fluids.

6.2 BARRIER CLASSIFICATION

There are two broad categories of barriers, or to be precise, barrier elements.

Normally open. These are barriers that are normally open, but can be closed to contain well fluids. For example, BOP rams, a downhole safety valve (SCSSSV), or Christmas tree valves.

Normally closed. A permanent barrier in the well that prevents fluid flow. For example, cemented casing, production packer, liner top seals.

Barriers may be further classified as:

Independent barrier. This is not reliant on another for integrity. A mechanical plug, properly tested, would constitute a single, independent mechanical barrier.

Dependent barrier. Relies on another barrier for integrity. For example, a check valve that requires a full column of kill weight fluid above it to remain closed.

Primary barrier. The first object that prevents flow from the well.

Secondary barrier. The second object that contains flow, the object that will contain flow from the failed primary barrier.

Tertiary barrier. The third object that prevents flow from a well. Only used if both primary and secondary barriers fail.

A properly implemented barrier philosophy will be based on the presumption that the primary barrier should remain intact. If it fails, operational priority will be to reinstate the primary barrier before work continues. For example, during the running of a completion string, the primary barrier is a column of kill weight fluid. Fluid loss to the formation results in a loss of overbalance and a kick. Well integrity is restored by closing the BOP annular preventer (secondary barrier), and installing a full opening safety valve on the tubing (secondary barrier). Restoration of the fluid barrier becomes the priority, and must be carried out before operations resume. Failure of the secondary barriers is contained by activating shear and blind rams (tertiary barrier).

6.2.1 Mechanical barriers

Mechanical barriers must be verified by pressure testing. Where possible, the test should be in the direction of flow. Closed barriers are usually tested when they are first installed. Normally, open barriers should be tested at the time of installation and then at intervals in accordance with regulatory requirements. For example, in the United Kingdom and Norwegian sectors of the North Sea, Christmas tree valves are normally integrity tested every six months.

There are several types of mechanical barrier elements, these include:

- Production packer (annulus barrier).
- Tubing hanger seals (annulus barrier). Hanger seal integrity is tested via ports in the wellhead and by performing an annulus pressure test.
- Wellhead annulus valves (or Valve Removal (VR) plugs in the side outlet bore).
- Drilling, coiled tubing, and wireline BOPs.
- Plugs. Some operating companies will only consider a mechanical plug equipped with chevron seals to be a barrier if it can be tested in the direction of flow. This is because a chevron seal (V-packing) can only hold pressure in one direction. Most plugs are dressed with two sets of seals, one for each direction of flow. Testing in one direction only tests part of the seal stack. Solid slab seals solve this problem, but are not available for all wireline set plugs.
- Tubing or wireline retrievable SCSSSV. There are many conflicting views about the use of downhole safety valves as a well control barrier. However, many operating companies permit the use of the SCSSSV, providing it can be inflow tested in the direction of flow. Safety valve acceptance is normally based on a zero leak rate rather than the API permissible leak rate of 400 cc/minute for liquids or 15 scf/minute gas.[2] Ideally, a small differential pressure from below is needed to keep the flapper seated. However, the risk of dropped objects must be considered, since the closed flapper valve in a safety valve does not have the resilience of, e.g., a wireline set bridge plug.
- Annulus SCSSSV.
- Surface controlled downhole "lubricator valve."

- Surface controlled "inflow control valves."
- Formation isolation valves (of the type commonly used during gravel packing operations).
- Cemented liner or casing.

6.2.2 Fluid barriers

A static column of mud or brine of sufficient density to overbalance the highest anticipated reservoir pressure can be classified as a fluid barrier.

Mud is classified as a barrier where:

- The mud can be conditioned (circulated) to keep the solids in suspension. If mud remains static for too long the mud solids (barite) will begin to settle, and the mud will lose density.

Brine can only be classified as a barrier if certain conditions are met:

- Since brine contains no solids, losses are probable unless lost circulation material (LCM) is held against the formation.
- Where LCM is required to prevent losses, the brine cannot be classified as an independent barrier, since it is dependent on the effectiveness of the LCM to maintain a hydrostatic overbalance.
- In low permeability reservoirs, e.g., shale, no LCM is required, and brine can be classified as an independent fluid barrier.
- Brine can only be classified as a barrier if the fluid level can be continually observed and the fluid level maintained.
- If a brine column is supported by a mechanical barrier, e.g., a plug, it can only be considered a single barrier, since it is dependent on the plug integrity.
- Where fluid losses are permissible, the loss rate must be less that the agreed limit.

6.3 BARRIER TESTING

Until successfully tested, no mechanical or fluid device can be considered a well control barrier. The normal method for testing mechanical barriers is to apply differential pressure. The main considerations are:

- *Differential pressure*: Test pressure (Δ_p) must be more than the anticipated maximum differential the barrier will be exposed to when in service. Typically, at least 10% more. There may be a requirement for some barrier elements to be tested to their working pressure.

- *Duration*: A barrier test must last long enough to allow pressure to stabilize, and for any pressure changes or flow to be observed. When testing a large volume component, e.g., a casing test, a small leak would not be noticeable on a short duration test.
- *Documentation*: Most operating companies require barrier tests to be recorded. Plots and print-outs should be retained for the well records.
- *Frequency*: Some well barrier elements will be tested at specified intervals, e.g., the SCSSSV, wellhead, and Christmas tree valves. Other elements might only ever be tested at the time of installation. Items of well intervention pressure control equipment and drilling BOP equipment are tested each time they are rigged up and following any repairs or break of containment. They are also tested to working pressure at regular intervals to obtain the necessary certification.
- *Verification*: Where downhole mechanical barriers are close together, it can be difficult to verify the integrity of the upper barrier. The lower barrier will normally be inflow tested. With the lower barrier in place, the only way of testing the upper barrier is to test from above. However, if the space between the plugs is small (low volume) it will be very difficult to determine if the upper plug is leaking. Plugs are now available that allow pressure and temperature between the plugs to be monitored.[3] This enables both plugs to be verified.

6.4 INFLOW TESTING

Inflow tests in high temperature wells can be difficult to interpret because of fluid expansion. It can be difficult to determine if flow from the well is a leak or fluid expansion, as the cold test fluid warms. In very high temperature wells it can take many hours for the fluid temperature to stabilize. One method of determining if flow is diminishing is to create a Horner plot of flow rate against time, with Horner Time on the x axis and flow rate on the y axis.

After plotting a series of rate against time points, draw a straight-line interpolation between the sample points. If the line intersects the *x* axis

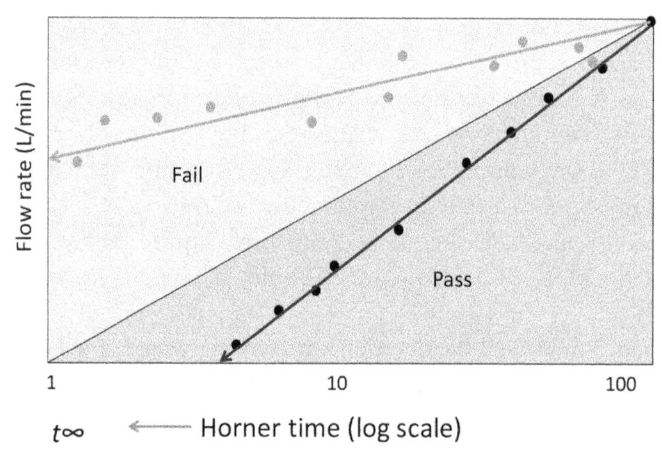

Figure 6.1 Horner plot example.

before infinite time (Horner time = 1) the leak is diminishing and the inflow test is good. If the plot makes an intersect of the *y* axis, it means there is flow at infinite time and the inflow test has failed (Fig. 6.1).

6.5 NONCONFORMANCE WITH BARRIER POLICY

Many workover and intervention operations are performed to repair or replace worn and damaged equipment that threatens the integrity of the well. In some cases, the condition of the well is such that installing two mechanical barriers may not be possible. Consider a situation where barriers are needed to enable the Christmas tree to be removed and the drilling BOP installed. Consider also that the reason for the intervention is a need to replace the production tubing that has parted just below the tubing hanger. In such a situation, a fluid barrier plus a mechanical plug set in the tubing hanger nipple profile may be all that can be achieved.

In exceptional circumstances, and after a rigorous examination of the risks, dispensation from normal barrier policy may be granted and permission to proceed given. Much will depend on the conditions at the well site. Factors influencing that decision will include reservoir pressure and the nature of the reservoir, GOR, the quantity of H_2S, and the general condition of the wellhead and Christmas tree.

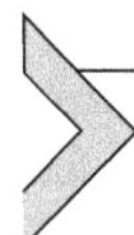

6.6 BARRIER REQUIREMENTS IN SUBHYDROSTATIC RESERVOIRS

Where a reservoir is known to be subhydrostatic and incapable of flow, then some operating companies and jurisdictions will allow work to be carried out against a single well control barrier. Note that subhydrostatic wells are very often capable of flowing gas to surface.

6.7 WELL INTERVENTION WELL CONTROL BARRIERS

Many well intervention operations are performed on live wells. The barriers are provided by pressure control equipment rigged up on the Christmas tree. Pressure control equipment specifically intended for use on live well interventions is described in some detail in the chapters of this book dealing with wireline, coiled tubing, and hydraulic workover (snubbing). The main barrier categories are summarized here.

6.7.1 Wireline

Primary (first) barrier.

- Stuffing box or grease head and lubricator system.
- When the wireline toolstring is at the surface, the tree valves are closed to allow the lubricator to be vented and opened. The Christmas tree valves form the primary barrier.

Secondary barrier.

- Wireline BOP that can close around and seal on the wire in the hole. For braided cable and e-line a dual ram BOP is required.
- Christmas tree valves if the wire parts and are ejected from the stuffing box (grease head). If the wire is ejected from the stuffing box, pressure should be contained in the lubricator by the BOP gland or ball check in the stuffing box/grease head.

During fishing operations, there may be occasions where wire is stripped through a closed BOP. In these circumstances, the BOP becomes the primary barrier.

Tertiary barrier.

- Wire cutting shear seal BOP.
- Wire cutting tree valve.

In the event of primary and secondary failure, and where no wire cutting BOP is fitted, it may become necessary to cut the wire with the tree valves. If the tree valves are not designed to cut wire, the gate may be damaged and it may not be possible to obtain a seal with the valve closed. If a tree valve must be used to secure the well, always use the upper master valve.

- If the lower master valve is damaged, two mechanical set barriers must be installed in the wellbore before repairs can be made.
- If the swab valve is damaged, it would not be possible to flow the well whilst maintaining double barrier isolation (swab valve and swab cap).

6.7.2 Coiled tubing

6.7.2.1 External pressure control

Primary Barrier.

- Stripper rubbers and riser.
- Christmas tree valves when deploying tools in and out of the riser. If a downhole lubricator valve is fitted, this would be the primary barrier when installing tools into or out of the well.
- Gate valves if a tool-string deployment system is being used.

Secondary Barriers.

- BOP rams.
- SCSSSV or lubricator valve if the coil tubing BHA is above.

Tertiary Barrier.

- Shear and seal capability in BOP. For many operations, a shear/seal combination BOP will be mounted immediately above the tree.
- Some trees are fitted with gate valves able to shear coil tubing.

6.7.2.2 Internal pressure control

Primary Barrier.

- Check valves (nonreturn valves) in the BHA.

Secondary Barriers.

- Fluid pumped into the coil to prevent hydrocarbon ingress and valves on the side of the reel.

Tertiary Barrier.

- Shear and seal capability in BOP. For many operations a shear/seal combination BOP will be mounted immediately above the tree.
- Some trees have been fitted with gate valves able to shear coil tubing.

Note: Some coil tubing operations use reverse circulation to clean out the wellbore. This means no check valves are included in the BHA. In these circumstances, the primary internal barrier is provided by maintaining overbalance on the formation to prevent hydrocarbons from entering the reel.

6.8 HYDRAULIC WORKOVER (SNUBBING) UNIT: LIVE WELL OPERATIONS

6.8.1 External pressure control

Primary Barrier.

- Stripper rubbers or annular preventer.
- If rigged up on top of a Christmas tree, the tree gate valves provide the primary barrier when rigging up and when lubricating tools in and out of the riser.
- If no Christmas tree is installed, tools are lubricated in and out of the well above closed BOP blind rams.

Secondary Barriers (pipe in the well).

- 2 x BOP pipe rams.

Tertiary Barrier.

- BOP shear and blind rams or a shear seal ram.

6.8.2 Internal pressure control

Primary Barrier.

- Check valves (nonreturn valves) in the BHA.

Secondary Barrier.

- A full opening stab-in safety valve (Kelly Cock).

Tertiary Barrier.

- Shear ram and blind ram or shear/seal ram on the BOP.
- Facility to kill or plug the well: pumps, kill weight fluid, LCM, cement, etc.

6.9 WELL BARRIER SCHEMATICS

Well schematics have been used by the oil industry for many years to provide a simple overview of the well architecture, and list critical features and depths. The concept of using a well schematic to identify well control barriers is said to have first begun in Norway in 1992.[4] Since then the use of barrier schematics has gained in popularity and is rapidly becoming standard practice for many operating companies.

A barrier schematic clearly illustrates both primary and secondary barrier envelopes, as well as listing the barrier elements. The barrier schematic can be used to detail how each barrier is tested. On a long and complex completion or intervention, the barrier envelope may change several times as the operation progresses. Embedding a barrier schematic for each phase of the operation in the work programme or procedures manual helps well site personnel keep track of well control barriers as the job progresses. This reduces the risk of errors in barrier identification, and consequently reduces the possibility of barrier failure. It also improves response time where barriers do fail, and will help in planning the actions necessary to reinstate the vital primary barrier.

Commercial barrier drawing software is now available,[5] and the drawings produced are similar to those found in NORSOK D 010. The example barrier drawings shown here are based on the NORSOK model. By convention, these drawings display the primary barrier in blue, and the secondary barrier in red.

When preparing barrier drawings, the correct identification of both primary and secondary barrier envelopes is important. A simple approach is to ask, between the reservoir and surface, what parts of the wellbore/completion are exposed to reservoir fluids, i.e., the primary well barrier envelope. If an element of the primary barrier envelope fails, where is the fluid going next? Anything exposed to fluid leaking from the failed primary barrier forms the secondary well barrier.

The next few pages are a series of well barrier drawings for some commonly performed completion, workover, and intervention operations.

6.9.1 Tripping with open ended completion tubing or work string – open perforations

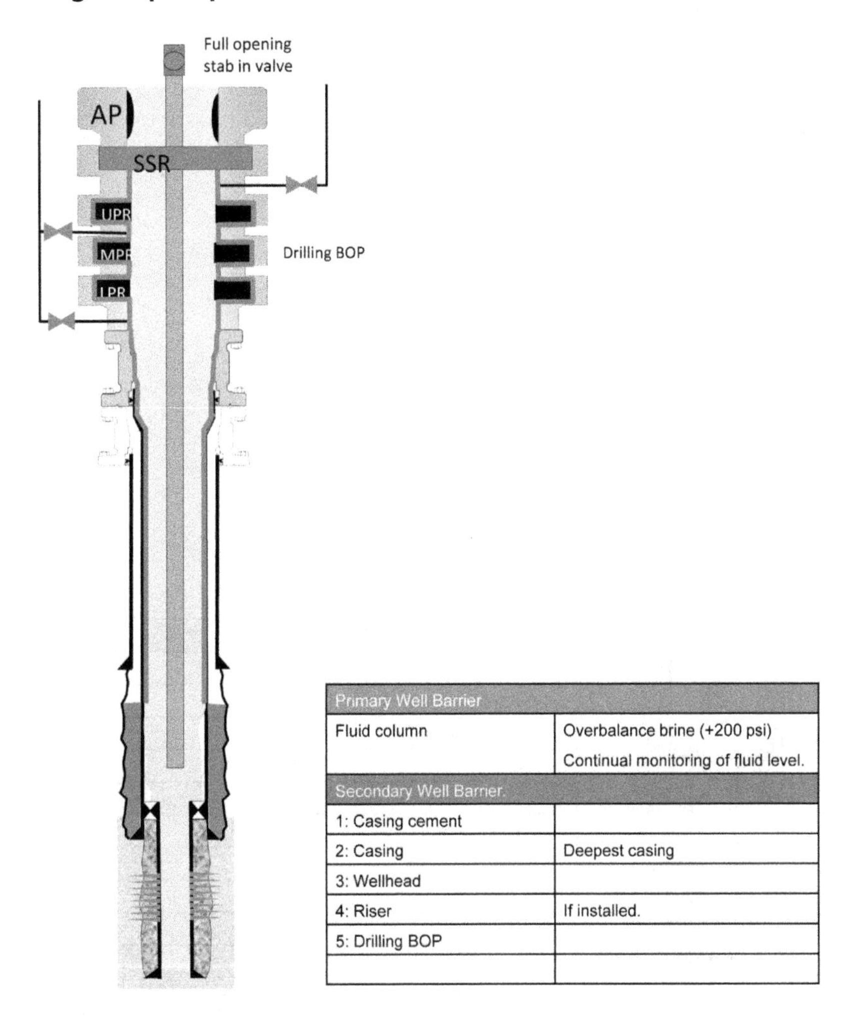

Primary Well Barrier	
Fluid column	Overbalance brine (+200 psi) Continual monitoring of fluid level.
Secondary Well Barrier	
1: Casing cement	
2: Casing	Deepest casing
3: Wellhead	
4: Riser	If installed.
5: Drilling BOP	

6.9.2 Removing BOP and installing Christmas tree

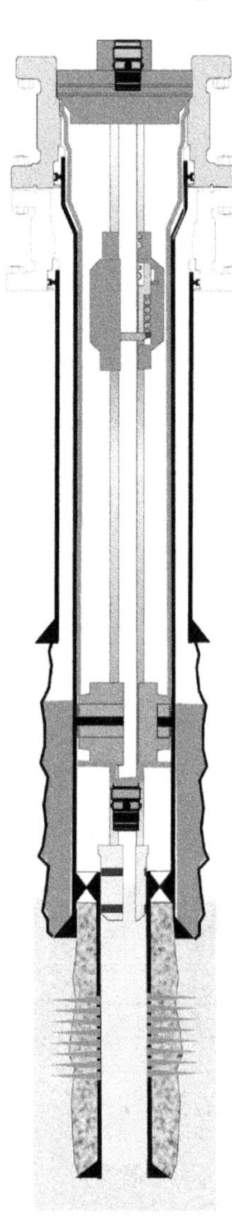

Primary Well Barrier	
1: Production packer	Inflow tested
2: Completion tubing	Positive test.
3: Deep set plug (tubing)	Inflow test.
Secondary Well Barriers.	
1: Cemented liner	
2: Production casing.	
3: Wellhead	
4: Tubing hanger	
5: Completion string	
6: Closed SCSSSV	
7: Tubing hanger plug	

6.9.3 Gas lifted production well with annular safety valve (ASV)

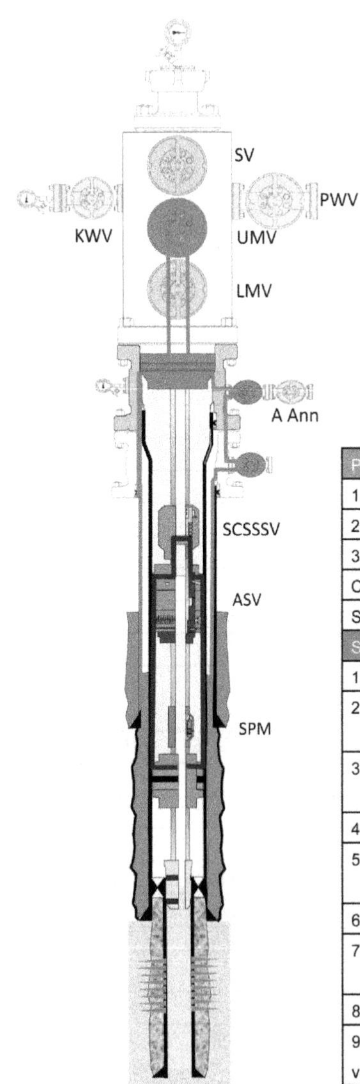

Primary Well Barrier	
1: Production packer	
2: Casing	Between ASV and packer
3: Annular safety valve	
Completion tubing	
SCSSSV	
Secondary Well Barriers	
1: Casing cement	Production casing cement
2: Casing	Production casing and intermediate casing.
3: Casing cement	Casing hanger, tubing Intermediate casing cement
4: Casing	Intermediate casing.
5: Wellhead	Intermediate casing hanger and seals
6: Tubing hanger	
7: Annulus inlet valve	Tubing head annulus gas injection line
8: Surface production tree	
9: Annulus access line and valve	B Annulus line and valve.

6.9.4 Barriers during wireline intervention in a live well

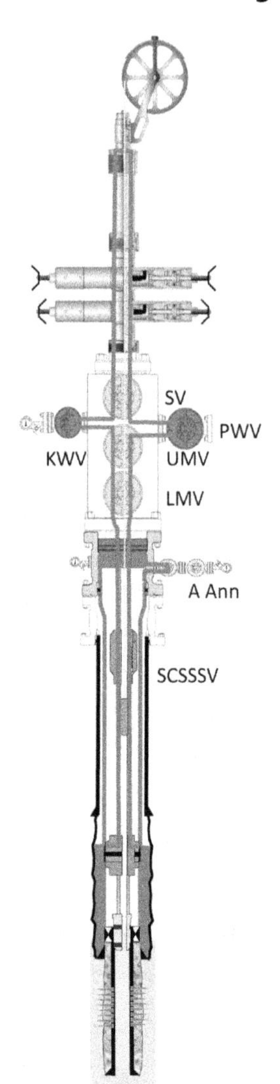

Primary Well Barrier	
1: Production packer	
2: Completion tubing	
3: Tubing hanger	
4: Surface tree body	
5: Kill wing valve	
6: Production wing Valve	
7: Wireline BOP	Body only
8: Wireline lubricator	
9: Stuffing box or grease head	
Secondary Well Barriers.	
1: Casing cement	
2: Casing	
3: Wellhead	
4: Tubing hanger	Common with primary
5: Christmas tree	Common with primary
6: Wireline BOP	Common with primary.

6.9.5 Barriers during wireline intervention in a live well

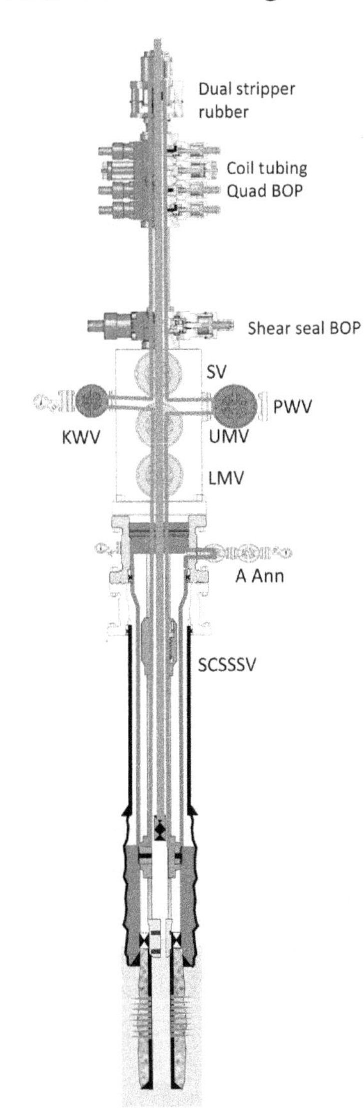

Primary Well Barrier	
1: Casing cement	
2: Casing	
3: Production packer	
4: Completion tubing	
5: Tubing hanger	
6: Surface tree body	
7: Kill wing valve	
8: Production wing Valve	
9: Shear seal BOP body	
10: High pressure riser	
11: Quad BOP body	
12: Stripper rubber	
13: Coil tubing	Below stripper
14: Check valves in reel	
Secondary Well Barriers.	
1: Casing cement	Production casing
2: Casing	Production casing and intermediate casing.
3: Wellhead	
4: Tubing hanger	
5: Surface tree	Common with primary
6: Quad BOP	

6.9.6 Hydraulic workover unit: running tubing in to a live well (shear rams able to function)

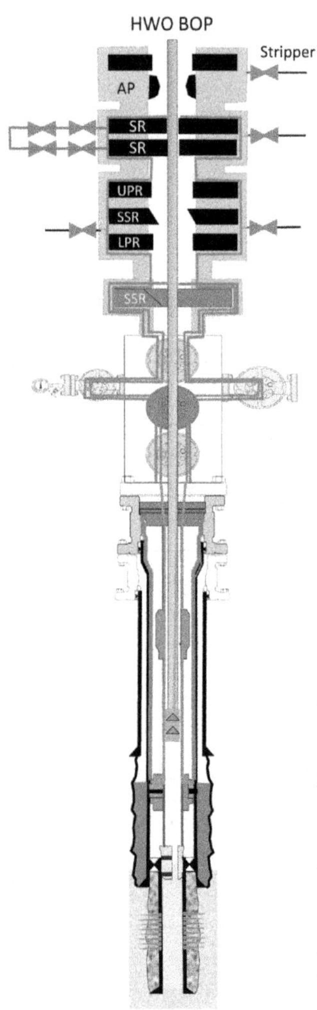

Primary Well Barrier	
1: Casing cement	
2: Casing	
3: Production packer	
4: Completion tubing	
5: Tubing hanger	
6: Surface tree body	KWV & PWV
7: Snubbing safety head	Body
8: HP riser	
9: Snubbing BOP	Body, kill and choke line, equalising loop and bleed off line.
10:Stripper rams	Either upper or lower ram when snubbing past tool-joint.
11: Annular preventer	If being used to strip/snub. LP application.
12: Tubing string	Below active stripper
13: Check valves in BHA	X 2.
Secondary Well Barriers.	
1: Casing cement	Common barrier with primary.
2: Casing	Common barrier with primary.
3: Wellhead	Including tubing hanger and annulus inlet valves
4: Tubing hanger	Common barrier with primary.
5: Surface tree	Common barrier with primary.
6: BOP pipe rams (rams closed)	

6.9.7 Subsea well test string—well shut in after flowing

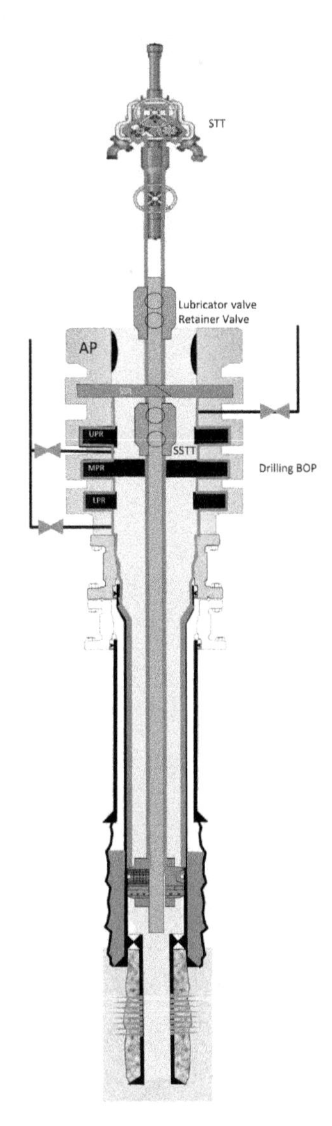

Primary Well Barrier	
1: Cemented liner	
2: Liner top packer	
3: Well test packer	Kill fluid in annulus
4; Downhole tester valve	
5: Downhole tester valve	Body only
6: Test string tubing	
7: Subsea test tree (SSTT)	Body only
8: Landing string	
9: Surface test tree (STT)	Tree body. Closed valves
Secondary Well Barriers.	
1: Casing cement	
2: Casing	
3: Wellhead	
4: BOP	Body, choke and kill valve and shear seal ram. SSTT disconnect before closing ram.

REFERENCES

1. NORSOK Standard D-010. Well Integrity in Drilling and Well Operations. Revision 3. August 2004.
2. Design, Installation, Repair and Operation of Subsurface Safety Valve Systems ANSI/API Recommended Practice 14b Fifth Edition, October 2005.
3. Stein A. Meeting the Demand for Barrier Plug Integrity Assurance. SPE paper 175489-MS. 2015.
4. Fjagesund T. Technology Update: Using Schematics for Managing Well Barriers. SPE paper-0915-0034-JPT.
5. https://www.wellbarrier.com.

CHAPTER SEVEN

Well Kill, Kick Detection, and Well Shut-In

7.1 INTRODUCTION

Killing a well prior to a workover or intervention is a necessary and planned part of the operation. There is normally time to decide upon the most appropriate kill method and design the operation, based on the best use of available data. A pre-intervention risk assessment can be carried out, and the kill procedure written into a detailed work program for use at the well site. This chapter describes the methods most commonly used to kill wells when preparing for a workover or intervention. It goes onto describe how well control can be restored following an unplanned kick during a completion, intervention, or workover on a dead well.

There are several important and distinct differences between drilling well control and well control during completions, workovers, and interventions.[1] With much of well control training focusing on drilling related issues, i.e., managing a kick whilst drilling ahead or tripping pipe, some of the well control complexities associated with completion and workover operations are overlooked. It is important that staff engaged in completion and workover understand, and can manage, these differences. The important ones are:

- When drilling, one of the main well control concerns is the integrity (fracture pressure) of the exposed formation at the previous casing shoe. This is normally the weak point and defines pressure limitations and the maximum allowable annulus surface pressure (MAASP). Exceeding the fracture pressure of the exposed formation would cause fluid loss. During completion and interventions, the weak point is almost always the exposed production formation. MAASP during a workover is either defined by the pressure limits on the production casing and wellhead, or by the fracture pressure of exposed formations.
- During a drilling operation, allowing a gas kick to migrate toward the surface without choke controlled expansion can be disastrous. Most

Well Control for Completions and Interventions.
DOI: https://doi.org/10.1016/B978-0-08-100196-7.00007-5

workovers and interventions are carried out on systems that are designed to withstand reservoir pressure to the surface.

- Well kill methods frequently used during workover operations are rarely used (post-kick) by drillers, i.e., bullheading and reverse circulation. If a bullhead were attempted in a well with an intact filter cake, it would almost certainly lead to a fracturing of the formation. Reverse circulation is widely used to kill the well pre-workover.
- Many drilling related kicks occur because higher than expected formation pressure is encountered, and the mud in the hole does not have sufficient density to maintain overbalance. To regain control, the mud in the hole must be replaced with a higher density mud that will return the well to an overbalanced condition. In most cases the new mud is forward circulated into the well using either the "drillers" or the "wait and weight" method. Kicks taken during workovers are rarely caused by higher than expected reservoir pressure; they are normally swabbed in or are a result of fluid loss, or loss of hydrostatic overbalance. Needing to increase brine weight during a completion, intervention, or workover is very unusual.
- During completion and workover operations, the producing zone is exposed for most of the duration of the operation. When drilling the well, zones with the potential to kick are only exposed for a relatively small percentage of the total drilling time.
- Overbalance pressure during drilling operations is sometimes higher than during workovers, as the mud filter cake creates an effective seal that stops fluid loss to the formation. The clear filtered fluid used during workovers has no sealing capability, and too much overbalance pressure against a permeable formation will cause fluid loss unless lost circulation material (LCM) is used. Overbalance with a clear fluid is normally limited to no more than 200-250 psi.
- A genuinely static fluid column is difficult to achieve when using clear fluids. In many workovers, fluid will be lost to the formation for the duration of the operation. Keeping the hole full becomes more complicated, as does monitoring the fluid level in the well. With even moderate, but manageable losses, accurate trip tank readings are more difficult to monitor. It also takes longer to identify flow or an increased loss rate.
- Many workovers are carried out on wells with depleted reservoirs. The use of brine as a kill fluid means that losses are sometimes difficult to control, and in very depleted sub-hydrostatic wells the fluid level cannot be monitored, rendering a fluid barrier invalid.

- A gas influx migrates towards the surface more quickly in completion brine than in mud.
- Many completion components, for example packers, are close to casing ID. Restricted fluid by-pass around these components increases the risk of swabbing and surging.
- Well integrity during workover operations is often compromised by corroded and damaged equipment; consequently, controlling the well becomes more difficult.
- Neither the annular preventer or pipe ram will seal around some types of completion equipment, for example sand control screens and slotted liners, nor will they seal around hydraulic control lines (flat pack), instrument cables, and Electric Submersible Pump (ESP) cables.
- During many workover operations, the well is killed with the Christmas tree still in place. If reverse circulation is used, hydrocarbons from the well can be routed through the flowline to the process facilities. This enables the well kill to be carried out in a closed, tested system; a very low risk operation.

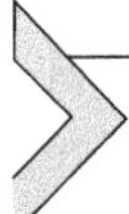

7.2 WORKOVER AND INTERVENTION WELL KILL PLANNING

Killing the well is a necessary part of many workover and intervention operations, and deciding how the well should be killed is a crucial step in the planning process. Any well kill decision must consider the type and condition of the completion, reservoir parameters, and the type of surface equipment available.

Initially, the choice will be between a circulating or non-circulating kill, with a circulating kill usually preferred since it is normally performed at lower pressure and is generally much less likely to damage the formation. If the Christmas tree is still in place, reverse circulation allows the tubing contents to be displaced through the process facilities, a very safe and environmentally-friendly way of removing the hydrocarbons from the tubing. If the blowout preventer (BOP) stack has been installed, reverse circulation is still normally preferred, as the hydrocarbons are kept in the tubing. Where it is not possible to establish a circulation path, a non-circulating kill method must be used; the two methods commonly used are:

- Bullhead
- Lubricate-and-bleed

During a bullhead (or deadhead) kill fluid is pumped down the production tubing, pushing any hydrocarbons back into the reservoir and leaving the wellbore filled with kill weight fluid. Its simplicity makes it the most popular method of killing a well prior to an intervention or workover. To succeed, the rate of injection into the well must be higher than the migration rate of any hydrocarbons in the tubing, meaning a relatively high pump rate is needed. Bullheading is only possible where the formation is permeable enough to permit the high rates of injection required without reaching formation fracture pressure. It is not suitable for easily damaged reservoirs, since any debris in the tubing (and some of the kill fluid) is swept into the formation during the kill.

If the completion design will not allow circulation, and if low permeability prevents a bullhead kill, coiled tubing can be used to circulate kill fluid into the well. Coiled tubing is often overlooked when planning a well kill, yet it has distinct advantages. The well kill can be performed at relatively low pressure, and equivalent circulating density (ECD) is low. Coil tubing circulation should be considered for any well that has an easily damaged formation, even if bullheading is possible.

Lubricate-and-bleed is used to kill gas wells where circulation cannot be established, or where it is not possible to obtain a high enough injection rate (fluid velocity) to overcome the migration velocity of gas during a bullhead kill. Lubricate-and-bleed involves pumping a measured quantity of fluid into the tubing, then bleeding gas pressure equivalent to the hydrostatic head of the pumped fluid. The process is repeated until the well is dead.

Killing the well prior to a workover or intervention is a planned event. During the running of a completion, or the execution of a workover or intervention, it may be necessary to respond to an unplanned kick.

7.2.1 Essential information

A great deal of information about the reservoir, completion, and surface equipment must be obtained to determine which kill method is best suited for the well. Essential data will include, but need not be limited to:

- Reservoir datum depth (usually top reservoir).
- Formation fracture pressure (P_{fr}) to set surface and downhole pressure limits.
- Reservoir pressure (P_r). This will be used to calculate the kill weight fluid density and static pressure during the kill.

- Formation permeability (K). This is needed to calculate potential losses and determine if LCM is going to be needed.
- Shut-in tubing head pressure (SITP).
- Shut-in casing head pressure (SICP).
- Tubing fluid type and density including oil/water contact (OWC) and gas/oil contact (GOC) if known.
- Tubing linear capacity.
- Annulus fluid type and density.
- Casing and liner linear capacity.
- Annulus linear capacity.
- Tubing and casing burst and collapse limits.
- Completion configuration—critical depths:
  - Tubing depth, including any changes in diameter.
  - Packer depth.
  - Depth of circulation ports.
  - Type of circulation device (if relevant).
  - Depth of seating nipples (if relevant) for barriers.
  - Casing configuration and shoe depth.
  - Liner top and liner shoe depth (if relevant).
  - In sand control screen completions: screen mesh size for sizing LCM.
- Surface well control equipment:
  - BOP configuration and working pressure.
  - Choke manifold configuration.
  - Wellhead configuration and condition.
  - Pump capacity, strokes per barrel, and working pressure.
  - Fluid storage.

If the well kicks during the completion or workover operation, the same data would be used to restore control.

7.2.2 Wellbore preparation

When planning a well kill in preparation for a rig assisted workover, consider killing the well before the rig is on location. Substantial cost savings can be made as the rig is required for less time. Preparatory work can include:

- Integrity testing of the Christmas tree valves (and sub-surface valve if fitted).
- Integrity testing of the wellhead valves and seals.
- Performing drift runs to locate any restrictions in the tubing.

- Running wireline deployed pressure gauges to obtain an accurate BHP measurement. A gradient survey can also locate water/oil and oil/gas contacts in the wellbore.
- Running caliper or sonic logs to determine the condition of the tubing and casing.
- Creating a flow path between production tubing and casing for circulating kills, for example opening a sliding sleeve or punching a hole in the tubing.
- Performing the well kill (circulating or non-circulating).
- Installing mechanical barriers in preparation for the removal of the Christmas tree and the installation of the BOP.

7.3 WELL KILL: REVERSE CIRCULATION

Although bullheading is the most commonly used method for killing a production well prior to workover, it is not necessarily the best. Reverse circulating is sometimes preferred, as surface pressure is generally very low and formation damage is less likely. Kill weight brine is pumped into the annulus, returning through the production tubing to the surface. The natural U tube effect created by the heavy fluid in the annulus helps push the lighter hydrocarbon in the tubing towards the surface and in most cases, a single circulation (combined volume of annulus and tubing) is enough to kill the well. The advantages of a reverse circulation kill are:

- Pump pressure is normally lower than a bullhead.
- Little or no formation damage occurs. If properly performed, fluid loss to the formation should be minimal and is eliminated if the well can be plugged below the circulation point.
- It is the shortest (quickest) route to circulate hydrocarbons in the tubing to the surface.
- If the Christmas tree is still in place and the flowline still connected, hydrocarbons can be produced to the production facilities for easy, low risk, disposal.

Note: Some remotely operated production chokes are not ideal for controlling flow from the well and may have to be by-passed, with flow routed through a manually operated choke. Nevertheless, the tubing contents can still be routed through the process facilities.

- Keeping the hydrocarbons contained in the tubing can be advantageous if the tubing has a higher burst rating than the casing, but can be a disadvantage of the tubing is badly worn or corroded.
- In some pre-workover or intervention situations, the packer fluid will be dense enough to overbalance reservoir pressure, thus simplifying the kill.
- In some deep-water subsea wells, ECD will be lower when reversing, a consequence of frictional pressure drop through the choke line (see Chapter 14, Well Control During Completion and Workover Operation).

Although the reverse kill is generally acknowledged to be beneficial, there are some concerns.

- In most cases, reverse circulation ECD is higher than during conventional forward circulation. Fluid being pumped to the surface through the tubing must overcome friction. Since tubing usually has a smaller cross-sectional area than the annulus, more frictional pressure drop occurs when reversing. For example, 4½ in. 12.6 lb/ft. tubing has a capacity of 0.0152 bbls/ft. Placed inside 9⅝ in. 47 lb/ft. casing the annulus has a capacity 0.0535 bbls/ft., 3.5 times more than the tubing.
- High ECD will result in more fluid loss to the formation unless the well is plugged below the circulation ports or effective LCM is used.
- High pump pressure in the annulus will result in increased collapse loading on the production tubing. Tubing weakened by corrosion is at risk of collapse.
- Debris that has settled above the production packer can block circulation ports.
- If the well is plugged below the circulation ports, reverse circulation will sweep debris from the annulus into the tubing. The debris will settle above the plug, making post intervention recovery more complicated and time consuming.
- In gas filled wells, small adjustments to the choke often result in large changes in circulation pressure owing to gas compression and expansion.
- Heavy fluid in the annulus. If reservoir pressure declined after the well was completed, the fluid in the annulus (packer fluid)

may be much denser than is needed to kill the well. When the tubing/annulus communication path is opened, heavy annulus fluid will drop down the well. There is a risk of this annulus fluid invading and damaging the formation, particularly if there is debris in the annulus. It will also be more difficult to displace the heavy annulus fluid out of the well with the lighter kill fluid unless a viscous pill is used. The use of a viscous pill will increase frictional pressure drop in the tubing.

- If the original completion was run before the casing (or liner) was perforated, a sub-hydrostatic fluid may have been left in the annulus. If this is the case, there will be a tendency for the lighter annulus fluid to migrate upwards during the well kill. To sweep the lighter fluid out of the annulus, the pump rate will need to be increased or a viscous pill used. This will increase frictional pressure drop in the tubing.
- If there is no packer in the well, or tubing leaks have allowed oil and gas to enter the annulus, the circulation rate must be higher than the hydrocarbon migration rate.

7.3.1 Static and circulating pressure

During a well kill, fluid is dynamic when circulation takes place. It is therefore necessary to calculate BHP whilst circulating, a more complex condition than when static. Friction needs to be considered and added to the hydrostatic pressure (HP). Frictional pressure drop is affected by:

- Fluid density
- Fluid viscosity
- Gel strength in viscos fluids
- Yield point
- Well depth (measured and vertical)
- Annular clearance
- Tubing ID
- Tubing and casing roughness
- Rate of circulation

When forward circulating, fluid is pumped down the tubing and returns taken from the annulus. Pump pressure when circulating is a result of friction through the circulation path and is a function of:

$$\text{Surface line friction} + \text{tubing string friction} + \text{friction through the BHA} \\ + \text{annulus friction} + \text{annulus hydrostatic} - \text{tubing hydrostatic.}$$

Similarly, when reverse circulating, fluid is pumped down the annulus and returns taken through the tubing. Pump pressure is a function of:

Surface line friction + annulus friction + friction through the BHA + tubing friction + tubing hydrostatic − annulus hydrostatic.

If the tubing and annulus HP is balanced (both have the same fluid density) then pump pressure is equal to friction whilst circulating or reverse circulating. If tubing and annulus have fluid of different density, the hydrostatic imbalance will register as an additional pressure increase over and above friction pressure, or decrease below friction pressure.

When circulating or reverse circulating, pressure acting against the formation is higher than under static conditions, owing to the addition of friction pressure. This can be calculated as follows.

BHP during conventional (forward) circulation:

$$\mathrm{BHCP} = \mathrm{HP} + \mathrm{SP} + \mathrm{FP}_{\mathrm{ann}} \tag{7.1}$$

where:

BHCP: bottom hole circulating pressure;
HP: hydrostatic pressure;
SP: surface pressure;
$\mathrm{FP}_{\mathrm{ann}}$: friction pressure in the annulus.

BHP during reverse circulation:

$$\mathrm{BHCP} = \mathrm{HP} + \mathrm{SP} + \mathrm{FP}_{\mathrm{tbg}} \tag{7.2}$$

where:

BHCP: bottom hole circulating pressure;
HP: hydrostatic pressure;
SP: surface pressure;
$\mathrm{FP}_{\mathrm{tbg}}$: friction pressure in the tubing.

During drilling operations, it is normal to carry out slow circulation rate tests to accurately establish frictional pressure drop. The information obtained would be used if a well kill had to be performed. When planning a workover, it is unlikely that empirical data will be available and so for planning purposes, hydraulic modeling software can be used to estimate bottom hole circulating pressure and frictional pressure drop.

7.3.2 Equivalent circulating density

Reservoir engineers will typically give reservoir pressure as a pressure, i.e., psi, kPa, or bar. Drill crews will normally convert a given pressure to a mud weight equivalent (ppg, SG, or kg/m^3). ECD includes any increase over and above HP resulting from fluid friction whilst circulating.

ECD during forward circulation can be calculated as follows.

Oilfield units:

$$\mathrm{ECD(ppg)} = \frac{\mathrm{FP_{ann}}}{0.052 \times \mathrm{TVD}} + \mathrm{FD(ppg)} \tag{7.3}$$

where:

FP_{ann}: annular friction pressure;
TVD: true vertical depth (ft.);
FD: fluid density in ppg.

SI units:

$$\mathrm{ECD(kg/m^3)} = \frac{\mathrm{FP_{ann}}}{0.00981 \times \mathrm{TVD}} + \mathrm{FD(kg/m^3)} \tag{7.4}$$

where:

FP_{ann}: annular friction pressure;
TVD: true vertical depth (m);
FD: fluid density in kg/m^3.

ECD during reverse circulation is calculated from:

Oilfield units:

$$\mathrm{ECD(ppg)} = \frac{\mathrm{FP_{tbg}}}{0.052 \times \mathrm{TVD}} + \mathrm{FD(ppg)} \tag{7.5}$$

where:

FP_{tbg}: tubing friction pressure;
TVD: true vertical depth (ft.);
FD: fluid density in ppg.

SI units:

$$\mathrm{ECD(kg/m^3)} = \frac{\mathrm{FP_{tbg}}}{0.00981 \times \mathrm{TVD}} + \mathrm{FD(kg/m^3)} \tag{7.6}$$

where:

FP_{tbg}: tubing friction pressure;
TVD: true vertical depth (m);
FD: fluid density in kg/m^3.

Fluid velocity in the tubing is normally much higher than in the annulus, as the cross-sectional area is smaller.

7.3.3 Slow circulation rate

Most circulating kills are carried out at a slow circulating rate. To enable slow circulation rate to be used effectively, slow circulation rate pressure (SCRP) must be recorded at different pump rates with the pipe on the bottom. SCPR is a record of pump pressure at any rate less than the normal circulating rate used to carry out work on the well and when the well is open. During drilling operations, SCRP is taken regularly, normally at least once per shift. However, for completion and workover operations, obtaining SCRP rarely becomes routine since the pipe is not normally on the bottom for any appreciable time. During most completions and workovers, the only operations requiring pipe to remain on the bottom are the pre-completion well bore cleaning trip, and remedial operations carried out as part of a workover. Examples include performing a cement squeeze, milling a packer, and cleaning fill from the bottom of the well. SCRP should be recorded any time there is a change to the geometry of the workstring (internal or external), or any time there is a change in the circulated fluid (density or viscosity). Normal practice is to record SCRP at several circulating rates (Table 7.1).

If a well control incident occurs and the well needs to be shut-in, the recorded SCRP is available to use when calculating circulating pressure for the removal of a kick.

There are several reasons why a slow circulation rate is preferred:

- ECD on exposed formation. The higher the circulation rate the higher the ECD.
- Disposal of kick fluids. When circulating out a large kick, disposing of a large volume of liquid hydrocarbon and contaminated brine will be problematic, especially on rigs with limited facilities. Slowing the pump rate allows the crew to deal with returning hydrocarbons in a controlled manner.
- Reaction time on the choke. A circulation kill relies on maintaining a constant BHP. Control of surface pressure (and by extension BHP) can be difficult as the kick approaches surface—especially with a gas kick. A slow circulation rate gives the choke operator more time to respond to changing surface pressure.

Table 7.1 Example of recorded slow circulation rate pressure

Strokes per minute (bbls/st = 0.0313)	bbls/min	psi
20	0.626	420
30	0.939	605
40	1.25	1050
50	1.56	1840

- Rig pumps/mud pumps. If a rig is on location for the well kill, the mud pumps will probably be used. These are high volume pumps, but are sometimes tricky to operate consistently if very small volumes are required. If small volume and consistency of output is required, a smaller capacity pump might be preferred, for example a cement unit.
- Surface equipment limitations. Be aware of any limitations on surface equipment downstream of the choke, i.e., the choke line, mud/gas separator. The maximum pump rate must be lower than the handling capacity of this equipment.
- Fluid filter throughput limits. Whilst not strictly a well kill concern, some circulating operations, for example gravel packing or wellbore cleaning, work in a closed loop where brine returns are filtered before they are returned to the pits for re-use. The pump rate will be limited to the capacity of the filtration system.

7.3.4 U tube pressure

When circulating, the tubing and annulus are in hydraulic communication. If the tubing and annulus contain fluids of differing density, any resulting difference in HP will cause the fluid level to change until equilibrium is reached. If the well is shut-in, higher pressure will be observed at the top of the fluid column of lower density (tubing or annulus).

Example:

The tubing is filled with base oil with a density of 6.5 ppg (0.78 sg) and the annulus is filled with a 10 ppg (1.2 sg) brine. The tubing end is at 5000 ft. (1524 m) and the resulting pressure at the surface is as follows.

Oilfield units:

HP of the brine column = $10 \times 0.052 \times 5000 = 2600$ psi.

HP of the base oil column = $6.5 \times 0.052 \times 5000 = 1690$ psi.

Tubing pressure at the surface = 2600-1690 = 910 psi.

SI units:

Brine hydrostatic = $(1.2 \div 10.2) \div 1524 = 179.294$ bar (17,929 kPa).

Base oil hydrostatic = $(0.78 \div 10.2)/1524 = 116.541$ bar (11,654 kPa).

Tubing pressure at the surface = $179 - 117 = 62$ bar (6200 kPa).

There are several occasions where U tube pressure will be present during workover, completion, and intervention operations:

- Circulation of a weighted-up fluid to kill a well following a kick.
- Circulation of a kill weight fluid into the well prior to a workover or intervention.

- Displacement of a heavy fluid with a lighter fluid to provide an underbalanced cushion prior to perforating.
- The opening of a communication port or sleeve above the production packer in a completed well where kill weight fluid has been left in the annulus and hydrocarbons remain in the tubing.

7.3.5 Opening the circulation path

In wells completed with a packer, it will be necessary to create a flow path between the annulus and tubing to allow circulation. Some completions have equipment that is designed to allow communication between tubing and annulus, usually a sliding sleeve, or side pocket mandrel.[a]

When using a sliding sleeve, it is normally opened using a slickline deployed mechanical shifting tool. Sliding sleeves in old wells are usually very difficult to open, and it may be less time consuming to simply run a tubing punch.

Side pocket mandrels are occasionally placed in a well specifically to provide a communication path between tubing and annulus. They are normally run with a dummy (blank) valve pre-installed. When tubing/annulus communication is needed, slickline is used to recover the dummy valve, leaving ports in the side pocket mandrel open. Side pocket mandrels are commonly found in gas lift wells, with most of these wells requiring more than one mandrel. The bottom mandrel is normally positioned a short distance above the packer, and is fitted with a retrievable orifice valve. The orifice valve allows fluid to flow from the annulus to the tubing, but flow in the opposite direction should be prevented by non-return valves. All other mandrels in the well are fitted with retrievable unloading valves. These allow annulus to tubing communication when open, but close-in response to a predetermined casing pressure. Like the orifice valve, the unloading valves have non-return valves to prevent fluid flow from tubing to annulus. To effectively kill a well, fluid must be circulated through the bottom mandrel, the one closest to the packer. To do this, the unloading valves in the higher mandrels must be replaced with blank valves that isolate the circulating ports. Failure to do so would result in fluid taking the path of least resistance and simply circulating through the upper side pocket mandrel. It is also best to pull the orifice valve in the bottom side pocket mandrel, since leaving it in place would restrict circulation rates. It would also prevent changing the

[a] The sliding sleeve and side pocket mandrel are described and illustrated in Chapter 3, Completion Equipment.

direction of circulation and invalidate any pressure observation made on the annulus.

If a well has not been completed with a circulating device, communication can be established by punching holes in the tubing. There are several considerations before punching the tubing:

- A mechanical punch run on slickline is simple and inexpensive, but only creates a single hole. Circulation rates will be restricted.
- An explosive punch using shaped charges will produce a flow area equal to or greater than the flow area through the tubing.
- If an explosive punch is used, it should be designed for limited penetration. This is to prevent casing damage when the tubing is perforated.
- An explosive tubing punch can be run on slickline using a time delay detonator, but if depth is critical it is better to run it on e-line and use Gamma ray and Casing Collar Locator (GR/CCL) to place the punch accurately on depth.
- To prevent any debris that has settled on the packer entering the tubing, a space is normally left between the packer and the location of the punch, typically, at least 50 ft.

Before opening or creating a communication flow path, it is important to minimize pressure differential between tubing and annulus. To do this, SITP + tubing HP and SICP + casing HP at the depth of the communication device must be calculated. If there is uncertainty about the depth gas/oil and OWCs, pressure gauges can be run. If this is not possible, tubing pressure at the point of circulation may have to be estimated.

- Pressure differential across a sliding sleeve (SSD) will make it more difficult to open.
- There is a risk of the wireline tools being "blown up the hole" if that differential is from annulus to tubing.
- When punching a hole in the tubing, there is a risk of the wireline tools being "blown up the hole" if there is too great a pressure differential from annulus to tubing.
- If a dummy valve needs to be pulled from a side pocket mandrel, there is a risk of the wireline tools being "blown up the hole" if there is too great a pressure differential from annulus to tubing.
- If a gas lift unloading valve or gas lift orifice valve need to be pulled from a side pocket mandrel, pressure from the annulus will equalize across the valve to the tubing. Pressure differential from tubing to annulus will make the valve more difficult to pull.

7.3.6 Plugging the well

There are significant advantages to plugging the well below the circulation ports. A plug prevents fluid loss to the formation, meaning the risk of losing the hydrostatic overbalance is significantly reduced. A plug also reduces the possibility of kill fluid damaging the formation. Furthermore, an inflow tested plug can be used as one of the two mechanical barriers that are required before the Christmas tree can be safely removed and the drilling BOP installed.

If left in place during a workover or intervention, the presence of a mechanical barrier greatly reduces any possibility of a well control event. Losses or a kick are prevented. Finally, having a plug below the circulation ports makes the kill more predictable, since losses are unlikely. However, the risk associated with placing a mechanical barrier in the well must be assessed:

- Debris settling above the plug can make recovery difficult. Two-piece plug and equalizing prong barriers are more debris tolerant than those equipped with an integral equalizing device. The only drawback is the need for two wireline runs to install the plug and prong type (and two runs to pull).
- Setting a debris catcher above a plug is recommended, but this does require an additional wireline run.
- Continued injection or production in offset wells often results in a change in reservoir pressure (increase or decrease) that might complicate plug recovery.
- It is not possible to monitor reservoir pressure in a plugged well.
- Knowing a plug is present can lead to complacency and a lack of vigilance with regard to monitoring fluid level. Plug integrity is not guaranteed for the duration of a workover, or even of the kill.

A plug can be set before or after the communication port is open:

- If the plug is set before the communication ports are opened or the tubing punched, debris from the annulus can move into the tubing and settle above the plug.
- If the plug is set before the communication ports are opened or the tubing punched, pressure above the plug can be increased, or reduced, to equalize between tubing and annulus.
- If the plug is set after the ports are open, debris contaminated heavy fluid from the annulus can be lost to the formation.

It is usually better to set the plug before opening the communication path. Debris above a plug can usually be removed, whereas formation damage is sometimes irreversible.

If the well is plugged, it must be killed to the point of circulation and not the reservoir depth. This assumes that the wellbore below the plug remains filled with (lower density) hydrocarbon, but allows for the consequence of a leaking plug.

If there is no plug, the well should be killed to the top of the reservoir. Any hydrocarbon between the point of circulation and the reservoir will migrate upwards (because of density differences) and be swept out during the kill, leaving the wellbore between the reservoir and circulation ports filled with kill weight fluid. However, the hydrostatic calculations for a kill should be based on a worst-case assumption; the interval between the circulation path and top reservoir remains filled with lower density hydrocarbon.

7.3.7 Pumping the kill fluid

The procedure described here assumes the well has been shut-in following production and is to be killed in preparation for a workover or intervention. In common with other circulating kills, the aim is to maintain a constant BHP throughout the kill, normally 200–250 psi above reservoir pressure. If circulating pressure drops below reservoir pressure there will be an additional influx; if circulating pressure is too high, fluid will be lost to the formation. Losses or an influx are less of a concern when the well has been plugged. However, even when a plug is in place it is good practice to maintain BHP within the parameters that would be used for an unplugged well. Plugs can leak!

As ECD is normally higher when reverse circulating, a slow circulation rate is normally a requirement. Fluid hydraulics modeling can be used to determine pump speed where SCRP data is not available

Lines are rigged up to enable kill fluid to be pumped into the annulus. A pressure relief valve should be fitted to the lines, and non-return valves placed close to the annulus valve. Returns from the tubing side will be routed through a choke, and from there, to disposal. If the flowline is still in place, returns can be handled by the process facilities. This is normally the preferred option, since there is no need for additional handling and disposal equipment.

Normal procedure for reverse circulation:

1. With the pumps lined up to the annulus and the wellhead valves open, bring pressure in the casing up to the calculated value whilst holding back pressure on the tubing using the choke.
2. Continue circulating controlling the casing pressure using the choke.

When the kill starts, tubing pressure will reduce quickly as the annulus (packer) fluid fills the tubing. There are a few points to note:

- When the kill fluid has the same weight as the fluid already in the annulus, pump pressure should be kept constant until the end of the kill.
- If the kill fluid is heavier than the fluid in the annulus, pump pressure will need to be stepped down to prevent over-pressure on the formation and potential losses.
- If the fluid needed to kill the well is lighter than the fluid already in the annulus, then pump pressure will have to be stepped up to overcome U tube pressure as the heavy annulus fluid fills the tubing. It will then be stepped back down as the lighter kill fluid begins to fill the tubing.

With a known reservoir pressure and required overbalance, creating a pump schedule is relatively simple. The surface pressure needed at the pump for any stage of the operation is:

$$(\text{reservoir pressure} + \text{overbalance}) \\ - \text{annulus hydrostatic to the plug (or reservoir)}.$$

Tubing head pressure should equal:

$$(\text{reservoir pressure} + \text{overbalance}) \\ - \text{tubing hydrostatic to the plug (or reservoir)}.$$

Casing pressure is predictable and can be plotted against volume pumped (or strokes). If the tubing contents are known, and GOC depth is known, a prediction of tubing pressure can be produced and used to assist the person operating the choke.

In vertical wells with single diameter tubing and single diameter casing, the critical points during the kill are:

- Gas cap (if present) has been removed and liquid reaches the surface.
- Oil/Water contact (if applicable) at surface.
- Annulus (packer) fluid fills the tubing.
- Kill fluid at the circulation ports.
- Full circulation is complete, at which point the well should be dead.

In deviated well and well with tapers in the tubing and/or casing, the pump schedule will be more complex. Three examples of reverse circulation well kills are demonstrated next.

7.3.8 Reverse circulation: worked example 1. Plugged vertical well

In this example, the well is vertical and equipped with a sliding sleeve a short distance above the packer. The sleeve will be opened to allow circulation. A plug has been installed in a seating nipple below the circulation ports of the sliding sleeve (Fig. 7.1).

Planning requirements:

1. Calculate the gradient of kill weight fluid required to overbalance well pressure at the SSD by 200 psi (for the purposes of this example, brine weight temperature and pressure corrections have been ignored).
2. Calculate all necessary volumes.

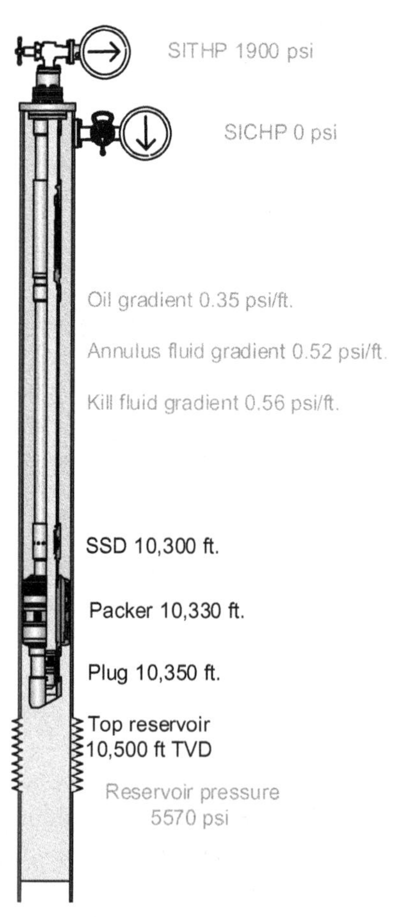

Figure 7.1 Well schematic. Worked example 1.

3. Calculate the surface pressure at the critical points in the kill, and plot a kill graph for tubing and annulus pressure. A 200 psi overbalance is required during and after the kill.

7.3.8.1 Calculate the gradient of kill weight fluid required to overbalance well pressure by 200 psi

Since the well is plugged, it must be killed to the SSD at 10,300 ft., and not the reservoir. This ensures the overbalance is maintained if the plug leaks.

Average gradient (psi/ft) of tubing fluid is given (Table 7.2 and Fig. 7.1) or can be calculated from:

$$\text{(reservoir pressure} - \text{SITP)} \div \text{reservoir depth}$$
$$(5570 - 1900) \div 10,500 = 0.3495\ \text{psi/ft} \rightarrow 0.35\ \text{psi/ft}$$

Tubing pressure at the SSD: SITP + (tubing fluid gradient × depth at SSD)

$$1900 + (0.3495 \times 10,300) = 5500\ \text{psi.}$$

Fluid density (PPG) needed to hold 200 psi overbalance at the SSD is:

$$\text{(pressure at circulating port + overbalance)}$$
$$\div\ \text{(depth of circulating port} \times 0.052)$$

$(5500 + 200) \div (10,300 \times 0.052) = 10.64$. By convention, kill weight is always rounded up.

$$10.64 \rightarrow 10.7 \quad 10.7\ \text{ppg} \times 0.052 = 0.5564\ \text{psi/ft.}$$

Table 7.2 Reverse circulation kill data

Top reservoir depth	10,500 ft.
Reservoir pressure (P_r)	5570 psi
Plug depth	10,350 ft.
Packer depth	10,330 ft.
Sliding sleeve or sliding side door (SSD) depth	10,300 ft.
5½ in 23 lb/ft tubing ID	4.670 in.
9⅝ in 53.5 lb/ft casing ID	8.535 in.
SITP	1900 psi
SICP	0 psi (before sleeve open)
Packer fluid gradient	0.52 psi/ft. (10 ppg)
Oil gradient	0.35 psi/ft. (6.73 ppg)
Planned overbalance	200 psi

7.3.8.2 Tubing and annulus volumes

Casing and annulus capacity can be calculated since ID and outer diameter (OD) are known. Alternatively, they could be looked up in tables.

$$\text{Tubing capacity: } 4.670^2 \div 1029.4 = 0.0212 \text{ bbls/ft.}$$
$$\text{The reciprocal gives ft./bbls} \quad 1 \div 0.0212 = 47.17 \text{ ft./bbls}$$
$$\text{Tubing volume to SSD: } 10,300 \times 0.0212 = 218.36 \text{ bbls}$$

$$\text{Annulus capacity: } (8.535^2 - 5.5^2) \div 1029.4 = 0.0414 \text{ bbls/ft.}$$
$$\text{The reciprocal gives ft./bbls} \quad 1 \div 0.0414 = 24.15 \text{ ft./bbls}$$
$$\text{Annulus volume to SSD: } 10,300 \times 0.0414 = 426.42 \text{ bbls}$$

Total volume (tubing and annulus) = 644.42 bbls.

7.3.8.3 Calculate pressures before and after opening the sliding side door

- Tubing pressure at the SSD before it is opened has already been calculated (5500 psi).
- Annulus pressure at the SSD: depth to SSD × annulus fluid gradient

$$10,300 \times 0.52 = 5356 \text{ psi.}$$

- Pressure across the SSD can be equalized by applying (5500 − 5356) = 144 psi to the casing.
- HP at the SSD after the kill will be: 10,300 × 0.5564 = 5731 psi. This is 231 psi above reservoir/well pressure at the SSD. It is also 31 psi more than the required 200 psi overbalance. The additional 31 psi is because the kill fluid density was rounded up from 10.64 to 10.7.
- After opening the SSD, bring the casing pressure up to 375 psi to give 231 psi overbalance at the SSD, 5731 − 5356 = 375 psi.
- As the circulation begins the casing pressure is 375 psi. Pressure at the SSD (kill depth) is overbalancing well pressure by 231 psi.
- Before the choke is opened the tubing head pressure will read 5731 − (10,300 × 035) = 2126 psi.
- The aim is to pump fluid into the annulus whilst maintaining at least a 231 psi overbalance at the SSD by controlling casing pressure on the choke. As the tubing fills with 0.52 psi/ft. packer fluid from the annulus, the tubing pressure will fall rapidly.

To complete the kill sheet pump schedule, tubing head pressure and casing head pressure at each stage of the kill must be calculated. Surface pressure at each stage of the kill is simply the required pressure at the plug

(5731 psi) minus the hydrostatic head of fluid in both the tubing and annulus. In this example the main steps are:

- Tubing filled with the original annulus (packer) fluid.
- Kill fluid filling annulus to the sliding sleeve.
- Full circulation (well dead).

7.3.8.4 Begin kill

With SSD open. Casing pressure applied, but no fluid pumped:

- Pressure at SSD: 5731 psi. This pressure must be maintained throughout the kill.
- SITP 2126 psi.
- SICP 375 psi.

Open the choke and begin to pump kill fluid at the slow circulation rate.

7.3.8.5 Tubing filled with annulus (packer) fluid

- With annulus (packer) fluid to surface the tubing pressure required to maintain overbalance will be:
    - $5731 - (10{,}300 \times 0.52) = 375$ psi.
- After pumping the tubing volume (218 bbls) annulus (packer) fluid will reach surface. At this point the kill fluid/packer fluid interface depth in the annulus is the volume pumped divided by the annulus capacity: 218(bbls) $\div$ 0.0414 bbls/ft = 5265.7 ft. This leaves $10{,}300 - 5265.7 = 5034.3$ feet of packer fluid remaining in the annulus.
- Annulus HP (to SSD)
- $5034.3 \times 0.52 = 2618$ psi.
- $5265.7 \times 0.5564 = 2930$ psi.
- Casing head pressure to maintain overbalance: $5731 - (2618 + 2930) = 183$ psi.

As the casing fills with kill weight fluid, the pressure can be stepped down from 375 to 183 psi. This is best done in stages. The choke operator should be provided with a schedule showing casing pressure versus volume (or strokes). In this example, after pumping 218 bbls, the pressure should have been reduced by 192 psi, or 1 psi for every 1.135 bbls pumped, or 0.881 psi/bbls. Casing pressure would continue to be stepped down until the annulus is filled with kill fluid (Table 7.3).

Table 7.3 Casing surface pressure schedule for first two stage of well kill

bbls pumped	Casing pressure
0	375 psi
50	330 psi
100	287 psi
150	243 psi
200	199 psi
218 packer fluid to surface	183 psi
250	155 psi
300	111 psi
350	67 psi
400	22 psi
426	0 psi

7.3.8.6 Annulus filled with kill fluid

- The tubing remains filled with annulus fluid (0.52 psi/ft.), therefore the tubing head pressure remains unchanged at 375 psi (as for the previous step).
- Because the annulus is full of kill weight fluid, static casing pressure will be 0 psi.
- $5731 - (10{,}300 \times 0.5564) = 0$ psi.

7.3.8.7 Tubing displaced to kill fluid

- As the kill fluid fills the tubing through the SSD, tubing head pressure will drop until it reaches 0 psi. When kill fluid reaches the surface, the HP at the SSD will be 5731 psi, giving the 231 psi overbalance.

Note: Although the choke should be used to control casing pressure, there will probably need to be adjustments made to the pump output. Good communications between the choke and pump operators are essential for maintaining the correct BHP (Fig. 7.2).

7.3.9 Reverse circulation: worked example 2—plugged vertical well with heavy fluid in the annulus

Packer fluid left in the annulus when the well was originally completed will often be significantly denser than would be necessary to kill the well after depleting the reservoir. Following on from the previous example, the effect of having a kill brine that is significantly lighter than the packer fluid is examined. The packer fluid used in this example has a gradient of 0.67 psi/ft. and reflects the density needed to control the well at the

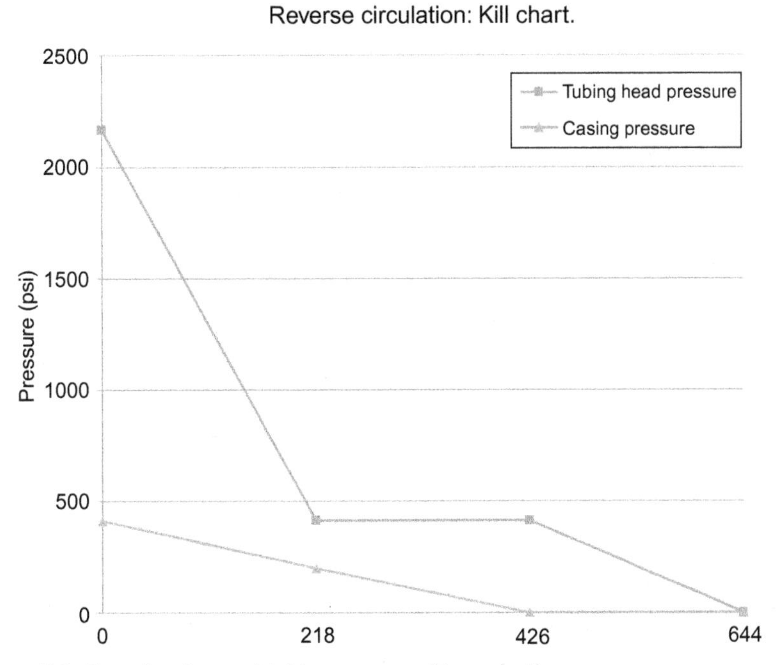

Figure 7.2 Plot of casing and tubing pressure (Example 1).

Table 7.4 Reverse circulation kill data

Top reservoir depth	10,500 ft.
Current reservoir pressure	5570 psi
Original reservoir pressure	6750 psi
Plug depth	10,350 ft.
Packer depth	10,330 ft.
Sliding sleeve or sliding side door (SSD) depth	10,300 ft.
$5\frac{1}{2}$ in. tubing ID	4.670 in.
$9\frac{5}{8}$ in. casing ID	8.535 in.
SITP	1900 psi
SICP	0 psi (before sleeve open)
Packer fluid gradient	0.67 psi/ft. (12.88 ppg)
Oil gradient	0.35 psi/ft. (6.73 ppg)
Planned overbalance	200 psi

original reservoir pressure. In all other respects, the well is identical to the previous example (Table 7.4).

Since the tubing head pressure, oil density values, and reservoir pressure are identical to those used in the first example, it follows that the kill fluid requirement (0.5564 psi/ft.) is also the same. Tubing and annulus volumes are also unchanged. However, having a heavier fluid present in the annulus will change the pressure profiles observed during the kill and the differential pressure across the SSD before it is opened.

7.3.9.1 Calculate tubing and casing pressure before and after opening the sliding side door

- Tubing pressure at the SSD before it is opened is 5500 psi (as per Example 1).
- Annulus pressure at the SSD:

$$10,300 \text{ ft.} \times 0.67 \text{ psi/ft.} = 6901 \text{ psi.}$$

- Differential pressure from annulus to tubing is therefore $6901 - 5500$ psi = 1401 psi.
- Since differential pressure across the SSD is high, opening the ports would risk blowing the wireline toolstring up the hole. To eliminate this risk, pressure must be applied to the tubing to balance casing and tubing pressure. Before opening the sleeve, tubing head pressure (1900 psi) must increase by an amount equivalent to the differential at the SSD, i.e., 1401 psi. Tubing head pressure is increased to 3301 psi.
- Tubing pressure at the sleeve becomes: (10,300 ft. × 0.3395 psi/ft.) + 3301 psi = 6901 psi, which is on balance with the annulus pressure at the sleeve before opening.

Increasing tubing pressure to equalize between tubing and casing will result in more overbalance than the 200 psi used to keep the well dead, in this case 1401 psi. The ability to apply tubing pressure is reliant on the integrity of the mechanical plug. A leak past the plug would result in fluid losses to the formation with the associated risk of formation damage. It could be argued that if there is confidence in the integrity of the plug there is no need to maintain the 200 psi overbalance at the SSD. Wellhead pressure could be bled-off after opening the SSD. However, should the plug leak, the resultant hydrocarbon influx would compromise the kill operation.

7.3.9.2 Calculate the fluid level (H) in the annulus if the tubing head pressure is bled to 0 psi

If, after opening the sleeve, the tubing pressure was bled down, heavy fluid in the annulus would U tube until hydrostatic pressure is equalised. The oil water contact depth in the tubing and fluid level in the annulus can be calculated (Fig. 7.3):

- Annulus capacity: $8.535^2 - 5.5^2/1029.4 = 0.0414$ bbls/ft.
- Tubing capacity: $4.678^2/1029.4 = 0.02125$ bbls/ft.

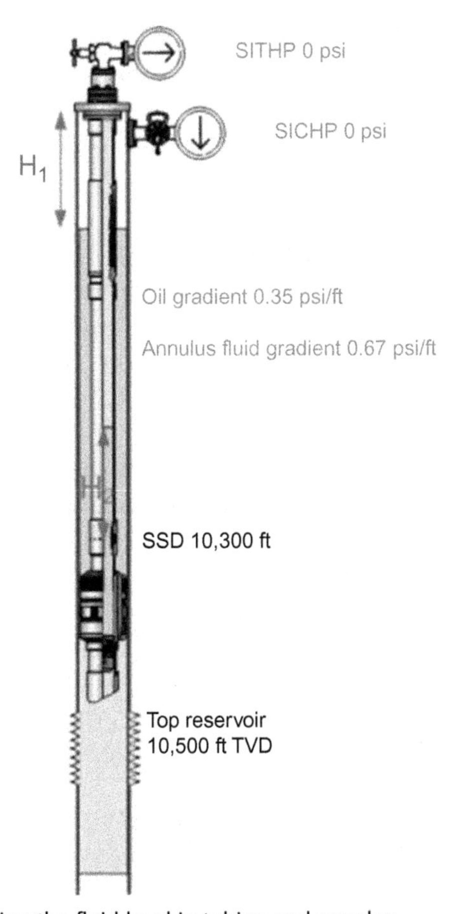

Figure 7.3 Calculating the fluid level in tubing and annulus.

- Annulus to tubing capacity ratio is: 0.0414/0.02125 = 1.95.
  - H2 = 1.95H1.

Balancing HP enables the fluid level (H1) in the annulus to be calculated.

- 0.67 x (10,300 − H1) = (1.95H1 × 0.67) + 0.35(10,300 − 1.95H1).
- 6901 − 0.67H1 = 1.31H1 + 3605 − 0.68H1.
- 6901−3605 = 1.31H1 − 0.68H1 + 0.67H1.
- 3296 = 1.30H1.
- H1 = 3296/1.30 = 2535 ft.
- H2 = 2535 ft. × 1.95 = 4943 ft.
- Tubing HP at SSD:
- 10,300-(4943 × 0.35) + (4943 × 0.67) = 5187.

Pressure at the SSD before bleeding off was 5500 psi. After bleeding off the tubing head pressure there is a 313 psi pressure differential across the plug (from below) (5500-5187). If the plug leaked a hydrocarbon influx would result. This could be prevented by holding back pressure on the tubing head.

7.3.9.3 Pumping the kill weight fluid

Before pumping begins, the fluid level in the annulus is 2535 ft. below the surface. It should mean that the volume required to completely fill the annulus is (0.0414 × 2535) 105 bbls. If the choke were closed this would be the case. However, with the choke open it is more complicated.

As kill fluid (0.5564 psi/ft.) begins to fill the annulus, HP initially begins to increase as the height of the fluid column increases. Increasing HP in the annulus will cause a U tube effect (Fig. 7.4), forcing annulus fluid though the SSD and into the tubing. As the pumping continues three changes happen simultaneously (Fig. 7.4):

- Kill fluid filling the annulus increases HP pushing packer fluid through the SSD into the tubing.
- The OWC in the tubing (originally at 5357 ft.) moves towards the surface as oil is displaced from the well.

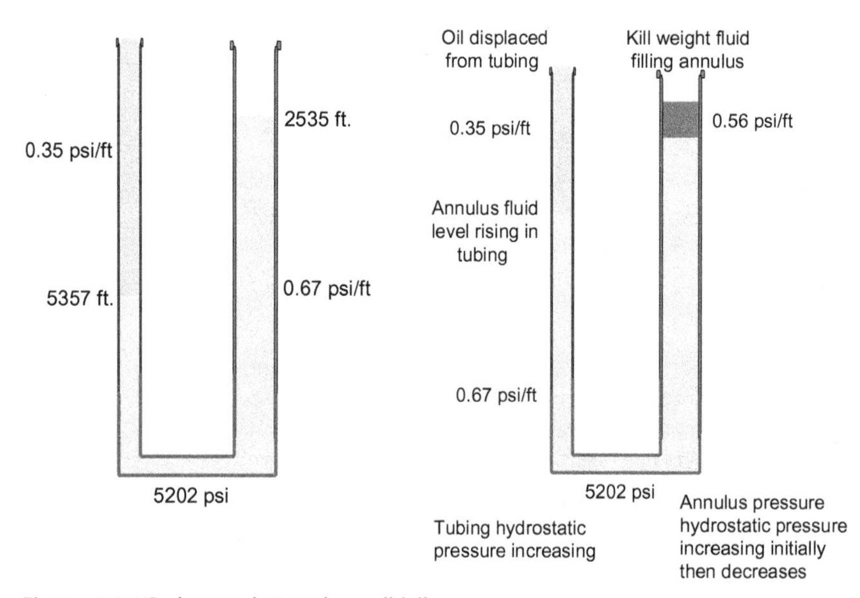

Figure 7.4 HP change during the well kill.

- Since oil is being displaced by packer fluid, HP in the tubing increases.

Tubing HP increases and reaches a maximum when the tubing is completely full of packer fluid: 10,300 ft. × 0.67 psi/ft. = 6901 psi. Annulus pressure must be increased to compensate since the kill fluid is lighter than the packer fluid.

- The oil water contact in the production tubing is at (10,300 − 4943 ft.) = 5357 ft. To get packer fluid to surface 114.21 bbls of oil must be displaced (5357 ft. × 0.02125 bbls/ft.).
- When the tubing is filled with packer fluid, HP at the SSD is 10,300 psi × 0.67 psi/ft. = 6901 psi.
- After pumping 114.21 bbls of kill fluid into the annulus, the kill fluid/packer fluid level is (114.21/0.0414 bbls/ft.) 2758 ft.
- Annulus HP is 2758 ft. × 0.56 psi/ft. + (10,300 − 2758) × 0.67 psi/ft. = 6597 psi.

Tubing HP exceeds annulus HP by 6901 − 6597 = 304 psi, meaning that 304 psi casing head pressure is required to balance the HP in the tubing. In fact, circulation would require a higher pump pressure as frictional pressure drop must be included (Fig. 7.5).

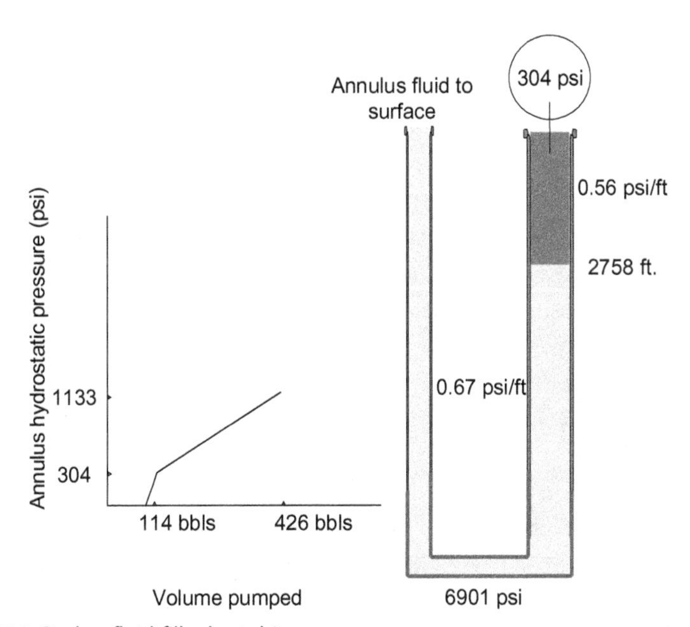

Figure 7.5 Packer fluid fills the tubing.

Although the point at which annulus HP peaks and begins to drop can be calculated, the variables involved mean that the calculations are complex and unlikely to be carried out at the well site. However, in this example, a casing pressure should be observed before 114.21 bbls have been pumped. Failure to observe wellhead pressure after pumping this volume would indicate that the plug was leaking and fluid was being lost to the formation.

7.3.9.4 Kill fluid at the sliding side door

- When kill fluid fills the annulus, pressure will be as follows:
  - Annulus HP: 10,300 ft. × 0.5546 psi/ft. = 5712 psi.
  - Tubing HP: 10,300 ft. × 0.67 psi/ft. = 6901.
  - Pump pressure required to circulate fluid: 6901-5712 = 1189 psi.
  - Annulus volume is 10,300 ft. × 0.0414 bbls/ft. = 426.42 bbls.

7.3.9.5 Displace tubing to kill weight fluid

As the lighter kill weight fluid passes through the SSD and replaces the heavy packer fluid in the tubing, hydrostatic pressure will drop. Once the tubing is full of kill weight fluid, the HP at the SSD should be the required 200 psi overbalance.

- Before kill fluid enters the tubing, HP is 1189 psi higher than the annulus HP:
  - 1189 psi/10,300 ft. = 0.11 psi/ft. density difference between the fluid in the tubing (0.67 psi/ft.) and the fluid in the annulus (0.5564 psi/ft.).
  - 1 bbl of fluid pumped will fill 45.05 ft. of tubing reducing the HP (and therefore casing head pressure) by 45.05 × 0.11 = 5.07 psi.
  - Tubing volume is 10,300 ft. × 0.02125 bbls/ft. = 218.87 bbls.
- Casing pressure against volume pumped can be plotted. Since the kill is carried out using overbalance fluids, and as the well is plugged, there is no tubing head pressure to record (Fig. 7.6).

7.3.10 Example 3 reverse circulation: deviated well with tapered string

Well kill calculations associated with vertical wells are relatively straightforward, since the relationship between pressure and depth remain constant throughout. In deviated wells, pressure change per volume pumped changes with changing well angle. As the well angle builds (increases) more kill fluid must be pumped in order to achieve pressure reduction at

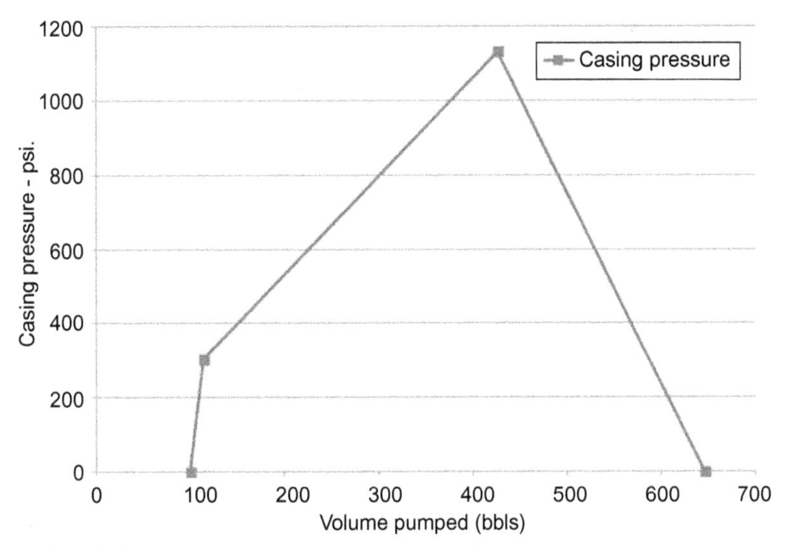

Figure 7.6 Casing pressure versus volume pumped.

the surface. It is important to emphasize that pressure must be calculated in TVD, whilst all volumes are calculated in measured depth (MD), sometimes called "along hole" (AH). When making numerous calculations that involve switching between vertical depth (pressure) and MD (volume), it is easy to make errors.

Additional complexity is introduced if there are diameter changes in either tubing or casing, as the fill rate will vary. For example, 1 bbl of fluid fills 65.78 ft. of 4½ in. 12.6 lb/ft. tubing, but only 43.1 ft. of 5½ in. 17 lb/ft. tubing. Where a kill graph is used, the slope of the pressure versus volume plot will change at each diameter change.

In a well that both deviates from the vertical and has one or more changes in casing and/or tubing diameter, a sequential approach is required to enable a kill graph to be prepared.

In this example, a deviated well is to be killed by reverse circulation. The well has been completed with a 5½ in. × 4½ in. tubing string inside 10¾ in. × 9⅝ in. casing. A plug has been set below the packer. The well is to be killed with a 200 psi overbalance at the SSD (Table 7.5 and Fig. 7.7).

To be able to prepare a kill schedule, it is necessary to determine the following:

- TVD for relevant points in the completion (components, diameter changes, and fluid interface).
- Reservoir pressure.

Table 7.5 Well data

SITP	2750 psi
Tubing gas oil contact	3500 ft. MD
Gas gravity	0.65 sg
Oil gradient	0.35 psi/ft.
Original packer (annulus) fluid	Inhibited water 0.433 psi/ft. 8.33 ppg
Casing 1 (upper)	10¾ in. 65.7 lb/ft. 9.560 in. ID
Casing 2 (lower)	9⅝ in. 53.5 lb/ft. 8.958 in. ID
5½ in. × 4½ in. cross-over	4800 ft. MD
SSD	13,060 ft. MD
Packer	13,100 ft. MD
Plug (in tail-pipe)	13,115 ft. MD
Reservoir datum (top reservoir)	13,520 ft. MD
TD	14,000 ft. MD

- Density (weight) of fluid is required to give a 200 psi overbalance at the SSD.
- Pressure differential at the SSD before it is opened.

7.3.10.1 Obtaining measured versus vertical depth data

For most wells, detailed trajectory data is easy to obtain (or at least it should be). Data may be obtained as a hard copy in the form of a spreadsheet. Multiple data points list MD in feet or meters, inclination (in degrees), azimuth (in degrees) and TVD in feet or meters.

More commonly, trajectory data is stored and manipulated using proprietary software such as Halliburton's "Compass." The use of software simplifies and speeds up the process of converting MD to TVD (and *vice versa*) (Fig. 7.8).

Completion schematics should list the MD of the key components and features in the well. Some, but not all, schematics will list the corresponding TVDs. If vertical depth is unknown, it can be obtained using software or from a deviation survey. If survey data is used, interpolation between two survey points is required unless known depth falls exactly on one of the data points. A deviation survey for this example is provided in Table 7.6.

For this example, the TVDs for each of the relevant points in the completion are provided in Table 7.7.

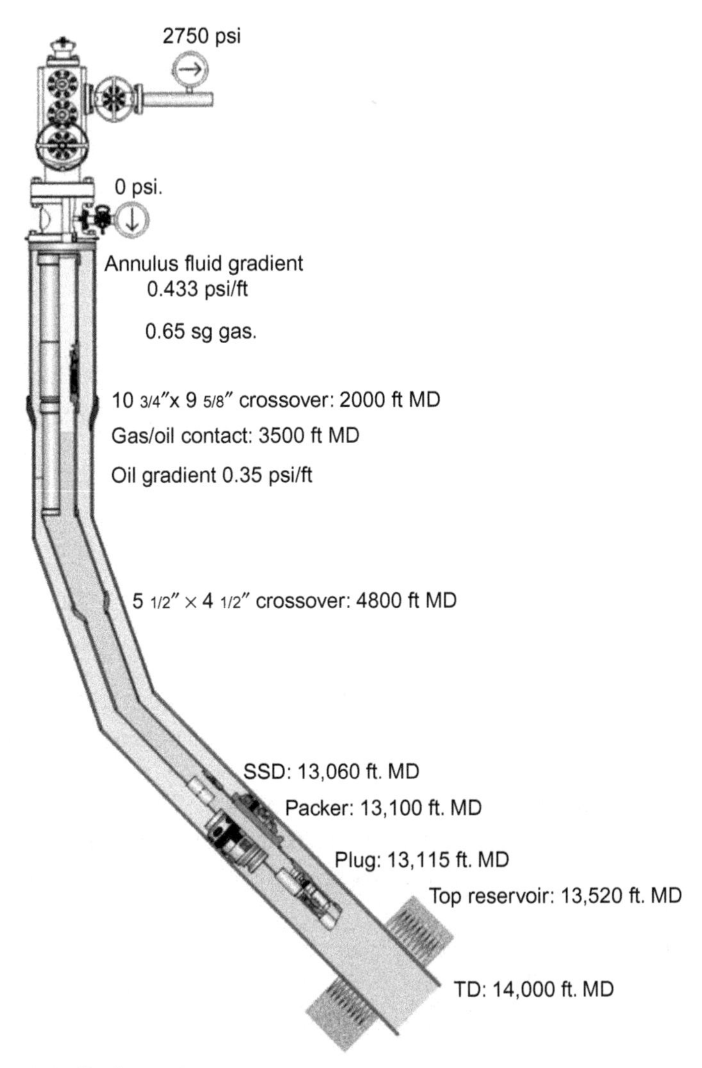

Figure 7.7 Well schematic.

7.3.10.2 Calculate reservoir pressure

Reservoir pressure is not given, and therefore must be calculated from the hydrostatic head of the gas and oil in the tubing string.

- Gas hydrostatic:
  - There is a column of 0.65 SG gas from surface to 3500 ft. TVD. The gas correction table provided can be used to determine the HP at the base of the gas column (Table 7.8).

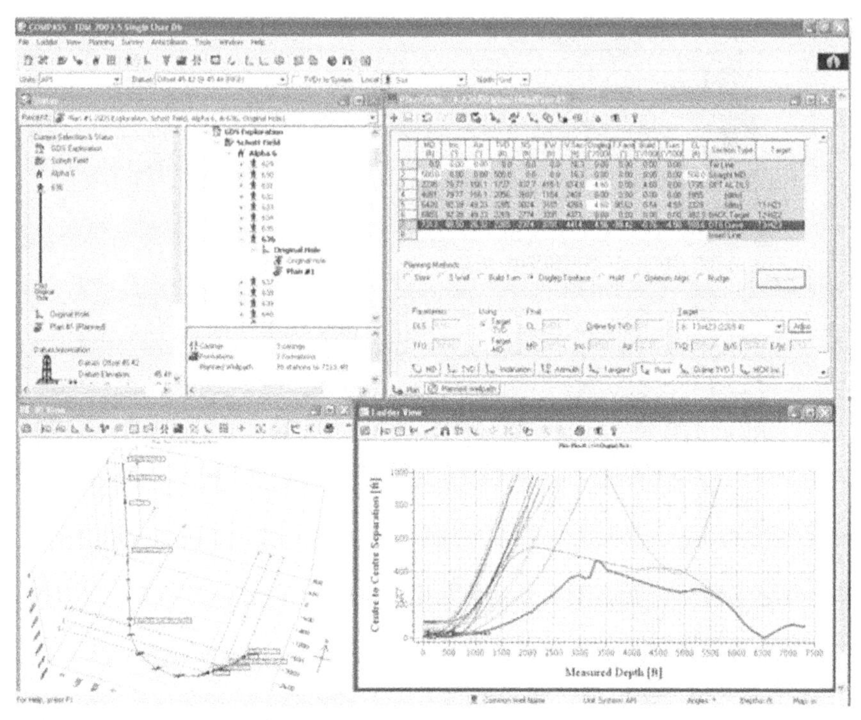

Figure 7.8 Screen-dump from "Compass," Halliburton's well trajectory software.

- There is no gas correction value for 3500 ft. The correction factor for 0.65 sg gas at 3000 ft. is **1.070** and **1.094** at 4000 ft. Interpolation between the two known factors gives a gas correction factor of 1.082.
- SITP is 2750 psi. Pressure at the bottom of the gas column (3500 ft. TVD) is SITP × the correction factor, 2750 × 1.082 = 2975.5 psi → 2976 psi.

- Oil hydrostatic:
  - The oil column is reservoir depth to GOC.
    - 11,256 ft. TVD-3500 ft. TVD = 7756 ft. TVD.
  - Oil gradient 0.35 psi/ft. × 7756 ft. = 2714.6 psi → 2715 psi.
  - 2715 psi (oil hydrostatic) + 2967 psi (gas hydrostatic) = 5691 psi.
- Reservoir pressure = 5691 psi.

7.3.10.3 Calculate kill weight fluid needed to give 200 psi overbalance at the sliding side door

- Gas hydrostatic is known already, 2976 psi.
- The oil column is 10,925 ft. TVD-3500 ft. TVD = 7425 ft. TVD.

Table 7.6 Deviation survey

Data-entry mode	MD (ft.)	INC (deg.)	AZ (deg.)	TVD (ft.)	DLS (deg./100 ft.)	Max DLS (deg./100 ft.)	Vsection (ft.)	Departure (ft.)
MD-INC-AZ	0	0.00	0.00	0			0	0
MD-INC-AZ	4000	0.00	0.00	4000	0.00	0.00	0	0
MD-INC-AZ	4100	2.00	0.00	4100	2.00	2.00	2	2
MD-INC-AZ	4200	4.00	0.00	4200	2.00	2.00	7	7
MD-INC-AZ	4300	6.00	0.00	4299	2.00	2.00	16	16
MD-INC-AZ	4400	8.00	0.00	4399	2.00	2.00	28	28
MD-INC-AZ	4500	10.00	0.00	4497	2.00	2.00	44	44
MD-INC-AZ	4600	12.00	0.00	4596	2.00	2.00	63	63
MD-INC-AZ	4700	14.00	0.00	4693	2.00	2.00	85	85
MD-INC-AZ	4800	16.00	0.00	4790	2.00	2.00	111	111
MD-INC-AZ	4900	18.00	0.00	4885	2.00	2.00	140	140
MD-INC-AZ	5000	20.00	0.00	4980	2.00	2.00	173	173
MD-INC-AZ	5100	22.00	0.00	5073	2.00	2.00	209	209
MD-INC-AZ	5200	24.00	0.00	5165	2.00	2.00	248	248
MD-INC-AZ	5300	26.00	0.00	5256	2.00	2.00	290	290
MD-INC-AZ	5400	28.00	0.00	5345	2.00	2.00	335	335
MD-INC-AZ	5500	30.00	0.00	5432	2.00	2.00	384	384
MD-INC-AZ	5600	32.00	0.00	5518	2.00	2.00	435	435
MD-INC-AZ	5700	34.00	0.00	5602	2.00	2.00	490	490
MD-INC-AZ	5800	36.00	0.00	5684	2.00	2.00	547	547
MD-INC-AZ	5900	38.00	0.00	5764	2.00	2.00	607	607
MD-INC-AZ	6000	40.00	0.00	5841	2.00	2.00	670	670
MD-INC-AZ	6100	42.00	0.00	5917	2.00	2.00	736	736
MD-INC-AZ	6200	44.00	0.00	5990	2.00	2.00	804	804
MD-INC-AZ	13,060	44.00	0.00	10,925	0.00	0.00	5569	5569
MD-INC-AZ	13,100	44.00	0.00	10,953	0.00	0.00	5597	5597
MD-INC-AZ	13,115	44.00	0.00	10,964	0.00	0.00	5608	5608
MD-INC-AZ	13,120	44.00	0.00	10,968	0.00	0.00	5611	5611
MD-INC-AZ	13,520	44.00	0.00	11,256	0.00	0.00	5889	5889
MD-INC-AZ	14,000	44.00	0.00	11,601	0.00	0.00	6222	6222

Table 7.7 True vertical depth of key completion components

Item	MD (ft.)	TVD (ft.)
Tubing hanger	0	0
10¾ in. × 9⅝ in. cross-over	2000	2000
Gas oil contact (before kill)	3500	3500
Start of build section	4000	4000
5½ in. × 4½ in. cross-over	4800	4790
End of build section	6200	5990
SSD	13,060	10,925
Packer	13,100	10,953
Plug	13,115	10,964
WLEG (tubing end)	13,120	10,968
Top reservoir	13,520	11,256
TD	14,000	11,601

Table 7.8 Gas correction factors

Depth (ft.)	Gas SG—relative to air										
	0.5	**0.55**	**0.6**	**0.65**	**0.7**	**0.75**	**0.8**	**0.85**	**0.9**	**0.95**	**1**
1000	1.018	1.019	1.021	1.023	1.025	1.026	1.028	1.030	1.032	1.034	1.035
2000	1.035	1.039	1.043	1.046	1.050	1.053	1.057	1.061	1.064	1.068	1.072
3000	1.053	1.059	1.064	**1.070**	1.076	1.081	1.087	1.093	1.098	1.104	1.110
4000	1.072	1.079	1.087	**1.094**	1.102	1.110	1.117	1.125	1.133	1.141	1.149
5000	1.091	1.100	1.110	1.119	1.129	1.139	1.149	1.159	1.169	1.179	1.190
6000	1.110	1.121	1.133	1.145	1.157	1.169	1.181	1.194	1.206	1.219	1.232
7000	1.129	1.143	1.157	1.171	1.185	1.200	1.215	1.229	1.244	1.260	1.275
8000	1.149	1.165	1.181	1.198	1.215	1.232	1.249	1.266	1.284	1.302	1.320

- Oil gradient 0.35 psi/ft. × 7425 ft. = 2598.75 psi → 2599 psi.
- 2599 psi (oil hydrostatic) + 2967 psi (gas hydrostatic) = 5566 psi.
- Tubing pressure at the SSD 5566 psi + 200 psi overbalance = 5766 psi.
- 5766 psi/10,925 ft. TVD = 0.528 psi/ft. → 0.53 psi/ft. (10.2 ppg).

7.3.10.4 Calculate pressure differential at the sliding side door before and after opening

- Annulus pressure at the SSD:
  - 10,925 × 0.433 = 4731 psi.
- Tubing pressure at the SSD has already been calculated (5566 psi).
- Pressure differential is 5566 − 4731 = 835 psi from tubing to annulus. Pressure differential can be equalized by applying 835 psi to the casing.
- Pressure at the SSD after the kill will be: 10,925 ft. × 0.53 psi/ft. = 5790 psi. This is 224 psi above well pressure at the SSD, 24 psi more than the required 200 psi overbalance.
  - The additional 24 psi is because of rounding up the brine fluid gradient.
- After opening the SSD, 1059 psi pressure needs to be applied to the casing to hold the required 200 psi overbalance.
  - 5790 psi (HP at the end of the kill) − 4731 psi (casing HP at the SSD) = 1059 psi.
- At the start of pumping, the pressure either side of the SSD is 5790 psi, as 1059 psi has been applied to the casing. Backing out the HP in the tubing, wellhead pressure would in theory read:
  - 7425 ft. oil at 0.35 psi ft. = 2599 psi.
  - The HP of 3500 ft. of 0.65 sg gas is 0.082 × 3500 = 287 psi.
  - 5790 − 2599 − 287 = 2904 psi.

7.3.10.5 Calculate tubing and annulus capacities

- 5½ in. tubing: $4.892^2/1029.4 = 0.0232$ bbls/ft.
- 4½ in. tubing: $3.958^2/1029.4 = 0.0152$ bbls/ft.
- 10¾ in. casing × 5½ in. tubing: $(9.560^2-5.5^2)/1029.4 = 0.0593$ bbls/ft.
- 9⅝ in. casing × 5½ in. tubing: $(8.535^2-5.5^2)/1029.4 = 0.0414$ bbls/ft.
- 9⅝ in. casing × 4½ in. tubing: $(8.535^2-4.5^2)/1029.4 = 0.051$ bbls/ft. (Table 7.9).

To enable tubing head pressure to be plotted against fluid pumped, it will be necessary to calculate tubing head pressure and casing pressure at several stages throughout the kill. These must include any point at which

Table 7.9 Tubing and annulus volume

Section	Volume
Annulus	
Wellhead to $10^3/_4$ in. × $9^5/_8$ in. cross-over	2000 ft. × 0.0593 bbls/ft. = 118.6 bbls
$9^5/_8$ in. × $5^1/_2$ in. annulus (2000 to 4800 ft.)	2800 ft. × 0.0414 bbls/ft. = 115.9 bbls
$9^5/_8$ in. × $4^1/_2$ in. annulus (4800 to 13,060) SSD	8260 ft. × 0.051 bbls/ft. = 421.2 bbls
Annulus volume	655.7 bbls
Tubing	
$5^1/_2$ in. tubing (0 to 4800 ft.)	4800 ft. × 0.0232 bbls/ft. = 111.36 bbls
$4^1/_2$ in. tubing volume (4800 to 13,060 ft.) SSD	8260 ft. × 0.0152 = 125.55 bbls
Tubing volume	236.91 bbls
Well volume (full circulation)	892.61 bbls

a density change reaches a change in tubing diameter. The steps calculated will be:

1. Oil reaches surface (all the gas has been removed from the tubing).
2. The 0.53 psi/ft. kill fluid in the annulus reaches the $10^3/_4$ in. × $9^5/_8$ in. casing cross-over.
3. The 0.433 psi/ft. packer fluid fills the tubing between the $5^1/_2$ in. × $4^1/_2$ in. cross-over and the SSD.
4. The 0.53 psi/ft. kill fluid in the annulus reaches the $5^1/_2$ in. × $4^1/_2$ in. tubing cross-over.
5. The 0.433 psi/ft. packer fluid is completely filling the tubing.
6. Kill fluid fills the annulus.
7. Kill fluid in the tubing reaches the $5^1/_2$ in. × $4^1/_2$ in. cross-over.
8. Full circulation.

Step 1, oil to surface:

- Tubing volume to GOC 3500 ft. × 0.0232 bbls/ft = 81.2 bbls.
- Kill fluid level in annulus when oil in tubing reaches surface.
  - 81.2 bbls/0.0593 bbls/ft. ($10^3/_4$ in. × $5^1/_2$ in. annulus volume) = 1369.3 ft. This is above the $10^3/_4$ in. × $9^5/_8$ in. cross-over and in the vertical section of the well.
- Packer fluid level in the tubing.
  - 81.2 bbls packer fluid is in the tubing.
  - 81.2 bbls/0.0152 bbls/ft. ($4^1/_2$ in. tubing volume) = 5342.10 ft. above the SSD.
  - OWC is at 13,060 ft. MD − 5342.10 ft. MD = 7718 ft. MD
- Vertical depth of oil/packer fluid contact in tubing when MD is 7718 ft. = 7082 ft.

- Required kill pressure at SSD = 5790 psi.
- Tubing hydrostatic:
  - Oil hydrostatic 7082 ft. TVD × 0.35 psi/ft = 2479 psi.
    Hydrostatic of annulus fluid in tubing (10,925 ft. TVD − 7082 ft. TVD) × 0.433 psi/ft. = 1664 psi.
- Wellhead pressure 5790 − (2479 + 1664) = 1647 psi.
- Annulus hydrostatic:
  - 1369.3 ft. × 0.53 psi/ft. (kill fluid) = 726 psi.
  - (10,925-1369.3) × 0.433 psi/ft. packer fluid = 4138 psi.
  - Casing pressure 5790 psi-(4138 psi + 726 psi) = 926 psi (Fig. 7.9).

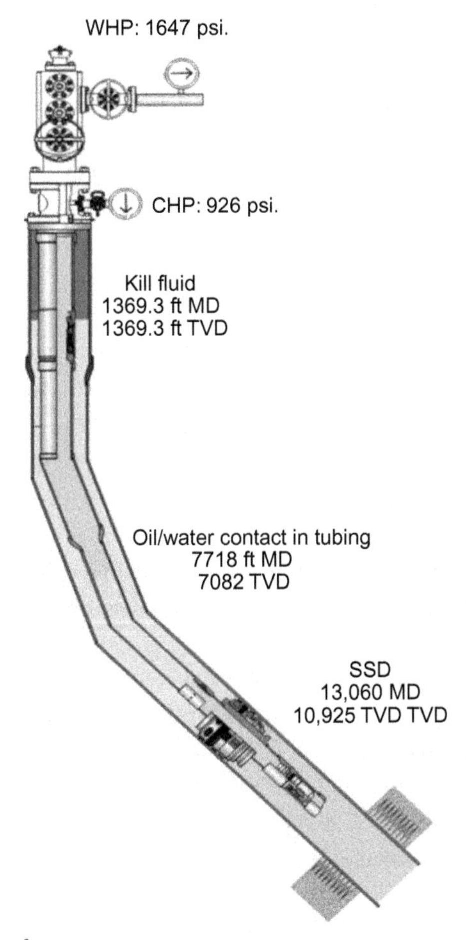

Figure 7.9 Oil to surface.

Step 2 Kill fluid in the annulus at the $10^3/_4$ in. × $9^5/_8$ in. *cross-over:*

- The kill weight fluid was at 1369.3 ft. MD when the oil reached surface. The cross-over depth is 2000 ft. MD.
  - 2000 − 1369.3 = 630.7 ft. × 0.0593 bbls/ft. = 37.4 bbls.
  - An additional 37.4 bbls needs to be pumped to get kill fluid to the cross-over.
- Oil/packer fluid contact in tubing:
  - 37.4 bbls/0.0152 bbl/ft. ($4^1/_2$ in. tubing capacity) = 2461 ft.
  - 7718-2461-5257 ft. MD.
- Annulus pressure:
  - 2000 ft. × 0.53 psi/ft. = 1060 psi.
  - (10,925 − 2000 ft.) × 0.433 psi/ft. = 3865 psi.
- Casing head pressure: 5790-(1060 + 3865) = 865 psi.
- Tubing pressure:
  - MD of the oil/packer fluid contact is 5257 ft. Interpolation using the data supplied in the well deviation survey gives a TVD of 5217 ft.
  - Oil hydrostatic 5217 ft. × 0.35 psi/ft. = 1826 psi.
  - Packer fluid hydrostatic (10,925 − 5217 ft.) × 0.433 psi/ft. 2472 psi.
  - 5790 − (2472 + 1826 psi) = 1492 psi.

Step 3, oil/packer fluid contact in the tubing reaches the $5^1/_2$ in. × $4^1/_2$ in. *Cross-over:*

- The $5^1/_2$ in. × $4^1/_2$ in. tubing cross-over is at 4800 ft. MD (4790 ft. TVD).
- Tubing head pressure:
  - Oil hydrostatic 4790 ft. × 0.35 psi/ft. = 1677 psi.
  - Packer fluid hydrostatic (10,925 − 4790 ft) × 0.433 psi/ft. = 2656 psi.
  - 5790 − (1677 + 2656) = 1457 psi.
- Annulus pressure:
  - Volume pumped 125.55 bbls ($4^1/_2$ in. tubing volume).
  - Annulus volume to $10^3/_4$ in. × $9^5/_8$ in. cross-over = 118.6 bbls.
  - 125.55-118.6 = 6.95 bbls.
  - 6.95 bbls/0.0414 bbls/ft = 168 ft.
  - Kill fluid at 2000 ft. + 168 ft. = 2168 ft. MD (2168 ft. TVD).
  - Kill fluid hydrostatic 2168 ft. × 0.53 psi/ft = 1149 psi.
  - Packer fluid hydrostatic (10,925 − 2168 ft.) × 0.433 psi/ft = 3792 psi.
- Casing head pressure: 5790-(1149 + 3792) = 849 psi.

Step 4, kill fluid in annulus at $5^1/_2$ in. × $4^1/_2$ in. *tubing cross-over:*

– Annulus pressure:
  - Kill fluid hydrostatic 4790 ft. × 0.53 psi/ft. = 2539 psi.
  - Packer fluid hydrostatic (10,925 − 4790 ft.) × 0.433 psi/ft. = 2656 psi.
  - 5790 − (2539 + 2656) = 595 psi.

- Tubing pressure:
  - Volume pumped 234.5 bbls (10 3/4" x 5 1/2" + 9 5/8" x 5 1/2" annulus volume).
  - $4\frac{1}{2}$ in. tubing capacity = 125.55 bbls.
  - 234.5 − 125.55 bbls = 108.95 bbls filling the $5\frac{1}{2}$ in. tubing.
  - 108.95 bbls/0.0232 bbls/ft. ($5\frac{1}{2}$ in. tubing capacity) = 4696 ft. of fluid above the cross-over.
  - Oil packer fluid contact in tubing is 4800 − 4696 ft. MD = 104 ft. MD.
  - 104 ft. MD = 104 ft. TVD.
  - Oil hydrostatic 104 ft. × 0.35 psi/ft. = 36 psi.
  - Packer fluid hydrostatic (10,925 − 104 ft.) × 0.433 = 4685 psi.
  - 5790 − (36 + 4685) = 1069 psi

Step 5: Tubing filled with annulus fluid:

- Tubing head pressure:
  - 10,925 ft. × 0.433 psi/ft. = 4731 psi.
  - 5790 − 4731 = 1059 psi.
- Casing head pressure:
  - Volume pumped 236.91 bbls (tubing volume).
  - Fluid level in casing. Volume to $5\frac{1}{2}$ in. × $4\frac{1}{2}$ in. cross-over = 234.5 bbls.
  - 236.91 − 234.5 = 2.41 bbl.
  - 2.41 bbls/0.051 bbls/ft. ($9\frac{5}{8}$ in. × $4\frac{1}{2}$ in. annulus) = 47 ft. MD.
  - Kill fluid at 4800 ft. MD + 47 ft. MD = 4847 ft. MD.
  - 4847 ft. MD interpolation (using the data supplied in Table 7.1—well deviation survey) gives a TVD of 4835 ft.
  - Kill fluid hydrostatic 4835 ft. × 0.53 psi/ft. = 2563 psi.
  - Annulus fluid hydrostatic (10,925 − 4835) × 0.433 psi/ft. = 2638 psi.
  - 5790 − (2563 + 2638) = 589 psi (Fig. 7.10).

Step 6: Kill fluid reaches SSD:

- Volume pumped = annulus volume (655.7 bbls).
- Annulus pressure:
  - 10,925 × 0.53 psi/ft. = 5790 psi.
  - Casing pressure 0 psi.
- Tubing head pressure:
  - 10,925 ft. × 0.433 psi/ft. = 4731 psi.
  - 5790 − 4731 = 1059 psi.

Step 7, Kill fluid reaches $5\frac{1}{2}$ in. × $4\frac{1}{2}$ in. cross-over in tubing:

- Volume pumped = 655.7 bbls (annulus volume) + 125.55 bbls ($4\frac{1}{2}$ in. tubing volume) = 781.25 bbls.

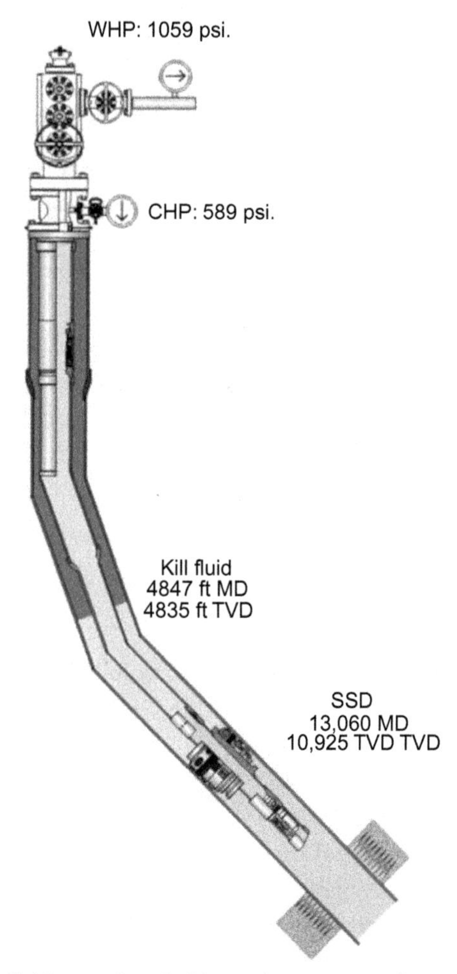

Figure 7.10 Packer fluid at surface (tubing volume pumped).

- Annulus pressure:
  - Kill fluid hydrostatic: 10,925 × 0.53 psi/ft. = 5790 psi.
  - Casing pressure: 0 psi.
- Tubing head pressure:
  - Packer fluid hydrostatic: 4790 ft. TVD × 0.433 psi/ft. = 2074 psi.
  - Kill fluid hydrostatic (10,925 ft. TVD − 4790 ft. TVD) × 0.53 = 3156 psi.
  - 5790 − (2074 + 3156) = 560 psi.

Step 8, Full circulation:

- Volume pumped = 892.61 bbls
- Annulus pressure:
  - Kill fluid hydrostatic: 10,925 × 0.53 psi/ft. = 5790.
  - Casing pressure: 0 psi.
- Tubing head pressure:
  - Kill fluid hydrostatic: 10,925 ft. × 0.53 psi/ft. = 5790 psi.
  - 0 psi tubing head pressure (Fig. 7.11 and Table 7.10).

Data from Table 7.10 is used to prepare a kill plot (Fig 7.12).

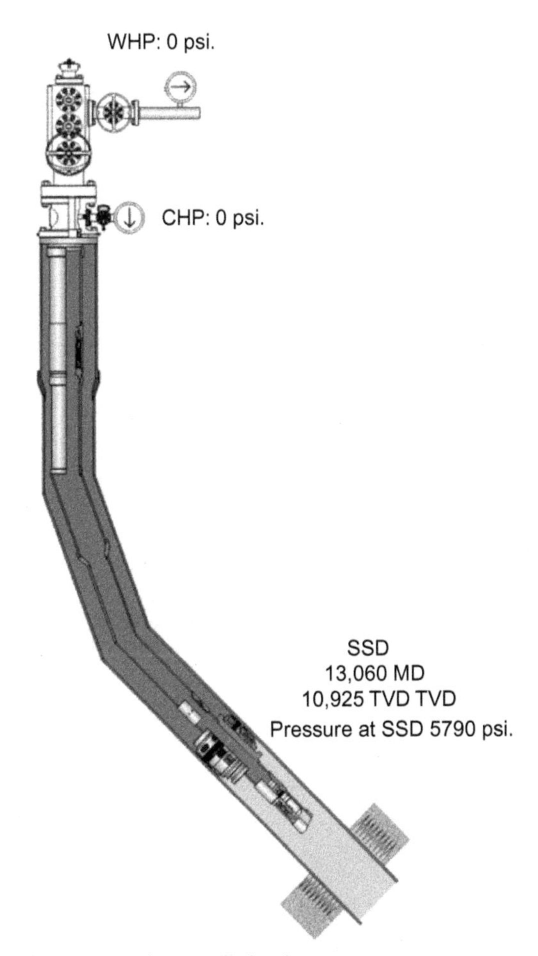

Figure 7.11 Circulation complete: well dead.

Table 7.10 Summary of well kill steps

Step	Volume pumped (bbls)	Cumulative volume (bbls)	Tubing head pressure (psi)	Casing head pressure (psi)
Before kill (SSD open) and casing pressure applied	0	0	2904	1059
1. Oil to surface	81.2	81.2	1664	926
2. Kill fluid in the annulus at 10¾ in. × 9⅝ in. cross-over	37.4	118.6	1492	865
3. Annulus fluid in tubing at 5½ in. × 4½ in. cross-over	6.95	125.55	1457	849
4. Kill fluid in annulus at depth of 5½ in. × 4½ in. cross-over	108.95	234.5	1069	595
5. Tubing filled with annulus fluid	2.42	236.91	1059	589
6. Kill fluid at SSD	418.79	655.7	1059	0
7. Kill fluid at 5½ in. × 4½ in. cross-over	125.55	781.25	560	0
8. Full circulation	111.36	892.61	0	0

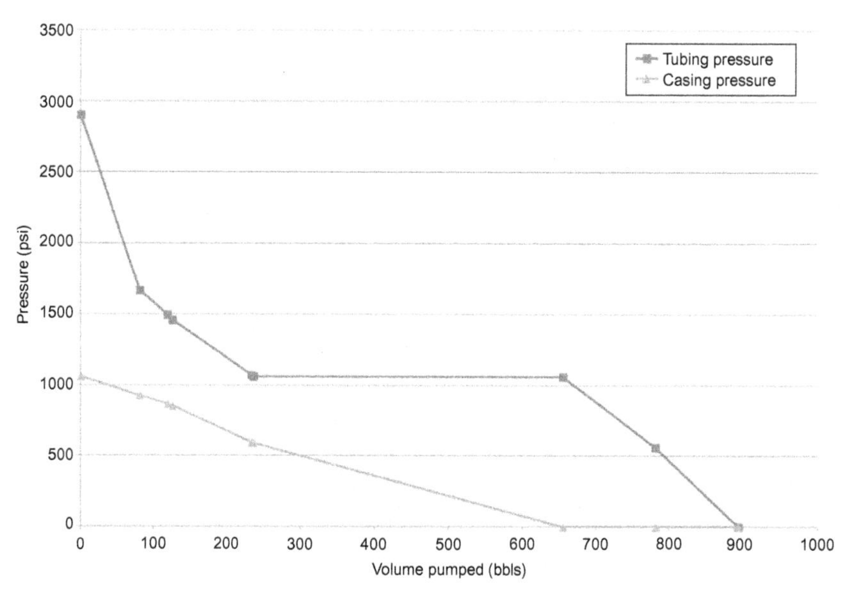

Figure 7.12 Plot of volume pumped versus tubing and casing pressure.

7.4 NON-CIRCULATING KILL: BULLHEAD

Bullheading is the most widely used kill method when preparing a well for a workover or intervention. Although principally intended for use where circulation is not possible, bullheading is frequently used where circulation is possible, as it is simple, easy to perform, and there is no need to handle hydrocarbons at the surface. During a bullhead, produced fluids are pushed back into the formation as kill fluid is pumped down the tubing (Fig 7.13).

Unlike a circulating kill, a bullhead does not need to be carried out at a constant BHP. Instead, BHP must be kept above pore (reservoir) pressure and high enough to inject the tubing contents (produced fluids) back into the formation, whilst not exceeding fracture pressure. For a bullhead to succeed, the reservoir needs to have good permeability, since the velocity of the kill fluid (as it is pumped down the tubing) must be higher than the upwards migration velocity of the lower density produced fluids. As gas migration velocity in clear brine can be as much as 6000 ft./min,[2] there is a widespread reluctance to use bullheading for killing gas wells. However, modeling shows that a combination of viscous pills and high rates of injection can overcome gas migration and sweep hydrocarbons back into the formation,[3] although as tubing size and well angle increase, the chances of a successful outcome diminish. The nature of the bullhead kill makes formation damage more likely than when using circulation. Despite these limitations, there are many occasions where a bullhead kill is preferred:

- The configuration of the completion prevents the creation of a circulation path.
- A well that has a leak high in the tubing, making it difficult to get circulated fluid deep enough into the well.
- A well with no tubing installed, or one that has kicked after the tubing has been removed.
- Any well that has a significant height difference between the lowest possible point of circulation and the top of the reservoir, i.e., a well with a long liner, or a well with a single string multiple zone completion across many different reservoir layers.
- A well with weak or corroded casing, or where there are concerns about wellhead integrity.
- An influx is known to contain H_2S.

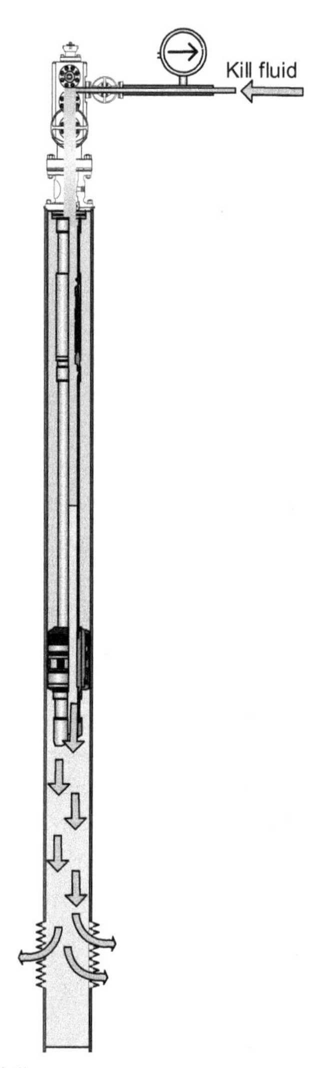

Figure 7.13 The bullhead kill.

- Prior to abandonment where there are no concerns about formation damage or formation fracture (fracture pressure of the cap rock becomes the limiting factor).
- In a well where circulation is possible, but the formation is both permeable and not easily damaged.

Whilst completion and intervention engineers routinely plan bullhead kills, drillers will normally only use it as a method of last resort since, with mud in the hole, a bullhead will almost always result in the fracturing of the formation.

7.4.1 Before a bullhead kill

In planning to kill the well by bullhead, there are several important considerations that can only be properly assessed if some essential information has been obtained. This will include, but need not be limited to:

- Formation fracture pressure. If the kill is prior to final abandonment of the well, the fracture pressure of the cap rock may be used.
- Reservoir pressure.
- SITP.
- Tubing fluid density, including oil water contact and GOC if known.
- Annulus fluid density and fluid level.
- Reservoir injection potential. Estimate or establish the maximum injection rate (bbls/min) that is achievable at below formation fracture pressure:
  - Theoretical values based on completion architecture, reservoir geometry, reservoir permeability, and fluid characteristics using inflow performance models such as "Prosper." Hand calculations can also be used.
  - Empirical data: in some fields, chemical treatments (scale and corrosion inhibitors) are routinely "squeezed" into the reservoir by bullheading down the completion.
  - Injection test: if no empirical data is available, and there are concerns about the validity of theoretical assumptions, an injection test should be performed, preferably in advance of the planned intervention.
- LCM: since a bullhead kill requires the formation to have good permeability, losses will occur when the well is filled with a clear, solids-free fluid that overbalances the formation. LCM will be needed to control losses after killing the well and can be pumped ahead of the kill fluid. LCM must be non-damaging, and compatible with the formation, reservoir fluids, and the completion equipment. In wells fitted with sand control screens, the LCM should be sized to bridge off across the inside of the screens and not the formation.
- Completion design limitations:
  - If the completion tubing is anchored to a packer, axial tension will increase during the bullhead due to a combination of thermal

contraction and ballooning. If the tubing is free to move (seal assembly and PBR) the combination of cooling, ballooning, and piston force will result in significant upwards movement of the tubing end, and there is a risk the end of the tubing will pull out of the seal bore. If the tubing remains within the seal bore, the combination of compressive piston force and high internal pressure will promote buckling.
  - Wellhead and tree pressure working pressure.
  - Safety valve opening pressure: the control line pressure required to keep a downhole safety valve open is normally about 1500–2500 psi (10,000–17,000 kPa) more than tubing pressure. Tubing pressure at the beginning of a bullhead may be significantly higher than the normal wellhead closed-in pressure. If the well is equipped with a tubing retrievable safety valve, control line pressure will have to be increased to compensate. Failure to do so risks damaging the downhole valve. If the well is equipped with a wireline retrievable valve, this can be pulled before the bullhead starts.
- Condition of completion:
  - Tubing burst limits: most operators recommend using 80% of API (American Petroleum Institute) burst as an upper limit during a bullhead kill. Burst pressures for API pipe grades are readily available from tables. If no tables are available, burst can be calculated using Barlow's formula (API values are calculated from Barlow's formula).

$$P_b = \mathrm{Tol}\frac{2Y_p t}{\mathrm{OD}} \tag{7.7}$$

where:
P_b: burst pressure, psi.
Tol: API wall thickness tolerance (0.875 for most API grades; 0.9 for some cold worked corrosion resistant alloys).
Y_p: tubing yield, psi.
T: tubing wall thickness, in.
OD: external diameter of tubing, in.

API burst value assumes new undamaged pipe. A lower burst value should be used if the pipe has been corroded or worn. API burst values should be treated with caution, as they will be altered by tri-axial loading during the kill. If the tubing is anchored to the packer injection of cold fluid under pressure, this will increase tension. Tubing under tension can have a tri-axial burst limit higher than API burst. Conversely, if the completion is equipped with a seal assembly and PBR (Polished Bore Receptacle) positioned above the packer, injection pressure will push the base of the tubing into compression. Tubing under compression can have a burst limit significantly below the API limit. Detailed stress analysis using proprietary software such as "WellCAT" can be used to establish tri-axial burst limits during a well kill.

- Where tubing integrity has failed and there is communication with the annulus, it may be necessary to bullhead down both tubing and annulus simultaneously to sweep hydrocarbons out of the annulus.
- Injection pressure may have to be limited in wells where past interventions have been carried out to isolate unwanted water or gas. High pressure can result in the failure of isolations such as bridge plug, straddles, cement squeeze, resins, and polymers.

7.4.2 Bullhead calculations: preparing the kill sheet

The following calculations need to be completed when preparing a bullhead kill schedule.

1 Calculate the kill fluid density

For most workovers and interventions an overbalance of approximately 200–300 psi is required. The kill fluid density is calculated as follows:

$$\text{Kill fluid (ppg)} = \frac{\text{reservoir pressure (psi)} + \text{over balance (psi)}}{\text{ft TVD} \times 0.052}$$

2 Calculate the volume of fluid required to kill the well (fluid pump to reservoir)

This volume must include surface lines from the pump to the wellhead, the tubing volume, any annular space below the packer, and the volume of the casing (or liner) below the end of the tubing.

3 Calculate the maximum surface pump pressure (formation fracture limit) at the start of the kill

At the beginning of a bullhead kill, the pressure acting on the formation is the HP of the reservoir fluid in the tubing (from the surface to the top reservoir) plus any applied surface pressure (pump pressure). The maximum (pump discharge) pressure that can be applied at the surface is formation fracture pressure minus the HP of the reservoir fluids in the tubing. A safety margin is normally applied.

The maximum surface pump pressure at the beginning of the kill formation fracture limit is

$$\text{Initial pump pressure (fracture limit)} = \text{formation fracture pressure} - \text{hydrostatic pressure to top reservoir} - \text{safety margin}$$

The safety margin varies, depending on location and company policy, but is normally between 100 and 500 psi. A common error when calculating the maximum pressure at the start of the kill is to include wellhead closed-in pressure with the hydrostatic calculation. This is a mistake and unnecessary, since no kill fluid can enter the well until pump discharge pressure is equal to or more than wellhead closed-in pressure.

4 Calculate the maximum surface pump pressure (formation fracture limit) at the end of the kill

At the end of a bullhead kill, the pressure acting on the formation is the HP of the kill fluid in the tubing (from the surface to the top reservoir) plus any applied surface pressure (pump pressure). The maximum (pump discharge) pressure that can be applied at the surface is formation fracture pressure minus the HP of kill fluid in the tubing. A safety margin is normally applied.

The maximum surface pump pressure at the end of the kill formation fracture limit is:

$$\text{Final pump pressure (fracture limit)} = \text{formation fracture pressure} - \text{kill weight fluid hydrostatic pressure at top reservoir} - \text{safety margin}$$

5 Tubing burst limit

Use 80% of the API burst limit.

6 Calculate maximum pump pressure (tubing burst mechanical limit) at the start of the kill

At the start of the kill, the highest pump pressure than can be applied at surface is tubing burst limit, minus the HP of the hydrocarbon fluids in the tubing at packer depth. This is a worst-case assumption, and makes no allowance for the HP in the annulus.

$$\text{Initial pump pressure (mechanical tubing burst limit)} = \text{tubing burst limit} - \text{tubing hydrostatic pressure at packer}$$

7 Calculate maximum pump pressure (tubing burst mechanical limit) at the end of the kill

At the end of the kill, the highest pump pressure than can be applied at the surface is tubing burst limit, minus the HP of the kill fluid in the tubing at packer depth. This is a worst-case assumption, and makes no allowance for the HP in the annulus.

$$\text{Final pump pressure (mechanical tubing burst limit)} = \text{tubing burst limit} - \text{tubing hydrostatic pressure at packer}$$

7.4.3 Bullhead procedure

If a rig is on location, the mud pumps are normally used to kill the well. However, a bullhead kill can be carried out using a mobile pumping unit before the rig arrives on location. This can bring about significant cost savings.

Outline procedure:

1. Hook up a temporary treating (Chiksan) line from the pumping equipment to the Christmas tree. Fluid is usually pumped via the kill wing or swab cap. The treating line should be fitted with non-return valves close to the tree, as well as a pressure relief valve. The line must be secured.
2. If there are holes in the tubing and simultaneous injection of kill fluid is required, or if pressure needs to be applied to the annulus to reduce tubing burst loading, a second temporary line will need to be made up to the production (A) annulus side outlet valve. The line should be fitted with a non-return valve close to the wellhead, as well as a pressure relief valve. The line must be secured.
3. The lines should be tested before pumping begins. Test pressure must exceed the maximum anticipated pump pressure by a suitable margin, typically a 10% excess, or 500 psi, whichever is greater.
4. Begin to pump kill fluid at a rate high enough to push produced fluids in the tubing back into the formation. In practical terms, the pump rate will be as high as possible whilst remaining below the fracture

pressure (with a safety factor), or the mechanical limit. If LCM is needed, this is normally pumped ahead of the kill fluid.

5. Continually monitor pump pressure. Pump pressure is stepped down as the tubing fills with kill weight fluid. Any reduction should be in accordance with the planned pump schedule.
6. If pumping into a well with an isolated annulus, pressure must be monitored. An initial increase caused by tubing ballooning should be expected, but it should stabilize.
7. Once the required amount of fluid has been pumped (surface to reservoir volume), stop pumping and observe for pressure build-up or losses. If LCM is pumped ahead of the kill fluid, anticipate a pressure increase at the surface when the LCM reaches the formation (Fig. 7.14 and Table 7.11).

7.4.4 Bullhead kill sheet example: vertical well

Kill sheet:

1. Calculate kill weight fluid required. Include 200 psi overbalance at top of formation:

$$\underset{\text{Form pressure + overbalance}}{5500 + 200} \;/\; \underset{\text{Top reservoir (feet TVD)}}{10{,}499} \;=\; \underset{\text{Kill fluid gradient (psi/ft.)}}{0.5429}$$

$$\underset{\text{Kill fluid gradient (psi/ft.)}}{0.5429} \;/\; \underset{\text{Factor (psi/ft. to ppg)}}{0.052} \;=\; \underset{\text{Kill fluid weight (ppg)}}{10.44}$$

Round up:

$$10.44 \quad \text{Rounds up to} \quad 10.5$$

2. Calculate the maximum tubing pressure (formation fracture limit)—at start of kill (tubing filled with oil and gas):

$$\underset{\text{Formation fracture pressure (psi)}}{7534\,\text{psi}*} \;-\; \underset{\text{Tubing hydrostatic Pressure}}{(722\times0.1)+((6.5\times0.052)\times(10{,}499-722))} \;=\; \underset{\text{Formation limit (start of kill)}}{4157\,\text{psi}}$$

**Formation fracture pressure is obtained by multiplying the fracture mud weight equivalent (13.8 ppg) by 0.052 to obtain a fracture gradient (0.7175 psi/ft.). The fracture gradient is then multiplied by the depth (TVD) to obtain the formation fracture pressure.*

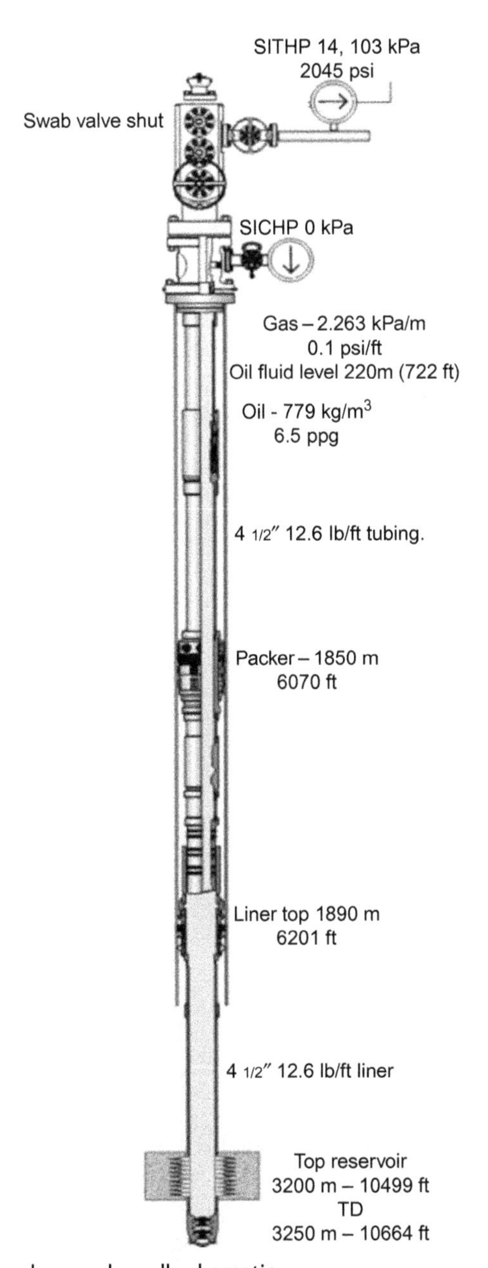

Figure 7.14 Bullhead example well schematic.

Table 7.11 Well data; bullhead kill example

Data	Value
Depth (top perforation)	10,499 ft. MD/TVD
Depth (TD)	10,664 ft. MD/TVD
Packer depth	6070 ft. MD/TVD
Liner top	6201 ft. MD/TVD
Tubing	4½ in. 12.6 lb/ft.
Liner	4½ in. 12.6 lb/ft.
Tubing capacity	0.0152 bbls/ft.
API burst pressure	8430 psi
Oil density	6.5 ppg
Gas gradient	0.1 psi/ft.
GOC depth	722 ft. MD/TVD
Annulus fluid density	9.2 ppg
Reservoir pressure	5500 psi
Shut-in tubing pressure (SITP)	2045 psi
Formation fracture equivalent	13.8 ppg
Required overbalance	200 psi

Formation fracture limit at the end of the kill—tubing filled with kill fluid:

$$\underset{\text{Formation fracture pressure}}{7534\,\text{psi}} - \underset{\text{Tubing (kill fluid) hydrostatic pressure}}{10.5 \times 0.052 \times 10{,}499*} = \underset{\text{Formation limit (end kill)}}{1802}$$

*Kill weight hydrostatic = 10.5 × 0.052 × 10,499 = 5732 psi.

3. Calculate the working tubing burst limit.

$$\underset{\text{API burst pressure psi (new)}}{8430} \times \underset{(80\%)}{0.8} = \underset{\text{psi}}{6744}$$

4. Calculate maximum initial and final tubing pressure (mechanical limits, no back-up).

This assumes the casing is evacuated to the packer.

Initial limit (tubing full of oil and gas):

$$\underset{\text{Working burst press}}{6744} - \underbrace{(722 \times 0.1) + ((0.052 \times 6.5) \times (6070 - 722))}_{\text{Tubing hydrostatic pressure (at packer)}}$$

$$= 4864_{\text{Initial limit (start of bullhead)}}$$

Final mechanical limit (tubing full of kill fluid):

$$\underset{\text{Tbg working burst press}}{6744} - \underset{\text{Tubing hydrostatic pressure (at packer)}}{10.5 \times 0.052 \times 6070} = \underset{\text{Final limit (tubing full of kill fluid)}}{3430}$$

5. Calculate bullhead volume. Surface to top of formation (bbls).
   Tubing:

$$\underset{\text{Tubing capacity factor (bbls/ft.)}}{0.01521} \times \underset{\text{Tubing length (MD)}}{6201} = \underset{\text{Barrels required}}{94.25}$$

Casing (or liner) to top of formation:

$$\underset{\textit{Liner}\text{ capacity factor}}{0.01521} \times \underset{\text{To top of formation}}{10,499 - 6201} = \underset{\text{Barrels required}}{65.32 \text{ bbls}}$$

Total volume:

$$\underset{\text{Tubing volume}}{94.25} + \underset{\text{Liner to formation}}{65.32} = \underset{\text{Total volume}}{159.57 \text{ bbls}}$$

Total strokes (if applicable):

$$\underset{\text{Tubing volume bbls}}{159.57} / \underset{\text{bbls/stk}}{0.0315} = \underset{\text{Strokes for total volume}}{5066}$$

With the kill sheet complete, a plot can be prepared showing volume pumped versus fracture and mechanical pressure limits. The plot should also show the lower pressure limit, i.e., the minimum surface pump pressure required to maintain a 200 psi overbalance. At the well site, the pump operator would be supplied with a copy of the plot. The pump is inherently conservative with regards to maximum pressure. Downhole pressure will be lower than calculated by a value equal to frictional pressure drop between surface and the reservoir (Fig. 7.15).

7.5 GAS LAWS AND GAS BEHAVIOR

Gas volume changes with pressure and temperature. As gas moves towards the surface, the reduction of HP in the fluid column allows the gas to expand. Boyle's law is used to calculate how much expansion will take place. In 1660, Robert Boyle established the relationship between gas volume and pressure for an ideal gas. The relationship between pressure and volume is:

$$P_1 V_1 = P_2 V_2$$

where:

P_1: initial gas pressure;
V_1: initial gas volume;

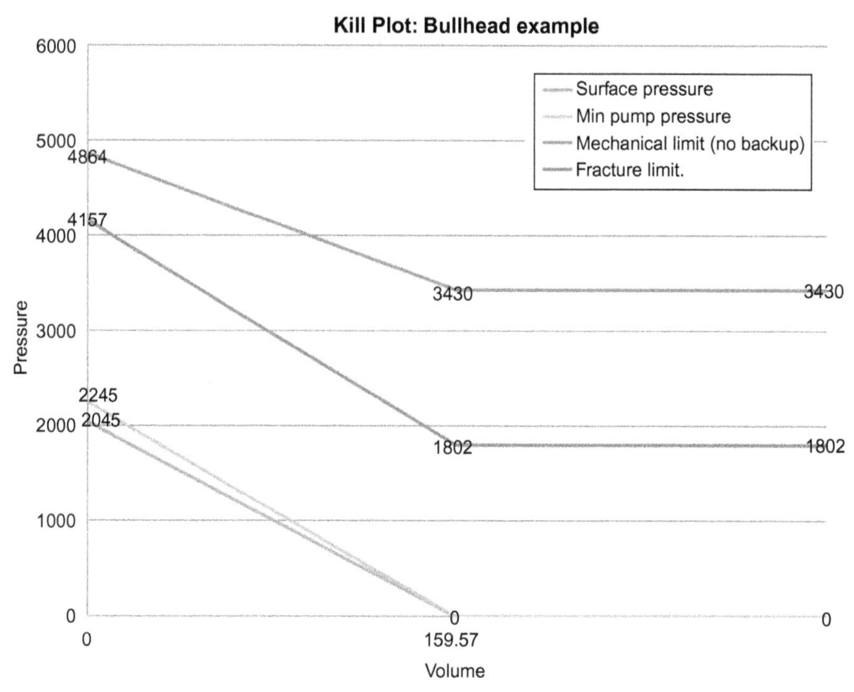

Figure 7.15 Kill plot. Bullhead example.

P_2: adjusted pressure;

P_2: adjusted gas volume.

In a well control (gas migration) context, P_1 is reservoir pressure and V_1 the size of a gas influx. P_2 is surface pressure and V_2 is the gas volume at atmospheric pressure (surface conditions).

Rearrange the equation to determine the volume of gas at the surface:

$$V_2 = \frac{P_1 V_1}{P_2}$$

Example (oilfield units):

A one barrel gas influx in a well with a reservoir pressure of 5000 psi would result in 340 bbls of gas at the surface:

$$V_2 = \frac{5000 \times 1}{14.7} = 340 \text{ bbls}$$

Note: 14.7 is atmospheric pressure in psi at standard conditions.

Example (SI units):

A 1 m^3 influx with a reservoir pressure of 3500 kPa would result in 34.65 m^3 of gas at the surface:

$$V_2 = \frac{3500 \times 1}{101} = 34.65\ \text{m}^3$$

Note 101 is atmospheric pressure in kPa at standard conditions.

7.5.1 Gas migration in a closed-in system

Gas entering a liquid filled well will migrate towards the surface, as it is less dense. The speed of gas migration depends on liquid viscosity, with gas generally moving more quickly through brine than mud. A rule of thumb widely used by drillers assumes a gas migration of 1000 ft./h. For clear fluids this is almost certainly an underestimate. Flow loop experiments have shown that gas can migrate at up 6000 ft./h.[2]

If the well remains shut-in at the surface, migrating gas is prevented from expanding. Pressure in the gas bubble will remain the same as it moves towards the surface, and consequently, a pressure increase will be observed at the surface. Unless measures are taken to reduce gas pressure as it migrates upwards, very high surface and BHP will result. There will be a risk of formation fracture, or of even greater concern, exceeding pressure limitations on critical completion components.

The effect migrating gas has on both surface and downhole pressure is demonstrated in the following example:

Well depth—5000 ft. (TVD).

Reservoir pressure 3000 psi.

Fluid in well—10 ppg brine.

Closed-in wellhead pressure 400 psi.

Gas enters the well and migrates towards the surface. By the time the gas bubble reaches 4500 ft. (500 ft. above the reservoir) the pressure at the surface will have risen to:

$$\text{WHCIP} = P_{\text{res}} - (0.053 \times \text{TVD}_{\text{gas}} \times \text{FD}) \tag{7.8}$$

where:

WHCIP: wellhead closed-in pressure in psi;

P_{res}: reservoir pressure when the influx occurred;

TVD_{gas}: the TVD of the gas bubble;

FD: fluid density in PPG.

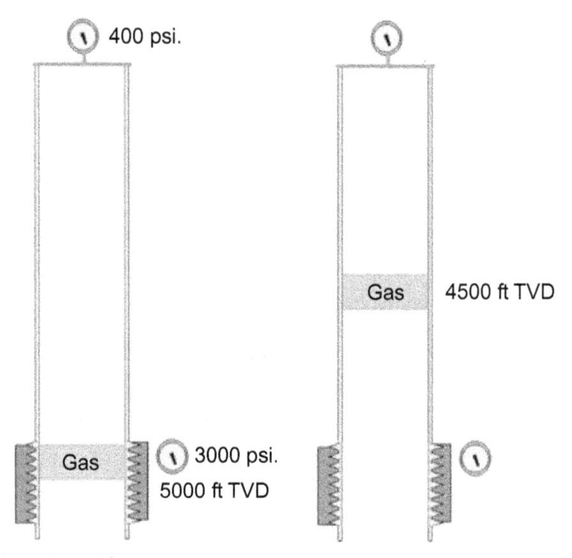

Figure 7.16 Gas migration.

$$\text{WHCIP} = 3000 - (0.052 \times 4500 \times 10) = 660 \text{ psi}$$

BHP will be:

$$\text{BHP} = P_{\text{res}} + ((0.052 \times \text{FD}) \times (\text{TVD}_{\text{res}} - \text{TVD}_{\text{gas}}))$$

where:

BHP: the bottom hole pressure in psi.

TVD_{res}: TVD of the reservoir (Fig. 7.16).

$\text{BHP} = 3000 + (0.052 \times 10) \times (5000 - 4500) = 3260 \text{ psi}$

If the gas bubble was allowed to migrate to the surface, shut-in pressure would rise to the original BHP of 3000 psi. Actual BHP would increase to:

$$3000 \text{ psi} + (0.052 \times 5000 \times 10) = 5600 \text{ psi}$$

7.6 PROCEDURE FOR CONTROLLING GAS MIGRATION

Since gas can migrate quickly, it is often necessary to reduce surface pressure whilst preparing to kill the well. Notable characteristics of gas migration include:

- Can occur any time the well is shut-in with gas present (either when the well is shut-in intentionally or the flow path is mechanically plugged or blocked).

- Indicated by a uniform increase in SICP and SITP.
- If uncontrolled, pressures will increase everywhere in the wellbore.
- Increased pressure can cause formation damage, loss of whole fluid into the formation, or damage to surface equipment.
- Occurs rapidly in clear, low viscosity, workover fluids.

Surface pressure is relieved by using either the "constant tubing pressure method" or the "volumetric method." Both methods are used when it is not possible to circulate out a kick, where circulation cannot be started immediately, or where bullheading is not possible or is inadvisable.

It should be noted that increasing pressure, caused by gas migration, is normally less of a concern during workover or intervention operation than during a drilling operation. With mud in the hole, increasing pressure acts against the filter cake. Since most filter cakes have extremely good fluid loss prevention properties, pressure can build to formation fracture pressure, resulting in fluid loss and formation damage. This is less likely to happen if clear (solids free) fluid is controlling the well. Increasing well pressure caused by gas migration normally results in increasing rates of fluid loss as pore pressure is exceeded. Formation fracture is unusual and only occurs when extremely robust LCM is used, permeability is exceptionally low, or the window between pore pressure and fracture pressure is unusually narrow. In some cases, gas expansion will self-regulate, increasing wellbore pressure being offset by an increase in leak-off to the formation.

7.6.1 Constant tubing pressure

This method, as the name implies, removes the gas whilst maintaining a constant tubing pressure. Although easier to implement than the volumetric method, it can only be used if these conditions are met:

- There is a communication path between the tubing and the annulus choke.
- Tubing pressure is able to be read.

7.6.1.1 Procedure for constant tubing pressure bleed method

1. Allow SITP to increase by a safety margin of about 50–100 psi (350–700 kPa). The safety margin prevents any additional influx caused by bleeding off too much pressure at the choke. This is the lower limit.
2. Allow SITP to increase by an additional margin, normally about 50–100 psi (350–700 kPa). This is the upper limit.

3. Using the choke, bleed the annulus until the tubing pressure drops to the lower limit. There will be a time lag between bleeding off pressure at the choke and seeing a response on the tubing pressure gauge. The time taken for the pressure signal to reach the tubing head is dependent on the depth of the tubing (or the circulation port) and the characteristics of the fluid in the well. To avoid bleeding off too much pressure on the annulus side, close the choke when the annulus pressure reads the lower limit and wait until the tubing pressure stabilizes.
4. Keep repeating steps 2 and 3—keeping the tubing pressure between the lower and upper limits for as long as is necessary or until gas reaches surface.

As the operation progresses, the casing gauge will show an increase even though the tubing pressure remains within the upper and lower limit. Do not try to keep the pressure on both sides the same by opening the choke. With successive bleed cycles, the gas expands as it rises in the annulus, creating an imbalance in HP between the annulus and the tubing.

7.6.2 Volumetric method

If there is no communication between the tubing and annulus, the volumetric method must be used. Control is managed by accurately recording the volume removed with each bleed cycle. A calibrated tank placed downstream of the choke is needed, and it should be capable of recording fluid volume in relatively small increments, from as little as ½ bbls or 0.8 m^3.

Outline procedure:

1. Select a safety margin. Normally 100 psi (700 kPa).
2. Select a range, also normally about 100 psi (700 kPa).
3. Calculate HP (P_h) per bbl or (m^3) fluid in the upper part of the annulus:

   P_h per bbl (psi/bbl) = fluid gradient (psi/ft.) ÷ annular capacity factor (bbl/ft.).

   P_h per m^3 (kPa/m^3) = fluid gradient (kPa/m) ÷ annular capacity factor (m^3/m).
4. Calculate volume to bleed to reduce pressure by the range:

   Volume to bleed (bbl/cycle) = range (psi) ÷ P_h per bbl.

   Volume to bleed (m^3/cycle) = range (kPa) ÷ P_h per m^3.
5. Construct casing pressure versus volume to bleed schedule.

6. Without bleeding any fluid, allow SICP to increase by the safety margin.
7. Without bleeding any fluid, allow SICP to increase by range.
8. Maintaining SICP, bleed small volumes of fluid until the volume in the tank is equal to that calculated in Step 4.
9. Keep repeating Steps 7 and 8 until gas is at the surface, or until a circulating or bullhead kill can begin (Fig. 7.17).

Small and gradual changes on the choke are recommended. The aim is to hold a constant casing pressure as fluid is bled from the well. Rapid opening of the choke, to speed up the process, can easily result in the casing pressure dropping enough to allow a second influx. It may take several hours to bring gas to the surface. Patience is required.

Example calculation (Table 7.12):

Procedure (field units):

Volume to bleed (using 100 psi steps) 100 ÷ 11.66 = 8.57 bbls.

1. Allow casing pressure to increase by range (100 psi) above safety margin (100 psi) above initial SICP: 500 psi + 100 psi safety margin + 100 psi range = 700 psi.

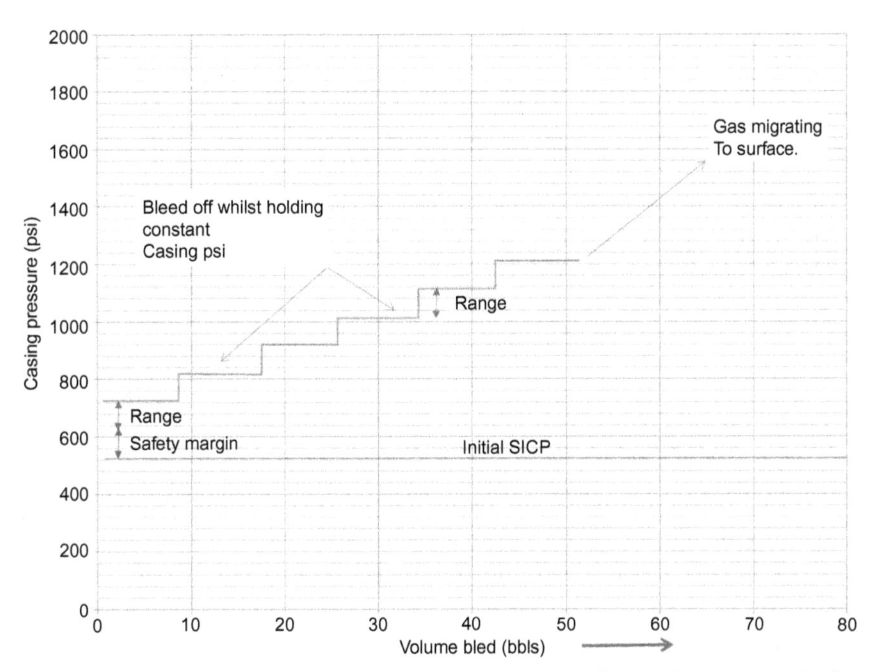

Figure 7.17 Stair-step schedule for bringing gas to the surface (volumetric method).

Table 7.12 Data for volumetric bleed calculation

Data	Field units	SI units
Fluid in annulus	12 ppg	1437.91 kg/m^3
9⅝ in. 47 lb/ft. × 4½ in. annular capacity	0.0535 bbls/ft. (18.69 ft./bbl)	0.0279 m^3/m (38.81 m/m^3)
P_h per bbl or m^3	11.66 psi (18.69 ft.)	504 kPa/m^3
Initial SICP following influx	500 psi	3448 kPa

2. Bleed pre-determined volume (8.5 bbls). Reduction in fluid hydrostatic means casing pressure will be 700 psi + 100 psi margin = 800 psi.
3. Allow casing pressure to build by 100 psi (range) 800 + 100 psi = 900 psi.
4. Bleed pre-determined volume (8.5 bbls). Reduction in fluid hydrostatic means casing pressure will be 900 psi + 100 psi margin = 1000 psi.
5. Keep repeating until gas is at the surface.

   Procedure (SI units):

   Volume to bleed (using 700 kPa steps) 700 ÷ 504 = 1.38 m^3.

1. Allow casing pressure to increase by range (700 kPa) above safety margin (700 kPa) above initial SICP (3448 kPa): 3448 + 700 kPa safety margin + 700 kPa range = 4848 kPa.
2. Bleed pre-determined volume (1.38 m^3). Reduction in fluid hydrostatic means casing pressure will be 4848 + 700 kPa margin = 5548 kPa.
3. Allow casing pressure to build by kPa (range) 5548 + 700 kPa = 6248 kPa.
4. Bleed pre-determined volume (1.38 m^3). Reduction in fluid hydrostatic means casing pressure will be 6248 + 700 kPa margin = 6948 kPa.
5. Keep repeating until gas is at the surface.

7.7 LUBRICATE-AND-BLEED

The lubricate-and-bleed procedure is used to kill gas wells, or liquid producing wells with a large gas cap, where circulation is not possible. Gas in the wellbore is replaced with kill fluid whilst keeping bottom hole pressure constant at above reservoir pressure, to prevent an additional influx from the formation. Lubricate-and-bleed is also used on wells where sustained annulus pressure is a problem, i.e., wells where casing

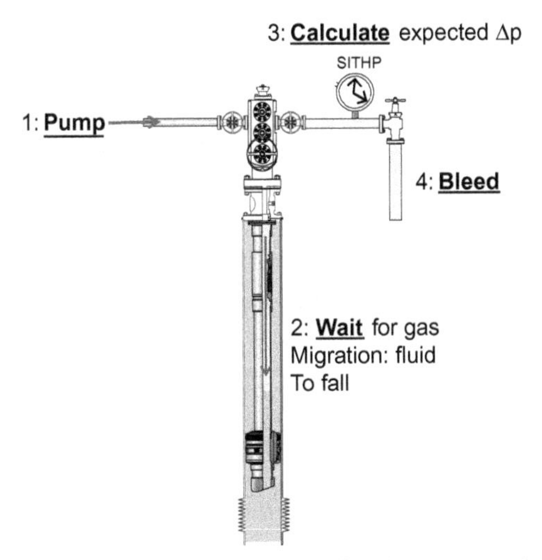

Figure 7.18 Pump–wait–calculate–bleed. The four basic steps for a lubricate-and-bleed well kill.

pressure repeatedly and quickly builds up after bleed-off. Three different lubricate-and-bleed methods are described here. In each, the fundamental principle is the same; fluid is pumped in at the surface and allowed to drop down the well, through the gas column. Gas pressure, equivalent to the hydrostatic head of the fluid pumped, is bled-off. The cycle of pump (lubricate) then bleed is repeated until the well is dead (Fig. 7.18).

As with the circulating kill, setting a mechanical plug close to the bottom of the well can be beneficial during a lubricate-and-bleed. Providing the plug maintains integrity, it will prevent losses to the formation. It will also prevent a second influx if the person operating the choke is too aggressive and allows BHP (at the plug) to drop below reservoir pressure. Although the presence of a plug undoubtedly aids the kill process, maintaining BHP above reservoir pressure is still recommended.

7.7.1 The constant volume method (lubricate and bleed)

The constant volume method is so called because the volume pumped at each lubricate and bleed cycle is the same, and therefore the hydrostatic pressure reduction should be approximately the same for each cycle (in a vertical well). Kill fluid is accurately measured (from a calibrated tank) before pumping, and bleed-off calculations are based on the hydrostatic pressure reduction of that measured volume. As the fluid is pumped into the well, gas in the wellbore is compressed, causing an increase in tubing pressure. The volume of fluid that can be pumped at each stage is limited by the combination of pressure increase due to gas compression and the increasing hydrostatic head. If the tubing has not been plugged, formation fracture pressure normally defines the upper limit; a safety margin should be included. Where the tubing is plugged (mechanical barrier), the maximum pressure limit will be the mechanical limit of the plug, the production tubing, the tree or surface pumping equipment; whichever is lowest.

7.7.1.1 Constant volume method. Calculations and procedure

1. Calculate the gas gradient in the tubing.

$$\text{Gas gradient (psi/ft.)} = \frac{\text{BHP} - \text{SITP}}{\text{reservoir depth TVD}} \tag{7.9}$$

2. Calculate pipe capacity and volume.
3. Calculate the kill weight requirement.
   In a plugged well: kill weight is calculated to the plug. Assumes a column of reservoir fluid from the plug to the reservoir.
   No plug: kill weight is calculated to the top of the reservoir.
4. Calculate the maximum volume that can be pumped without exceeding the maximum allowable surface pressure using Boyle's law. If no plug is installed, formation fracture pressure is likely to determine the maximum pressure that can be applied at surface.

$$V_2 = \frac{P_1 V_1}{P_2}$$

   where
   - V_2 is the maximum volume that should be pumped (to reach maximum surface pressure).
   - P_1 is closed in pressure.
   - V_1 is the tubing volume (to the plug or reservoir).
   - P_2 is the maximum allowable surface pressure.
5. Pump fluid until the wellhead pressure reaches the maximum allowable pressure (or the calculated amount has been pumped).
6. Record the volume of fluid pumped.
7. Allow time for fluid to drop down the well.
8. Bleed dry gas from choke to reduce casing pressure to the previous recorded wellhead pressure plus the calculated hydrostatic pressure increase. Allow the well to stabilize.

9. Note the wellhead pressure and pump the same volume of kill weight fluid as used at the first stage (step 5).
10. Allow time for the fluid to fall in to the well.
11. Bleed dry gas until the pressure is reduced to that recorded before the second stage was pumped (beginning of step 9). Further reduce the pressure equivalent to the hydrostatic head of fluid pumped into the well. The final pressure will be equivalent to the pressure recorded at the end of step 8/beginning of step 9, plus the increase in hydrostatic head.
12. Continue to lubricate and bleed until the well is dead. The volume (and hydrostatic pressure increase) should be the same for each step.

7.7.1.2 Example lubricate and bleed kill using the constant volume method

A well is to be killed by lubricate and bleed, constant volume method. A 200 psi overbalance is required at the end of the kill. A plug has been placed close to the end of the tubing. All the necessary data are presented in (Table 7.13 and Figure 7.19).

1. Calculate the gas gradient.

$$\text{Gas gradient } (\text{psi/ft.}) \frac{5747 - 4911}{10,450} = 0.08 \text{ psi/ft}$$

Table 7.13 Data for lubricate and bleed—constant volume

Data	Value
WHCIP	4911 psi
Plug depth	10,250 ft.
Packer depth	10,240 ft.
Reservoir depth	10,450 ft. (TVD)
Reservoir pressure	5747 psi
Maximum surface pressure (wellhead mechanical limit)	5500 psi
Tubing size	4½ in. 13.5 lb/ft.
Tubing ID	3.92 in.
Tubing capacity	0.0149 bbls/ft.
Tubing end (tail-pipe)	10,260 ft.
Casing size	7 in. 29 lb/ft.
Casing ID	6.184 in.
Casing capacity	0.0371 bbls/ft.
Kill fluid to give 200 psi overbalance	10.94 → 11 ppg
Fracture pressure	6950 psi
Safety factor (below fracture pressure)	500 psi

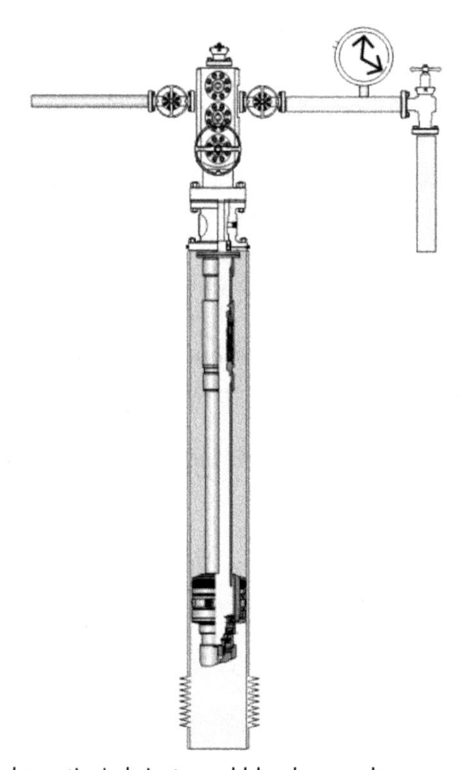

Figure 7.19 Well schematic. Lubricate-and-bleed example.

2. Calculate kill weight fluid density and gradient (to kill to the plug).

   Pressure at the plug is reservoir pressure minus the gas hydrostatic from the top of the reservoir to the plug depth. Gas gradient is 0.08 psi/ft. (10,450 − 10,250) × 0.08 = 16 psi. Pressure at plug is 5747 − 16 = 5731 psi.

$$\text{Kill fluid (ppg)} = \frac{5731 + 200}{10250 \times 0.052} = 11.127 \rightarrow 11.2 \text{ ppg}$$

$$11.2 \text{ ppg} \times 0.052 = 0.5824 \text{ psi/ft.}$$

3. Calculate the tubing volume (to the plug)

$$\text{Tubing volume} = 0.0149 \times 10,250 = 152.725 \text{ bbls} \rightarrow 153 \text{ bbls.}$$

4. Fluid is pumped into the well until the maximum wellhead pressure (5500 psi) is reached. Increasing pressure will compress the gas. The amount of gas compression, and therefore the amount of fluid that can be pumped is calculated using Boyle's aw.

$$\text{Volume to reach } P_{max} = \frac{4911 \times 153}{5500} = 136.61 \text{ bbls}$$

5. The volume of fluid that can be pumped is the original tubing volume (153 bbls) minus the volume occupied by the compressed gas (136.61 bbls) = 16.39 bbls.
6. Pump the calculated volume (16.39 bbls), or until the wellhead pressure reaches the maximum allowable value (5500 psi). Do not exceed this value even if not all the calculated amount of fluid has been pumped.
7. Calculate the reduction in hydrostatic head from the volume of fluid pumped.
   $\text{Height of fluid column (ft.)} = \dfrac{\text{volume pumped (bbls)}}{\text{tubing capacity (bbl/ft)}}$ becomes

   $$\frac{16.39}{0.0149} = 1100 \text{ ft.}$$

   $$\text{Hydrosatic pressure} = (\text{kill fluid gradient} - \text{gas gradient}) \times \text{height of fluid column}$$

   $(0.5827 - 0.08) \times 1100 = 553$ psi.
8. Bleed back to the wellhead pressure at the start of the kill (4911 psi). Continue to bleed off pressure equal to hydrostatic increase from the fluid pumped (553 psi). 4911 − 553 = 4358 psi.
9. Carry out a second lubricate and bleed cycle using the same volume that was pumped at step 6 (16.39 bbls). The expected pressure increase is again calculated using Boyle's law. While not essential, this calculation assures that the operation is proceeding as planned and allows an accurate kill graph to be prepared.

   $$\text{Wellhead pressure after pumping 16.31 bbls} = \frac{4358 \times 136.61}{136.61 - 16.39} = 4952 \text{ psi.}$$

10. Bleed off until the wellhead pressure is 553 psi lower than the starting pressure (before the second stage was performed). 4358 − 553 = 3805 psi.
11. The process of lubricate and bleed is repeated until the tubing is full of kill weight brine and no gas remains at surface.

The best approach is to create a kill sheet with a table of expected pressure against volume pumped (Table 7.14). A spreadsheet helps in this respect. Properly constructed it can be configured to make the calculations for each step of the lubricate and bleed process (Figure 7.20).

Table 7.14 Lubricate and bleed spreadsheet—constant volume

Stage	Volume to pump (bbls)	Pressure at start (psi)	End pressure $(P_1 \times V_1)/V_2$	Cu Vol (bbls)	Head (fluid pumped) (psi)	Bleed back to (psi)	P_1 (psi)	Vol 1 (bbls)	Vol 2 (bbls)
1	16.38	4911	5500	16.38	553.00	4358.00			
2	16.38	4358	4952	32.77	553.00	3805.00	4358.00	136.62	120.23
3	16.38	3805	4405	49.15	553.00	3252.00	3805.00	120.23	103.85
4	16.38	3252	3861	65.54	553.00	2699.00	3252.00	103.85	87.46
5	16.38	2699	3321	81.92	553.00	2146.00	2699.00	87.46	71.08
6	16.38	2146	2789	98.31	553.00	1593.00	2146.00	71.08	54.69
7	16.38	1593	2274	114.69	553.00	1040.00	1593.00	54.69	38.31
8	16.38	1040	1817	131.08	553.00	487.00	1040.00	38.31	21.92
9	16.38	487	1300	147.46	553.00	0.00	487.00	21.92	5.54
10	5.54	0	783	153.00	186.00	0.00			

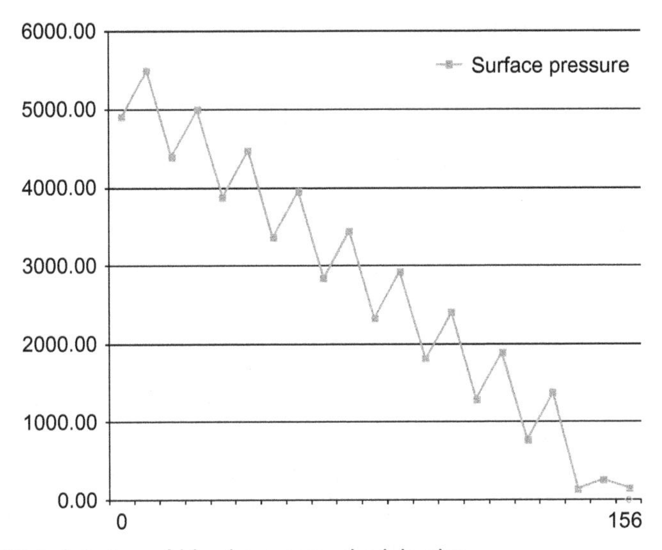

Figure 7.20 Lubricate-and-bleed pressure schedule plot.

7.7.2 Lubricate and bleed. Constant volume method (no plug)

If there is no plug in the tubing, the well must be killed to top reservoir. As the reservoir is exposed, the combination of increasing tubing pressure due to gas compression and increasing hydrostatic head must not be allowed to exceed fracture pressure. It is advisable to include a safety margin when setting an upper limit for bottom hole pressure (BHP). Overbalance must be maintained throughout the kill, and since overbalance will be increased by a significant margin during lubrication, there is an additional risk of losses.

7.7.2.1 Example calculation

Using the data from Table 7.13 (as per the previous example), prepare a lubricate and bleed kill schedule. BHP must not exceed fracture pressure. A 200 psi overbalance is required at the end of the kill.

1. Calculate the gas gradient.

$$\text{Gas gradient (psi/ft.)} = \frac{5747 - 4911}{10,450} = 0.08\ \text{psi/ft}$$

2. Calculate the tubing/casing volume (to the reservoir)
   Tubing volume = 0.0149 × 10,260 = 152.874 bbls → 153 bbls.

Casing volume (from tail-pipe to the perforations) = $0.0371 \times 190 = 7$ bbls.
Total volume $153 + 7 = 160$ bbls.

3. Calculate the kill fluid density and gradient needed to achieve a 200 psi overbalance at the end of the kill.

$$\text{Kill fluid (ppg)} = \frac{5747 + 200}{10,450 \times 0.052} = 10.94 \rightarrow 11 \text{ ppg.}$$

$$11 \text{ ppg} \times 0.052 = 0.572 \text{ psi/ft.}$$

4. Calculate the maximum volume of fluid that can be pumped without exceeding the formation fracture pressure (including the designated safety margin). The method used in this example first calculates the approximate pressure increase (downhole) with each barrel pumped. Separate calculations for the casing and tubing are required since the hydrostatic head increase will be different.
   a. Approximate pressure increase per bbl of fluid pumped due to gas compression.

$$\text{Pressure after pumping 1 bbl} = \frac{5747 \times 160}{159}$$
$$= 5783 \text{ psi. } (5783 - 5747 = 36 \text{ psi increase})$$

   Note: The increase at surface after pumping 1 bbl is 31 psi due to the lower starting (P_1) pressure 4911 psi.
   b. Hydrostatic pressure increase, per bbl of kill fluid pumped.
   Tubing: 1/0.0149 = 67 ft/bbl. 67 × (0.572–0.08) = 33 psi/bbl.
   Casing: 1/0.0371 = 27 ft/bbl. 27 × (0.572–0.08) = 13 psi/bbl.
   Approximate pressure increase (at reservoir) per bbl pumped.
   Tubing: 36 (gas compression) + 33 (hydrostatic head) = 69 psi.
   Casing: 36 (gas compression) + 13 (hydrostatic head) = 49 psi.
   Maximum pressure at reservoir depth is fracture pressure minus the safety margin (6950–500 = 6450 psi). The maximum permitted pressure increase is maximum allowable pressure minus reservoir pressure: 6450–5747 = 703 → 700 psi.
   – 700/69 = 10.14 → 10 bbls of kill fluid in the tubing.
   – 700/49 = 14.28 → 14 bbls of kill fluid in the casing.

5. Since the casing volume (from tubing end to reservoir depth) is 7 bbls, pumping the maximum calculated volume of 14 bbls would place 7 bbls of kill fluid in tubing, exceed the allowable maximum

pressure downhole. For the sake of simplicity and consistency the first lubrication will use the calculated maximum tubing volume (10 bbls):

$$\text{Surface pressure after pumping 10 bbl} = \frac{4911\text{x}160}{(150)} = 5238 \text{ psi.}$$

The casing (7 bbls capacity) has been filled with kill fluid. The remaining 3 bbls pumped is filling the tubing, placing the fluid level at 3/0.0149 = 201 ft. above the tubing end (10,260 − 201 = 10,059). The head of fluid pumped is (10,450 − 10,059) × (0.572 − 0.08) = 192 psi.

BHP after pumping 10 bbls is:

$$5238 + (10,059 \times 0.08) + (10,450 - 10,059) \times 0.572 = 6266 \text{ psi.}$$

6. Bleed off to the starting pressure of 4911. Bleed an additional 192 psi (the head of fluid pumped) = 4719 psi. The BHP is after bleeding off is:

$$4719 + (10,059 \times 0.08) + (10,450 - 10.059) \times 0.572) = 5747 \text{ psi.}$$

7. Pump a second 10 bbl batch of fluid. The pressure increase due to gas compression is:

$$\text{Surface pressure} = \frac{4719 \times 150}{140} = 5056 \text{ psi}$$

10 bbls of fluid fills 10/0.0149 = 671 ft. of tubing. The fluid level is at 10,059-671 = 9388 ft. The BHP is 5056 + (9388 × 0.08) + (10,450 − 9388) × 0.572 = 6414 psi.

8. Bleed back pressure to the starting WHCIP of 4719 bleed an additional 330 psi (the head of fluid pumped) = 4389 psi. The new pressure downhole is

$$4389 + (9388 \times 0.08) + (10,450 - 9388) \times 0.572 = 5747 \text{ psi.}$$

9. Continue pumping 10 bbl batches of fluid. The pressure reduction should be approximately 330 psi with each cycle of lubricate and bleed. As with the first (plugged) example, it is useful to build a spreadsheet (Table 7.15). This keeps track of surface and downhole pressure as the kill progresses.

Table 7.15 Lubricate and bleed (no plug) pressure schedule plot

Surface

Stage	Volume to pump (bbls)	Vol 1 (bbls)	Vol 2 (bbls)	Pressure at start (psi)	End pressure (P1 × V1)/V2	Fluid column height (ft.)	Fluid level (ft. TVD)	Head of fluid (psi)	BHP after pumping (psi)	BHP after bleed off (psi)
1	10.00	160.00	150.00	4911.00	5238	391.00	10059.00	192.37	6266.77	5747.00
2	10.00	150.00	140.00	4718.63	5056	671.14	9387.86	330.20	6414.25	5747.00
3	10.00	140.00	130.00	4388.43	4726	671.14	8716.72	330.20	6414.77	5747.00
4	10.00	130.00	120.00	4058.23	4396	671.14	8045.58	330.20	6415.39	5747.00
5	10.00	120.00	110.00	3728.02	4067	671.14	7374.44	330.20	6416.11	5747.00
6	10.00	110.00	100.00	3397.82	3738	671.14	6703.30	330.20	6416.98	5747.00
7	10.00	100.00	90.00	3067.62	3408	671.14	6032.15	330.20	6418.05	5747.00
8	10.00	90.00	80.00	2737.42	3080	671.14	5361.01	330.20	6419.38	5747.00
9	10.00	80.00	70.00	2407.22	2751	671.14	4689.87	330.20	6421.09	5747.00
10	10.00	70.00	60.00	2077.02	2423	671.14	4018.73	330.20	6423.37	5747.00
11	10.00	60.00	50.00	1746.82	2096	671.14	3347.59	330.20	6426.56	5747.00
12	10.00	50.00	40.00	1416.61	1771	671.14	2676.45	330.20	6431.35	5747.00
13	10.00	40.00	30.00	1086.41	1449	671.14	2005.31	330.20	6439.34	5747.00
14	10.00	30.00	20.00	756.21	1134	671.14	1334.17	330.20	6455.31	5747.00
15	10.00	20.00	10.00	426.01	852	671.14	663.03	330.20	6503.21	5747.00
16	10.00	10.00	0.00	95.81	0	671.14	0.00	330.20	5977.40	5977.40

10. If the pressure increase is not consistent with the volume pumped, i.e., P_2 is less than the anticipated value, the most likely explanation is that fluid is being lost to the formation as the downhole pressure increases. Since fluid loss will result in a reduced hydrostatic head, bleeding off pressure equivalent to the head of fluid pumped will likely result in an influx. The revised fluid level can be estimated from the pressure at surface. For example, Table 7.15 predicts a pressure rise from 3728 to 4067 psi during the 5th stage of the kill. Consider a situation where, after pumping 10 bbls, the pressure had not risen to the expected 4067 psi but had stabilized at 3920 psi. Boyles law can be used to estimate the volume of fluid in the tubing, and therefore the revised hydrostatic head.

$$V_2 = \frac{3728 \times 120}{3920} = 114.12 \text{ bbls.}$$

The compressed gas volume remaining in the tubing is 114.12 bbls, not 110 as planned. Even though 10 bbls of fluid were pumped, only $120 - 114.21 = 5.88$ bbls remain in the tubing, the rest (4.12 bbls) is lost to the formation. The head increase is therefore $5.88/0.0149 = 395$ ft. $\times$ $(0.572 - 0.08) = 194$ psi, not the 330 psi anticipated. Bleeding off the planned amount of pressure (330 psi) would result in an influx. If losses are suspected, then pressure adjustments will need to be made "on the fly." More stages will be required, the kill will take significantly longer and more kill fluid will be needed.

7.7.3 Lubricate and bleed. Constant pressure method

Like the constant volume method, the constant pressure method also uses four steps at each stage of the kill; pump, wait, calculate, and bleed. Where the pressure method differs is that each time fluid is pumped, the wellhead pressure is brought back up to maximum allowable pressure. This technique will, in theory, kill the well more quickly as fewer stages are required. However, it must be treated with caution. As the operation progresses, the combination of high wellhead pressure and increasing hydrostatic head results in increasingly high BHP. BHP must not be allowed to go above formation fracture pressure, nor should pressure be allowed to reach the tubing burst limit. The most conservative approach when defining a tubing burst limit is to assume that the casing is evacuated. Differential pressure is based on the tubing hydrostatic at the packer + applied surface pressure against an annulus pressure of 0 psi—complete evacuation. This very stringent approach can be revised if verifiable data on casing fluid density and casing fluid level is available.

Clearly the pressure method is not suitable for wells where the condition of the completion is suspect, where losses are likely to occur or where fracture pressure is low. The pressure method should only used where the condition of the tubing is known to be good and where there is a plug above the reservoir.

7.7.3.1 Procedure for the constant pressure method (lubricate and bleed)

1. Calculate the gas gradient in the tubing.

$$\text{Gas gradient psi/ft.} = \frac{\text{BHP} - \text{SICHP}}{\text{reservoir depth TVD}}$$

2. Calculate pipe capacity and volume.
3. Calculate the kill weight requirement.
   Plugged well: Calculated kill weight to the plug. Assumed a column of gas from the plug to the reservoir.
   No plug: Calculate kill weight to the top of the reservoir.
4. Calculate the maximum volume that can be pumped without exceeding the maximum allowable surface pressure using Boyle's law. If no plug is installed, formation fracture pressure is likely to determine the maximum pressure that can be applied at surface.

$$V_2 = \frac{P_1 V_1}{P_2}$$

5. Pump fluid until the wellhead pressure reaches the maximum allowable pressure (or the calculated amount has been pumped).
6. Allow time for fluid to drop into the well.
7. Record the volume of fluid pumped.
8. Bleed dry gas from choke to reduce casing pressure to the previous recorded wellhead pressure plus the calculated hydrostatic pressure increase. Allow the well to stabilize.
9. Pump fluid until the wellhead pressure reaches the maximum allowable pressure.
10. Allow time for the fluid to fall in to the well.
11. Bleed dry gas from choke to reduce casing pressure to the recorded wellhead pressure at the start of the stage, plus the calculated hydrostatic pressure increase. Allow the well to stabilize.
12. Repeat steps 9–11 until the well is dead.

7.7.3.2 Example "pressure method" lubricate and bleed well kill

To enable comparisons to be made between the two lubricate and bleed techniques, the data set used for the constant volume method (Table 7.13) will be used for this constant pressure example. The well is plugged.

The first part of the constant pressure method is identical to constant volume method (Section 7.7.1.2)

1. Calculate the gas gradient.

$$\text{Gas gradient}\left(\text{psi/ft}\right) = \frac{5747 - 4911}{10,450} = 0.08\ \text{psi/ft}$$

2. Calculate kill weight fluid density and gradient (to kill to the plug).

   Pressure at the plug is reservoir pressure minus the gas hydrostatic from the top of the reservoir to the plug depth. Gas gradient is 0.08 psi/ft. (10,450 − 10,250) × 0.08 = 16 psi. Pressure at plug is 5747 − 16 = 5731 psi.

$$\text{Kill fluid (ppg)} = \frac{5731 + 200}{10,250 \times 0.052} = 11.127 \rightarrow 11.2\ \text{ppg.}$$

$$11.2\ \text{ppg} \times 0.052 = 0.5824\ \text{psi/ft.}$$

3. Calculate the tubing volume (to the plug)

$$\text{Tubing volume} = 0.0149 \times 10,250 = 152.725\ \text{bbls} \rightarrow 153\ \text{bbls.}$$

4. Stage 1: Fluid is pumped into the well until the maximum wellhead pressure (5,500 psi) is reached. Increasing pressure will compress the gas. The amount of gas compression, and therefore the amount of fluid that can be pumped is calculated using Boyle's Law.

$$\text{Volume to reach } P_{\max} = \frac{4911 \times 153}{5500} = 136.61\ \text{bbls.}$$

5. The volume of fluid that can be pumped is the original tubing volume (153 bbls) minus the volume occupied by the compressed gas (136.61 bbls) = 16.39 bbls.
6. Pump the calculated volume (16.39 bbls), or until the wellhead pressure reaches the maximum allowable value (5500 psi). Do not exceed this value even if not all the calculated amount of fluid has been pumped.
7. Calculate the reduction in hydrostatic head from the volume of fluid pumped.

$$\text{Height of fluid column (ft)} = \frac{\text{volume pumped (bbls)}}{\text{tubing capacity (bbl/ft.)}} \text{ becomes}$$

$$\frac{16.39}{0.0149} = 1100 \text{ ft.}$$

$$\text{Hydrosatic pressure} = (\text{kill fluid gradient} - \text{gas gradient}) \times \text{height of fluid column}$$

$$(0.5824 - 0.08) \times 1100 = 553 \text{ psi.}$$

8. Bleed back to the wellhead pressure at the start of the kill (4911 psi). Continue to bleed off pressure equal to hydrostatic increase from the fluid pumped (553 psi). 4911 − 553 = 4358 psi. Up to this point the kill is identical to the constant volume method. The constant pressure method differs from this point on in that the volume pumped regulated by the maximum allowable surface pressure.
9. Stage 2: Pump fluid until the wellhead pressure reaches the maximum allowable value of 5500 psi. Calculate the volume pumped and consequent reduction in pressure.

$$\text{Volume to reach } P_{max} = \frac{4358 \times 136.61}{5500} = 108.24 \text{ bbls}$$

$$\text{Volume pumped} = 136.61 - 108.24 = 28.37 \text{ bbls.}$$

$$\text{Height of fluid column} = \frac{28.37}{\text{Tubing capacity } 0.0149} = 1904\text{ft.}$$

$$\text{Hydrosatic pressure} = (\text{kill fluid gradient} - \text{gas gradient}) \times \text{height of fluid column}$$

$$(0.5824 - 0.08) \times 1904 = 957 \text{ psi.}$$

10. Bleed back to the original wellhead pressure (4358 psi) plus the hydrostatic increase from the fluid pumped (957 psi). 4358 − 957 = 3401 psi.
11. Stage 3: Pump fluid until the wellhead pressure reaches the maximum allowable value of 5500 psi. Calculate the volume pumped and consequent reduction in pressure.

$$\text{Volume to reach } P_{max} = \frac{3401 \times 108.24}{5500} = 66.93 \text{ bbls}$$

$$\text{Volume pumped} = 108.24 - 66.93 = 41.31 \text{ bbls.}$$

$$\text{Height of fluid column} = \frac{41.31}{\text{Tubing capacity } 0.0149} = 2772\text{ft.}$$

$$\text{Hydrosatic pressure} = (\text{kill fluid gradient} - \text{gas gradient}) \times \text{height of fluid column}$$

$$(0.5824 - 0.08) \times 2772 = 1392 \text{ psi.}$$

12. Bleed back to the original wellhead pressure (3401 psi) plus the hydrostatic increase from the fluid pumped (1392 psi). 3401 − 1392 = 2009 psi.

The process of pump, wait, calculate, and bleed continues until the well is dead. In this example, five stages should be sufficient to kill the well. As demonstrated, 10 stages are needed to kill the well using the constant volume method.

As with the other examples, the best approach is to construct a spreadsheet to plan the kill. The same spreadsheet can be used during the operation if adjustments need to be made to pressure or volume. The spreadsheet can also be used to calculate bottom-hole pressure (or pressure at the plug). As shown here, increasing pressure to the maximum allowable at each step will result in very high BHP toward the end of the kill. In this example, pressure at the plug is over 11,000 psi at the final stage Table 7.16.

7.7.4 Alternative (simplified) pressure method

This method uses a simple pressure calculation where the outcome depends on recorded surface pressure before and after pumping kill fluid. Kill fluid is pumped and the wellhead pressure allowed to climb by a predetermined amount. Whilst, strictly speaking, there is no need to record the volume of fluid pumped at each stage, keeping a record is strongly recommended.

Calculations for the bleed-down pressure are based on pressure readings that are taken both before and after the fluid is pumped. The following equation is used:

$$P_3 = P_1^{\,2} \div P_2$$

where:

P_1: shut-in pressure before pumping.
P_2: stabilized shut-in pressure after pumping.
P_3: tubing pressure after bleeding off (pressure to bleed-down to).

Table 7.16 Constant pressure kill schedule

Step	Start volume (bbls)	End volume (bbls)	Cumulative pumped (bbls)	Volume pumped (bbls)	Start pressure (psi)	End pressure (psi)	Hydrostatic reduction (psi)	Bleed to (psi)	Pressure at plug. (psi)
1	153	136.62	16.38	16.38	4911.00	5500.00	552.30	4358.70	6873.78
2	136.62	108.27	44.73	28.35	4358.70	5500.00	955.91	3402.79	7829.69
3	108.27	66.99	86.01	41.28	3402.79	5500.00	1392.04	2010.75	9221.72
4	66.99	24.49	128.51	42.50	2010.75	5500.00	1432.89	577.86	10,654.62
5	24.49	2.57	150.43	21.92	577.86	5500.00	738.98	0	11,393.59

The obvious benefits of this system are its simplicity and adaptability. Since P_2, the stabilized shut-in pressure after pumping, is an actual value, the method becomes self-adjusting. If kill fluid is being lost to the formation, the gauge reading will fall and not stabilize until equilibrium is reached.

Note: This method should not be used for controlling the well where a kick has occurred because the fluid in the well is underbalanced. The calculation $P_3 = P_1^2 \div P_2$ *only works if the fluid in the well is sufficiently dense to overbalance the formation by the required amount when all of the gas has been removed from the tubing. If a weighted-up brine is needed to restore control to the well, then use the volume method.*

7.7.4.1 Alternative pressure method (example)

1. Calculate the kill weight necessary to overbalance the formation by the required amount.
2. Calculate the volume of kill fluid required (from the top reservoir to the surface).
3. Record the SITP.
4. Decide on a pressure range (pressure increase for each cycle).
   In this example, the SITP before the kill starts is 1750 psi.
   A range of 100 psi will be used (Table 7.17).
5. Initial SITP is 1750 psi. Pump fluid into the well until the SITP increases by the set amount, in this case 100 psi.
6. Wait for the pressure to stabilize.
7. Calculate the pressure to bleed back to using the equation $P_3 = P_1^2 \div P_2$
8. Repeat until the well is dead.

This method will only work if the pressure has stabilized after the fluid is pumped. If the formation begins to take fluid, the pressure will drop. For example, in this case after pumping enough fluid to increase

Table 7.17 Lubricate-and-bleed: self-adjusting pressure method

$P_3 = P_1^2 \div P_2$: Pressure range: 100 psi		
P_1	P_2	P_3
1750 psi	1860 psi	1646 psi
1646 psi	1750 psi	1549 psi
1549 psi	1670 psi	1437 psi

the SITP to 1860 psi the pressure was to drop to 1820 psi, the calculation $P_3 = P_1^2 \div P_2$ would simply be repeated using the new stabilized pressure and P_3, and the bleed-off pressure revised, becoming 1683 psi.

If the pressure drops on subsequent stages, then the range should be reduced.

7.8 CAUSES AND DETECTION OF KICKS

As described at the beginning of this chapter, many, if not most completion and workover operations are conducted above an open permeable formation. A clear solids-free brine provides the overbalance pressure needed to control the well. Minimizing fluid loss to the formation means operating with low overbalance pressure and that in turn means continual and careful monitoring of fluid level, since nearly all well control problems during completion and workover operations are caused by fluid loss, rather than an unexpected increase in pressure. The most common causes of a kick during completion, workover, and intervention operations are as follows.

- Fluid loss to the formation
- Loss of hydrostatic overbalance
- Fluid level changes caused by pipe displacement
- Swab and surge pressure
- Incorrect brine weight
- Trapped pressure
- Mechanical failure
- Disabled or malfunctioning alarms

7.8.1 Fluid loss to the formation

Well control incidents during completion and intervention operations are usually caused by loss of hydrostatic overbalance. It is unusual for an influx to occur because of higher than expected pressure in the reservoir. In common with drilling operations, losses are more likely to occur under dynamic (circulating) conditions as friction increases BHP. Nevertheless, static losses still pose a risk and are more likely than during a drilling operation since solids-free fluids are used. Where LCM is used, it is not normally as effective at controlling losses as a mud filter cake.

The main causes of fluid loss to the formation and consequent loss of overbalance are:

1. *Decay or breakdown of solids-free LCM.*

   Solids-free LCM is often used to control losses. Fluid viscosity is increased by mixing polymers with the base fluid to slow or stop fluid leak-off. High temperature downhole reduces viscosity, and over time can break the polymer. It is important that viscous fluids are tested at bottom hole conditions to enable the break to be anticipated.

2. *Loss of LCM as a consequence of remedial operations.*

   There is often a need to carry out remedial work before a tubing string is installed. Commonly performed operations include the removal of unwanted debris or scale, and casing scraping operations in preparation for packer setting. Fluid circulated across an open formation can wash away LCM.

3. *High ECD.*

   High ECD will lead to losses. Some of the operations associated with running a completion or performing an intervention require high circulation rates and a consequently high ECD.
   - Wellbore cleaning, especially in deep, high angle wells.
   - HPHT wells where very dense (and therefore more viscous) brines must be used.
   - During remedial cementing operations.
   - Forward circulation—wells where there is limited clearance between the tubing and open hole or tubing and casing.
   - Reverse circulation—wells with small ID tubing.
   - When circulating in gravel during an open hole gravel pack (water pack). Particularly during the propagation of the "Beta" wave.

4. *Cross-flow between zones.*

   Many wells are completed across more than one reservoir layer, and not all completions are configured to provide isolation between the layers. With the well is shut-in, cross-flow from the high-pressure zone into low pressure zones takes place. Losses into the lower pressured zone can promote losses from the wellbore and loss of overbalance. Cross-flow will also tend to sweep LCM out of the higher pressure zone.

5. *Formation fracture.*

   Some operations performed during completions and workovers increase the risk of fracturing the formation and consequent losses. Remedial squeeze cementing, if poorly executed, can lead to formation fracture and fluid loss. Fracturing of the formation can also occur during the "beta wave" phase of an open hole gravel pack (circulating water pack).

6. *Declining reservoir pressure caused by production in an adjacent well.*

   During a workover or completion bottom hole (reservoir) pressure may decline if nearby wells are produced. Declining reservoir pressure means the calculated over-pressure provided by the brine will increase, potentially leading to fluid loss. Pressure decline will depend on well spacing, connectivity between the wells (how they communicate), and the rate of offtake from nearby wells. With a good understanding of the reservoir properties and an accurate reservoir model, the amount of depletion can be anticipated. In some cases, adjacent wells may have to be closed-in as a precautionary measure, although there is an understandable reluctance to do this for commercial reasons.

7. *Mechanical failure of a downhole barrier.*

   Mechanical barriers are used in many completions and workovers to isolate the reservoir. For example, in gravel packed wells it is routine to use a closed mechanical barrier such as a formation isolation valve to isolate the reservoir whilst the upper completion (production tubing) is run. Similarly, it is common practice to install a retrievable bridge plug in a liner to allow a workover to take place. This has the advantage of isolating the reservoir from potentially damaging completion fluids, and makes post workover clean-up of the reservoir unnecessary. Mechanical barriers are very useful in wells where fractured and permeable formations make the control of losses problematic, and for many workover and completion operations will be the preferred method of well control if completion design and conditions allow. However, mechanical barriers fail from time to time, and when they do losses are more likely to occur because it is unlikely that LCM will have been used.

   Some workover operations are entirely reliant on mechanical barriers and are performed with underbalance conditions above the barrier(s). Barrier failure means an immediate and sizable influx.

7.8.2 Loss of hydrostatic overbalance

There are several circumstances that can lead to loss of overbalance, even where the well is properly monitored, and the hole kept full:

1. *Thermal effects.*

   As temperature increases brine expands and density decreases (Chapter 5, Completion, Workover, and Intervention Fluids). If no correction is made for wellbore temperature, particularly when temperature is high, brine pumped into the hole will not provide sufficient overbalance.

2. *Brine dilution.*

   Something as basic as an unattended and forgotten water hose left running into a brine pit can dilute brine sufficiently for overbalance to be lost. In monsoon affected areas (and Scotland) where heavy rainfall is common, brine in uncovered pits can be diluted. Brine in the well can also be diluted by cross-flow between zones.

3. *Brine crystallization.*

   When crystallization of the brine salts occurs, the brine leaving the pits will be less dense than planned (Chapter 5, Section 5.4, Completion, Workover, and Intervention Fluids). Regular checks on brine density and visual checks on the pits are needed if ambient temperature is close to crystallization point.

4. *Reservoir pressure higher than anticipated.*

   If a well is perforated with kill weight fluid in the well, an unexpectedly high reservoir pressure will overcome the kill fluid and the well will kick.

5. *Reservoir pressure increase.*

   Overbalance will be lost if reservoir pressure increases during the workover. This can happen if nearby water or gas injection wells remain on line. A pressure increase caused by continued injection depends on well spacing, connectivity between the wells, and injection rate. Once again, a good understanding of the reservoir properties and an accurate reservoir model are useful. There may be a requirement to reduce or cease injection until the completion is secure.

6. *Trapped pressure.*

   When a well is killed prior to a workover, gas pockets are sometimes left trapped in the wellbore; for example, below a packer. If the packer is released, or milled out, the gas bubble will be free to rise to the surface. If gas expansion is not controlled, it will displace fluid from the well and reduce the overbalance. Pockets of high pressure can also become trapped below bridge plugs and obstructions in the wellbore, e.g., a scale bridge.

7.8.3 Fluid level changes caused by pipe displacement

When pipe is tripped into the well, displacement will cause the fluid level to rise. Similarly, pulling pipe causes the fluid level to drop. If the hole is not regularly topped up as tubing is pulled, overbalance will be lost. A widely used rule of thumb is to top up the hole when the overbalance has been reduced by 75 psi. This is based on the 200–300 psi overbalance widely used for completion operations. That overbalance will reduce more quickly when pulling closed end tubing, since it has a greater displacement volume than open ended pipe. Displacement volume for open ended and closed ended tubing is readily available from tables or can be calculated.

Open ended pipe displacement volume bbls/ft. = $OD^2 - ID^2/1029.4$
Closed end pipe displacement volume bbls/ft. = $OD^2/1029.4$

Allowances must also be made for nonstandard sized equipment where the displacement volume is different from the rest of the string. External control lines and flat packs, a feature of many completions, will add to the displacement volume (per unit length). Whilst a single ¼ in. hydraulic control line may not displace much fluid and might reasonably be ignored, multiple "flat-packs" or power cables for electric submersible pumps displace a significant volume and must be included. For example, Reda 3 × 6 AWG flat ESP cable measures 1.40 in. × 0.55 in. and displaces 4 US gallons per 100 ft. By comparison, open ended 3½ in. 9.2 lb/ft. tubing will displace 13.4 US gallons per 100 ft.

To calculate hydrostatic pressure drop per (vertical) foot of dry (open ended) pipe pulled:

$$\Delta P_{\text{psi/ft.}} = \frac{0.052 \times \text{brine weight}_{\text{ppg}} \times \text{tubing displacement}_{\text{bbl/ft.}}}{\text{annulus capacity}_{\text{bbl/ft.}} + \text{tubing capacity}_{\text{bbl/ft.}}} \tag{7.10}$$

To calculate hydrostatic pressure drop per (vertical) foot of wet (closed end) pipe pulled:

$$\Delta P_{\text{psi/ft.}} = 0.052 \times \text{brine weight}_{\text{ppg}} \frac{\text{tubing capacity}_{\text{bbl/ft.}} + \text{tubing displacement}_{\text{bbl/ft.}}}{\text{annulus capacity}_{\text{bbl/ft.}}} \tag{7.11}$$

To calculate length of open ended (dry) pipe pulled to reduce overbalance by a given value:

$$\text{Feet} = \frac{\Delta P_{\text{psi}} \times (\text{annulus capacity}_{\text{bbl/ft.}} + \text{tubing capacity}_{\text{bbl/ft}})}{0.052 \times \text{brine weight}_{\text{ppg}} \times \text{tubing displacement}_{\text{bbl/ft.}}} \quad (7.12)$$

To calculate length of closed end (wet) pipe pulled to reduce overbalance by a given value:

$$L_{\text{ft.}} = \frac{\Delta P_{\text{psi}} \times \text{annulus capacity}_{\text{bbl/ft.}}}{0.052 \times \text{brine weight}_{\text{ppg}} \times (\text{tubing capacity}_{\text{bbl/ft}}. + \text{tubing displacement}_{\text{bbl/ft.}})} \quad (7.13)$$

Example calculation:
Open ended pipe displacement:

Tubing size	4½ in. 12.6 lb/ft. (ID = 3.958 in.)
Casing size	7 in.29 lb/ft. (ID = 6.184)
Fluid in wellbore	9 ppg brine
Hydrostatic overbalance with hole full	250 psi

How much pipe can be pulled from the well before the hydrostatic overbalance is reduced by 75 psi?

How much can be pulled from the well before the overbalance is lost? Assume the well is static (no losses or gains):

1. Calculate tubing capacity : $\frac{3.958^2}{1029.4} = 0.01521$ bbls/ft.
2. Calculate tubing displacement : $\frac{4.5^2 - 3.958^2}{1029.4} = 0.004453$ bbls/ft.
3. Calculate annulus capacity. $\frac{6.184^2}{1029.4} = 0.03174$ bbls/ft.

   $\Delta P_{\text{psi/ft.}} = \frac{0.052 \times 9 \times 0.004453}{0.03174 + 0.01521} = 0.04438$ psi pressure reduction per foot of tubing pulled.
4. To reduce HP by 75 psi, 75/0.04438 = 1690 ft.
5. To reduce HP by 250 psi, 250/0.04438 = 5633 ft.

   Or

1. $L_{\text{ft.}} = \frac{75 \times (0.03174 + 0.01521)}{0.052 \times 9 \times 0.004453} = 1690$ ft.
2. $L_{\text{ft.}} = \frac{250 \times (0.03174 + 0.01521)}{0.052 \times 9 \times 0.004453} = 5632$ ft.

Pipe displacement data should be included on a pipe tally or trip sheet in an easy-to-use format. This value can be incorporated into a pipe tally spreadsheet and used to calculate the displacement volume of each joint.

Since the displaced volume of open ended pipe is relatively small, the trip tank should be used to accurately monitor fluid level when tripping (a trip tank description and illustration can be found in Chapter 4, Well Control Surface Equipment). If the fluid level in the well is static, gains or losses from the trip tank should be equal to the displacement volume of the tubing as it is run or pulled from the hole. However, if the level in the trip tank is changing by more than the expected volume, it should be treated as an early warning sign:

- *Running pipe*: Trip tank gain more than pipe displacement volume—the well is flowing.
- *Running in*: Trip tank gain less than pipe displacement volume—fluid loss to the formation.
- *Pulling pipe*: Hole fill-up volume less than tubing displacement—the well is flowing.
- *Pulling pipe*: Hole fill-up volume more than pipe displacement—fluid loss to the formation.

Whilst the response to a trip tank/pit gain (an indication of flow) should always be to shut-in the well and record pressure, the response to losses during completions and interventions is not as straightforward. In many cases some fluid losses are expected and tolerated. However, the maximum loss rate should be defined and agreed beforehand. If losses exceed the agreed amount, measures must be taken to stem the losses, usually the pumping of LCM. In formations where formation damage is not a problem, and where there is a sizable reserve of kill fluid, quite high loss rates can be tolerated and managed.

7.8.4 Swabbing and surging

If the tubing in the well is picked up too quickly, it creates a piston effect, lowering BHP. If the pressure reduction is more than the overbalance provided by the kill weight fluid, the well will kick. The pressure reduction caused by upward pipe movement is called swabbing and is a common cause of kicks. Similarly, when tubing is run into the well, the piston effect will increase pressure downhole. This is called surging and can lead to fluid losses or in extreme cases, fracture the formation.

Swab and surge pressure is primarily a function of trip speed pipe size, casing (or open hole) size, and the viscosity of the fluid in the well. Surge and swab pressure can be significant, as illustrated Table 7.18. This shows the BHP reduction for 4½ in. pipe inside three different hole sizes across several trip speeds.

The table clearly indicates that even a quite moderate trip speed can create swab pressure if there is limited clearance between the workstring and the bore hole, i.e., a 4½ in. pipe inside a 6½ in. hole creates a swab pressure of 256 psi if a stand (typically 90 feet of pipe) is pulled in 30 seconds. This is more than the overbalance pressure used for many completion and workover operations, and would cause an influx. During the running and pulling of completions, trip speed is generally slower than during operations with drill pipe or workstring. Nevertheless, caution must be exercised and the tubing run at a speed that keeps surge and swab pressure within safe limits. Most completions are run in clear, solids-free brines that generally have lower viscosity than drilling mud. However, viscous brines are widely used to control losses and as carrier fluids during gravel packing and fracturing operations. The presence of viscous fluid in the wellbore will increase surge and swab pressure, and trip speed must be modified accordingly. Debris in the wellbore will also increase the surge and swab pressure, as it reduces clearance around the pipe and completion components. Often, the greatest risk from increased surge and swab pressure during completion and workover operations is caused by the very narrow clearance between the production packer and the casing. This can be especially problematic when pulling a retrievable packer that has been left in the well for a long time. After release, the packer seal element will remain in a partially extruded state, and may well be in physical contact with the casing wall around the full circumference. Swabbing is likely, even at moderate trip speed, unless the seal element is given time to relax. A useful rule of thumb is to allow at least 30 min before attempting to pull pipe. If concerns about swabbing remain, the worst effects can be alleviated by punching holes in the tubing immediately above the packer. This reduces swab pressure by allowing fluid to by-pass the packer through the tubing.

Table 7.18 Swab pressure reduction for a range of trip speeds (swab pressure in psi)
4½ in. OD pipe, 14.0 ppg mud

Hole size (in.)	Pulling speed (seconds per stand)					
	15	22	30	45	68	75
8½	267	167	124	98	84	75
6½	589	344	256	192	159	140
5¾	921	524	294	289	231	200

During the running of a completion equipped with a sliding sleeve (SSD) or gas lift mandrel above (and close to) the packer, the communication ports can be left open to reduce surge pressure. Surge and swab pressure can be estimated using hydraulics modeling packages such as WellPlan.

7.8.5 Mechanical failure

Surface and downhole equipment can fail. Analysis suggests that as many as 20% of blowouts are, at least in part, caused by equipment failure. If surface equipment fails, then a kick becomes a blowout unless secondary and tertiary barriers are in place.

When newly installed, a completion should be able to withstand the maximum anticipated pressure anywhere in the wellbore. Equipment failure does occur in newly completed wells, usually as a result of design errors or equipment malfunction. In older wells, the failure of downhole equipment is common, and to be expected, since metallic components will erode and corrode, and elastomer seals will degrade. Many workovers are carried out because of failed downhole equipment. The poor condition of tubing, and other completion components, adds to well control complexity during intervention or workover operations. Well kill is often more difficult and secondary well control (BOP) is not as effective when trying to seal around corroded and holed pipe (Fig. 7.21).

- Corroded or holed tubing, or a leaking packer, will result in a live annulus; a condition that is not normally tolerated.
- Liner cement and liner lap failure—reservoir pressure will reach the production casing. A failure of the liner lap can be very serious, as reservoir pressure can migrate behind pipe and reach the production casing. An abnormally heavy kill fluid may be needed to control the well.
- Breakdown of isolated (squeezed) perforations resulting in cross-flow between layers.

7.8.6 Disabled alarms

Some rigs have audible and/or visual alarms that are activated by changes in return flow or tank fluid level. If improperly calibrated or maintained, they will give spurious warnings. A system that repeatedly gives false alarms is likely to be disabled by the crew, and if not disabled, may be ignored.

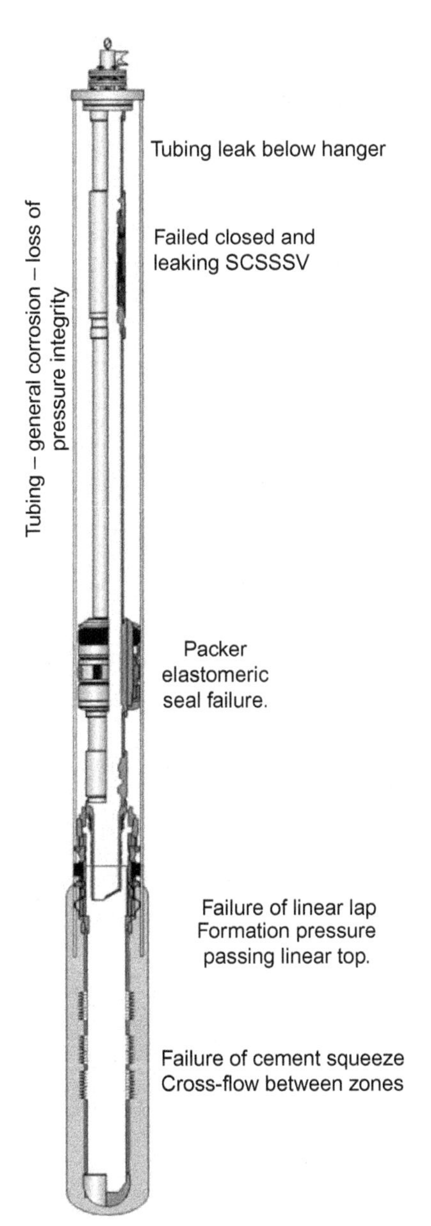

Figure 7.21 Some commonly seen failure modes that can both require a well intervention or workover to remedy, and will complicate well control and the well kill.

7.9 KICK DETECTION

There are several indications that a kick is imminent, and the ability to recognize these indicators is an essential skill for a rig crew. Early detection ensures the well is shut-in quickly, preventing or limiting the size of a kick. Early kick indicators are as follows.

- Gas or oil shows whilst circulating
- Trip tank volume inconsistent with tubing displacement while tripping
- A decrease in pump pressure whilst circulating
- Increased flowline temperature whilst circulating
- Increase in weight indicator reading

Gas or oil shows whilst circulating.

Trace gas and oil contamination whilst circulating is a good early indication of a kick. Gas cut fluid often has a foamy appearance, and gas should be detected by monitoring equipment. Oil contamination will reduce the density of the returning fluid, and will also leave a visible sheen on any water-based fluid once it has been returned to the pits. Even a small quantity of oil will be visible.

In addition to hydrocarbon shows, an influx of reservoir water will affect the density and chloride concentration of returning fluid. Regular inspection and testing of the returning fluid for hydrocarbons, density changes, and changes in chemical composition will aid in early kick detection.

Inconsistent fluid level during tripping.

When tripping pipe, the fluid level change should be consistent with the tubing displacement volume (see Section 7.8.3). Rig crews need to be aware of displacement volume and expected fluid level changes. If fluid level change is not consistent with expectations, stop tripping and observe.

Decreasing pump pressure during circulation.

If there is an influx of low density oil or gas, the HP of the returning fluid will reduce. This will cause a reduction in overall circulating (or reverse circulating) pressure. Any reduction in pump pressure that does not correspond to a change in rate must be investigated. Stop circulation, shut-in, and observe.

Increased flowline temperature during circulation.

When a kick occurs, the temperature of the invading fluid is normally higher than the circulated fluids. Any unexpected temperature increase in returns should be investigated. Stop circulation, shut-in, and observe.

7.9.1 Positive kick indication

There are three positive indicators of a kick. If any of the following conditions are observed, immediate steps must be taken to control the well:

- Increase in pit volume.
- Increase in return flow, or flow with the pumps off.
- Surface pressure when the well is shut-in.

Increase in pit volume.

Since the circulating system is a closed loop, any increase in volume must be entering from the formation. An increase in pit level (or trip tank level) not consistent with pipe displacement should be treated as a positive indication of a kick. Close-in immediately.

Increase in return flow rate (or flow with pumps off).

If fluid out of the well is more than fluid in, the well is flowing. Any indication of flow should be confirmed by shutting down the pumps and observing the well under static conditions. If a kick is suspected it is common practice to pick up off bottom, stop the pumps and observe. Continued pit gain is a positive indication of a kick, close-in immediately.

Surface pressure with the well shut-in

Observed surface pressure after closing-in the well is a positive indication that a kick has occurred. Once a well has been closed-in, the pressure control equipment should be checked for leaks. The crew will continue to monitor tubing and annulus pressure whilst preparing to kill the well.

On rigs where daylight only work is being carried out, and the well is left closed-in overnight, checks should be made for surface pressure before resuming operations.

7.10 MINIMIZING THE INFLUX

A large influx is more difficult to remove, since it results in higher pressure in the wellbore. In addition, a larger volume of contaminated fluid will need to be handled at surface. Early recognition and a rapid response (close-in) will limit the size of the influx and the amount of kill fluid that flows from the well. After shut-in, recording surface pressure and the pit gain is used to establish the size and type of influx.

Factors affecting shut-in pressure after a well has kicked are:

- The size of the influx
- The geometry of the well (annulus capacity)
- The type of influx
- Formation (reservoir) pressure

7.10.1 Establishing reservoir pressure, influx size, and influx type

During a kick, lower density fluid from the formation displaces the kill weight fluid from the top of the well, hence the pit gain. If the overall density of the fluid column is reduced sufficiently, overbalance will be lost and pressure will be observed at the surface after closing-in the well. The density of the kill fluid in the well is known, as are the dimensions of the tubing and casing. Using the recorded shut-in pressure and the recorded volume of pit gain, it is possible to calculate the BHP as well as the height of the influx. It is also possible to estimate the type of influx (gas, oil, or formation water).

To calculate formation (bottom hole) pressure with the well shut-in following a kick:

$$\text{BHP} = \text{SITP} + (\text{mud wt ppg} \times 0.052 \times \text{TVD}_{\text{ft.}}) \tag{7.14}$$

Example:

Reservoir depth:	8000 ft. TVD (vertical well)
Brine weight (pre-kick)	13.7 ppg
7 in. 29 lb/ft. casing ID	6.184 in.
Tubing OD	4½ in.

Following a 20 bbl pit gain, the well was closed-in. Initial SICP = 760 psi and initial SITP, 320 psi.

New BHP:

$$\text{BHP} = 320 + (13.7 \times 0.052 \times 8000) = 6019 \text{ psi}$$

The length of an influx is calculated from:

$$\text{Influx length}_{\text{ft.}} = \frac{\text{influx size}_{\text{bbls}}}{\text{lower annulus capacity}_{\text{bbls}}} \tag{7.15}$$

Continuing the previous example, the 20 bbl pit gain inside the 4½ in. × 7 in. annulus would have a length of:

$$\text{Annulus capacity is} \frac{6.184^2 - 4.5^2}{1029.4} = 0.0174 \text{ bbl/ft.}$$

$$\text{Influx length}_{\text{ft.}} = \frac{20}{0.0174} = 1149 \text{ ft.}$$

Since a smaller capacity means a longer influx for a given volume, it follows that the HP reduction and surface pressure increase will be greater in a small capacity well. HP reduction is also a function of well angle. In a vertical well, the length and vertical height of the influx are the same. As well angle increases the length of the influx is unchanged, but the vertical height, and hence the reduction in HP, gets smaller. For example, at a 45° well angle the 1149 ft. long influx would have a vertical height of 812 ft. As the kick is circulated out of the well, the vertical height of the kick will increase as it moves into the vertical section of the well (ignoring gas expansion).

It is useful to establish the type of kick (gas, oil, or water), before attempting to bring it to the surface. Removal of a gas influx will give high casing and wellbore pressure during removal. Oil kicks, especially large ones, can contaminate the kill fluid and are difficult to dispose of. Formation water is the least problematic, but still has the potential to dilute kill weight brine. The precipitation of scale can be a problem if the formation water is not compatible with the kill weight brine.

In proven reservoirs, recent production records will help identify the nature of the influx. In newly developed fields, not enough information will be available to properly assess the influx. However, the size of the kick (from the pit gain) combined with the shut-in pressures, allow the pressure gradient (psi/ft.) of the influx to be calculated. If the gradient of the influx is known, assumptions can be made about its type (Table 7.19).

To calculate the pressure gradient (psi/ft.):

$$\text{Influx gradient}_{\text{psi/ft.}} = \text{kill fluid gradient}_{\text{psi/ft.}} - \frac{\text{SICP} - \text{SITP}}{\text{influx length}_{\text{ft.}}} \tag{7.16}$$

Table 7.19 Kick fluid pressure gradient: approximate values

Kick fluid	psi/ft.
Formation water	>0.43
Formation water/oil/gas	0.2–0.43
Oil	0.25–0.35
Oil/gas	0.2–0.3
Gas	<0.2

Continuing the example:

$$\text{Influx gradient}_{\text{psi/ft.}} = 0.7124 - \frac{760 - 320}{1149} = 0.33\ \text{psi/ft.}$$

In the example, an oil influx is suspected.

7.11 SHUT-IN PROCEDURES

As soon as a kick is detected, the most important single step in preventing a blowout is closing the chosen preventer to shut-in the well. The procedures described here assume that there is pipe (workstring or completion tubing) across the BOP. Some of the equipment run through the BOP during completions and workovers will compromise the ability of both pipe rams and annular preventers to seal against well pressure. Procedures for dealing with this type of equipment are covered towards the end of this chapter.

7.11.1 Close-in procedure whilst circulating

This procedure assumes that a circulating head is in place, or the top drive is made up to the top of the tubing string and circulation is underway. It further assumes that the tubing is at, or close to, reservoir depth. Under forward circulation fluid returns flow from the top of the well through the return line and back to the pits. The BOP kill line is closed. The line to the choke is open, but the choke is closed. At the first indication of a kick:

- Alert the crew
- Position the string so that couplings are clear of the BOP sealing elements and a connection(s) is accessible at the rig floor
- Stop circulation
- Close the uppermost applicable preventer and confirm the well is shut-in

- Record:
  - Shut-in tubing pressure
  - Shut-in annulus pressure
  - Pit gain volume
  - Time the well was shut-in

After shut-in the well should be monitored for any pressure increase associated with gas migration.

The drilling industry and the API describe a hard shut-in as one where the choke is in the closed position before the BOP preventer is closed (API RP59).[4] A soft shut-in is one where the choke is closed after the BOP preventer is closed. When drilling, the soft shut-in is sometimes preferred as the wellbore is exposed to lower pressure during the shut-in. For completion and workover operations, the hard shut-in is normally used since it is simple, allowing well closure in the shortest possible time. During completion and workover operations the higher pressure associated with the hard shut-in is not normally a concern.

7.11.2 Shut-in procedure whilst tripping pipe

The risk of a kick is normally greater when pipe is moving, partly because of swab and surge pressure, and partly because it is more difficult to detect changes in trip tank level when having to include tubing displacement.

Rig crews need to be especially vigilant when tripping a completion with minimal clearance between large diameter components, such as a packer and the casing wall. The effect of swabbing can be cumulative, with each joint pulled causing a small influx until hydrostatic control is lost. If this is suspected, the pipe should be run back to the bottom and the well circulated bottoms-up to remove any reservoir contamination of the hydrostatic column. To close-in the well when tripping pipe:

- Alert the crew.
- Bring the string to a working height at the rotary table and stab the full opening safety valve:
  - If drill pipe is being tripped, then the full opening safety valve will normally have a thread compatible with the workstring.
  - If production tubing or a drill stem test (DST) string is being run, the tubing in the rotary table (box up) will be different from the pin down connection on the full opening safety valve. A cross-over must be available on the rig floor.
- Position the string so that couplings are clear of the BOP sealing elements and a connection(s) is accessible at the rig floor.

- Close the uppermost applicable preventer and confirm the well is shut-in.
- Record:
  - Shut-in tubing pressure (SITP).
  - Shut-in annulus pressure (SICP).
  - Pit gain volume.
  - Time.

If the string weight is light (not much pipe in the well), there is an additional hazard associated with shutting in; force exerted by well pressure can exceed string weight. If this is the case and the annular preventer was used to close-in, piston force will eject the pipe from the well; an extremely hazardous occurrence. Most production tubing has upset connections. Closing the pipe rams will hold the pipe in the well against piston force as the upset is too large to pass through the closed ram.

If the tubing positioned across the BOP is not the correct size for the pipe ram, or does not have upset connections, then the crew has two choices:

- If time permits, a "kick stand" of drill pipe (with a full opening safety valve) can be made up on top of the tubing in the rotary table (a cross-over will be needed). Once made up, the drill pipe can be lowered across the BOP and the pipe rams closed.
- As a last resort, it may be necessary to drop the tubing:
  - Lower the string until the elevators are at working height above the rotary table.
  - Close the annular preventer to take the weight of the string and enable the elevators to be opened. Once the elevator is unlatched, retract the annular BOP, dropping the string dropping the string.
  - Close the blind ram.
  - Record SICP.

It is not unusual to install non-return valves in the tubing string during completion and workover operations. Where this is the case, additional steps are needed to determine tubing pressure after a shut-in:

1. Line up to pump into the tubing.
2. Pump down the tubing at a low rate (approximately ½bbls/min).
3. Tubing head pressure will increase until the BPV opens, at which point there will be a momentary hesitation in the rate of pressure increase.
4. Note the pressure at which the hesitation occurs.

If possible, plot pressure against volume pumped. The point at which the BPV opens should be easy to read from the plot. It will be similar in appearance to the break-over-pressure during a formation leak-off test (Fig. 7.22).

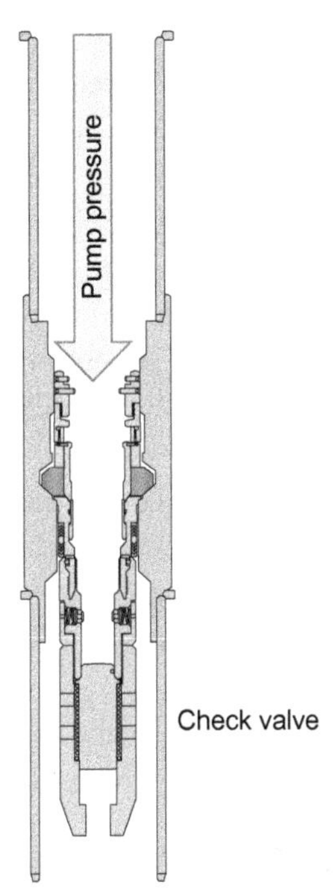

Figure 7.22 Pressure against volume with a BPV in the tubing.

7.12 REGAINING WELL CONTROL FOLLOWING A KICK

During completion, workover, or intervention operations on a dead well, the potential for a kick is always present. If a kick occurs, the well will need to be closed-in and the kick removed. The methods used to remove a kick are fundamentally the same as those used to kill the well before workover or intervention operations began. In general, if circulation was used to make the initial kill, then it is likely that it will remain the most appropriate method for regaining control after a kick. What may change is the direction of circulation. In anything other than a truly vertical well, density differences will mean that hydrocarbons entering the wellbore during a kick will tend to migrate towards the high side of the hole. Unless centralized, the tubing will rest on the low side of the hole, meaning most, perhaps all, of the influx, will occupy the annulus. Given the potentially rapid migration rate of gas kicks in brine (up to 6000 ft./h) the best way to remove the kick is normally with forward circulation. Using a non-circulating technique (lubricate-and-bleed or bullhead) is more dependent on the nature of the kick. Most completion or intervention well control incidents are caused either by a failure to keep the hole full, or by swabbing in the well. Crucially, reservoir pressure remains unchanged, meaning the well can be killed using the brine already in use. However, there will be times when the brine in use does not have sufficient density and must be replaced with heavier brine. The need to increase the brine weight can result from:

- Exposure to higher than expected pressure in the wellbore; e.g., perforating a new zone, drilling a side track, milling through a scale bridge, or washing out a sand plug.
- Contamination of kill weight brine with lighter formation fluid. Often this is the result of cross-flow within the reservoir.
- Poor management of the completion brine:
  - Brine dilution at surface, e.g., heavy rain or the mistaken addition of drill water in the pits.
  - Reduced density resulting from brine crystallization.
  - Reducing density due to the hygroscopic properties of some heavy brines.[5]

If the well can be killed by circulation, then replacing the underbalanced fluid in the well with a suitably weighted-up brine can be achieved using either the drillers' method or wait and weight. If circulation is not possible, filling the wellbore with kill weight brine becomes more problematic. Bullheading can be used, but requires the entire well contents to be pushed into the formation to restore overbalance. If bullheading is not possible, a kick can be brought to surface using the volumetric method and then removed by lubricate-and-bleed. However, the well would still be in an underbalanced condition. Circulation would be possible if a workstring could be snubbed into the well. Coiled tubing would also allow circulation of kill weight fluid.

Where the brine in the well is still kill weight and a circulation path is available, forward or reverse circulation can be used. If no circulation path is available, a gas kick can be controlled and brought to the surface using the volumetric method. The gas cap can be removed using the lubricate-and-bleed technique.

If an unexpectedly high reservoir pressure has been encountered, then the brine weight will need to be increased before the well can be killed. The revised kill weight is calculated as follows:

Oilfield units:

$$\text{Kill fluid weight (overbalanced)} = (\text{SITP} + \text{overbalance})/\text{TVD}_{\text{perfs}}/0.052 + \text{tubing fluid weight} \quad (7.17)$$

SI units:

$$\text{Kill weight fluid} = (\text{SITP (kPa)})/(0.00981 * \text{TVD(m)}) + \text{original brine weight (kg/m}^3) \quad (7.18)$$

7.12.1 The drillers method

The drillers' method is widely used. It is simple and often the most appropriate technique for workover well control situations. Like all constant BHP methods, the aim is to circulate out the kick whilst holding enough pressure against the formation to prevent a second influx. Depending on the nature of the kick, it will be removed with one or two circulations:

- If the kick has been caused by swabbing, not keeping the hole full, or by fluid loss, the brine in use will be of sufficient density to kill the well. Control can be restored by circulating the kick to the surface. A single circulation (or less) is all that is normally required.
- If the kick was caused by an unexpected increase in BHP, or by dilution of the kill fluid in the well, two circulations are required. On the first circulation, the kick is brought to the surface. On the second circulation, the fluid in the well is displaced with the new kill weight fluid.

The advantage of the drillers' method is that it is simple. The first circulation can start soon after shut-in, since only minimal calculations are required and there is no need to wait for kill weight fluid to be made available. However, this method will take longer if two circulations are required, the well remains under pressure for longer, and annulus pressure is normally higher than if the well is killed by wait and weight.

7.12.1.1 Outline procedure

During drilling operations, it is standard practice to record slow circulation rates and pressures (SCRP) on each shift. This is less likely to occur during workover operations, since the pipe is not normally on the bottom for an extended duration. However, where possible, circulation rates and pressures should be recorded, as described in Section 7.3.3.

7.12.1.1.1 First circulation

1. After closing-in the well, record SITP and SICP, and the volume of the pit (or trip tank) gain.
2. Line up to pump down the tubing, raking returns from the annulus through the choke.
3. If a SCRP was recorded, calculate the pressure required on the tubing for the first circulation:

$$\text{ICP} = \text{SCPR} + \text{SITP}.$$

4. Open the choke, whilst at the same time bringing the pump up to kill speed.
5. As circulation begins, the choke operator should keep the casing pressure constant at the SICP reading.
6. Once circulation is underway, the choke should be used to hold the tubing pressure constant at ICP on the tubing pressure gauge.
7. ICP should be held constant on the tubing pressure gauge until the kick has been circulated out of the well. Pump rate must also be held constant at kill rate throughout. It will not always be necessary to complete a full circulation to remove the kick from the well.
8. Once the kick is out of the well, stop the pumps and close-in on the choke.
   a. If the influx was caused by swabbing, or by a failure to keep the hole full, then bringing the influx to surface should have restored the well to an overbalanced condition.
   b. If the influx was caused by a pressure increase, or contamination (dilution) of the kill weight brine, a second circulation will be required (Fig. 7.23).

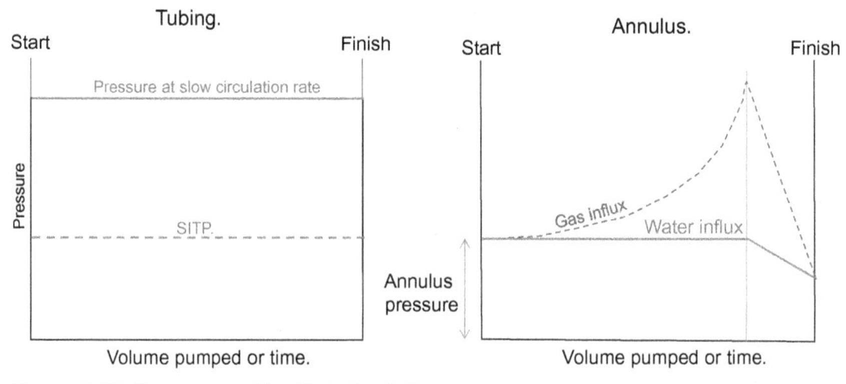

Figure 7.23 Pressure profile: First circulation.

7.12.1.1.2 Second circulation (if required)

9. Prepare enough kill weight fluid to displace the well.
10. Once the kill weight brine has been prepared, open the choke, whilst at the same time bringing the pump up to kill speed.
11. As the pumps are coming up to speed, the choke should be used to hold the casing pressure steady at SICP (this is the pressure recorded after the first circulation).
12. As the tubing fills with kill weight fluid, there are two methods for keeping a constant BHP:
    a. Use the choke to maintain a constant casing pressure.

       Note: If the influx was gas, and it was not all removed during the first circulation, this option might mean higher annulus pressure.
    b. Prepare a plot of pump pressure versus barrels (or strokes) pumped with time, starting with ICP and finishing with FCP. To calculate the pressure drop per increment for plotting on a kill graph a useful formula is:

$$\frac{\text{ICP} - \text{FCP}}{10} = \text{pressure drop per increment}$$

13. As the tubing fills with kill fluid, pressure at the surface will reduce. If the kick was fully removed during the first circulation, and if pump rate is constant, there should be little or no requirement to adjust the choke until the kill fluid reaches the base of the tubing.
14. Final circulating pressure (FCP) is achieved when kill fluid reaches the base of the tubing. Pump pressure will be the slow circulating rate adjusted for the heavier fluid:

$$\text{FCP} = \text{SCRP} \times \frac{\text{kill weight fluid}}{\text{original fluid weight}}$$

15. As the kill fluid fills the annulus, the choke is used to hold tubing pressure constant at FCP.
16. Once the second circulation is complete, the well can be shut-in. Tubing pressure and casing pressure should both read 0 psi.
17. Observe the well for flow or losses. If losses are observed, they must be within acceptable (pre-agreed) limits (Fig. 7.24).

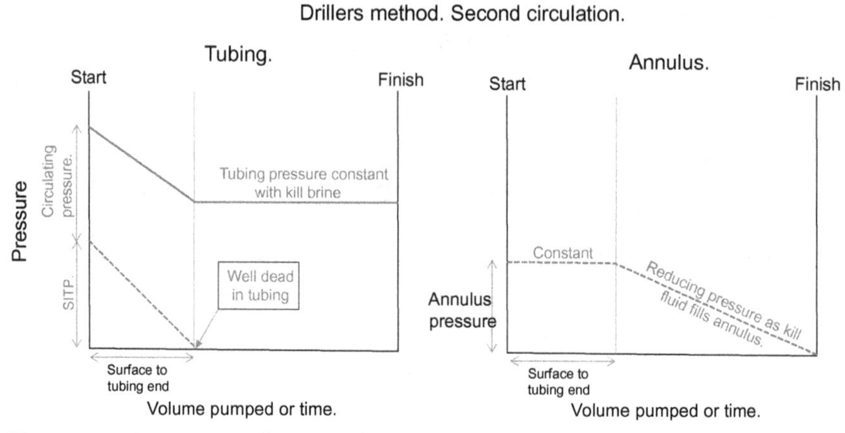

Figure 7.24 Pressure profile: Second circulation.

7.12.2 Forward circulation: wait and weight

The wait and weight, or engineer's method, removes a kick and kills the well in a single circulation. The name is suggestive of the method; there is a "wait" whilst the brine "weight" is increased sufficiently to kill the well to the required overbalance, using SITP and SICP to calculate the required brine weight increase. Although it requires slightly more complex calculations than those used in the drillers' method, it is generally preferred by well control specialists, since lower pressures make it safer and require less time "on the choke." Since time is required to weight up the brine, migrating gas will need to be controlled. It is possible for gas to be migrated to the surface before the brine weight has been increased.

In addition to preparing brine, a well kill circulating "pressure schedule" should be prepared. It will be used to monitor and control tubing pressure as it is displaced to the new kill weight fluid.

7.12.2.1 Outline procedure

1. After shutting in the well, record the SITP, SICP, and the volume of pit or trip tank gain.
2. During the preparation for the well kill (calculations and brine mixing) gas migration must be monitored and controlled using the volumetric method.

3. Calculate the brine weight required to kill the well:

$$\text{Kill weigh (ppg)} = \frac{\text{SITP (psi)}}{0.052 \times \text{reservoir depth (TVD).}}$$
$$+ \text{ original brine weight (ppg)}$$

Note: This only restores the well to balance. If an overbalance margin is required, it can be added to the SITP, i.e.:

$$\text{Kill weigh (ppg)} = \frac{\text{SITP (psi)} + \text{required overbalance (psi)}}{0.052 \times \text{reservoir depth (TVD).}}$$
$$+ \text{ original brine weight (ppg)}$$

4. Calculate the initial circulating pressure (ICP).

$$\text{ICP} = \text{SCRP} + \text{SITP}$$

5. Calculate the final circulating pressure.
6. Calculate surface to tubing tail-pipe volume and strokes.
7. Calculate time to pump surface to tubing tail-pipe.

$$\text{Time} = \frac{\text{strokes (surface to bit)}}{\text{strokes per minute}} \text{ or Time} = \frac{\text{bbls (surface to bit)}}{\text{bbls per minute}}$$

8. Plot pump pressure against pump strokes and time for the tubing on a graph with the ICP on the left-hand vertical axis and FCP on the right-hand axis. A simple way of determining the pressure drop per increment is to use the formula:

$$\frac{\text{ICP} - \text{FCP}}{10} = \text{pressure drop per increment}$$

For example, if 825 strokes are needed to fill the tubing, the kill rate is 25 strokes per minute, ICP is 900 psi, and the FCP is 450 psi, the pumping schedule would appear as shown in Fig. 7.25.
9. Having prepared the pump schedule and weighted-up enough brine to bring the well back under control, the kill can commence.
10. Ensure there is good communication between the choke and pump operator. Zero the stroke counter (if applicable).
11. Bring the pump up to the slow circulation rate whilst holding casing pressure constant.

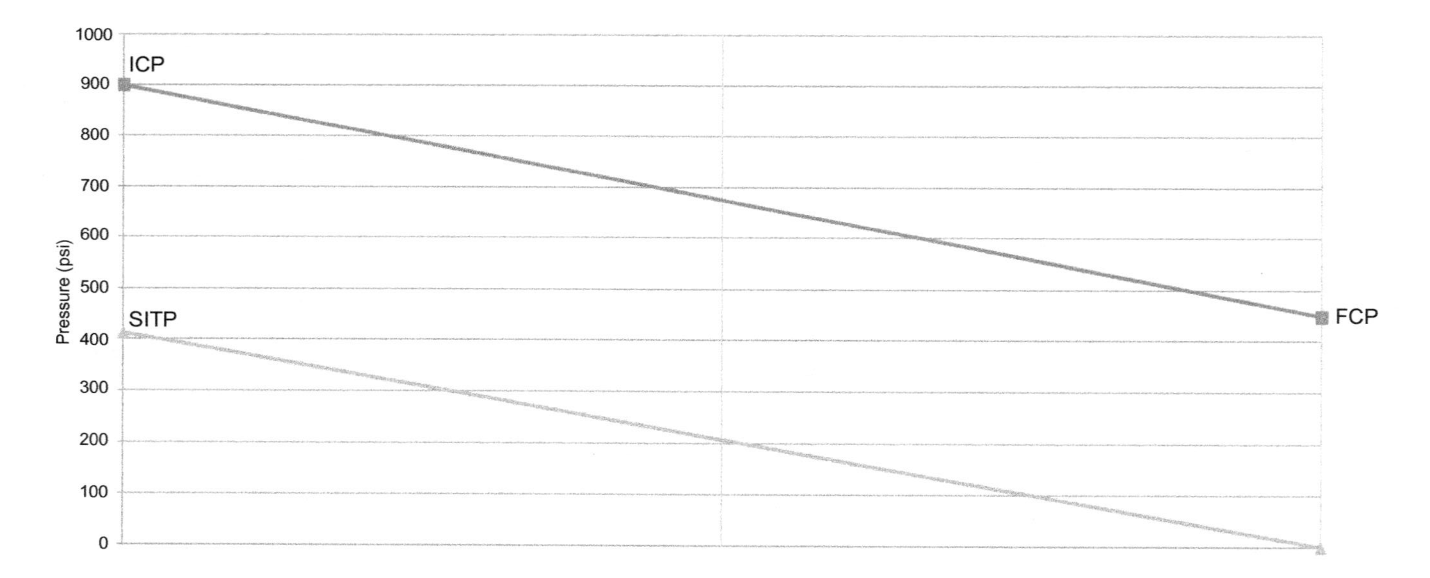

Time (mins)	0	3.5	7	10.5	14	17.5	21	24.5	28	31.5	35
Strokes	0	87	175	262	350	437	525	612	700	787	825
Pressure (psi)	900	855	810	765	720	675	630	585	540	495	450

Figure 7.25 Tubing pressure versus volume pumped.

12. Once the pumps have reached kill speed and pressures have stabilized, record the actual circulating tubing pressure:
   a. If the actual circulating pressure is equal to, or close to, the calculated ICP continue to pump, adjusting the tubing pressure in accordance with the pump schedule. Slight differences between the calculated ICP and the observed ICP are normally because the SCPR used to calculate the ICP is inaccurate. The FCP can be adjusted as follows:
      i. Actual (updated) SCRP = actual ICP − SITP.
      ii. Adjusted FCP = (actual SCRP) × kill brine weight/original brine weight.
   b. If the difference between the calculated ICP and the observed circulating pressure is significant, the kill should be stopped and the well closed-in for further investigation. Discrepancies are nearly always because of trapped pressure.

13. Once the kill fluid reaches the tubing end and begins to fill the annulus, the choke operator holds the tubing pressure constant until kill fluid reaches the surface.

14. Once the full circulation is complete, the well can be shut-in. Tubing pressure and casing pressure should both read 0 psi.

15. Observe the well for flow or losses. If losses are observed, they must be within acceptable (pre-agreed) limits (Fig. 7.26).

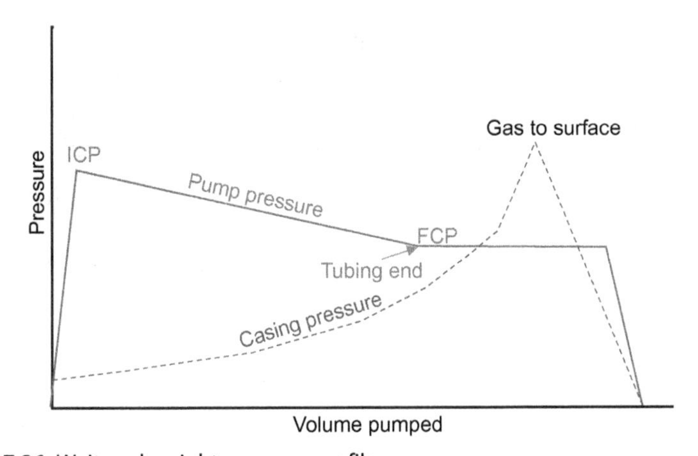

Figure 7.26 Wait and weight pressure profile.

7.12.3 Bullhead and lubricate-and-bleed

If circulation is not possible, or if there is no tubing in the well, regaining control of the well following a kick might be possible using bullheading or lubricate-and-bleed. The cause of the kick is critical when determining how to regain control of the well.

A kick caused by increased pressure downhole means the brine remaining in the well is no longer kill weight and must be replaced. If formation permeability allows, bullheading could be used to return the well to an overbalanced condition. However, this could cause significant formation damage since all the old brine in the wellbore will have to be displaced into the formation. In addition, the large volume of brine lost to the formation will make production startup difficult. If the formation has low permeability, bullheading may only be achievable if the formation is fractured.

Killing the well by lubricate-and-bleed is not possible if the brine remaining in the well is below kill weight. Where bullheading is not possible or where there are concerns about fracturing the formation and formation damage, establishing a circulation path may be the only way to kill the well. Two options are available:

- Snubbing in a workstring
- Using coiled tubing

A kick caused by swabbing, fluid loss to the formation, or failing to keep the hole full, means the brine in the well is still kill weight.

Providing a reasonable rate of injection can be established without risking formation fracture, a limited volume bullhead can be used to push the kick back into the formation. The volume of fluid needed to kill the well will depend on how soon after the kick pumping begins, and the migration velocity of the influx. If a gas influx occurs, it can migrate a significant distance towards the surface before pumping begins, meaning a larger volume of kill fluid must be displaced to the formation, with the attendant risk of damage. Moreover, it may not be possible to pump at a rate high enough to overcome the gas migration velocity. This is especially problematic in high angle wells. When dealing with a gas influx, it may be preferable to use the volumetric method to bring the gas to surface the replace the gas cap with kill fluid by a process of lubricate and bleed.

7.12.4 High angle and horizontal wells

A horizontal wellbore maximizes reservoir exposure, resulting in higher flow rates and increased recovery. In very permeable formations, productivity index (PI) values can be very high, with wells producing at a high rate with very little drawdown.

The well control techniques and principles used in horizontal wells are basically the same as those used in other wells. The calculation for kill weight fluid must be carried out using TVD, irrespective of MD. However, the kill sheet must be adjusted to account for well angle. In a vertical well, pressure change per unit volume pumped is uniform throughout the kill. In a deviated well, the pressure change per unit volume pumped varies as the well angle changes.

Fig. 7.27 shows a deviated well where the angle builds from the vertical then remains at a constant angle (volume on the X axis and pressure on the Y axis). The red line represents a theoretical pressure plot for the second circulation of a weight and wait kill; a straight-line decrease between ICP and FCP. The blue line accounts for the trajectory and represents true pressure. This plot clearly shows that ignoring well angle and constructing a kill sheet that only accounts for volume will result in high overbalance pressure, with the consequent risk of formation fracture and fluid loss.

In an S shaped trajectory both overbalance and underbalance will occur if the theoretical pressure plot is followed (Fig. 7.28).

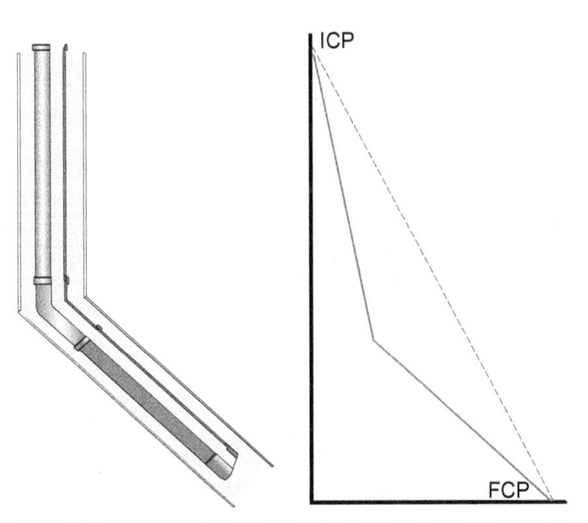

Figure 7.27 Theoretical pressure (red line; dashed line in print version) versus actual pressure (blue line; solid line in print version). Build and hold deviated well.

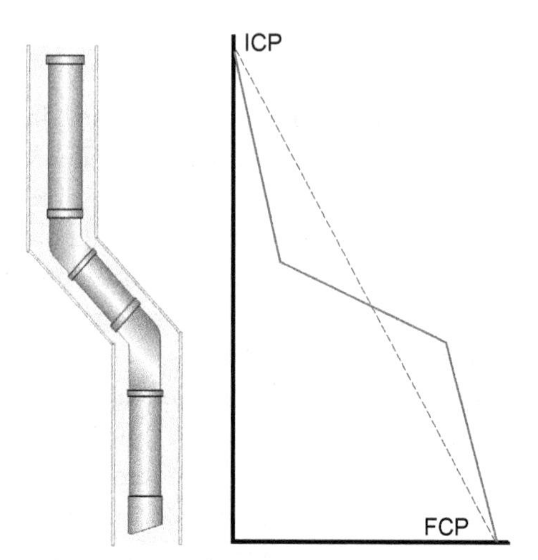

Figure 7.28 Theoretical pressure (red line; dashed line in print version) versus actual pressure (blue line; solid line in print version). S shaped trajectory.

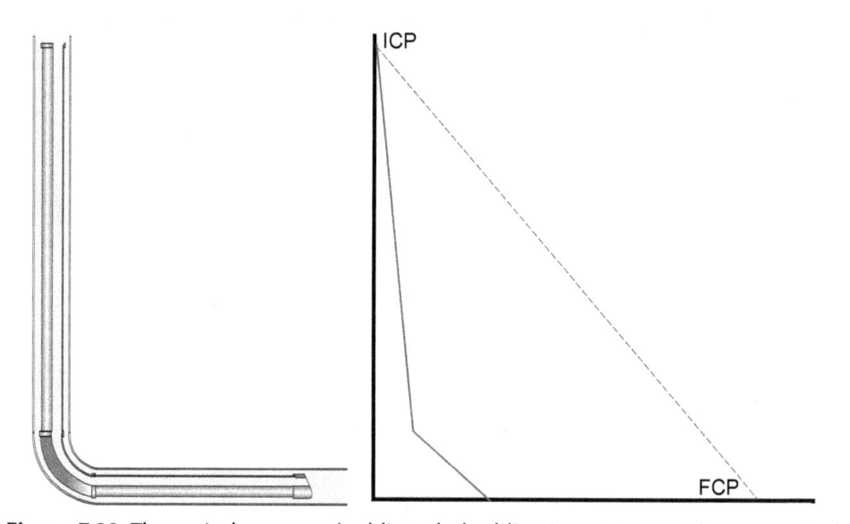

Figure 7.29 Theoretical pressure (red line; dashed line in print version) versus actual pressure (blue line; solid line in print version). Horizontal well.

Horizontal wells will experience very high overbalance pressure unless the kill is adjusted to account for well angle. The longer the horizontal section, the worse the effect will be (Fig. 7.29).

Other well control considerations when working on high angle and horizontal wells are:

- Since a horizontal well can produce relatively large amounts of reservoir fluid at low drawdown, swabbing has the potential to pull in a large influx. The influx volume will normally be considerably larger than in a vertical well.
- It is possible to have a large influx entirely within the horizontal section. This could go unnoticed at the surface since it would have no measurable effect on SITP or SICP. With the influx in the horizontal section, overbalance is still intact; the well is still dead. However, if circulation continued, the influx entering the vertical section would result in reduced hydrostatic. When tripping onto a horizontal well, carry out a partial circulation once the pipe is on the bottom. This should enable any influx to be detected.
- When tripping out of a vertical well, a flow check should be performed when outsized completion components (such as packers) are pulled above the end of the horizontal section, i.e., into the heel of the well.
- Some single string multi-packer completions are particularly prone to swabbing/surging in horizontal wells.
- Effective cleaning of horizontal casing/liners is difficult. Debris left behind in the horizontal section will increase swab pressure.
- ECD will be higher than in a vertical well, since there is more frictional pressure drop. Since circulating pressure will be significantly higher than static pressure (necessary to overcome friction) when circulation stops there will be a correspondingly large pressure reduction. A flow check should be carried out, especially prior to tripping out.

7.12.5 Multi-lateral wells

With a few additional precautions, the well control methods used for single bore wells can be used for multi-lateral wells.[6] Kick prevention follows the same basic principles. In high angle or slim hole laterals, the problems associated with ECD and swab pressure must be considered. Similarly, the methods of kick detection used for conventional wells should work equally as well in multi-lateral wells. However, cross-flow between laterals can affect pit gain and mask a kick; in addition, it will not always be possible to determine if the kick has originated in the active wellbore or the static lateral. There are some signs that can help determine which wellbore has kicked:

- Active wellbore—influx below the junction:
  - SICP should read higher than SITP (if the wellbore is not horizontal).
  - If volumetric measures are used to control gas migration, SITP will remain relatively stable whilst SICP increases.
- Static wellbore—influx below the junction.
  - SITP should be equal or close to SICP (if the wellbore is not horizontal).
  - Both SITP and SICP increase during gas migration until the gas reaches the junction.

If tubing is in the active wellbore, and the kick originated from there, a conventional weight and wait or drillers method kill can be used. The effects on the lateral must be considered when planning to kill the active bore.

If the kick originated in the static wellbore, or if there is uncertainty about which bore kicked, and if the tubing string is above the junction, it may be possible to bullhead. If this is not possible and if it is not possible to run tubing into the bore where the kick originated, then the well can be killed to the junction. This will almost certainly mean having to increase brine weight. The may also be a requirement to spot LCM above the junction to slow losses.

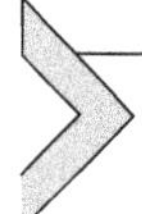

7.13 COMPLETION AND WORKOVER: WELL CONTROL CONTINGENCIES

There are several widely used and well documented procedures for shutting in the well and removing a kick (some of which have already been described). However, most of those procedures are heavily biased towards drilling operations. During completion and workover activities, there are occasions when standard procedures for shutting a well are inadequate, often because BOP preventers will not seal around some components run as part of the completion. Described here are some of the more common completion activities where additional steps are required to shut-in a well. Also described are completion and workover operations where the risk of a kick can increase.

Almost without exception, operating companies and drilling contractors advocate and implement a program of well control drills; only regular practice can provide a drill crew with the knowledge and experience necessary to cope with an emergency. Drills traditionally concentrate on recommendations made by the API in "Recommended Practice for Well Control Operations" (API RP 59). Since many of the procedures necessary to shut-in a well during a completion or workover operations are not routinely practiced, additional "job specific" drills are strongly recommended, especially where the rig crew has had limited completion or workover experience. A well-trained crew will know how to react to every eventuality during the completion.

7.13.1 Running the liner

The OD of liner hangers and liner packers is normally very close to casing drift. During liner running and cementing of conventional liner hanger systems, the tight clearance between the hanger and the casing can increase the risk of well control problems:

- The liner, liner hanger, liner hanger packer (if run), and the running tool are run on drill pipe. For the rig crew, running drill pipe is routine, and there may be a tendency to run at speed. The trip speed must be moderated to reduce surge/swab pressure.
- In order to obtain a good cement bond between the liner and formation, turbulent flow in the annulus must be achieved during cement displacement. Achieving turbulent flow requires a high circulation rate; the result is high ECD, which combined with dense fluid (cement slurry) can lead to fluid loss. Long liners in high angle wells are particularly problematic.
- In recent years some liners have been run in combination with swell packers. If liner running is taking longer than planned, or if the swell packer elastomeric chemistry is not the correct specification, the elastomeric seal element could begin to swell (OD expansion) whilst the liner is still being run, increasing the swabbing and surging effect.
- Most liners are run with a float shoe made up as part of the shoe track, requiring the liner to be filled as it is being run. Closed end (dry) liner displacement will be significantly more than for open ended pipe.

Expandable liner hanger systems can be used to lessen some of the problems associated with conventional liner hangers. The small pre-expansion OD of the hanger reduces surge and swab pressure when running and ECD when circulating.

7.13.2 Running slotted liners and sand control screens

When sand control screens or slotted liners are positioned across the drilling BOP, closing the pipe rams or annular preventer will not prevent flow from the well. The following steps are needed to secure the well:

1. A drill pipe kick stand should be racked in the derrick with a full opening safety valve made up on top.
2. A cross-over to connect the kick stand to the screens in the rotary table must be available on the rig floor.

   Note: It is not advisable to attach the cross-over to the kick stand, particularly if premium connections are in use. Trying to stab a premium tubing connection on the bottom of a stand of drill pipe is likely to result in thread damage. The risk increases if high chrome (22% or 25% Cr) screens are in use.
3. At the first indication of a kick, lower the screen joint being run to a working height above the rotary table and set the slips.
4. Make up the cross-over and kick stand to the screen joint in the rotary table.
5. Close the full opening safety valve.
6. Lower the drill pipe across the BOP and close-in the well on the uppermost relevant preventer.
7. Record pressures.

If there is no time to make up the cross-over and kick stand the only option may be to cut and drop, or just drop the screens. The multi-layer construction of many screens and the need to run wash pipe means that not all BOP shear rams are able to cut the pipe. To drop the screens:

1. Lower the screen until the elevators are just above the rotary table.
2. Close the annular BOP to support the weight of the screen.
3. Slack off weight to enable the elevator to be unlatched. Unlatch the elevator.
4. Open the annular preventer, allowing the screens to drop into the well.
5. Close the blind rams.
6. Record pressure.

7.13.3 Pre-completion wireline perforating

Some wells are perforated on wireline before the completion is run (with the drilling BOP still in place). It is normal to perforate with the well in an overbalanced condition. There are well control risks when perforating in this way:

- Once the well is perforated, fluid loss can be expected. Spotting LCM above the reservoir will reduce, but not eliminate, the risk. If hydrostatic overbalance is lost the well will kick.
- If assumptions about the reservoir pressure are wrong and it is higher than anticipated, the well will kick.
- With no tubing, well kill is limited to bullhead or lubricate-and-bleed. It may not be possible to achieve the required rate of injection to bullhead successfully.

The use of wireline pressure control equipment is recommended. If it is not used, the only possible response to a kick when wire is in the hole is to cut the wire and close-in on the blind rams. If pressure control equipment is used, the kick can be contained while the wire and guns are recovered from the well.

With the BOP in place, the wireline pressure control equipment must be rigged up in combination with a "Dutch Cap" or "Shooting Nipple." The misuse of a Shooting Nipple has been the cause of several blowouts (*see* Chapter 9, *Wireline Operations, for a full description of the Dutch Cap, the Shooting Nipple, and the problems they can cause*).

7.13.4 Tubing conveyed perforation guns in an overbalanced well

Perforating with underbalance is generally preferred for well performance reasons. However, when tubing conveyed perforating guns (TCP) are used, cost and operational complexity are both reduced if perforating takes place with kill weight fluid filling the well. However, even with an overbalance fluid in the well, the risk of a kick cannot be overlooked (Fig. 7.30). The following should be considered:

- If the guns are pressure activated, use a time delay firing head to enable the gun activation pressure to be bled-off before the gun fires. If the wellbore pressure is high when the guns fire, there is a risk of fracturing the formation. In addition, formation damage will be considerably worse because of the additional overbalance.
- Make sure a suitable LCM pill is circulated into the well and spotted above the perforations.
- If there is a kick because of losses, or a higher than expected reservoir pressure, then the annular bag or pipe rams must be closed.

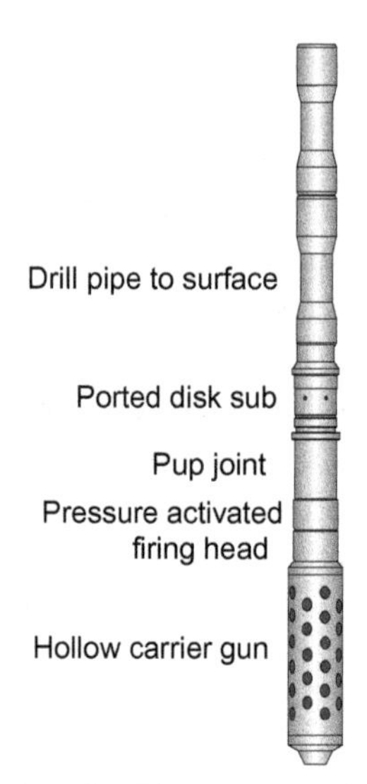

Figure 7.30 Typical pipe deployed tubing conveyed perforation configuration for overbalanced perforation.

- BOP pipe rams will not seal on spent (fired) perforating guns. Many pipe conveyed gun assemblies are long—in the case of horizontal wells many thousands of feet. If guns are straddling the BOP when the well kicks, they will have to be sheared and dropped so that the blind rams can be closed, making the well safe.

7.13.5 Underbalanced perforation; tubing conveyed perforation guns run on a drill stem test (DST)

To perforate the well in an underbalanced condition using tubing conveyed guns, a properly designed, fully functional DST running string is required. After perforating, the well is produced to remove perforating debris and completion fluids; the flow period may include a well test. At the end of the flow period the well must be killed to enable the DST string to be retrieved before the completion can be run. Well testing and the use of a DST string are described in Chapter 12, Well Control During Well Test Operations. The main well control concerns when underbalance perforating using a DST string are:

- A packer is required to isolate the annulus from the reservoir pressure.
- Two reversing valves are recommended, each one operated by a different method (e.g., annulus pressure, tubing pressure, and string movement).

- Many operating companies insist on the use of gas tight premium connection tubing for any TCP operation where the well is to be flowed.
- It is good practice to run a safety joint above the packer in case it cannot be released.
- Prior to conducting any DST, the BOPs and the gas detection system should be tested.
- The use of a surface test tree that enables the drill stem to be closed-in is recommended.

7.13.6 Wellbore clean-out

Completion engineers understand the benefits of properly cleaning the wellbore before running a completion. This is especially important when completion fluids contact exposed formation, and where complex completion components are vulnerable to malfunction because of debris left in the well. It has become routine to make a dedicated clean-out trip to prepare the wellbore using a range of specialized tools and wellbore cleaning chemicals (Fig. 7.31).

Usually the clean-out string will be run with mud in the well which is then displaced with the clean-out chemicals (viscous pills, solvents, flocculating agents, surfactants, etc.) and replaced with clear, clean,

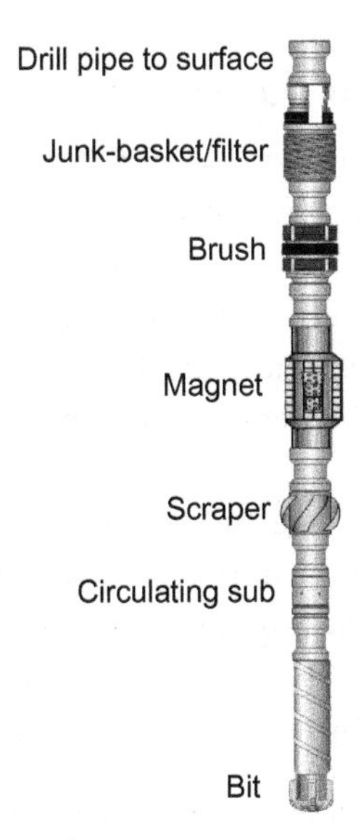

Figure 7.31 Wellbore clean-out string.

solids-free brine. The main well control concerns during the wellbore clean-out trip are:
- The high flow rate needed to clean the well coupled with the use of viscous fluids will increase ECD.
- Some of the most effective wellbore clean-out tools, such as the MI (Schlumberger) "Well Patroller" and Well Commissioner have seal elements and swab cups that have minimal clearance inside the casing. Running speed should be moderated to reduce the risk of surging and swabbing.

Many wellbore clean-out trips are run in wells that have been cased and cemented but not perforated. Providing the casing (and liner if relevant) has been integrity tested, the risk of a well control incident is significantly reduced. However, some concerns remain. In unperforated wells it is common to displace the kill weight mud with underbalanced completion brine. This is normally done to improve perforating performance. If no inflow test has been carried out on the liner prior to the clean-out trip, this will be the first time that the liner lap is exposed to a differential pressure from below. The well must be carefully monitored as the kill fluid is displaced from the well. At rig locations where pit space is limited (offshore), mud displaced from the well is often transferred straight to a boat rather than going back to the pits. This can make it difficult to identity flow and a kick can go unnoticed.

7.13.7 Fluid loss control valves

Fluid loss control valves[b] are widely used to isolate the reservoir in open hole completions. They are normally placed a short distance below a gravel pack or liner hanger. They provide a closed tested mechanical barrier whilst the upper completion is being run.

With the completion in place, the fluid loss valve must be opened to allow the well to produce. Opening is normally accomplished by pressure cycles above the valve.

Some operators will displace the wellbore above the valve to underbalanced fluid before the completion is run (during the wellbore clean-out trip). Where this is done, the rig crew needs to be especially vigilant. Well control incidents, caused by the fluid loss control valve opening prematurely, have been documented. On one occasion a valve was mistakenly opened

[b] Fluid loss control valves, also called formation isolation valves and formation saver valves, are described in more detail in Chapter 3, Completion Equipment.

during the wellbore cleaning trip when the bottom of the workstring snagged the opening mechanism in the valve. Fluid loss control valves are also vulnerable to premature opening if subjected to repeated surge pressure.

7.13.8 Running electric submersible pumps

The combination of motor, motor seal, pump intake, gas separator (if run), pump, and discharge head can be very long—easily more than 100 ft. (30 m). Correct assembly takes time. If the well kicks whilst the pump is hanging across the BOP, it will not be possible to close the pipe rams. Closure of the annular preventer might, in theory, be possible, but will have no benefit if the pump intake is below the ram (fluid will blow out through the pump), or if the power cable has been attached. If the pump intake is above the ram, the piston force acting on the pump would probably force it out of the annular preventer, a potentially dangerous situation.

- If time (and derrick height) allows, pick up until the pump is above the blind rams and close-in the well.
- If the pump assembly is long and pick up will not clear the blind rams, it will need to be dropped:
  - Position the elevators at a safe working height above the rotary table.
  - Close the annular preventer around the pump to support the weight.
  - If the power cable has already been attached, it will have to be cut.
  - Unlatch the elevator.
  - Retract the annular preventer, dropping the ESP assembly.
  - Close the blind ram.

7.13.9 Running production tubing

The main well control considerations when running production tubing are:

- Establish whether the shear rams will cut the production tubing. Some BOP shear rams may not be able to cut heavy wall, high yield tubing.
- Control lines, instrument cables (singly or in flat pack form), and ESP cable will prevent pipe rams from sealing.
- At low wellhead pressure the annular preventer will normally provide a usable seal against a single (non-encapsulated), control line.
- Multiple control lines and cables will have to be cut before the chosen preventer (annular or pipe ram) can be closed on the tubing. A suitable cutting device must be kept on the rig floor. Bundled control lines (flat pack) and ESP cables are tough and need to be cut quickly.

Hand portable hydraulic cutters are reliable, easy-to-use, and fast. Once the lines have been cut, the string must be lowered to place bare pipe across the BOP before the uppermost applicable preventer is closed. In some circumstances, it may be faster to make up a kick stand to the tubing in the rotary table (whilst the cables are being cut) and lower it across the BOP (Fig. 7.32).

7.13.10 Running dual string completions

In some fields, completions with dual and occasionally triple strings of tubing are run. Dual string completions are usually installed for reservoir management reasons, allowing different zones in the reservoir to be produced independently. In most cases the two tubing strings are run simultaneously, requiring specialist handling equipment, for example dual slips and elevators (Fig. 7.33).

With two strings of tubing across the BOP, the annular preventer cannot shut-in the well, so pipe rams must be used. They must to be equipped with a dual string ram block of the appropriate size. Shearing blind rams designed to cut multiple tubing strings should also be used.

Figure 7.32 Self-contained hydraulic cutter (left) and multiple flat-packs (right).

Figure 7.33 Dual string slips and elevators. *Image courtesy of Bilco Tools Inc.*

7.13.11 Completion components

Special consideration needs to be given to configuring some of the components that are run with the completion:

- When running any completion component, be aware of where it is, relative to the BOP rams. In the event of a kick, the string might need to be raised or lowered to move completion components away from the pipe rams and annular preventer. Shear rams will not be able to cut some completion components, for example, the packer.
- Surface controlled sub-surface safety valves should be run with control line pressure applied to hold the valve in the open position. If the valve is closed, reverse circulation is not possible, nor is it possible to monitor tubing head pressure as the completion is run.
- If gas lift mandrels are run with live valves installed (a common practice), the ability to reverse circulate out any kick is compromised. Fluid will take the path of least resistance—usually the uppermost mandrel in the tubing string. A kick will therefore have to be removed by forward circulation.
- If the completion is equipped with a sliding sleeve positioned above the packer, leaving it in the open position whilst the tubing is run can help reduce surge/swab pressure.

7.13.12 Daylight only operations

In some locations, completion and workover operations are restricted to daylight only working, with minimal overnight crew cover. At the end of each day operations are suspended and the crew secures the well before leaving. Cases have been recorded where well control has been lost whilst the crew was absent. Before leaving a well with minimum crew cover:

- At minimum, circulate bottoms-up, checking the returned fluid for gas. This step requires the tubing to be on the bottom.
- Do not leave the work string in "open hole." Pull back above the casing shoe.
- If the well has been taking losses, consider spotting LCM above any open formation.
- Make up a pup joint on the top of the tubing string. Lower the string, close the pipe rams on the pup joint, and lock the pipe rams. (The pup joint collar below the rams will prevent upward movement of the tubing string in the presence of unforeseen well pressure that might build overnight.)
- Install a full opening tubing safety valve and a pressure gauge on top of the pup joint.
- Close the safety valve.

7.13.12.1 Procedure for opening the well in the morning

Gas entering the well during the hours of inactivity is not unusual, and a sizable accumulation is possible. Opening the well without a proper check would result in flow:

- Check the tubing string pressure gauge by opening its needle valve. If no pressure registers on the gauge, check for flow with the full opening safety valve open.
- Check the annulus pressure gauge. If no pressure registers, check for annular flow. The flow check is normally through the choke manifold.
- If there is no pressure or flow from either tubing or the annulus, it is safe to open the well.
- If pressure or flow are observed, close-in and manage the kick.

7.13.12.2 Other considerations for daylight only operations

In a well with an open reservoir and no tubing, it is good practice to run (and test) a mechanically set retrievable packer. Most packers of this type require a minimum amount of weight suspended below them to prevent unseating where there is a pressure differential from below. When the packer is set and the running string recovered, close the blind rams to secure the well.

7.13.13 Failure to meet barrier policy

Inevitably, many workovers are carried out on wells where the condition of the tubing and completion equipment is poor. Those tasked with planning the workover are sometimes placed in a position where it is impossible to meet the normal requirement to have the two independent (tested) mechanical barriers needed for safe Christmas tree removal and BOP installation.

Where barrier policy cannot be met, the decision to proceed will depend on company policy and local regulatory requirements. Most operating companies will have a policy granting dispensation from the standard two mechanical barrier policy if certain stringent measures are met. Much will depend on what barriers can be established, and how reliable those barriers are judged to be:

- A detailed risk assessment would normally be required before any operation can proceed.
- In many operating companies, dispensation can only be granted by experienced senior engineers, and must be endorsed by management.
- For wells that can flow naturally (i.e., without artificial lift), the minimum requirement is normally to have a fluid barrier and one tested mechanical barrier.

Additional considerations when assessing barrier integrity/reliability:

- During a workover, the newly installed BOP is normally tested against the uppermost mechanical barrier. The barrier must be able to hold BOP test pressure. Where possible, the upper barrier is set in the tubing hanger.
- There are risks associated with setting bridge plugs in corroded and damaged tubing. Slips can further damage already weakened pipe. In extreme cases, bridge–plug slips can hole or deform the tubing to such an extent that integrity is compromised. Plugs set in damaged tubing are normally more difficult, and sometimes impossible, to pull.
- In the event of a tubing leak above the packer, well integrity is reliant on a single mechanical barrier (the tubing hanger). It may be possible to plug the tubing below the packer to fulfill the requirement for a second barrier.

7.13.14 Pulling damaged and corroded tubing

Pulling badly corroded or holed tubing will limit the ability to control the well in the event of a kick:

- The BOP pipe rams and annular preventer will be ineffective if closed around holed pipe.

- Corrosion will reduce the yield of the tubing. If the pipe is badly corroded (but not holed) it could fail following a BOP shut-in—a combination of point loading damage from closing the rams (or annular) and/or shut-in pressure.
- If the well kicks and is successfully closed-in at the BOP stack, holes in the pipe below the stack will complicate the subsequent kill operation.
- Prior to pulling damaged and corroded tubing, it is good practice to run a caliper or ultrasonic log. This will establish the amount of corrosion, and possibly enable an estimate of yield reduction to be made.
- Corrosion in water injection wells tends to be concentrated in the upper part of the tubing (Fig. 7.34).

If the yield of the tubing has been reduced to the point where it is likely to part under its own weight, then serious consideration should be given to making one or more controlled cut(s) in the tubing string using a radial cutting torch, jet cutter, or some similar tubing cutting technique.

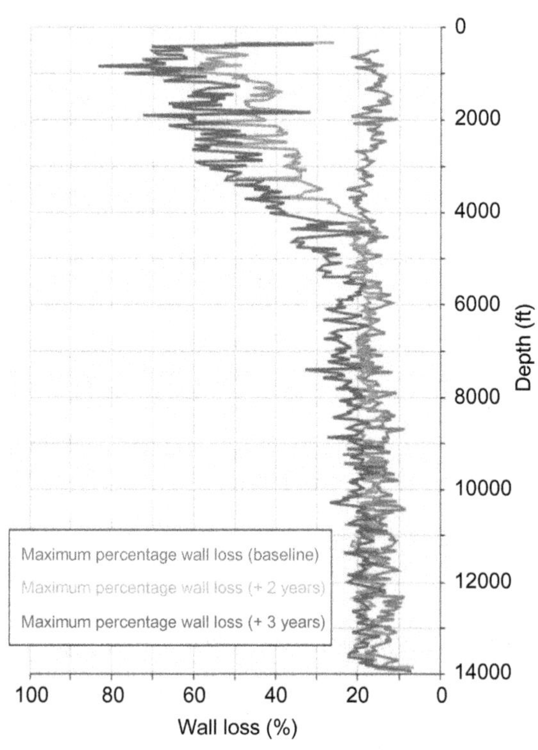

Figure 7.34 Caliper log data from a water injection well—the upper part of the string has significant corrosion.

Disintegration of the tubing string could lead to protracted fishing operations. If the well kicks, the presence of a partially fished string in dubious condition will complicate matters significantly.

7.13.15 Remedial operations

Many workover programs will include remedial operations that take place between the recovery of the old completion and the installation of the new one. Some of these activities will increase the risk of a well control incident:

- Forward circulation routinely used to remove sand fill and other unwanted debris from the wellbore. To clean effectively, high velocity circulation is necessary, with a correspondingly high ECD. As the sand or debris is removed, there is a risk of increased fluid loss to the newly exposed formation.
- Sulfate scales are dense, for example of 4.5 g/cc (37.5 ppg) for barium sulfate ($BaSO_4$). During the milling operations sometimes used to remove these scales from the wellbore, high annular velocity is required to lift the cuttings. Consequently, ECD will be high and losses can occur.
- Running casing patches, bridge plugs, and straddles to repair leaks or isolate watered out production zones risks surging and swabbing the formation.
- Any operation where circulation takes place across open formation or open perforations risks disturbing and sweeping away LCM.

7.13.16 Workover in steam flood fields

Many heavy oil reservoirs are relatively shallow, often less than 2000 ft. For example, the main formation in the Kern River field near Bakersfield California is between 400 and 1300 ft. TVD. Oil in this reservoir is very viscous, approximately 3000 cp at 100°F. Steam injection is used in the Kern River field and other similar reservoirs, to reduce oil viscosity. The liquid hydrocarbon then flows more easily.

Steam injection raises the temperature in the reservoir significantly, and under specific downhole temperature/pressure conditions water-based kill fluid can flash off into steam, leading to an explosive evacuation of wellbore. This phenomenon is known as a boiling liquid expanding vapor explosion, or BLEVE.

At atmospheric pressure (14.7 psi), water boils at 212°F (100°C). Under pressure, water becomes superheated since the boiling point has increased. For example, at 250 psi the boiling point is raised to 400°F. If the pressure of the 400°F water were suddenly reduced to atmospheric pressure, liquid would instantly flash to steam. One gallon of water would expand to 1660 gallons of steam (Fig. 7.35).

Steam tables are used to determine the minimum pressure necessary to prevent steam formation for any given temperature, or conversely, the maximum temperature that can be allowed for any given pressure (Fig. 7.36).

For example, if reservoir pressure is 200 psi, bottom hole temperature must remain below 381.8°F to prevent steam forming; at 381.8°F, steam will begin to form if the pressure drops below 200 psi. A BLEVE is prevented by keeping pressure above, and temperature below, values obtained from steam tables.

If temperature increases to the saturation temperature, steam begins to form. This reduces HP in the well, since expanding steam will displace fluid from the well. As pressure reduces, more steam flashes. Once started, the process is almost impossible to stop, and will continue until all the fluid in the wellbore has flashed off. The sudden and dramatic reduction in HP will inevitably lead to a kick, and it is worth emphasizing that the BLEVE is not a kick, but is the cause of one.

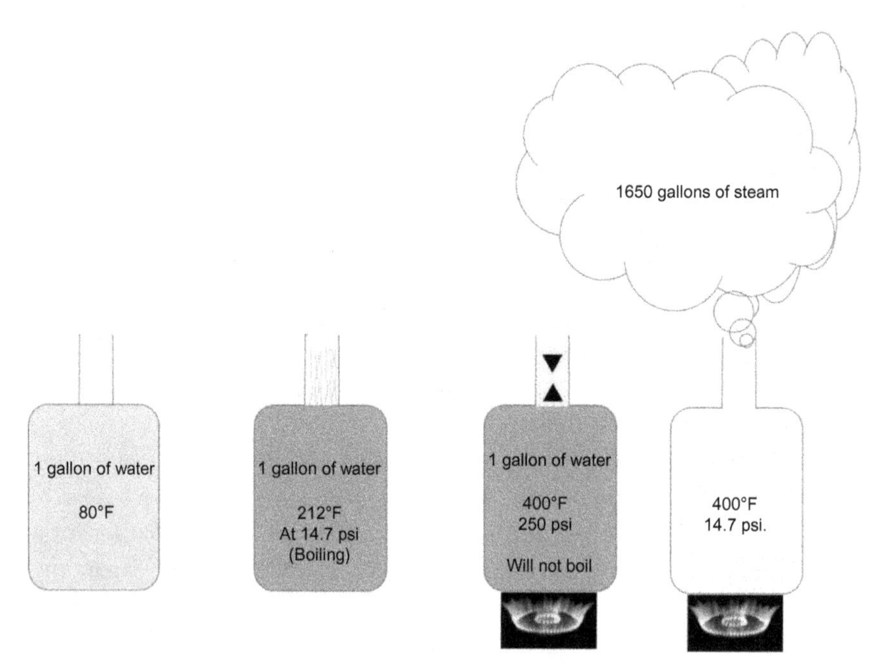

Figure 7.35 One gallon of water produces 1660 gallons of steam.

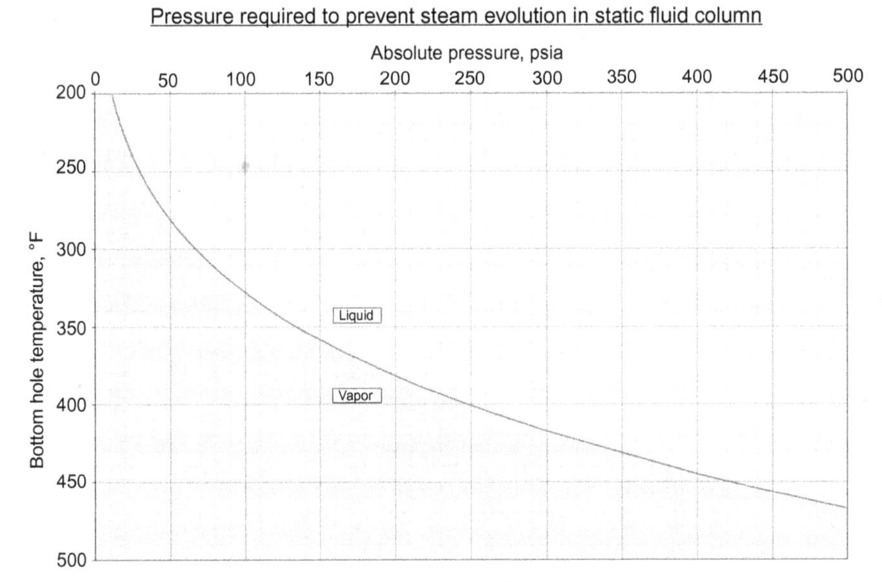

Figure 7.36 Pressure required to prevent steam forming.

During most workovers, the density of the kill fluid will be determined by the requirement to overbalance reservoir pressure by a desired amount. Having calculated BHP, the maximum bottom hole temperature, and the consequent risk of a BLEVE, is determined by referencing steam tables.

Example:

Formation pressure (oil/gas or steam):	254 psi at 460 ft. TVD
Highest expected reservoir temperature:	400°F
Fracture pressure at top reservoir	302 psi (12 ppg EMW)

In shallow wells, the window between formation fracture and pore pressure is narrow, and the "normal" 200 psi overbalance is not possible. In this example, an 11 ppg kill fluid has been selected and gives a BHP of 278 psi at 460 ft. TVD.

The minimum pressure needed to stop a BLEVE at 400°F is 250 psi. The margin between the HP the pressure at which flashing can begin is small, only 28 psi. This equates to only 49 ft. of head with the 11 ppg brine. Increasing BHP by increasing the brine weight is not a viable option. as there is a risk of exceeding fracture pressure. Cooling the wellbore would reduce the risk of a BLEVE.

Preventing a BLEVE is important since:

- It is almost impossible to stop once started. The event will be sudden and violent. The crew will be exposed to the rapid discharge of very hot fluids.
- There is very little time between the first indication a BLEVE has started, and evacuation of the tubing. It will be difficult, probably impossible, to shut-in the well conventionally. In most cases, the fastest and safest response is to close the shearing blind ram.

During any workover or intervention activities on wells where the risk of a BLEVE is present it is important to:

- Keep pressure in the wellbore high enough to prevent the formation of steam.
- If heavy enough kill fluid cannot be used, the wellbore must be cooled to below boiling point (with reference to a steam table).

7.13.17 Unconventional (shale) reservoirs

Many unconventional reservoirs produce from extremely low permeability formations. For example, the Eagle Ford Shale (Southwest Texas) has a permeability range of between 50 and 1500 nano-darcys.[7] Long horizontal reservoir sections with multi-stage hydraulic fractures are needed to make wells produce at an economic rate. Wells in the Eagle Ford Shale can have horizontal sections in excess of 5000 ft., with more than 25 fractures spaced between 200 and 250 ft. apart. In most areas, multiple wells are drilled from a single pad. The horizontal sections are drilled parallel to maximum horizontal stress direction in the formation (Fig. 7.37).

At present, most wells completed in shale formations use one of three different techniques to complete the reservoir section: plug and perf, drop ball activated sliding sleeves, and coiled tubing operated sliding sleeves. Fracturing takes place after perforating or opening a sliding sleeve adjacent to the zone of interest. High fracture pressure is needed to obtain the fracture geometry required for a productive well, typically in the range of 12,000–14,000 psi. Surface equipment rated at 15,000 psi is normally used. Isolation between the fractures is achieved by either cementing the lower completion, or using external casing packers.

Well control incidents are not uncommon in unconventional wells. In a study of 3533 Pennsylvanian wells monitored between 2008 and 2011, there were 85 examples of cement or casing failures, four blowouts, and two examples of gas venting.[8] In 2013 it was reported that the rate of

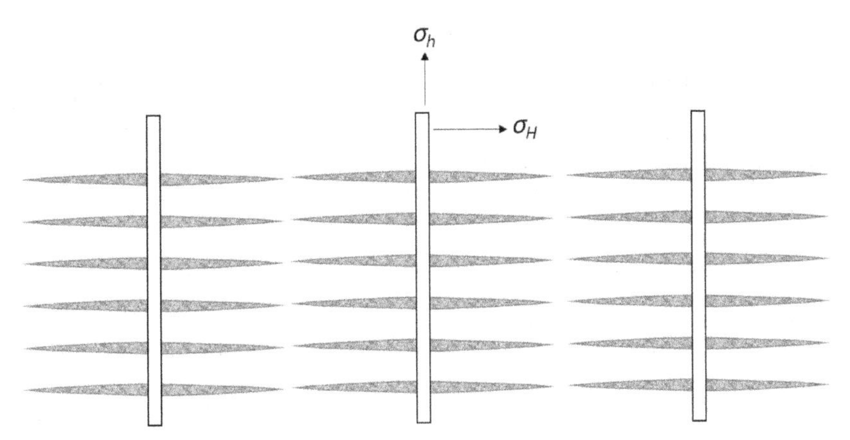

Figure 7.37 Multiple wellbores, parallel to minimum stress (σ_h). Fracture propagation is perpendicular to minimum stress.

blowout incidents for shale gas was twice that of conventional wells and shale oil was three times higher.[9] At least some of these well control difficulties have been attributed to drilling into fractures created in already completed wells.[10] The current practice of drilling several wells from a single pad, all having relatively close spacing, makes the continuation of these problems likely. Where fractures in offset wells have been intersected, problems arise in some wells, since the higher mud density needed to control formation pressure has led to fluid loss in shallower permeable formations. These permeable formations remain exposed until the lower completion is in place.

During the running of the casing, the normal well control measures must be implemented. The fracturing operation is conducted with a frac-head in place, reducing the risk of a well control incident.

Post-completion, the same well control precautions used for workovers and interventions on conventional wells can be applied to unconventional reservoirs. However, particular attention should be given to the selection of LCM, since damage to fractures can have a severe impact on productivity.

REFERENCES

1. Rike JL, Whitman DL, Rike ER, Hardin LR. *Completion and workover well control needs are different*. SPE paper 90–22; June 1990.
2. Johnson A, Tarvin J. New model improves gas migration velocity estimates in shut-in wells. *Oil Gas J* 1994.

3. Oudeman P, ter Avest D, Grodal EO, Asheim HA, Meissner RJH. *Bull heading to kill live gas wells*. SPE Paper 28896-MS; 1994.
4. Recommended practice for well control operations. API Recommended Practice 59. 2nd ed.; May 2006.
5. Formate Technical Manual. Version 3; September 2013. http://www.cabotcorp.com/solutions/products-plus/cesium-formate-brines/formate-technical-manual.
6. Luo Y, Gibson A, Mountford C, Hibbert T, Weddie C. Well control procedures developed for multilateral wells, oil and gas. *Oil Gas J* 1996.
7. Danney D. Improve unconventional reservoir completion and stimulation effectiveness. *J Petrol Technol* 2012;**64**(10). Available from: http://dx.doi.org/10.2118/1012-0115-JPT.
8. Davies RJ, Almonda S, Ward RS, Jackson RB, Adams C, Worralla F, et al. Oil and gas wells and their integrity: implications for shale and unconventional resource exploitation. *J Mar Petrol Geol* 2014;**56**:239–54.
9. Elshehabi T, Bilgesu I. *Well integrity and pressure control in unconventional reservoirs: a comparative study of marcellus and utica shales*. SPE paper–184056–MS; 2016.
10. Ridley K, Jurgens M, Billa RJ, Mota JF. *Eagle ford shale well control: drilling and tripping in unconventional oil and gas plays*. SPE paper 163984; 2013.

CHAPTER EIGHT

Pumping and Stimulation

Most of the pumping operations carried out on wells are not for the purposes of well kill. Instead they are performed as part of an intervention. Interventions that involve the use of pumped fluids include:

- Well stimulation carried out at below fracture pressure (matrix acid stimulation).
- Well stimulation carried out at above formation fracture pressure.
    - Propped fracture treatment in sandstone and shale reservoirs (Fig 8.1).
    - Acid fracture treatments in carbonate (limestone) reservoirs.
- Scale inhibitor and corrosion inhibitor squeeze treatment (bullhead of scale or corrosion inhibitor into the producing formation at below fracture pressure).
- Remedial cementing.
- Pressure testing of the completion and completion components.
- Pumping associated with coiled tubing operations—described in Chapter 10, Coiled Tubing Operations.
- Pumping associated with hydraulic workover operations (snubbing)—described in Chapter 11, Hydraulic Workover (Snubbing) Operation.
- Pumping associated with well kill operations—described in Chapter 7, Well Kill.

Figure 8.1 Some pumping operations require a large amount of equipment. Propped frac in progress, Wyoming, USA.

Well Control for Completions and Interventions.
DOI: https://doi.org/10.1016/B978-0-08-100196-7.00008-7

8.1 PUMPING EQUIPMENT

A pump is a device that is used to raise, compress, or transfer fluids. Pumps used in well interventions are either centrifugal pumps, positive displacement pumps, or displacement pumps.

8.1.1 Centrifugal pumps

Centrifugal pumps are widely used for high volume low pressure applications, such as transferring fluids between tanks at the wellsite during pumping and stimulation interventions.

At the center of a centrifugal pump is an impeller, a series of vanes arranged around a central axis. The impeller is driven (rotated) using a motor (Fig. 8.2).

Fluid enters the pump through the intake and reaches the impeller along, or near to, the rotating axis. It is accelerated by the impeller before flowing outwards through a diffuser or volute chamber.

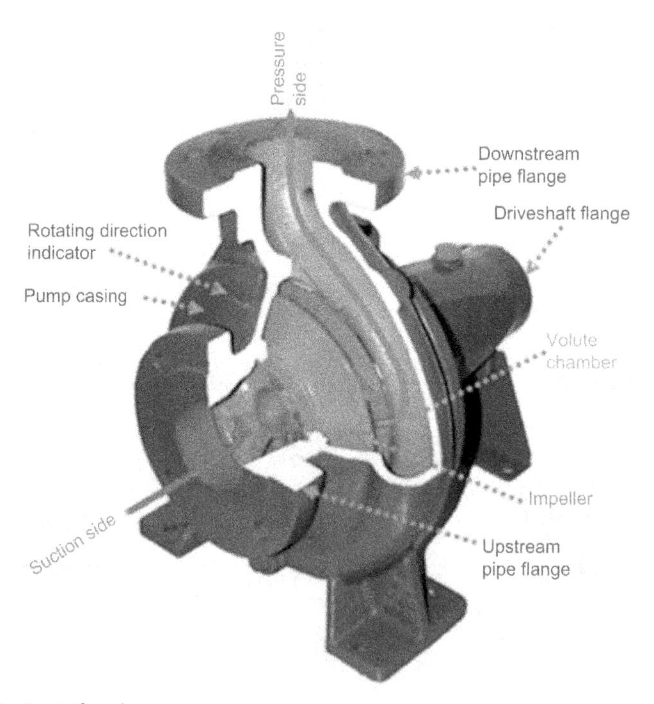

Figure 8.2 Centrifugal pump.

Figure 8.3 A diesel driven centrifugal pump—in use during a gravel packing operation to transfer fluids between vessels.

The pressure increase or head of fluid that can be lifted by a single impeller is limited. Some centrifugal pumps will use several impellers and diffusers in series, with each impeller/diffuser stage increasing fluid pressure by a set amount. When combined the head of fluid that can be lifted becomes significant. Perhaps the best know oilfield application for multistage centrifugal pumps is the electric submersible pump.

8.1.2 Reciprocating pumps

Mud pumps, the cement unit, and the high pressure fluid pumps used for fracturing and stimulation operations are almost all positive displacement reciprocating pumps (Fig 8.4). A rotating shaft turns a crank that converts rotational movement into reciprocation of pistons, in much the same way the pistons in an internal combustion engine drive the crankshaft.

Elastomeric seals on the piston sit inside ceramic or stainless steel liners. The seals and liners are high wear parts, and most pumps are designed to allow easy access for maintenance and replacement of the worn parts.

Nearly all of the pumps used in oilfield applications are triplex—three cylinder. Reciprocation of the cylinders means discharge from the pump is erratic and needs to be damped to reduce vibration and wear on the components downstream.

Pump discharge pressure is determined by the rated horsepower of the motor, and the liner and piston size. The smaller the liner, the

Figure 8.4 High pressure triplex frac pump.

higher the discharge pressure, but the smaller the displacement per stroke. Supervisors must know the following basic information about the pump:

- Pump pressure rating
- Maximum and minimum pump speed—strokes per minute
- Volume pumped per stroke—it is convenient to report volume in units that are in common use at the rig-site, e.g., strokes per barrel or strokes per m^3
- Setting of pressure relief valve.

8.1.3 Guidelines for the use of mud pumps and cement pumps

In most cases, dedicated high pressure pumps will be brought to a location specifically for an intervention operation. On offshore platforms and drilling rigs, there may be limitations on the space available for intervention pumps. As a consequence, the mud pump, or more usually the cement unit, will be used. Even where dedicated intervention pumps are used, the cement unit is often used to pressure test surface lines and as a back up to the intervention pump.

8.1.3.1 Mud pumps

- Before pumping operations begin, the drilling, mud pumps, mud manifold, valves, and main discharge lines should be pressure tested with water to the circulating system working pressure.
- For some critical operations a minimum of two pumps should be available to allow for redundancy.
- Hydraulic output should be sufficient to circulate maximum anticipated weight of fluid at planned well profile and worst case geometry.
- Functional stroke counters should be installed to enable displacement to be properly monitored.

8.1.3.2 Cement pump

- The emergency high pressure kill pump (and/or the cement pump), manifolds, valves, and lines should be tested to the maximum rated working pressure of the lowest rated part of the system.
- Cement pumps may be required as back up during well intervention operations where fluid is to be pumped, e.g., coil tubing clean out.
- Ideally the cement pump will have an independent power source (diesel) in the event of total power loss at the host facility.
- Functional stroke counters or flow/volume measuring devices must be installed prior to displacement.

8.2 TEMPORARY HIGH PRESSURE LINES

Pumping fluid into a well for intervention purposes requires the use of high pressure lines. These lines are usually temporary, and are rigged up for each operation. It is unusual to use equipment that is part of the permanent production facility, such as flowlines or production manifolds.

Temporary pipework is often mobilized and assembled by a specialist vendor, and will include a range of components. A temporary pumping system might include, but need not be limited to:

- Straight pipework in various lengths.
- T-pieces and laterals (Y-pieces).
- Swivel joints.
- Treating loops.
- Crossovers.
- High pressure (coflex) hoses.

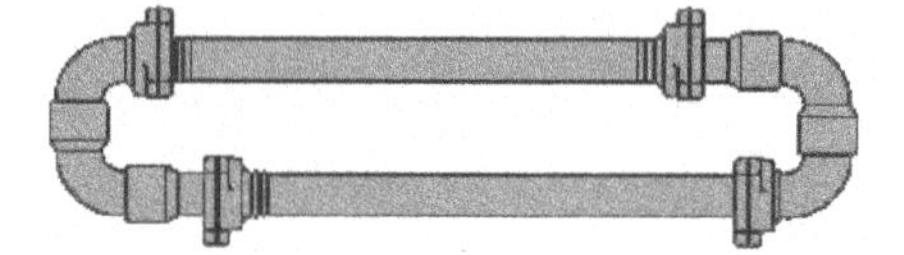
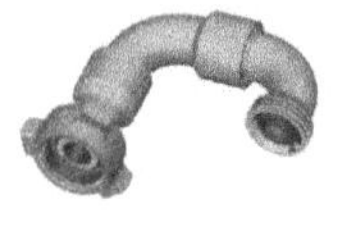

Figure 8.5 Chiksan components.

- Pressure relief valves (PRV)
- Lo-torque (plug) valves.
- Back pressure valves.

The type of temporary high pressure lines supplied by many operating and vendor companies is Chiksan. Chiksan (frequently misspelt "Chicksan") is a registered trademark of FMC, but has become the generic name for any temporary pipework of the same type. It is also referred to as a treating iron. The original Chiksan design patent has long expired, and there are many independent manufacturers making similar products. Not all are made to the same exacting specification as the FMC original (Fig. 8.5).

8.2.1 Temporary high pressure lines: connections

When constructing temporary high pressure lines, a connection is needed that provides pressure integrity, robust mechanical strength, and is easy to make up and break out. By far the most common connection uses a course Acme thread that is hammered up to tighten and seal. It is usually referred to as a Weco, or hammer lug connection.[a] Weco connections feature a winged nut (female) that fits over a male sub. Protruding lugs are used to hammer the connection tight. All Weco connections have a metal-to-metal seal, with some connections having an additional elastomeric seal.

- Low pressure service (1000–2000 psi). Weco unions for low pressure service use a metal-to-metal seal.
- Medium pressure service (2000–4000 psi). Weco unions use a resilient o-ring seal in the male sub and a metal-to-metal seal. The o-ring protects the metal seal face from corrosion.
- High pressure service (6000–20,000 psi). Weco unions for high pressure service are fitted with a lip type elastomer seal ring in the female sub (Fig. 8.6).

[a] Weco is a registered trademark of FMC.

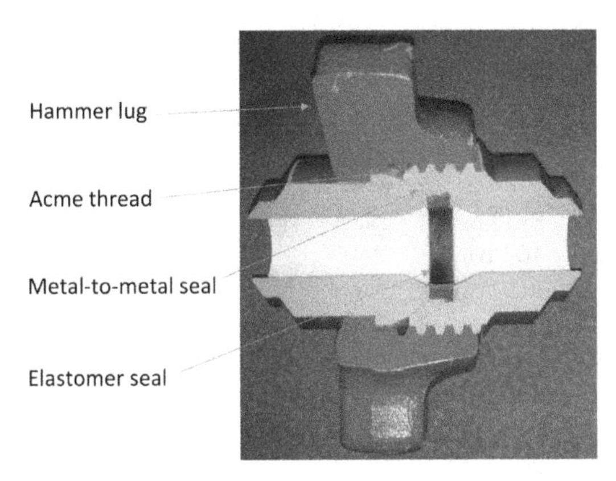

Figure 8.6 A Weco connection.

Figure or "Fig" numbers are used to denote pressure rating and seal type for hammer union connections. For example, the widely used 1502 connection has a working pressure of 15,000 psi for standard service, as indicated by the first three digits (150). Similarly 206 connections would be rated to a working pressure of 2000 psi. The final digit relates to the seal type. In a 1502 connection the "2" indicates a lip type elastomer seal, and for 206 the "6" indicates an o-ring seal. Some manufacturers use color codes to indicate connection type. However, this should not be relied on. Vendors can supply temporary lines in a wide range of pressure rating and pipe diameters. These are listed in Table 8.1.

8.2.1.1 Connection mismatch

Anyone preparing temporary lines needs to be aware of the potential fatal mismatch between connections of the same size, but with different pressure ratings. For example, it is possible to connect a 2″ 602 to a 2″ 1502 connection. Whilst the treads will engage, and the connection can give the appearance of being fully made up, they can undergo catastrophic failure when pressure is applied. This can, and has, caused fatalities. Anyone using temporary flowlines must have knowledge of the problem areas. An excellent source of information is the FMC website. http://www.fmctechnologies.com/FluidControl/Technologies/Flowline/WingUnions.aspx (Fig. 8.7).

8.2.2 Valves for temporary flowlines

Plug valves are the most commonly used type of valve in temporary high pressure flowlines. However, it is unusual to hear anyone refer to them as

Table 8.1 "Chiksan" pressure rating and size

Series	Standard pressure rating		Sour service pressure rating		Nominal size (inches)
	Cold working	Test	Cold working	Test	2–2½, 3–4, 6–8
100	1000 psi 69 bar	1500 psi 103 bar	N/A		1–1¼, 1½–2, 2½–3, 4
200	2000 psi 138 bar	3000 psi 207 bar	N/A	N/A	1–1¼, 1½–2, 2½–3, 4
206	2000 psi 138 bar	3000 psi 206 bar	N/A	N/A	1–1¼, 1½–2, 2½–3,46, 8–10
207	2000 psi 138 bar	3000 psi 206 bar	N/A	N/A	3–4, 6–8, 10
211	2000 psi 138 bar	3000 psi 206 bar	N/A	N/A	1–2
400	4000 psi 276 bar	6000 psi 414 bar	2500 psi 172 bar	3750 psi 259 bar	5–6, 8–12
602	6000 psi 414 bar	9000 psi 621 bar	6000 psi 414 bar	9000 psi 621 bar	1–1¼, 1½–2, 2½–3, 4
1002	10,000 psi 690 bar	15,000 psi 1034 bar	7,500 psi 517 bar	11,250 psi 776 bar.	1–1¼, 1½–2, 2½–3, 4–5, 6
1003	10,000 psi 690 bar	15,000 psi 1034 bar	7500 psi 517 bar	11,250 psi 776 bar.	2–3, 4–5
1502	15,000 psi 1034 bar	22,500 psi 1551 bar	10,000 psi 690 bar	15,000 psi 1034 bar	1–1½, 2–2½, 3–5
2002	20,000 psi 1379 bar	30,000 2068 bar	N/A	N/A	2–3
2202	N/A	N/A	15,000 psi 1034 bar	22,500 psi 1551 bar	2–3

a plug valve. They are usually called Lo Torq valves. Lo Torq is the Halliburton trade name for their own design of plug valve, however, the name has become generic (Fig. 8.8).

Most of the Lo Torq valves used with temporary pipework are opened and closed manually. Operation of the valve requires a quarter turn (90 degrees) of the plug using a bar. Larger sizes of temporary line can be equipped with manually or hydraulically operated gate valves.

8.2.3 Nonreturn valves

When temporary pipework is hooked up to the well, a nonreturn valve should be located as close as possible to the outlet on the well. This will minimize the escape of wellbore fluids if the lines fail. Two types of check valves are used for temporary pipework, the dart and the flapper (Fig. 8.9).

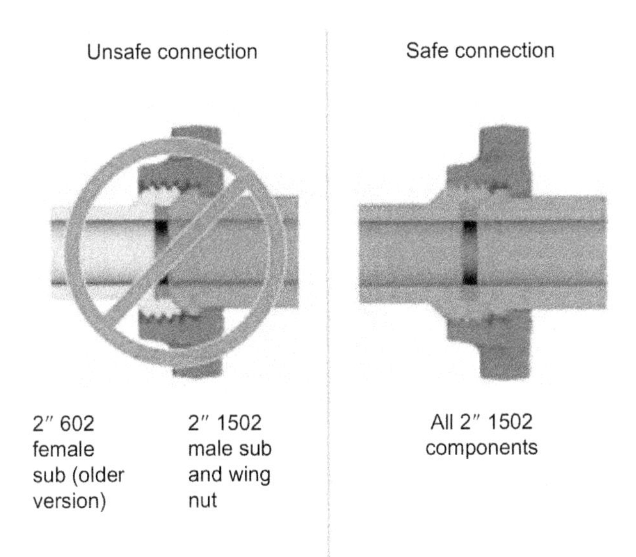

Figure 8.7 Unsafe: female 602 sub made up to a 1502 male sub. *Image courtesy of FMC.*

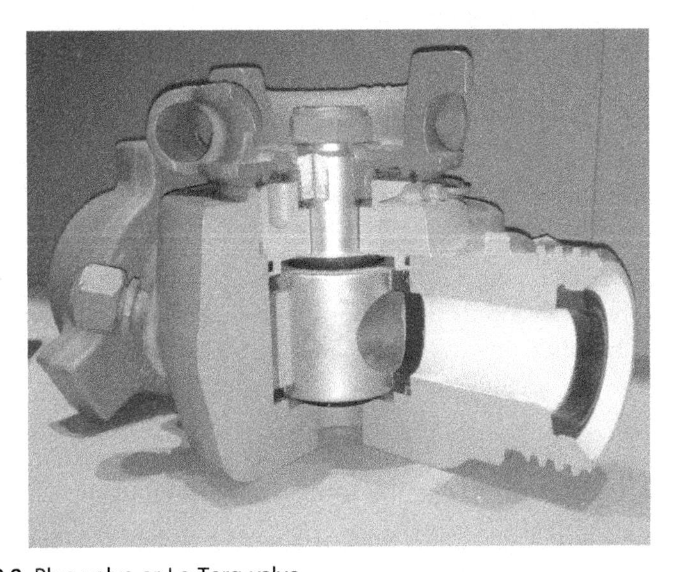

Figure 8.8 Plug valve or Lo Torq valve.

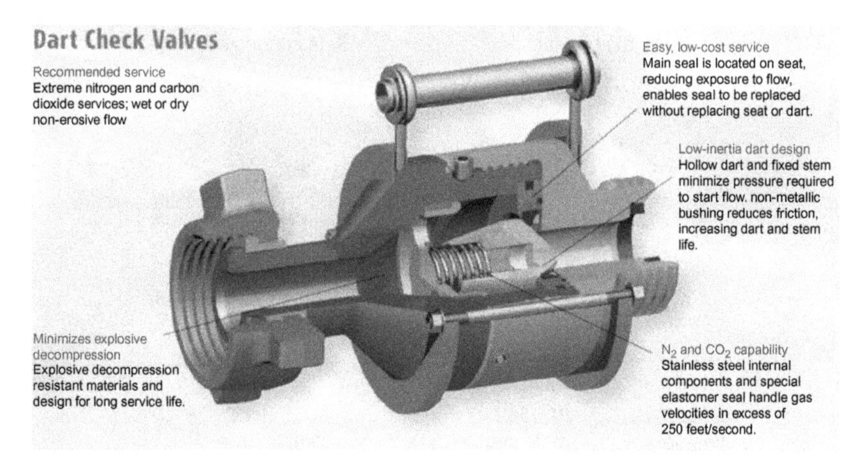

Figure 8.9 Check valve. *Image courtesy of FMC.*

8.2.4 Pressure relief valves

During pumping operations, a pressure relief valve must be used. These should be set to relieve at a pressure slightly higher than the maximum anticipated pump pressure, but lower than the maximum working pressure of the lowest rated component in the system. Before operations begin, the valve should be function tested in the workshop to confirm it is operating at the set pressure (or within an acceptable margin). It is common practice to isolate the PRV when pressure testing the lines, but it must be reinstated before pumping operations begin.

If hazardous or harmful chemicals are being pumped, the exhaust from the PRV must be piped to a suitable (safe) area or container (Fig. 8.10).

8.2.5 Wellhead or tree connection

In most cases, the temporary line will be connected to the tree or wellhead using a flange crossed over to a Weco type hammer union connection. In some installations, the outlet on the kill wing valve will have a Weco type hammer union connection as a permanent fixture. At some locations the A annulus outlet will also be equipped with a Weco connection.

8.2.6 Securing high pressure temporary flowlines

During stimulation operations, there is always a risk of a temporary line failing. If this occurs and the ends of the pipe are unsecured, they will flail

Figure 8.10 Pressure relief valve. *Image courtesy of FMC.*

around until the stored energy in the line dissipates. This can, and has, caused fatalities. When energized fluids and gases are used, considerable force is involved.[1] Conventional methods of tying down the lines (chain and wire stops) may not be sufficient to restrain the ends of a parted line. In recent years, specialized restraint systems have been developed[2] by a number of vendors, including Weir and FMC. These newer systems are designed to absorb the forces associated with lines failing under pressure, and provide a much safer alternative. Their use is recommended (Fig. 8.11).

Supervisors must always think about the consequences of line failure during pumping operations. If any part of the high pressure system fails, people will be at risk.

- Flying debris and shrapnel—radius of exposure: hundreds of feet away from the point of failure.
- Flailing lines—radius of exposure: tens of feet.
- Direct exposure to high pressure fluid or gas: a few feet.

Exposure to the risks of line failure must be limited by minimising the time people have to spend inside the radius of exposure. Automation and remote monitoring of the pumping operation is advisable.

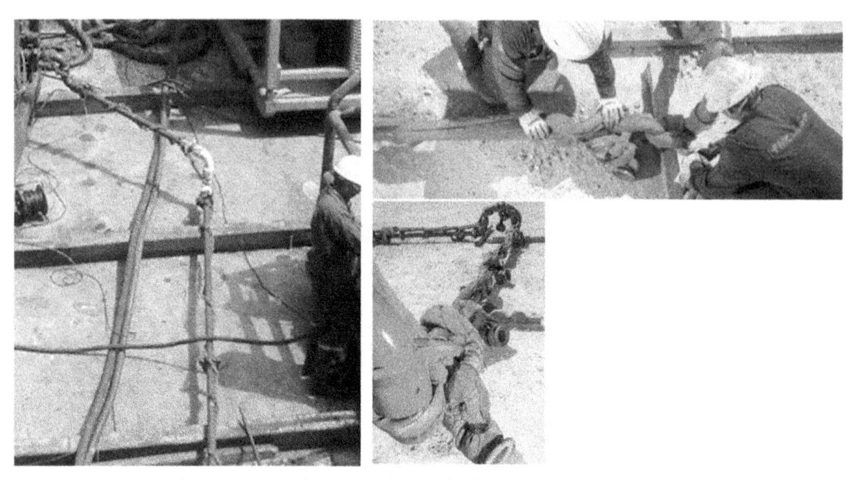

Figure 8.11 Chicksan secured using chain (left) and a properly designed restraint system (right top, right bottom). *Restraint system images courtesy of FMC.*

8.3 PUMPING OPERATIONS

8.3.1 Acid wash (carbonate scale removal)

Hydrochloric acid (HCl) is frequently used to remove carbonate scale from the production tubing. Carbonate scaling becomes worse with reducing pressure (and increasing temperature). Since well pressure is lowest at the top of the well (wellhead and tree), scaling tends to become worse as the produced fluids move towards the surface. Scale is often deposited where there are sudden pressure drops, e.g., at crossovers, above downhole safety valves, across the tree, and downstream of the choke. Scale deposition is always a matter of concern. It can have a significant economic impact, as flow from the well will be choked, or may even stop. Of greater concern is the impact scale can have on the safety of the well. Scale deposits can prevent the tree valves and downhole safety valves from functioning correctly.

The most common method of scale removal is to pump acid through coiled tubing as part of a live well intervention (see Chapter 10: Coiled Tubing Well Control). This method is very effective, since acid can be placed exactly where it is most needed.

Although acid can be placed by bullhead this is rarely done as there is a significant risk of formation damage.

8.3.2 Chemical inhibitor squeeze treatment

Bullheading chemical inhibitors into the formation is routine in many fields. Most chemical squeeze treatments are carried out to slow or prevent the build-up of carbonate or sulfate scales in the well.

Squeeze treatments vary in size, with large treatments requiring many hundreds of barrels of fluid. The chemical inhibitors are designed to be adsorbed onto the formation matrix, and are slowly released along with the produced fluids. As they flow back, they afford a degree of protection against scale or corrosion.

Regular sampling of the produced fluids needs to be carried out post-squeeze. Inhibitor in the produced fluids needs to be at above a minimum concentration (MIC). When the inhibitor concentration falls below the MIC necessary to effectively combat scale (or corrosion), the treatment needs to be repeated. Treatment frequency depends on the chemistry of the reservoir fluid (hydrocarbon and water). Scale squeezes for carbonate scale can last from a few weeks to a year. Sulfate scale treatments are typically more frequent, and do not normally last for more than 6 months. On one North Sea well, the treatment had to be repeated every 6–8 days[3]; not a very efficient method of scale management.

Nearly all inhibitor squeeze operations conducted during the life of a well are pumped through the production tree.

8.3.3 Cement squeeze

There are a number of reasons why remedial cementing may be required in a well.

- Approximately 15% of all primary cement jobs are unsuccessful in achieving the desired isolation. Cement will have to be squeezed into the compromised zone until integrity is ensured. The industry average is three squeeze jobs to ensure isolation after a poorly executed primary cement job.
- Hydrocarbon production zones water out, or begins to produce excessive and unwanted gas. These are squeezed to reduce water or gas production.
- Cement can be used to repair leaking packers.
- Cement plugs are used for zone or well abandonment.

Some squeeze operations are carried out during the drilling of the well. Shoe integrity is a crucial well integrity concern. Cement at a casing shoe must be strong enough to contain pressure in deeper zones as the well is drilled. If it is not, the cement must be repaired (squeezed) before drilling ahead. These squeeze operations are carried out with an overbalanced kill weight mud still in the well. Many other squeeze operations are performed on live production wells. Cement can be placed using coiled tubing or a hydraulic workover unit deployed workstring. Bullheading cement through the completion is not normally done, although there are exceptions.

8.3.4 Hydraulic propped fracture treatment

With a very few exceptions, propped "fracs" are used to increase productivity from sandstone and shale reservoirs. Fracturing a formation creates a high permeability pathway from the formation into the wellbore. Fracture permeability will be many times the permeability of the formation. For example, some of the new shale gas developments have reservoir permeability measured in nanodarcys, whilst permeability in the fractures is multi Darcy—many orders of magnitude more.

To create a fracture, a fluid "pad" is pumped into the well at above the reservoir formation fracture pressure. This hydraulically fractures the formation. Fracture direction will be determined by the (in situ) stress in the formation, and the fracture will always open perpendicular to the minimum stress in the formation.

Fracture size (height, width, and length) are a function of pump rate, pressure, formation permeability, and fluid pad viscosity. Once the fracture is open, proppant is blended into the fluid as it is pumped into the well. The proppant is carried into the fracture, packing it, and preventing it from closing when the pumping stops.

Proppant selection will depend on the size of the fracture and closure stresses in the formation. Where closure stress is low, natural sand will be used. Medium and high closure stresses will require ceramic and bauxite, respectively. Proppant concentration will be low at the beginning of the treatment, but is gradually increased as the job progresses. As a result, a propped frac operation normally sees the highest fluid (slurry) density and the highest pressure at the end of the treatment, when "screen out" is reached. Fracture pressure and screen out pressure can be very high. It should be noted that offset wells could be at risk during a fracturing operation. In one incident in Canada, a fracturing operation resulted in overpressure of an adjacent rod pump well shearing the polished rod at the pump jack and destroying the stuffing box.[4] Similar problems have been reported at multiwell pads during the hydraulic fracturing of shales.

8.3.5 Acid stimulation of carbonate reservoirs

Productivity from carbonate reservoirs can be improved by acid stimulation, since carbonate reservoir rock (limestone, dolomite, and chalk) is acid soluble. For example, 1000 (US) gallons of 15% HCl will dissolve 1840 lbs (900 kg) of carbonate material. Acid fracturing uses a similar technique as a propped frac to improve inflow performance, creating a highly permeable pathway extending from the wellbore out into the formation. Acid is pumped into the well at above formation fracture pressure. As the acid contacts the formation, it dissolves carbonate material in the rock matrix, etching the face of the fracture. Fracture geometry is controlled by pump rate, net pressure, fluid leak-off, and the speed of the acid reaction. When pumping stops, and pressure drops below fracture pressure, the fracture cannot close completely because some of the formation limestone has been dissolved, leaving open a conductive channel. Most limestone and dolomite reservoirs have sufficient compressive strength to keep the etched channels, so no proppant needs to be pumped (Fig. 8.12).

Unlike a propped frac, maximum pressure during acid fracturing is normally seen at the beginning of the operation and occurs as the fracture is propagated. With a fracture open, acid improves injection into the formation, and surface pump pressure declines. If diversion of the acid to different layers in the formation is needed, then the pressure profile will be different. For example, a ball sealer diversion, or diversion using pH sensitive gels, will case an erratic (rising and falling) pressure profile until final "screen out" (Fig. 8.13).

Many acid jobs are carried out at below fracture pressure and are called "matrix treatments." Pumping acid into the formation at below fracture pressure will dissolve the matrix, although the distribution of acid in the near wellbore region can be uneven. Acid creates dendritic

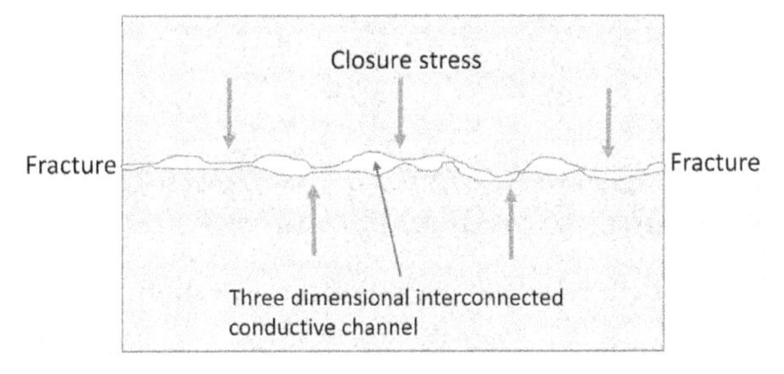

Figure 8.12 Acid fracture profile.

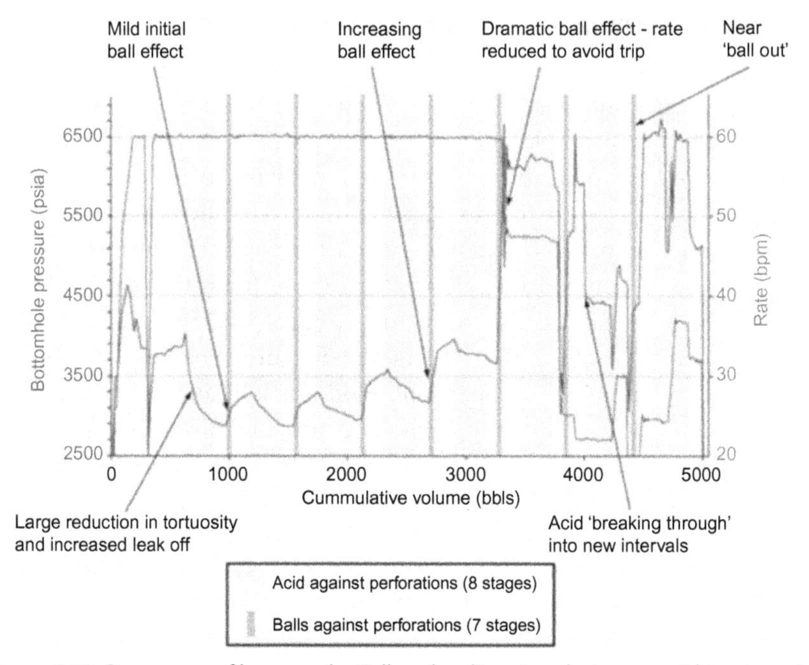

Figure 8.13 Pressure profile example. Ball sealer diversion during an acid treatment.

(branching) pathways sometimes referred to as "wormholes." Matrix treatments are preferred where fracturing risks contact with nearby water or gas intervals.

Matrix treatments are also carried out on sandstone reservoirs to remove formation damage. A mixture of hydrofluoric acid (HF) and HCl, known as mud acid, is used as the main fluid to remove mud damage from a sandstone matrix. HCl is also used to remove carbonate lost circulation material. HF is never used on a carbonate reservoir, because it produces an insoluble precipitate (calcium fluoride).

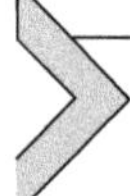

8.4 WELL CONTROL CONSIDERATIONS DURING PUMPING AND STIMULATION OPERATIONS

Pumping fluids into a well through the production tree is a routine intervention technique. The number of pumps, the horsepower requirement, capacity, and pressure rating depends on the type of operation planned and well pressure. Operational requirements and well pressure will also dictate the size and pressure rating of temporary flowlines.

Although the pump and line specifications will vary, there are some basic principles that will be common to all operations and will help ensure that well integrity and well control is maintained at all times.

- Christmas tree valve integrity should to be confirmed before rigging up, either by reference to wellhead and tree maintenance records, or by integrity testing.
- Confirm that any and all chemicals are compatible with surface and downhole components; tree, tubing, and completion equipment, both metal and elastomers. If there are any concerns, the chemicals should be tested on a representative sample; preferably at simulated well conditions (pressure and temperature).
- Obtain and read recent well history. Pay particular attention to any trends that are indicative of potential well integrity problems, such as persistent annular pressure build-up.
- All of the high pressure lines should be tested to a pressure in excess of the maximum planned pump pressure necessary for the operation, but no higher than the working pressure of the lines.
- Ensure that all of the temporary high pressure line connections are of the correct type and pressure rating.
- Under no circumstances should attempts be made to tighten leaking connections when the lines are under pressure. People have been killed doing this. If a leak appears, the operation should be stopped, the well closed in, and the line depressurized before repairs are carried out.
- Ensure that the high pressure lines (and any associated elastomer seals) are compatible with any chemicals that are to be pumped.
- The high pressure temporary lines, valves, and other accessories should be accompanied by up-to-date and valid pressure test certification.
- If the Christmas tree is equipped with actuated valves, these should be isolated from the host facility hydraulic supply and controlled using a single well control panel (Fig. 8.14 and as described in Chapter 9: Wireline Operations). This will prevent the valve closing unexpectedly whilst pumping is underway. A pump "deadheading" against a closed valve risks over-pressuring the lines, even if a PSV is used.
- A nonreturn valve must be placed in the lines as close as possible to the tree. This will prevent back-flow from the well if there is a leak or failure of the high pressure line.

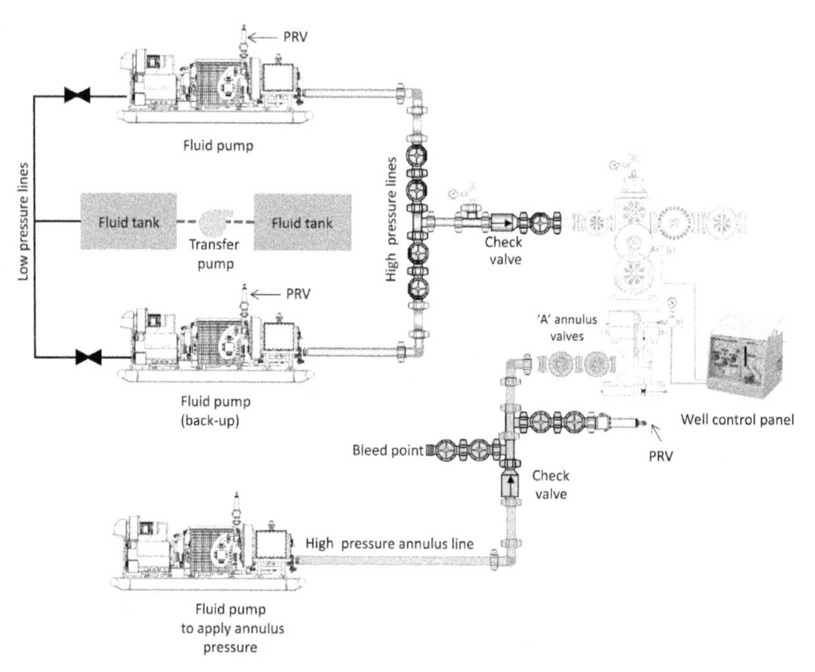

Figure 8.14 Rig up configuration for pumping operations.

- All high pressure temporary lines should be secured with properly designed restraints.
- Most operations will be pumped through the kill wing valve (Fig. 8.14), or through the swab valve. Double valve isolation must be maintained at all times when rigging up the lines. If rigging onto the swab cap, the master valve and swab valve remain closed while the lines are connected. If the kill wing is used, then the kill wing and (upper) master valve should remain closed during the rig up.
- Many operations use multiple pumps manifolded together. Even on operations where only a single pump is needed, it is common to have a back-up pump at the location. Where more than one pump is used, the lines should be rigged up so that any pump can be isolated from the high pressure manifold with double valve isolation (Fig. 8.14). This will enable repairs to be carried out whilst pumping continues.
- Some operations will require annulus pressure to be applied to reduce burst load on the tubing. Annulus line minimum requirements (Fig. 8.14) are:

- Double valve isolation on the "A" annulus outlet while the temporary lines are installed.
- A check valve (NRV) placed as close as possible to the A annulus outlet on the wellhead.
- A pressure relief valve placed in the line downstream of the NRV. The set pressure for the PRV must be lower than calculated tubing collapse.
- A bleed point with double valve isolation positioned between the "A" annulus side outlet and the NRV.
- The annulus line can be tied into the main treatment line manifold and annulus pressure applied using the main fluid pump. However, this is not recommended. It is much better to have a dedicated annulus pump (Fig. 8.14). This will allow annulus pressure to be topped up without having to interrupt operations.
- The annulus pressure history should be checked for pressure build-up, especially persistent build-up.

- If pump pressure is expected to reach or exceed the pressure limitations on the tree or wellhead, then a "tree-saver" valve should be used.
- Ensure reliable communications are established between the pump operator, the well site supervisor, and the host facilities operators.
- Many treatments use water-based chemicals or are mixed with water. In low pressure reservoirs, a stimulation or inhibitor treatment is likely to kill the well. Measures may have to be in place to underbalance the well following treatment.
- Bullheading will be difficult in wells with low permeability reservoirs.
- A controlled bullhead will be problematic in high permeability, low pressure reservoirs. A column of water-based fluid will, in effect, fall down the tubing, meaning that contact time with the chemical wash may be insufficient.

8.4.1 Special considerations when using frac boats

Many rigs and platforms do not have enough deck space or fluid storage capacity to enable large volume treatments to be carried out. In these circumstances, a frac boat will be used. The main points to consider when using a frac boat are:

- Fluids are pumped from the vessel to the treating iron on the rig via high pressure flexible hoses (Coflexip). Once the vessel arrives on

Figure 8.15 Coflex hose reels clearly visible on the stern (back) of a frac boat.

location, the end of the hose needs to be picked up from a reel at the stern of the vessel and hung from a support bracket on the side of the rig. Specially designed "frac hangers" are normally required, and are not standard on all rigs. These may need to be made and sent offshore for fitting in advance of the operation.

- Communications. All of the pumps, mixing, and monitoring equipment are on the boat. Before any pumping can begin, it is essential that good lines of communication are in place between the boat and the rig (Fig. 8.15).

8.4.2 Formation fracture pressure

When performing a propped or acid fracture treatment, formation fracture pressure is deliberately exceeded. In all other treatments, fracturing the formation must be avoided. Fracturing the formation will be detrimental to future production, and it can lead to well integrity problems. Data obtained during the drilling of the well in the form of formation leak-off tests or formation integrity tests will establish fracture pressure. As part of the planning process, well engineers should calculate the maximum surface pump pressure and rate that can safely be used whilst remaining below formation (reservoir) fracture pressure. Most engineers will include a safety margin, normally about 500 psi (3450 kPa).

Pump rate and discharge pressure are interdependent. Increasing the rate will cool the formation, lowering the fracture pressure. WellCAT (WS-Prod), Prosper, or similar commercially available software can be used to model temperature profiles during injection. They can also be

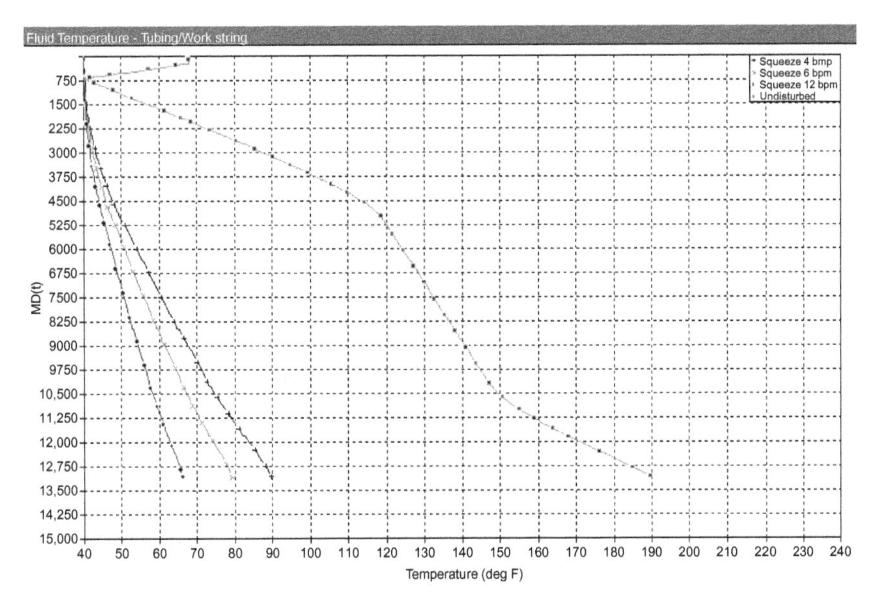

Figure 8.16 "Screen dump" from a WellCAT temperature model showing temperature against depth at different injection rates.

used to estimate frictional pressure drop in the tubing during injection (Fig. 8.16).

Temperature predictions can be combined with data from rock mechanics studies to estimate formation fracture pressure during the treatment (at the point of maximum cooling) (Fig. 8.17).

A plot of formation fracture pressure extrapolated back to the surface is then produced. Hydrostatic pressure of the treatment fluid and frictional pressure drop in the tubing need to be accounted for. Fig. 8.18 illustrates the relationship between the rate and surface pressure for two different tubing sizes. In the larger (7″) tubing, surface pressure must be reduced as the rate increases to remain below fracture pressure. In the smaller 5½″ tubing, a much higher surface pressure is required to obtain an equivalent rate. However, unlike the 7″ tubing, this does not translate to higher pressure downhole, as there is significantly more frictional pressure drop in the smaller tubing.

8.4.3 Surface controlled subsurface safety valve

Most safety valves are held open with hydraulic pressure of between 1500 and 2500 psi above well pressure. Well control panels are set to deliver

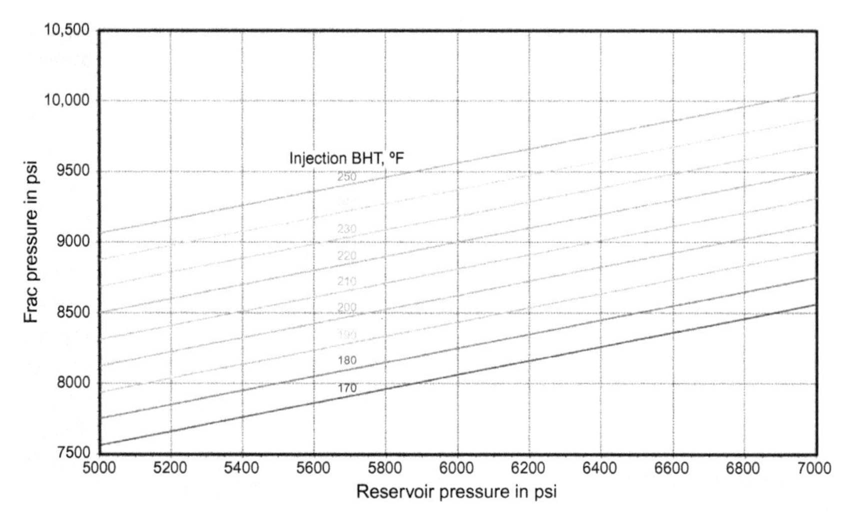

Figure 8.17 Fracture pressure, temperature reservoir pressure relationship. Example from a North Sea reservoir.

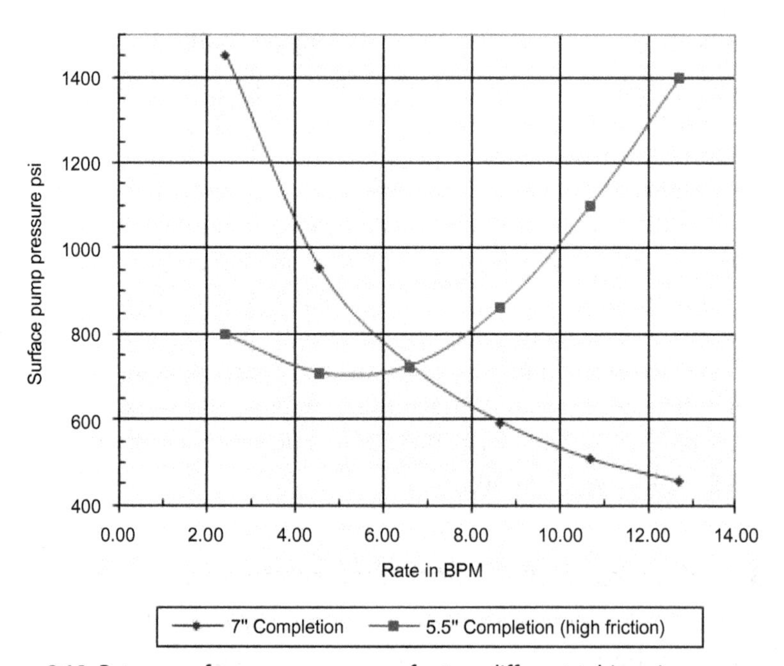

Figure 8.18 Rate vs surface pressure curves for two different tubing sizes.

enough pressure to keep the safety valve open when wellhead pressure is at the maximum anticipated value, normally closed in pressure.

Pumping operations are often performed at a pressure in excess of the maximum anticipated wellhead closed in pressure. If no mitigating measures are taken, the increased well pressure can cause the surface controlled subsurface safety valve (SC-SSSV) to close. If this happens there is no visible indication at the surface, and it is still be possible to continue to pump fluid down through the closed valve. Damage to the valve is almost certain. Where wellbore pressure is high enough to close the SCSSSV, the inner mandrel (flow-tube) is retracted to the closed position, exposing the flapper and flapper seat. Injected fluid washes past the flapper, causing vibration and erosion damage (from an abrasive slurry). To prevent this happening the following measures need to be taken:

- Take control of the SCSSSV on the single well control panel. Set the control line pressure to a value that is high enough to keep the valve open at the highest planned pumping surface pressure anticipated during the intervention. This can be calculated as follows:

$$Pc = Pvo + Pt + Pm - Ph \tag{8.1}$$

where:

Pc = control line pressure at surface (panel pressure).

Pvo = spring force—pressure required to depress the main spring.

Pt = tubing pressure acting on the valve piston (i.e., maximum pump pressure at the depth of the valve).

Pm = manufacturers safety margin—normally 200–300 psi.

Ph = hydrostatic head of control line fluid above valve.

If the required pressure (based on the above calculation) is higher than the working pressure of the valve piston chamber, then the following action will be necessary.

- Wireline Retrievable SCSSSV.
  - Pull the safety valve.
  - If a fluid-only treatment is to be performed, then a pack-off should be set in the SCSSSV nipple to isolate the control line and protect the seal bore.
  - If a propped frac is planned, running a pack-off will protect the nipple from erosion and the control line from blockage. However, there is a risk of the pack-off becoming stuck in the well. Frac sand can pack the gap between the fish neck of the straddle and the inside of the tubing.

- Tubing retrievable SCSSSV.
    - Run a hold-open sleeve/pack-off. This must be long enough to hold the flapper out of the flow-path and protect the metal flapper seat.

8.4.4 Gas lift wells

There are additional considerations needed when performing pumping operations on a well equipped with gas lift mandrels.

- If the well has been on gas lift, there will be a gas gradient in the annulus. During pumping the tubing will be filled with much denser fluid. As there is a density difference, hydrostatic pressure at the packer will be much higher in the tubing than the annulus. The potential for bursting the tubing must be addressed. If necessary, measures must taken to eliminate any risk. These would include:
    - Displacing the annulus to a dense fluid to cancel out any hydrostatic imbalance.
    - Increasing annulus pressure.
    - If the well has live gas lift valves fitted, it will not be possible to apply pressure to the annulus before any pumping operation begins, as any pressure increase in the annulus would simply equalize through the gas lift valve. Annulus pressure needs to be applied as the tubing pressure increases. This needs careful coordination between the crew on the frac pump and those operating the fluid pump connected to the annulus. The risk is that tubing pressure is brought up too quickly, before the pressure in the annulus is increased to compensate. Coordination is likely to be difficult in a situation where high output pumps (e.g., on a frac boat) are connected to the tubing, and the cement unit is being used to control annulus pressure. A more time consuming, but safer option, is to replace the live gas lift valves with blank (dummy) valves. This eliminates the problem caused by annulus-to-tubing communication, and the annulus can be pressured before pumping begins.
- Gas lift valves are equipped with a check valve to prevent tubing-to-annulus communication. These have a tendency to leak if they have been in the well for a long time. Before pumping starts, tubing-to-annulus integrity must be confirmed. This can be accomplished by performing an annulus bleed-off test, or by plugging and pressure testing the tubing.

- Frac proppant is likely to create problems for gas lift valves. If it gets into the valve it will block the orifice or damage the check valve. Proppant can also build up in side pocket mandrels, making access difficult. It makes sense to replace any unloading valves and the gas lift valve with dummies before treating the well.

8.4.5 Tree saver valves

If pump pressure at the surface is expected to reach, or even exceed, the working pressure of the tree or wellhead, then protective measures are needed. A tree saver protects the tree and wellhead from damage and failure resulting from exposure to high pressure. It will also protect the tree from exposure to abrasive slurry and corrosive chemicals. Replacing an existing tree with one having a high enough working pressure is both time consuming and expensive. In addition, a potentially damaging well kill would be needed. Replacing the wellhead is impracticable.

A tree saver is mounted on the existing tree and an integrity test is made. Once integrity has been ensured, the valves on the production tree are opened, and hydraulic pressure is used to extend a mandrel through the tree bore and down into the top of the production tubing (just below the hanger). Elastomer cups on the bottom of the mandrel seal inside the production tubing isolate the production tree from stimulation chemicals and stimulation pressure.

Once in place, a tree saver can increase the working pressure of the tree and wellhead up to as much as 20,000 psi (138,000 kPa). When a tree saver is used, it will usually be the production tubing that limits surface pump pressure. Once the stimulation treatment is finished, the mandrel is retracted and the production tree valves are closed. The tree saver can then be removed safely from the tree (Fig. 8.19).

8.4.6 Tubing loads during pumping operations

Pumping large volumes of cold, sometimes dense fluid, at high pressure will stress the completion. Relatively simple hand calculations can give an idea of how much stress is caused by a pumping operation. However, to accurately assess the stresses associated with any treatment, full tri-axial analysis must be carried out. This is most easily accomplished using tubing stress analysis software such as WellCAT. Failure to properly understand and determine tubing stress could result in a catastrophic failure of the completion.

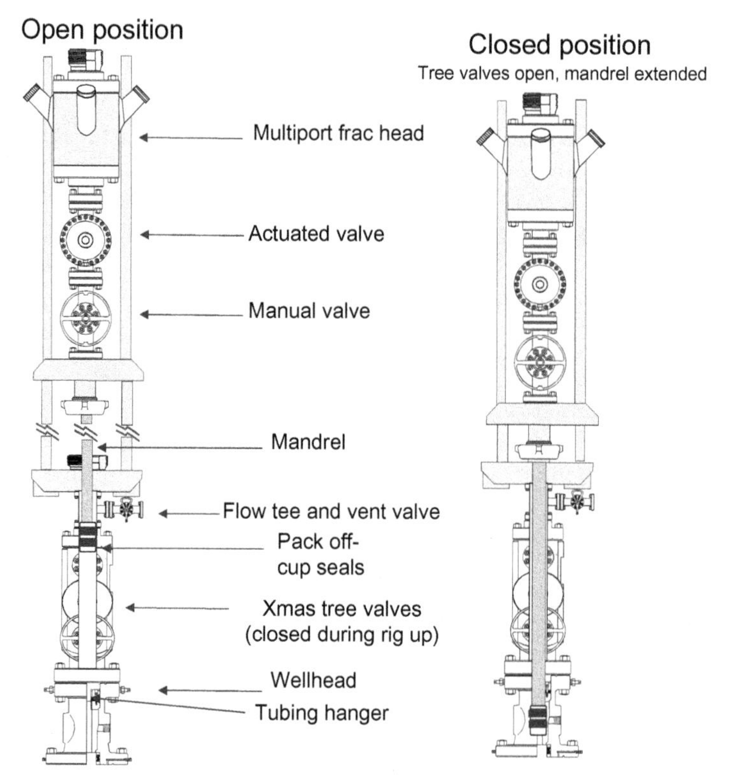

Figure 8.19 Tree/wellhead isolation tool.

8.4.6.1 Thermal effects

Steel will expand when heated and contract when cooled. If the tubing is free to move, pumping cold fluids into the well will cause the tubing to contract. There is a risk that seals could pull out of the top of a seal assembly if it is of insufficient length. Modeling can be carried out to determine the temperature change (average temperature in the string before, during, and after the treatment). If the amount of temperature change is known, movement can be calculated using the formula:

$$\Delta L_{\mathrm{TEMP}} = C_T \Delta T L \tag{8.2}$$

where:

C_T is the coefficient of thermal expansion of steel (approximately $6.5 \mathrm{x} 10^{-6}\ {}^{\circ}\mathrm{F}^{-1}$)

ΔT is the change (up or down) in temperature (F)

L is the length of the tubing string (ft).

For example, a 10,000 ft long tubing string cooled by 100°F would contract by 6.5 ft.

$$0.0000065 \times -100 \times 10,000 = -6.5$$

If the tubing is anchored to the packer (not free to move), strain (fractional length change) becomes stress, resulting in an increase in tension (cooling) or compression (heating). For example, a 5½″ 17 lb/ft string (tubing wall area 4.96 square inches) would be subjected to 96,720 lbs of additional tension during the injection of cold fluid down the well.

−6.5 (ft) × 30,000,000 (modulus of elasticity of steel) × 4.96 (tubing wall area)/10,000 (string length in feet) = −97,720 lbs.

8.4.6.2 Piston forces

Many wells are equipped with dynamic (moving) seals. Tubing pressure (internal or external) will result in piston forces that can lengthen or shorten the tubing. Polished bore receptacle (PBR) type seal assemblies normally have seal bores that are significantly larger than the ID of the tubing, and this will result in large piston forces. For example, a PBR run with 5½″ tubing will have a seal bore ID of approximately 6 to 6¼″. By way of comparison, 5½″ 17 lb/ft tubing has an ID of 4.892″. The main risk during a pumping operation is high axial compression and/or the tubing seal assembly being pushed out of the top of the polished bore receptacle. Piston force can be estimated using the following equation:

$$\text{Piston force} = P_o(A_b - A_o) - P_i(A_b - A_i) \tag{8.3}$$

where:

P_o is annulus (external) pressure
A_b is seal bore area
A_o is the area of the tubing OD
P_i is tubing pressure (internal)
A_i is the area of the tubing ID (Fig. 8.20).

8.4.6.3 Ballooning

Tubing will balloon if internal pressure is higher than external pressure. If external pressure is higher than internal pressure, reverse ballooning will occur. Where the tubing string is free to move, ballooning will cause the end of the string to move up and reverse ballooning will move it down. If the tubing is anchored, then ballooning will increase tension, and

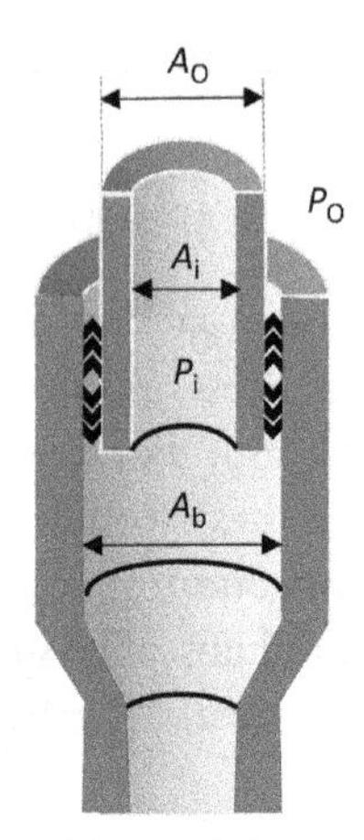

Figure 8.20 Polished bore receptacle critical dimensions.

reverse ballooning will increase compression. The amount of length change caused by ballooning can be calculated from the formula:

$$\Delta L_{BAL} = \frac{-2\mu L_P}{E(A_o - A_i)}(\Delta p_i A_i - \Delta p_o A_o) \tag{8.4}$$

where:

μ is Poisson's ratio (0.3 for steel)

E is the modulus of elasticity of steel (3×10^6)

L_p is the length of the tubing string to the packer.

Where tubing is fixed to a packer, the amount of force caused by ballooning is calculated from:

$$F_{ballooning} = 2\mu(A_i \Delta p_i - A_o \Delta p_o)$$

Example: a 10,000 ft deep well with 5½″ 17 lb/ft tubing (ID 4.892″), 5000 psi internal pressure would cause the tubing to move up 3.8 ft if it were free to move.

$$\Delta L_{BAL} = \frac{-2 \times 0.3 \times 10,000}{30 \times 10^6\ x\ (4.96)}(5000 \times 18.8 - 0)$$

$$= -3.8 \text{ ft}$$

or result in 56,388 lbs of additional tension if it were fixed to the packer.

$$F_B = 2 \times 0.3\left(4.892^2 \frac{\pi}{4}\right) \times 5000 - 0$$

$$= 56,388 \text{ lbs.}$$

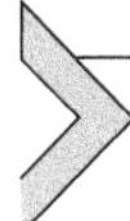

8.5 OPERATION SPECIFIC WELL INTEGRITY AND WELL CONTROL CONCERNS

8.5.1 Acid wash

- Scale and other debris detached from the tubing wall will be pushed downhole as the treatment progresses. If this is allowed to enter the formation it could be extremely damaging.
- Acid can cause emulsions to form when in contact with hydrocarbons. All treatment chemicals should be tested for compatibility with reservoir fluids.
- Acid can mobilize fines if allowed to contact sandstone formations.
- All of the completion above the maximum depth of the treatment is exposed to unspent acid. Some materials used to make tubular and completion components are vulnerable to acid attack unless the acid is properly inhibited. Chrome at 13% is particularly easily damaged.
- During flowback/clean up following an acid wash, check the pH of the produced fluids.
- Acid neutralizing agents should be available.

8.5.2 Chemical inhibitor squeeze

- Inhibitors must be compatible with the formation and formation fluids.

8.5.3 Hydraulic propped fracture treatment

- Ensure measures are in place to manage flow-back of excess proppant. Normally, excess proppant will be circulated out of the well using a workstring or coiled tubing, but once the well is put on production some proppant production can be expected. Are the host facilities able to handle the solids?

8.5.4 Acid stimulation in carbonate reservoirs

- Many carbonate reservoirs contain nonacid soluble material. During acid stimulation this insoluble material is released from the carbonate matrix and will have to be removed from the well. This can cause problems for the topside facilities. If fluid produced from the well is not dirty the well is not cleaning up.
- Fluid density increases as the acid reacts. HCl dissolving carbonate material in the formation produces calcium chloride ($CaCl_2$), carbon

dioxide, and water. The calcium chloride increases the density of the mix water in the well.
  - Spent 10% HCL = 9 ppg (1078.43 kg/m^3)
  - Spent 15% HCL = 10 ppg (1198.26 kg/m^3)
  - Spent 15% HCL = 11 ppg (1318.29 kg/m^3)
- Spent acid has a tendency to foam during flow-back. Foams cause problems in separator vessels.

8.6 THE PRICE OF GETTING IT WRONG

How many of us consider the potential for a well to blow out when pumping fluid into the wellbore during, e.g., a fracture stimulation operation?

A number of years ago, an operator of considerable experience was performing a stimulation operation on a well in Colombia. As the treatment was underway, a high pressure line parted close to the swab cap on the Christmas tree, killing one of the service company employees. Hydrocarbons blowing back through the parted line ignited and burned until the crew were able to close in the upper master valve.

An investigation into the failed line, fatality, and subsequent fire found a number of errors had been made:

- Although a pressure relief valve had been included in the rig up, it was set at too high a pressure, and thus failed to prevent overpressure in the treating lines.
- The high pressure lines had not been secured.
- There was no check valve between the well and the relief valve.
- The upper master valve on the Christmas tree had been locked open with a fusible lock-out cap.
- The report also noted that the safety valve had been removed. However, removal of wireline retrievable safety valves during stimulation operations of this type is not unusual.

The incorrect PRV setting, failure to secure the lines, and the absence of a nonreturn valve are self-explanatory, but it is worth expanding a little on the use of the fusible lock-out cap. Fusible caps are used to hold open actuated valves when they have been disconnected from their hydraulic supply, in this case the upper master valve (UMV). A fusible cap contains an element made from a low melting point alloy. If there is a fire, alloy melts, allowing the valve to close. In

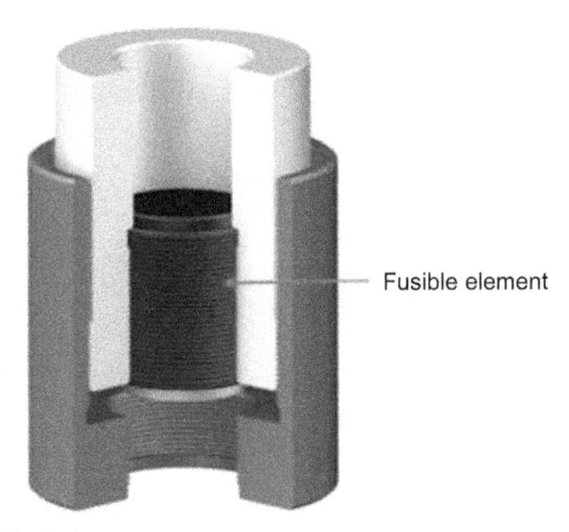

Figure 8.21 Fusible lock-out cap.

this incident, the fire failed to melt the element. The base of the fire was at the top of the tree (above the swab valve). Heat rises. The fire was eventually extinguished when an oxy-acetylene cutting torch (on the end of a very long pole) was used to melt the element in the cap (Fig. 8.21).

A much safer alternative to the fusible cap is the use of a hydraulic single well control panel (as described in Chapter 9: Wireline Operations). Fusible lock-out caps have been banned in many places, but in some locations their use persists. As this incident sadly showed, they are not safe.

REFERENCES

1. Stromberg JL, Surjaatmadja JB. (1992). Restraining System To Help Contain Well Flowlines and Equipment During Rupture for Increased Safety. SPE-24619-MS.
2. Matzner M, Hayes G. (2002). Safety Restraint System for High-Pressure Oilfield Piping. OTC-14165-MS.
3. Bourne HM, Williams G, Hughes CT. Increasing Squeeze Life on Miller with New Inhibitor Chemistry. SPE—60198.
4. Enform Safety Alert. Issue #13-2011. Induced Hydraulic Fracture Results in Blowout.

CHAPTER NINE

Wireline Operations

Non-conductive wireline is one of the oldest methods of well intervention. The first wireline was a simple flat steel measuring tape spooled in to a well to obtain a rudimentary depth measurement. The modern drawn wire slickline was originally developed in the 1930s by the Otis company (now part of Halliburton) to cope with deeper wells. However, it was the development of pressure control equipment, enabling slickline to be run into a well with pressure at the surface that guaranteed its widespread use.

Over the years, the type of interventions that can be performed using slickline have grown considerably, as more and varied downhole tools have been developed. Today, slickline is by far the most common method for intervening in live wells. Costs are generally lower and operational durations shorter than other intervention methods. In addition, the equipment is compact, easy to transport, and widely available (Fig. 9.1).

Electric line logging began in 1927 when Conrad and Marcel Schlumberger carried out the first electrical resistivity well log in Pechelbron, France. Although originally developed to log "open hole", e-line today is routinely run inside the casing, tubing, and liner of live wells. The range of operations performed using e-line continues to grow as more downhole tools are introduced to the market. The main advantages of slickline and electric line interventions are:

- Low cost.
- Live well intervention capability.
- Easy to transport equipment.
- Small crew numbers (two or three people).
- Generally, much faster than other intervention techniques.

The main disadvantages are:

- Relatively low strength of the wire (limits on the weight of the tools that can be deployed).
- Inability to rotate or circulate.
- Difficult to use in high angle and extended reach wells.

Well Control for Completions and Interventions.
DOI: https://doi.org/10.1016/B978-0-08-100196-7.00009-9

Figure 9.1 Slickline on location in the Netherlands. *Photograph courtesy of Bob Baister.*

9.1 WIRELINE INTERVENTIONS IN LIVE WELLS

For most intervention engineers, wireline will be the first choice of intervention method. If a problem can be remedied using wireline then more complex, costly alternatives can be ignored. Common applications for slickline and non-conductive braided cable include:

- Tubing inspection i.e. running a drift.
- Removal of wax and soft scales using broaches and cutters.
- Removal of overbalance fluids from a well; swabbing.
- Tubing leak detection.
- Running and retrieving downhole flow control equipment such as plugs, chokes, straddles, pack-off, etc.
- Service gas-lift systems.
- Open and close circulation paths.

- Repair tubing retrievable SCSSSV[a] and replace wireline retrievable SCSSSV.
- Remove debris and fill.
- Fish lost equipment.

The development of robust electronic components and improved battery technology, coupled with the seemingly exponential increases in the capacity of computer memory, has seen an increase in the number of interventions carried out using slickline. Operations that, in the past, were only possible using e-line are now routinely performed with slickline, using tools powered by lithium batteries with large amounts of data stored on memory cards, these include:

- Cased hole production logging.
- Perforating.
- Explosive setting of bridge plugs and straddle equipment.
- Casing and tubing caliper surveys.
- Downhole camera.

Logging (e-line) interventions (real time surface read out):

- Reservoir monitoring (production logging).
- Condition monitoring (caliper survey and image logging).
- Perforating.
- Bridge plug and straddle setting.
- Tubing cutting/punching.

9.2 THE WIRE

Three types of wireline are commonly used for interventions.

- Slickline: A single strand drawn wire.
- Non-conductive braided cable (multi-strand wire).
- Electric line (e-line): A braided cable with one or more electrical conductors.

9.2.1 Slickline

Slickline is a solid steel drawn wire that is available in a range of diameters with 0.108″ and 0.125″ being the most widely used. The wire is manufactured from carbon steel or a range of different alloys, depending on the service conditions. Slickline for use in oilfield applications will normally

[a] SCSSSV, surface controlled sub-surface safety valve.

Table 9.1 Slickline breaking loads (lbs)

Diameter (in.)	Carbon steel	UHT carbon	316 stainless	Nickel alloy 400	Nickel alloy 700	Nickel alloy 750
			14% Ni 18% Cr	6% Ni 23% Cr	18% Ni 21% Cr	26% Ni 21% Cr
0.092″	1547	2050	1420	1630	1610	1560
0.108″	2109	2730	1940	2190	2120	2115
0.125″	2837	3665	2330	2850	2650	2860
0.140″	3505	4600	3130	3500	3130	3280
0.160″	4580	6005	4050	4200	3920	

be manufactured and tested in conformance with API 9 A[1] and ISO 2232:1990.[2]

Plain carbon steel is cheap, readily available, and has good strength. However, it is not suitable for sour (H_2S) service. Slickline suitable for sour service is made from corrosion resistant (austenitic) alloys. Composition varies according to well conditions. Several different elements are added to the steel to improve strength and corrosion resistance.

- Chromium (Cr) improves corrosion resistance, particularly CO_2 (sweet) corrosion.
- Nickel (Ni), when combined with chromium, toughens the steel and improves resistance to H_2S corrosion.
- Manganese (Mn) is used to prevent free sulfur and increase hardenability of the steel.
- Molybdenum (Mo) increases high temperature strength and helps prevent pitting.
- Silicon (Si) is used to increase strength.
- Nitrogen (N) is used in very low concentrations to increase strength.

Whilst these alloys enable use in sour wells, the wire is not as strong as UHT carbon steel, and is more prone to fatigue cracking and failure. Stainless steel wire costs significantly more than carbon steel wire.

Table 9.1 shows the approximate minimum breaking loads for a range of wire sizes and grades. The exact composition of the corrosion resistant grades varies, depending on the manufacturer.

9.2.2 Braided cable (non-conductive)

Braided cable is constructed from multiple strands of wire laid together to form a single wire rope. It is normally used instead of slickline where higher cable strength is needed, for example during fishing or swabbing

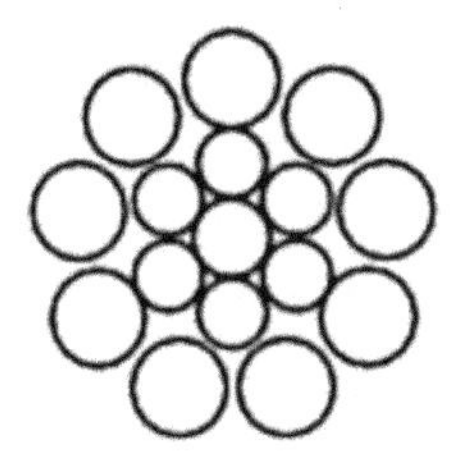
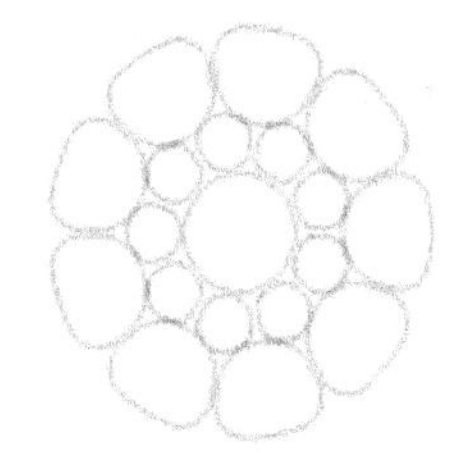

Figure 9.2 Braided cable construction. Standard cable (left) and Dyform cable (right).

Table 9.2 Minimum breaking load for braided cable (lbs)

Diameter (in.)	Galvanized carbon steel		316 Stainless steel		Nickel alloy 26% Ni 21% Cr	
	Conventional	Dyform	Conventional	Dyform	Conventional	Dyform
$^{3}/_{16}$	4960	6170	3990	4940	4320	4960
$^{7}/_{32}$	6610	8370	5400	6500	5842	6500
$^{1}/_{4}$	8640	11,200	7030	8640	7600	8530
$^{5}/_{16}$	13,460	17,550	11,000	13,560	11,880	13,380

operations. Although braided cable is significantly stronger than slickline, there are drawbacks associated with its use. Pressure control equipment is more complex. The delicate manipulation of many downhole tools is more difficult, as the heavier braided cable lacks the sensitivity of slickline. Prolonged downwards jarring can cause the wire to unwrap, leaving gaps in the braid and weakening individual strands.

There are currently two types of braided cable used for well intervention work, conventional (standard) wire, and "Dyform". Dyform is a registered trademark of Bridon PLC. During the manufacturing process, the wire is formed by drawing it through a die—hence the name "Dyform" (Fig. 9.2).

Dyform wire has several advantages over conventional braided cable.

- Increase in breaking load (see Table 9.2); there is more steel for the same diameter.
- Smooth external profile and closer tolerance of outside diameter—reduced leakage at the grease head.
- Higher crush resistance because of the increased steel content of the cable.
- Low twist tendency because of the Dyform process.

Other manufacturers are also now supplying a Dyform type wire (Table 9.3).

Table 9.3 Braided cable—essential information

Diameter	Recommended flow tube (in.)	Approximate weight lbs/1000ft	Approximate weight lbs/1000ft	Recommended pulley diameter (in.)
	New cable	Conventional	Dyform	
$^3/_{16}$	0.196	71	85	12
$^7/_{32}$	0.228	96	111	14
$^1/_4$	0.263	126	148	16
$^5/_{16}$	0.330	196	232	20

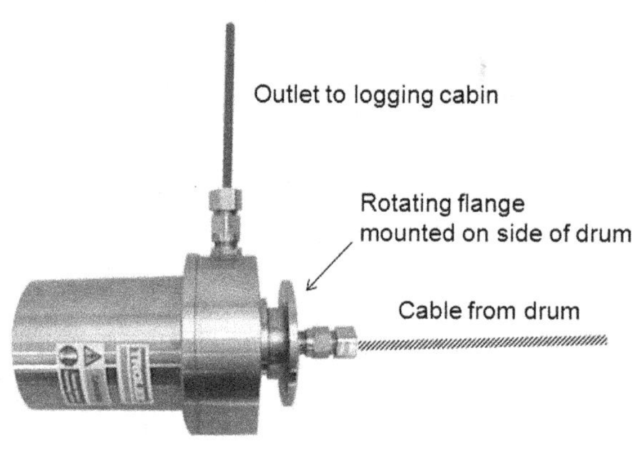

Figure 9.3 Slip ring connector.

9.2.3 Electric line (e-line)

Logging cable (e-line) is outwardly indistinguishable from standard braided cable. E-line is in fact a braided cable with between one (mono-cable) and seven (hepta-cable) conductors running through the core. The conductor is used to send electrical power to, and receive digital data from, logging tools. A slip ring mounted on the side of the e-line storage drum allows uninterrupted data and power transfer whilst the drum is turning (Fig. 9.3).

Like slickline and braided cable, logging cable is made using range of material grades from basic carbon steel through several different corrosion resistant alloys.

9.2.4 Care and handling of wireline

When wireline is in use, it is bent and straightened repeatedly. It straightens coming off the drum, bends round the counter head, and then straightens coming off the counter head. It is bent once more as it goes through the sheave at the bottom of the pressure control equipment, straightens, and then bends through 180° over the top sheave before straightening to run through the stuffing box and into the well. As the wire is worked back and forth through these sheaves, the repeated bending and straightening will lead to fatigue and ultimately failure of the wire. If the wire is under tension it will fatigue more quickly. Wireline crews must take the appropriate measures to limit fatigue. Wire failure normally happens at the surface, where tension loads and fatigue are greatest. Broken wire falling into the well risks an escape of hydrocarbons from the stuffing box or grease head. Broken wire needs to be recovered from the well, a difficult task with an increased risk of hydrocarbon release.

Slipping and cutting wire frequently changes the fatigue points and significantly reduces the chance of a break. Where wire has been worked at the same depth for a prolonged time, for example jarring to free a stuck tool, it is good practice to shear release the pulling tool, pull it back to surface, and then cut and slip wire. If the toolstring cannot be released then, where possible, the fatigue points should be changed by moving the top sheave, floor pulley, and the winch itself. Jarring can also be suspended for a period to allow the worked area of the wire to cool. After prolonged jarring at high tension, inspect the wire for diameter reduction.

Most operating companies and wireline service companies will enforce a limit on slickline and braided cable tension for routine operations; typically, 60%–70% of the breaking load.

The tension limit on e-line cable is normally 50% of the breaking load. Whilst the mechanical strength of the cable is more than capable of withstanding more than 50% of the breaking load, the electrical conductor is not. Some vendor companies charge the customer for the full cost of the cable if the 50% limit is exceeded, even if the conductor appears to be unharmed.

There are several precautions that can be implemented to prolong and preserve wire properties:

- When in use, slickline should be lubricated to prevent it from binding in the stuffing box. Braided cable and e-line are lubricated by grease injection.
- Do not allow the wire to rub against a metal surface—for example the top of the v-door. Use wood, matting, or additional pulleys to eliminate run points.

- Any surface damage or kinking of the wire will weaken it. The damaged section should be cut from the drum. Damage deep in the drum will mean wire replacement.
- Ensure the sheaves are the correct diameter for the wire in use.
- Ensure sheaves grooves are not worn or damaged and can rotate freely.
- Avoid harsh breaking when running in to the well.
- Lubricate the wire before storage.
- Store wireline spools vertically.

9.2.5 Spooling wire

Spooling wire correctly will improve wire life and help reduce operational problems. It is important that all wire (slickline, braided cable and e-line) is spooled onto the winch under tension. Spooling e-line under tension reduces the risk of a phenomena commonly known as "drum crush". Drum crush can cause electrical failure in logging cables and occurs as a result of the cable being crushed to such an extent that the armor wires press and distort the plastic insulation around the conductor. In some instances, distorted insulation degrades signal transmission, in other cases the outer armor cuts through the insulation and contact the conductor creating a short. These failures are caused by spooling cable at high tension over cable (lower down the drum) spooled at abnormally low tension.

Pre-packing e-line and braided cable with grease as they are being spooled on to the drum helps maintain pressure integrity at the grease head as the wire is run in to a live well for the first time. New, "unseasoned" wire is very prone to leaks, especially in gas wells.

When spooling slickline, it is standard practice to "smooth wrap" the first few layers on the drum. This gives a solid bedding for the remainder of the wire, and lets the wireline operator know they are close to the end of the wire. In recent years, the practice of smooth wrapping slickline all the way to the top of the drum has been adopted by some companies. This has the advantage of reducing minor bends and surface damage as the wire is pulled out of the well, but does slow down trip speed.

All wire manufacturers recommend spooling from the top of the storage drum to the top of the wireline unit drum. This means that the curvature of the wire is not reversed during the spooling process. Spooling bottom to top, or top to bottom, is detrimental to the wire. Most wireline service companies now subcontract wire spooling to specialist vendors.

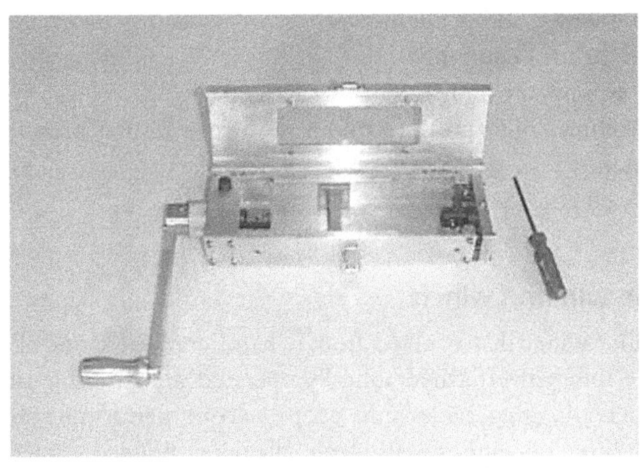

Figure 9.4 Slickline torsion tester. *Image courtesy of NOV Elmar.*

9.2.6 Ductility testing

Testing slickline for ductility is normally carried out by a torsion test or a wrap test. Torsion testing is the recommended method for testing ductility in carbon steel slickline; wrap testing is the recommended method for stainless steel slickline.

The reason for using two different methods for testing ductility is purely mechanical and relates to the properties of the steel: carbon steel has the same ductility properties in all directions. This means it will tolerate being twisted. Stainless steel has antistrophic properties and will not tolerate being twisted. Instead, wrapping of the wire around itself several turns is the preferred method (Fig. 9.4).

9.3 WIRELINE SURFACE EQUIPMENT

Being able to perform interventions on live wells makes wireline a popular option when choosing an intervention method. Part of the reason wireline is so widespread is that the surface equipment is compact, easily portable, and easily rigged up on the well, even by a small crew (usually two or three people).

To perform a wireline intervention requires the following items of surface equipment:

- Wireline winch and prime mover.
- Pressure control equipment.
- Lifting equipment (hoist).
- High volume, high pressure pump for hydro testing pressure control equipment.
- Single well control panel.

9.3.1 The wireline winch

The wireline winch has evolved from a hand cranked spool of flat measuring tape into powerful hydraulically operated units capable of spooling large diameter logging cables into deep wells or performing heavy duty wireline fishing operations with large diameter braided cable. If a real time, surface read-out, logging unit is used in a hazardous (zoned) area, it must have a pressurized cabin to prevent hydrocarbon gas contacting live electrical components.

Units used for land operations are normally mounted on a vehicle, whereas those used on offshore rigs and platforms are skid mounted and have crash frames to protect against shipping damage (Fig. 9.5).

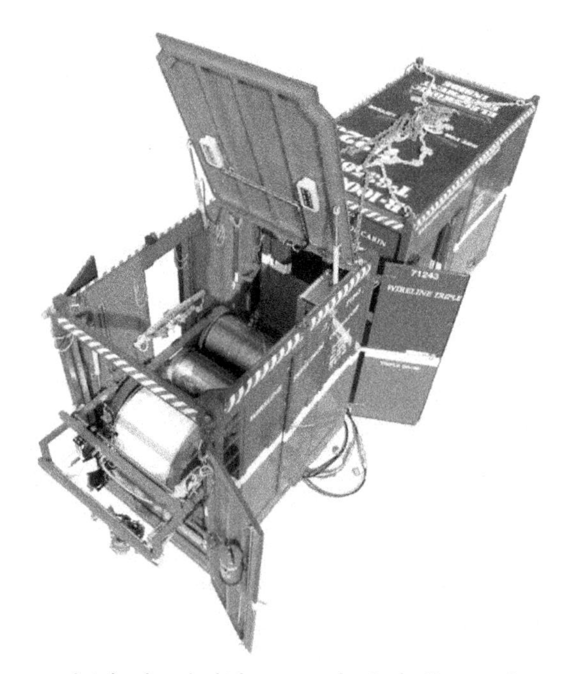

Figure 9.5 Multipurpose (triple drum) skid mounted winch. Source: *Image courtesy NOV Elmar.*

9.3.1.1 Winch controls

All wireline units (slickline, braided cable, and logging) have a control panel that allows the operator to manipulate the drum, spooling wire into or out of the well. As a minimum, the operator will have:

- Direction control leaver—connected to a hydraulic valve and pushed forward to spool wire into the well and pulled back to pull out of the well.
- Brake—on most units a mechanical brake leaver tensions a band around the circumference of the drum.
- Weight indicator—reads line tension.
- Depth indicator—reading in meters or feet.
- Hydraulic pressure adjustment valve and hydraulic pressure gauge.
- Level wind—for even spooling of the wire on to the drum.
- Gear selector leaver (Fig. 9.6).

Depth and weight indicators should be positioned to allow simultaneous reading of both. During slickline operations it is these two instruments that the operator relies on most to determine where and what is happening to the toolstring. Similarly, the direction control leaver and reel brake should be ergonomically positioned to allow easy manipulation of the controls. This is particularly important to allow prolonged jarring during slickline operations.

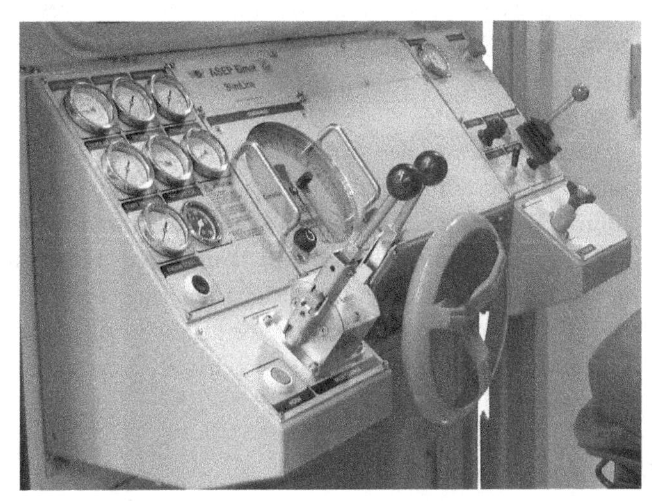

Figure 9.6 Wireline unit control panel (double drum unit). *Image courtesy of NOV Elmar.*

Figure 9.7 Modern wireline units have touch screen displays that control multiple functions including the winch, power pack, pressure control equipment, and in some cases well monitoring. *Image courtesy of NOV Elmar.*

On more sophisticated units, the operator will have access to engine controls and instrumentation on, or adjacent to, the main control panel. These can include.

- Engine shut-down (ESD) button.
- Engine throttle control.
- Diesel power pack—rev counter, oil temperature and pressure, water temperature (for water cooled units), hours run counter, fuel gauge, amp meter.
- Hydraulic oil temperature.

Although normally controlled from a remote, stand-alone panel, some wireline units will be configured to give the operator access to the hydraulic controls operating the pressure control equipment (blow out preventer (BOP) and stuffing box) and automated well shut-down functions on the Christmas tree (hydraulic master valve and SCSSSV). In very modern units, touch screen displays give access to all of the controls necessary to conduct safe and efficient operations from a single point (Fig. 9.7).

9.3.1.2 The weight indicator

An accurate measurement of wire tension is essential. Older slickline units have the weight indicator load cell positioned on the pulley wheel (hay

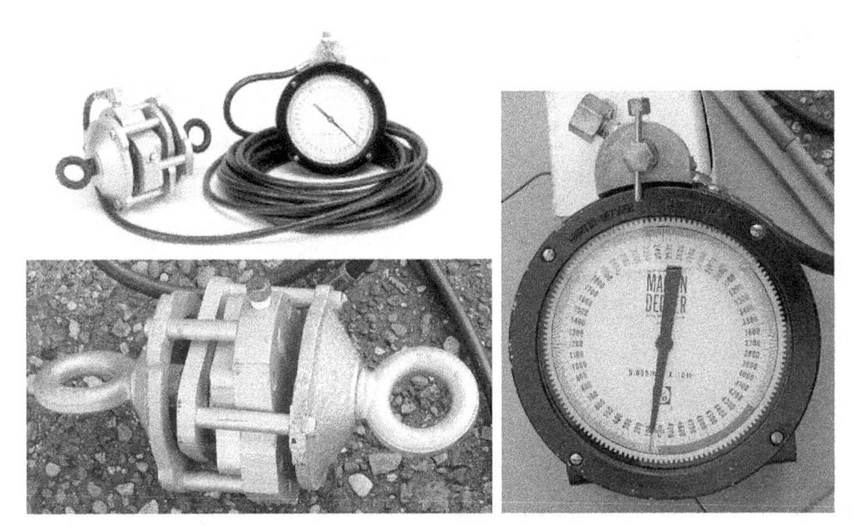

Figure 9.8 Martin Decker type weight indicator. *Photograph courtesy of Bob Baister.*

pulley) at the base of the pressure control equipment. A hydraulic hose connects the load cell to the gauge on the operator's console. Modern units usually have the load cell mounted on the unit level wind.

Properly interpreted, a weight indicator reveals a great deal about what is happening downhole. A skilled slickline operator can carry out delicate manipulation of a toolstring thousands of feet into a deviated well using just this simple instrument. A poorly maintained or incorrectly calibrated weight indicator can be an indirect cause of broken wire (Fig. 9.8).

9.3.1.3 Depth measurement

Most modern logging units use a "straight through" depth indicator where the wire passes between pressure wheels. Slickline units normally have a counter wheel type depth indicator where the wire is wrapped once around the measuring wheel (Fig. 9.9). All mechanical depth indicators are prone to error caused by wear and tear on the parts. Errors can also occur when converting units from metric measurement (meters) to field units (feet) and vice versa. The consequences of depth reading errors are obvious and can easily be avoided by ensuring the measuring equipment is properly serviced and maintained.

9.3.1.4 The power pack

Wireline units are powered by internal combustion engines (usually diesel) or electric motors. Electric motors have the advantage of a more

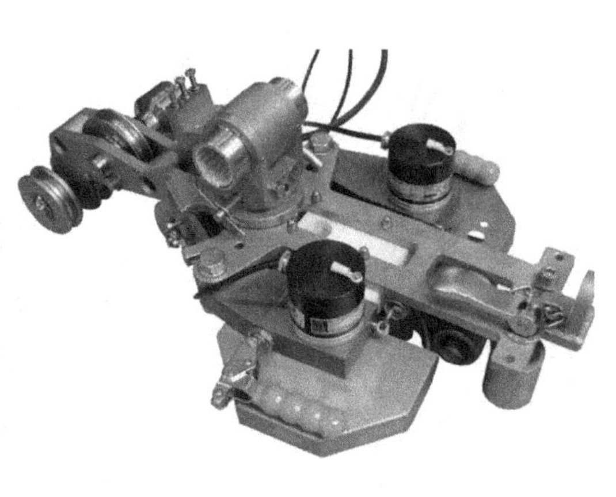

Figure 9.9 Slickline measuring head (left) and logging measuring head/weight indicator (right). Source: *Images courtesy of NOV Elmar.*

favorable power-to-weight ratio, are much quieter to operate, and normally require less in the way of day-to-day upkeep (refueling, oil changes). However, they can only be used where there is a readily available and reliable power supply. This means they are only widely used on large offshore platforms, and even here there are limitations since an electrical shutdown on the platform will render the unit inoperable. Electric motors must be explosion proof and comply with industry zoning regulations. Electric power packs for use with well intervention equipment should have a Zone 1 classification.

Internal combustion has the advantage of being able to operate independently of any external power source. Diesel engines are normally used because they are more reliable than petrol, and can be made to function more safely in hydrocarbon hazardous areas (no spark plugs, contact breakers, distributors, etc.). The exhaust can be fitted with an efficient spark arrester. Diesel fuel is widely available offshore, whereas petrol is normally not allowed.

As the engine is running in potentially hazardous areas where the release of hydrocarbons is a possibility, an automated shut-down system must be fitted. Under normal operating conditions, engine oil pressure is supplied to the following equipment:

- Over speed valve.
- Exhaust temperature valve.
- Fuel shut-off valve.
- Control cylinder.
- Water temperature valve.

If engine oil pressure is lost, or seriously reduced, the fuel shut-off valve and air intake valve close, stopping the engine. Oil pressure losses at the fuel shut-off valve can be caused by any of the following:

- Shortage of engine oil.
- Damaged or broken oil line.
- Oil pump failure.
- High exhaust gas temperature causing valve to open, thus dumping oil.
- High water temperature causing valve to open, dumping oil to the sump.
- Engine over revving causing over-speed to dump oil to sump.

If the engine is over speeding due to incorrect operator control or to flammable gas entering the inlet manifold, a valve will close off the inlet, preventing further entry of gas.

Note: Even if the fuel is shut off, the engine could continue to run on the flammable gas entering the inlet manifold if the inlet manifold is not closed off.

The power pack must be positioned and operated only in areas designated as safe, in accordance with IP "model code of safe practice[3] in the petroleum industry" which classifies areas as:

- Zone 0—In which flammable atmosphere is continuously present or present for long periods (more than 1000 hours per year).
- Zone 1—In which a flammable atmosphere is likely to occur in normal operation (about 10–1000 hours per year).
- Zone 2—In which a flammable atmosphere is not likely to occur in normal operation, and if it occurs will exist only for a short period (less than 10 hours per year).

9.3.2 Wireline pressure control equipment

Most wireline operations are carried out on "live" wells, i.e., wells with pressure at surface where a system of pressure containment is necessary.

Pressure control equipment contains well pressure whilst wire is run into and pulled out of the well.

Wireline crews must have a detailed knowledge of how this equipment works, and how to respond to unplanned events such as leaks, equipment malfunction, and broken wire.

Wireline pressure control equipment must never be left unattended when in use on a live well.

9.3.2.1 Equipment configuration

Wireline pressure control equipment consists of five basic components:

- A stuffing box (slickline) or grease injection head (e-line and braided cable) that can seal around moving wire, enabling it to be run in to a live well.
- Lubricator: sections of high pressure pipe of sufficient length to accommodate the longest toolstring in use.
- Wireline BOP (more properly called a wireline valve).
- Riser: on offshore platforms and satellite towers, the working deck is often located some distance above the wellhead and Christmas tree. At these locations a riser is needed to bridge the gap between the top of the Christmas tree and the working deck. Some operators insist on using flanged connections for the riser below the level of the wireline BOP, but this is by no means universal.
- Tree connection: a method of connecting the bottom of the pressure control equipment to the Christmas tree.

In addition to the pressure retaining components, ancillary equipment is needed to complete the rig up.

For slickline operations:

- Hydraulic hand pump and high-pressure hose to operate the stuffing box pack-off.
- Hydraulic pump and high-pressure hoses to operate the wireline valve (BOP).

For e-line and braided cable operations:

- Grease pump (with high pressure hoses). This maintains a seal around the wire as it passes through the grease head and to pack-off between the rams on the wireline valve (BOP).
- Hydraulic hand pump and hose to operate the line wiper and/or pack-off in the grease injection head.

For all wireline operations:

- High pressure pump for hydro testing the pressure control equipment.

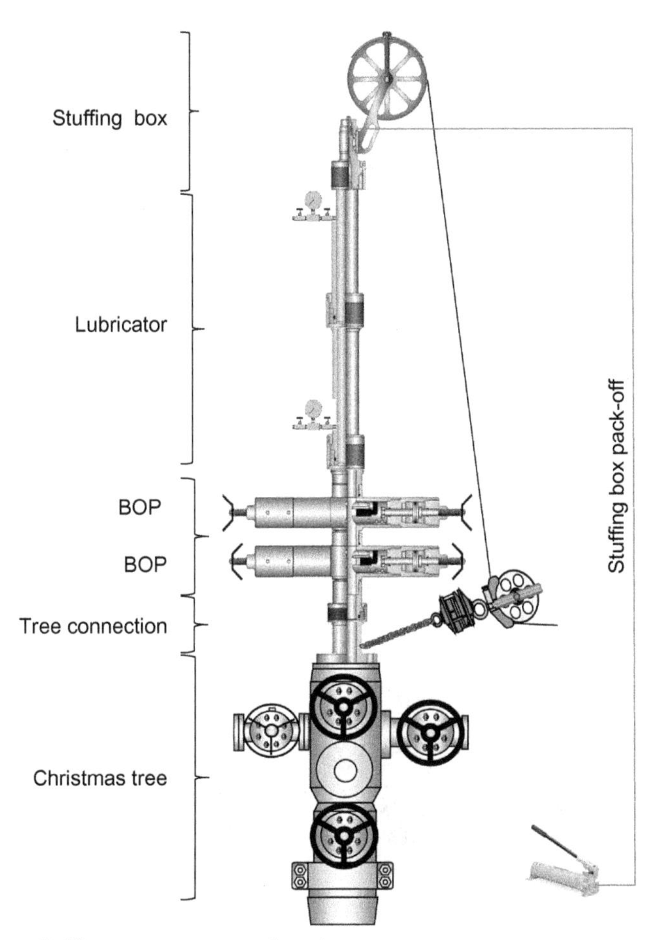

Figure 9.10 Slickline pressure control equipment.

- High pressure manifold—mounted on the lubricator to enable pressure to be monitored and the lubricator depressurized.
- Single well control panel (location dependent) (Figs 9.10 and 9.11).

9.3.2.2 Slickline stuffing box

When the Christmas tree valves are open and the pressure control equipment is exposed to well pressure, the stuffing box is a critical component in the pressure containment envelope. It is a primary barrier, sealing around moving and stationary slickline. Conventional stuffing boxes use a stack of five or six hard rubber packing elements (Fig. 9.12) to create the seal with the wire running through the center of the seal stack.

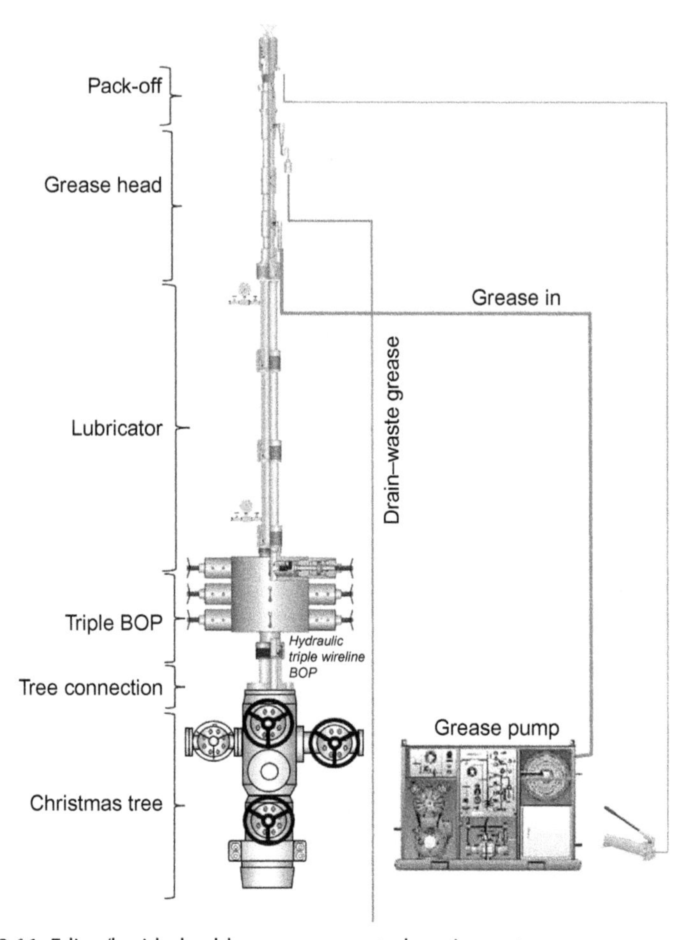

Figure 9.11 E-line/braided cable pressure control equipment.

Moving wire wears the seal elements, and worn seals lead to a release of hydrocarbons unless preventative measures are taken. Most slickline stuffing boxes use a hydraulic piston to compress the seal and pack off around the wire. The piston is operated using a hand pump that is connected to the stuffing box via a hydraulic hose. Older stuffing boxes use a simple hand-tightened nut to squeeze the packing. Stopping a leak requires access to the stuffing box; someone has to reach the top of the lubricator. This is no longer acceptable, as it exposes crew members to unnecessary risk.

Intermittent and minor leakage from the stuffing box is not unusual, and can normally be remedied by squeezing the packing as described.

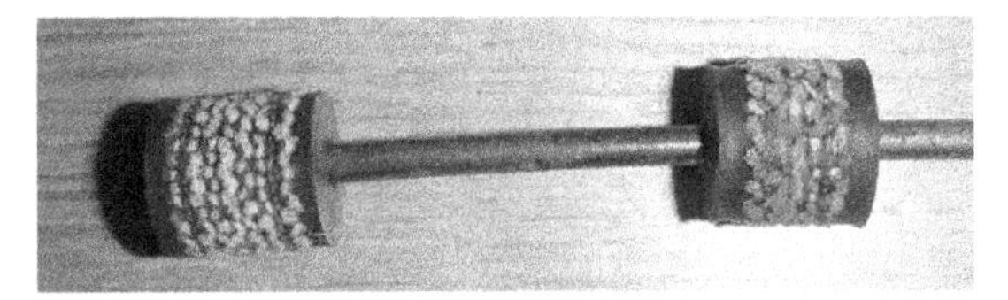

Figure 9.12 Stuffing box packing.

Figure 9.13 Blow out preventer plunger.

However, excessive packing wear can lead to incurable stuffing box leaks. This will occur when the piston and packing gland is fully bottomed out and when this happens, applying additional hydraulic pressure achieves nothing. Preventing this happing is simple:

- Replace stuffing box packing regularly. High pressure wells and gas wells will require more frequent replacement.
- Keep the wire lubricated.
- Do not apply more hydraulic pressure than is necessary to prevent leaks. Too much pressure simply accelerates the amount of wear during wire running. It will also make it more difficult to move the wire.

If wireline breaks close to the surface, well pressure will eject the wire from the stuffing box. Oil and gas will escape unless a secondary sealing mechanism is fitted. Two methods are commonly used. Older designs of stuffing box are fitted with a plunger (Fig. 9.13). As the wire is ejected, well pressure forces the plunger up into a taper in the bottom of the lower gland nut. The rubber element in the plunger is compressed into

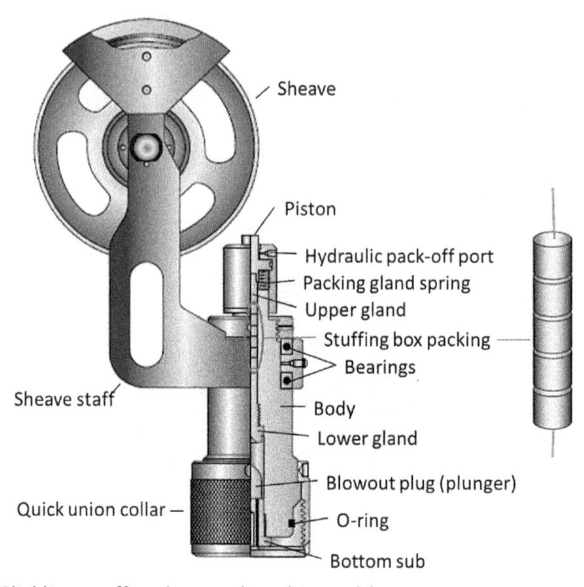

Figure 9.14 Slickline stuffing box and packing rubber.

the taper, sealing off the leak. This type of seal has a poor record of reliability, as they are seldom replaced often enough.

Most modern stuffing boxes use a ball check to prevent the unwanted escape of hydrocarbons. These are generally more reliable.

Systematic pre-job checks can reduce unwanted leaks and malfunctions:

- Is the stuffing box packing in good condition? If not change it out.
- Check that the hydraulic packing nut is moving freely and retracts when pressure is bled off.
- No leaks in hydraulic packing piston, hose, or pump.
- BOP plunger in good condition and correctly sized for the wire. (If a ball and seat are used, check the condition of the seat and the ball.)
- Top sheave correctly sized for the wire and rotating freely. Bearings in good condition.
- Sheave staff rotating freely around stuffing box body.
- Plunger stop nut secure.
- Quick union O-ring in good condition (Fig. 9.14).

9.3.2.3 Liquid seal stuffing box

The liquid seal stuffing box (or control head) is increasingly used for high pressure applications, particularly when working on high pressure gas wells. It is normally used in combination with a conventional hydraulically operated stuffing box. The liquid seal control head is positioned

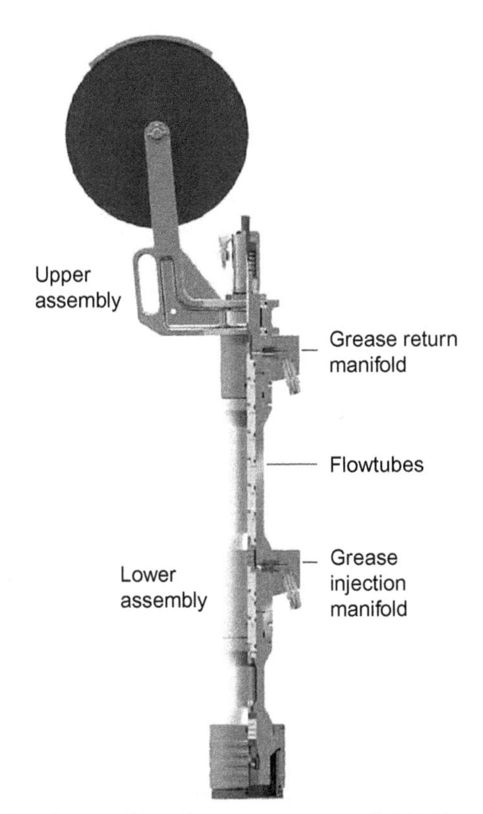

Figure 9.15 Liquid seal control head. *Image courtesy of NOV Elmar.*

immediately below a conventional stuffing box, or both functions can be combined in a single assembly.

When the slickline is moving, a seal is maintained by pumping grease into the annulus between the slickline and a flow tube in much the same way as the seal around braided cable is maintained using a grease head. Grease pressure needs to be slightly above well pressure for the seal to be maintained. The stuffing box functions as a line wiper, keeping any excess grease extruding from the top of the flow tube from escaping.

There is the need for additional equipment in the form of a grease pump. An adequate supply of grease is also required (Fig. 9.15).

9.3.2.4 Grease injection head

Braided cable and e-line have void space between the individual strands of wire that make up the cable. Well pressure will percolate through the gaps and escape to the atmosphere unless these void spaces are sealed.

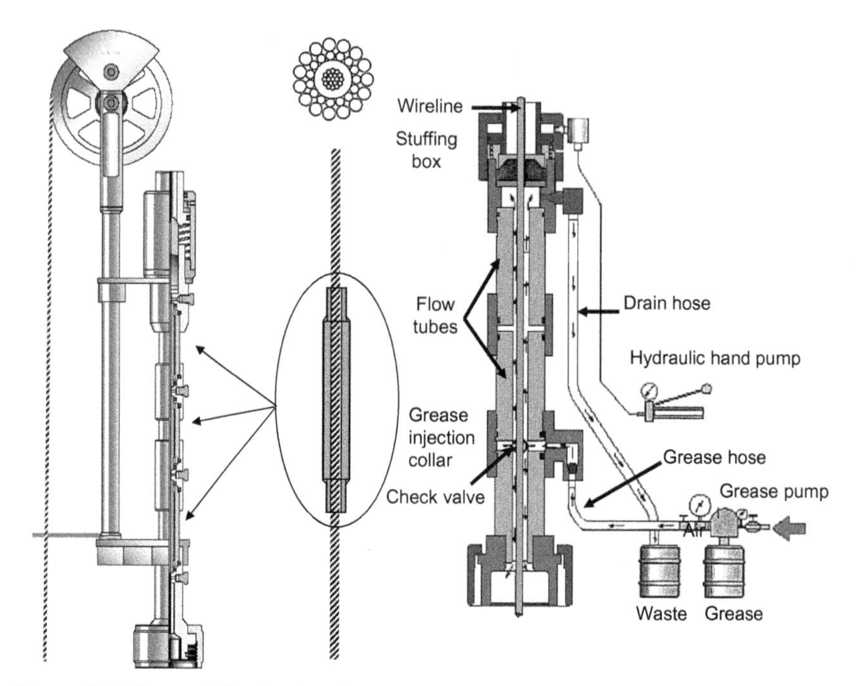

Figure 9.16 Grease injection head.

To create a high-pressure gas tight seal, the cable is passed through close fitting steel tubes called "flow tubes". The number of flow tubes required is pressure dependent, but it is rare to find a greased head with fewer than three. A viscous, stringy, grease (sometimes called honey oil) is pumped into the grease head at above wellhead pressure from an injection point above the bottom flow tube. Grease packs the void spaces between the wire strands, as well as the annular space between the wire and the inside of the flow tubes. Since the grease is viscous, and pumped at above wellhead pressure, an effective seal is created (Fig. 9.16).

Grease seal pressure integrity is dependent upon many factors:

- Wellhead pressure.
- Ambient temperature (grease viscosity is temperature dependent).
- Line speed. It is more difficult to maintain a seal at high line speed.
- Direction of travel. It is more difficult to maintain a seal when pulling out of the well.
- Number of flow tubes.
- Flow tube diameter. The inner diameter (ID) of the flow tube should be between 0.004″ and 0.006″ larger than the outer diameter (OD) of the cable. Any larger and it will be difficult to maintain a seal. In

Table 9.4 Flow tube configuration

Well pressure	Fluid type	Number of flow tubes
0–5000 psi	Liquid	3
0–5000 psi	Gas	3
5000–10,000 psi	Liquid	4
5000–10,000 psi	Gas	4–5
10,000–15,000 psi	Liquid	6
10,000–15,000 psi	Gas	6 or more

addition, greater quantities of grease will be used. If the tolerance is too tight the cable will not run.

- Wellbore fluids. It is more difficult to maintain a seal when there is a gas cap.
- Wire diameter—the bigger the cable the more difficult it is to maintain the seal.
- Grease pressure supplied by the pump.
- The number of grease injection points—the standard system uses a single injection point.

The number of flow tubes is mainly a function of well pressure and fluid type. Table 9.4 is a guide to the number of flow tubes needed. Cable diameter should also be considered. The larger the diameter the more difficult it will be to maintain a seal.

Nearly all grease heads have a pack-off at the top of the assembly (above the flow tubes). This traps any grease extruded above the top flow tube, allowing it to vent through a drain line where it is captured in a waste drum. The pack-off creates an effective seal around the wire in low pressure wells. Although the pack-off will wipe the wire reasonably effectively, many grease heads are equipped with additional line wiper.

Most grease heads will have a ball and seat below the lower flow tube. This is designed to prevent hydrocarbon escape if the wire breaks and is forced out through the top of the head. Like the BOP gland on the slickline stuffing box, the ball and seat should not be relied on, having a poor record of reliability.

9.3.2.5 Grease injection pump

A grease injection pump is an essential component part of a braided cable rig up. Without high pressure grease, the grease head cannot maintain a seal. Pump design varies depending on the location, pressure requirements, and the number of functions. A simple grease pump will have at least one high pressure feed to the grease head flow tubes and a second

Figure 9.17 Grease injection pump. *Image courtesy of Lee Specialties.*

feed to the wireline BOP to enable the void space between the closed dual rams to be pressurized (see Section 9.3.2.6).

Since pump failure means loss of pressure integrity, most modern systems are equipped with a back-up pump. Many grease pump assemblies are equipped with additional hydraulic pumps for operating the BOP, line wiper, and pack-off.

Most grease pumps are powered by diesel or electric motors. Grease pumps used on offshore platforms and rigs are normally air operated, and are therefore reliant on continual, rig supplied, compressed air. Since any failure of the rig supply would lead to a loss of pressure integrity, a suitable back-up must be provided. Compressed nitrogen (in portable cylinders) is normally used. There should be enough compressed nitrogen available to ensure the toolstring can be recovered to surface, and the well shut in and made safe (Fig. 9.17).

9.3.2.6 Wireline blow out preventer (wireline valve)

A wireline BOP is an essential part of the wireline pressure control rig up. To be classified as a BOP, the API requires rams to close in less than 30 seconds (equipment up to and including 18¾″ diameter).[4] NORSOK, the Norwegian regulatory authority, has the same requirement for BOPs used in wireline interventions.[5] However, in most jurisdictions, closure time for wireline BOPs is not stipulated. Moreover, many wireline BOPs would fail to meet the 30 s recommendation, as they are hand operated, or operated using simple hand pumps. Although the wireline BOP shares a name with its drilling counterpart, it performs a different function. It is mainly used to facilitate wireline fishing operations. It can also be used to repair leaks in the pressure control equipment above the BOP, when wireline is in the well. However, it is not normally used as a secondary well control barrier if the wire breaks and is ejected from the stuffing box or grease head. Instead, the tree valves are closed to restore integrity. Nor would the BOP be closed should there be a sudden and catastrophic failure of the pressure control equipment. Again, the safest course of action is normally closure of the tree valves. This ambiguity and the slightly different role for the wireline BOP, mean that some prefer to call it a "wireline valve".

Wireline BOPs are normally positioned low down in the pressure control rig up (above the Christmas tree and below the lubricator). Like a drilling BOP, the wireline BOP has movable rams fitted with elastomeric seal elements that are designed to close and seal around slickline or braided cable without causing damage. This allows well pressure to be contained below the rams whilst the lubricator above is depressurized. Closing the BOP is normally done to allow:

- De-pressuring of the lubricator to enable a leak to be repaired, i.e., stuffing box, grease head, or lubricator leak when wire is still in the well.
- Making up and dropping of wireline cutter bars or drop bars during wire fishing operations.
- Isolation of well pressure during the fishing of wire from the well. In some circumstances, a limited amount of wire can be stripped through the BOP. The BOP is not designed for stripping long lengths of wire, since that wears the elastomer seal causing it to leak.

BOPs are fitted with an equalizing device to enable pressure to be equalized across the rams before they are opened (Fig. 9.18).

Wireline BOP rams are operated either manually or hydraulically. Manual BOPs are closed by rotating the handle clockwise. The rotational movement turns a stem that pushes the ram into the closed position. The rams are opened by counter-clockwise rotation. Whilst this type of arrangement is simple, it is slow to operate. It also means members of the crew have to be positioned unnecessarily close to pressurized equipment; equipment that may be leaking (Fig. 9.19).

Risk is reduced if a hydraulically operated BOP is used. This enables the crew to open and close the rams remotely using hydraulic pressure. Ram travel time (open and close) depends on pump output, piston size, and well pressure, but it is normally faster (and safer) than using a manual

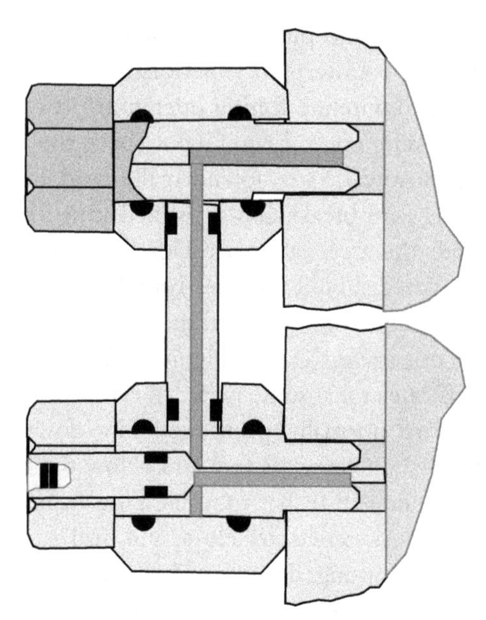

Figure 9.18 Wireline blow out preventer equalizing assembly.

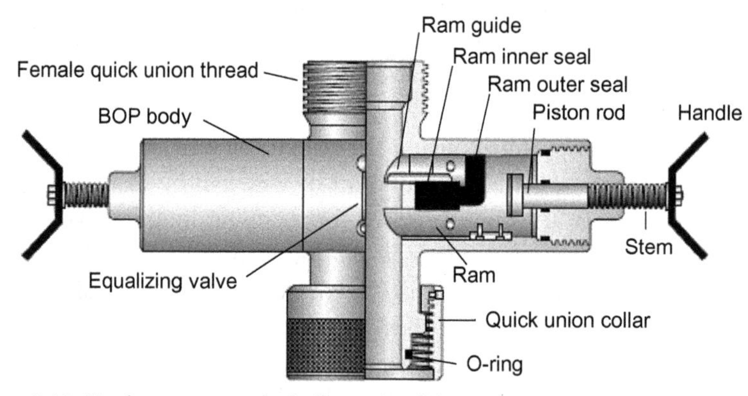

Figure 9.19 Single ram manual wireline valve (blow out preventer).

BOP. Most hydraulic BOPs have a manual lock in addition to the hydraulic piston, and in some jurisdictions manual locks are mandatory. Once the BOP rams are closed, manual handles are wound in, locking rams closed.

Some hydraulically operated BOP pump skids have accumulator bottles to enable the rams to be opened and closed if the hydraulic pump fails, or air supply to pneumatically operated pumps is lost. Again, in some jurisdictions, accumulators are mandatory (Fig. 9.20).

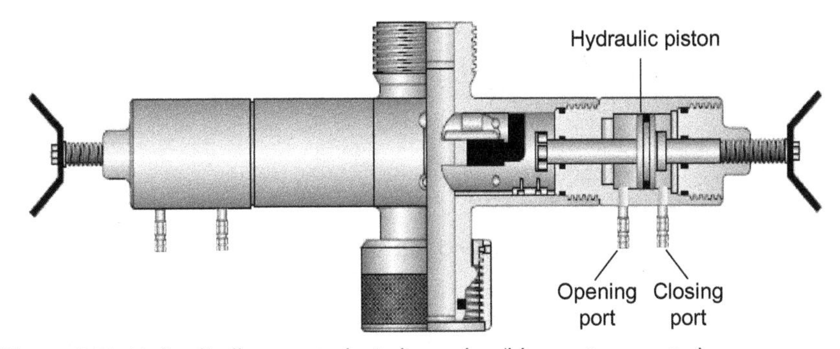

Figure 9.20 Hydraulically operated wireline valve (blow out preventer).

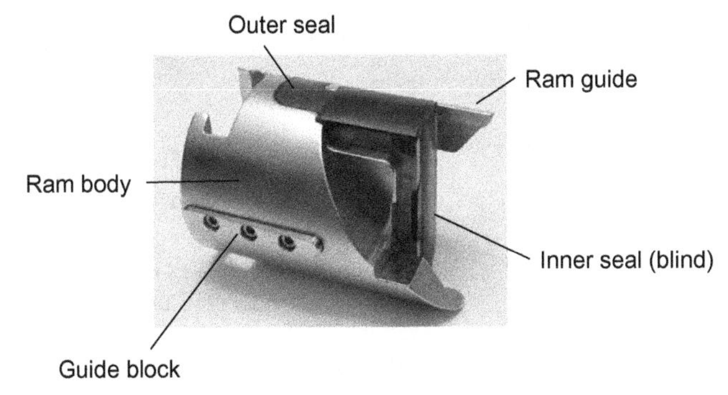

Figure 9.21 Wireline valve blind ram.

9.3.2.7 Wireline blow out preventer: ram configuration

Slickline BOPs are fitted with blind rams; slabs of rubber with a flat seal face (Fig. 9.21). When the rams are closed the pliable elastomer (usually nitrile or hydrogenated nitrile) deforms around the wire creating a seal. A single ram BOP forms a single barrier. To conform with a two-barrier well control policy, a dual ram BOP or two single ram BOPs are required.

For braided cable and e-line operations, two rams are needed to create a single barrier. BOP rams are fitted with an elastomer seal, each cut with a semi-circular profile matched to the cable OD (Fig. 9.22). If a single ram BOP is used with braided cable, well pressure will migrate through the void spaces in the wire. Braided cable BOPs therefore use two rams. The upper ram is configured to hold pressure from below, but the lower ram is inverted, to hold pressure from above (Fig. 9.23). With both sets of rams closed, grease is pumped (at above wellhead pressure) into the cavity between the closed rams. The grease packs the voids between the

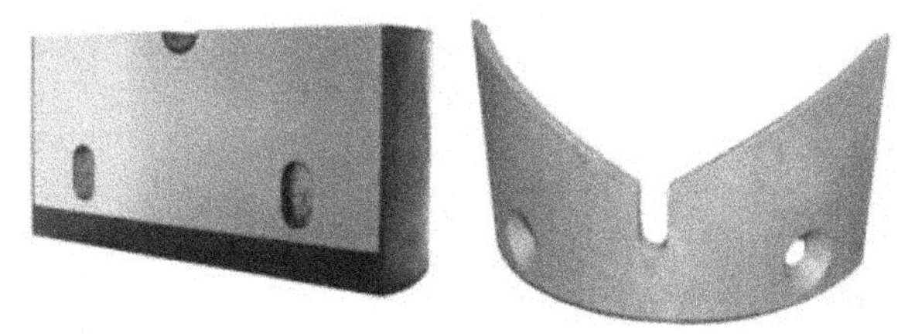

Figure 9.22 Braided cable ram insert and wire guide.

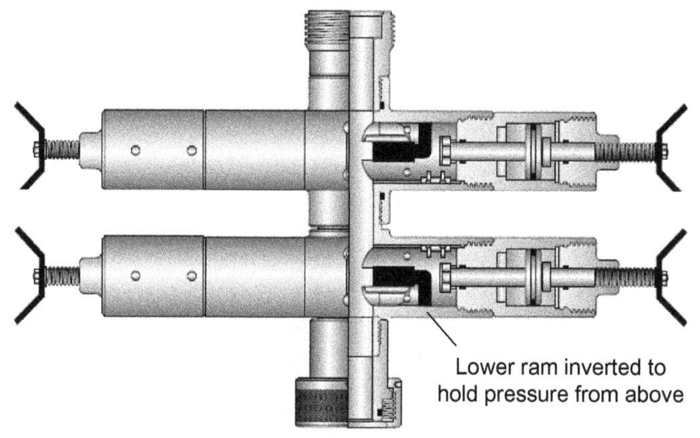

Figure 9.23 Dual ram wireline valve configured for braided cable.

wire strands, preventing gas migration. The grease used in the BOP is the same as that used for the grease head.

Since a dual ram BOP is a single barrier, conformance with two-barrier policy requires a second dual ram BOP.

Many vendors supply a three ram BOP with the lower set of rams inverted and the middle and upper rams configured to hold pressure from below, providing two mechanical barriers. However, as most operating companies require two "independent" mechanical barriers, it can be argued that a three ram BOP may not conform with a two-barrier policy. The two uppermost rams only form a barrier if the lower (inverted) ram maintains pressure integrity. If this fails, so do both upper rams. The two upper rams are therefore reliant for their integrity on a single barrier (Fig. 9.24).

9.3.2.8 Shear seal wireline valve (blow out preventer)

Shear seal BOPs designed to cut wire and seal the well and are particularly useful if multiple strands of wire or cable are pulled back to surface during a fishing operation. They are normally rigged up directly above

Figure 9.24 Triple ram wireline valve. *Image courtesy NOV.*

Figure 9.25 Wire cutting blades for a shear seal blow out preventer.

the Christmas tree (or as close to the tree as feasible). This ensures that cut cable drops below the tree valves, reducing the potential for valve gate and seat damage when closing the well in.

Shear seal BOP rams have a combined cutting blade and blind ram seal. Ideally, the cutting blade should be able to slice through multiple strands of wire (Fig. 9.25).

At some locations, the use of a shear seal BOP is mandatory for some prescribed operations, for example when fishing wire.

9.3.2.9 Wireline lubricator

A lubricator allows long wireline toolstrings to be deployed into and out of the well. It must be a long enough to accommodate the longest toolstring likely to be used during an intervention (Fig. 9.26).

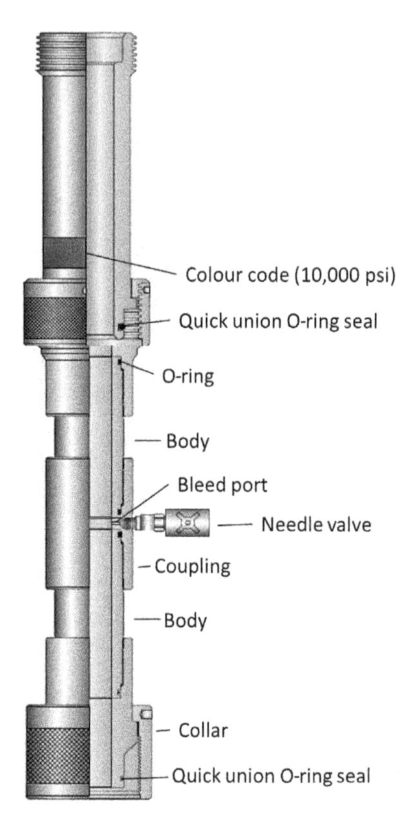

Figure 9.26 Wireline lubricator.

Wireline lubricator is made up from number of sections of pipe connected together using quick union connections. Lubricators can be tapered, with a wider lower section sized to accommodate large diameter wireline tools. The main selection criteria for a wireline lubricator are:

- Pressure rating. This should be more than the highest wellhead pressure expected during the operation.
- Type of service: H_2S considerations.
- Length. Long enough to accommodate the longest toolstring (with mechanical jar and power jars open). This must include the length of any equipment that might have to be pulled from the well, for example a wireline retrievable safety valve.
- Diameter: Greater than the OD of the largest component to be run/pulled.
- Bleed off and pressure monitoring. At least one bleed off point must be included in the lubricator—it is normal to fit the bleed off port

Table 9.5 Working pressure color coding for pressure control equipment

1500 psi	10,342 Kpa	White
3000 psi	20,684 Kpa	Green
5000 psi	34,474 Kpa	Red
10,000 psi	68,948 Kpa	Blue
15,000 psi	103,442 Kpa	White
20,000 psi	137,896 Kpa	Tan

Table 9.6 Service rating color code

Standard	**No band**
H_2S	Yellow
Cold	Purple

with a manifold configured for double block and bleed and a pressure gauge. In most rig ups the bleed off manifold is positioned for easy access close to the bottom of the lubricator. Many lubricators will be fitted with a second bleed point at the top of the lubricator. This enables air to be expelled from the lubricator when pressure testing.

Some lubricators are color coded to enable quick identification of pressure rating, service type (sour/non-sour service), and temperature limitations.

Although by no means universal, the tables above (Tables 9.5 and 9.6) are widely used by vendor companies.

9.3.2.10 Quick union connections

Wireline pressure control equipment components are joined together using "quick union connections". Quick unions have a coarse (Acme) square thread that can be quickly made up and backed off by hand. This saves a great deal of time during operations where lubricator continually has to be disconnected and connected to allow wireline tools to be installed and removed. Two thread forms are in common use; Otis and Bowen. Both use a threaded collar and box connection to join components together, and whilst similar in appearance, Otis and Bowen connections are incompatible. Pressure integrity at each connection relies on an O-ring in the pin connection sealing inside a polished bore in the box profile (Fig. 9.27).

Where possible, a quick union should be an integral part of the parent assembly, as is the case with high pressure lubricator. Some quick union

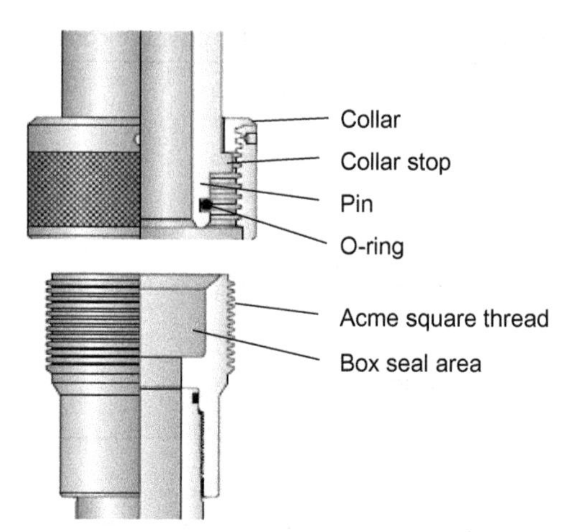

Figure 9.27 Quick union connection.

connections are welded to the parent assembly. Some quick unions are threaded onto the parent assembly, but this is only acceptable for low pressure non-sour service operations.

When making up quick unions, the O-ring must be inspected for damage each time a connection is made. The thread should make up easily and quickly, and it is good practice to back the thread off one quarter turn from fully made up. This makes the connection easier to undo after the equipment has been exposed to pressure. Rocking the lubricator and keeping it vertical above the Christmas tree will make it easier to fully make up and break out a connection.

When backing off a quick union, resistance should always be treated with caution—if a quick union will not turn by hand, it may be because there is pressure trapped in the lubricator. Always ensure pressure control equipment is fully vented. It is bad practice to use pipe wrenches or chain tongs. A quick union should never be hammered as this causes damage. Hammering a quick union that is still under pressure could be fatal.

Before use:

- Inspect for general damage and corrosion.
- Inspect the internal bore for corrosion or wire cutting.
- Check the condition of the O-ring groove and the O-ring.
- Check the condition of the box and pin sealing surface.
- Carefully check the pin and box dimensions.

9.3.3 Pressure control equipment: accessories

In addition to the main pressure control components, there are several additional items of equipment that can be included in a rig up. The selection of these additional items will be determined by operational requirements.

9.3.3.1 Pump-in tee

A pump-in tee enables large volumes of fluid to be pumped in to the well through the wireline pressure control equipment. Most pump-in tee sections are equipped with a 2″ 1502 Chicksan connection (Fig. 9.28).

9.3.3.2 Chemical injection sub

A chemical injection sub is usually positioned between the stuffing box or grease head and the top section of lubricator. It allows the wire to be coated with corrosion inhibitors, lubricant, or friction reducers as it is being run. Chemicals are supplied to the sub from a high pressure pump skid via a flexible high pressure hose. The sub should have a non-return valve to prevent hydrocarbon escape should the hose or pump fail (Fig. 9.29).

Figure 9.28 Pump-in tee.

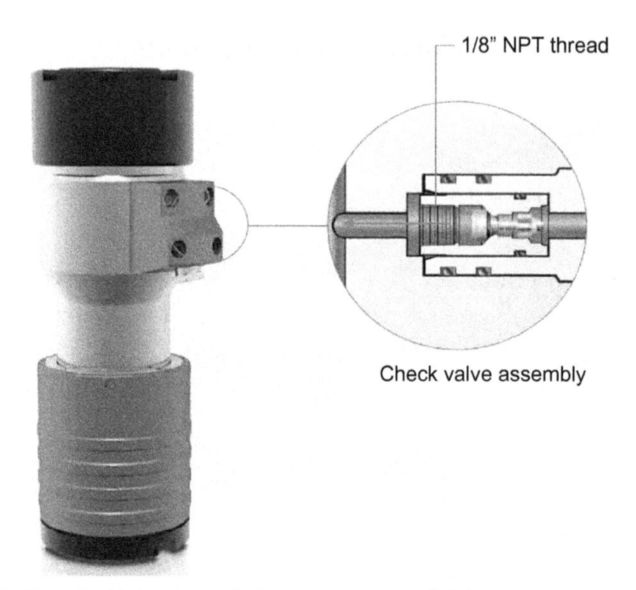

Figure 9.29 Chemical injection sub. *Image courtesy NOV Elmar.*

9.3.3.3 Tool catcher (tool trap)

Two types of tool catchers are in widespread use. A simple mechanical tool trap can be placed at the bottom of the lubricator. A hinged flapper holds the toolstring in the lubricator until it is manually released. Hydraulically operated tool traps are also available (Fig 9.30).

Hydraulic tool catchers are located at the top of the lubricator. When a toolstring is picked up to the top of the lubricator, a latch mechanism in the tool catcher engages the rope socket (cable head). The toolstring is released by applying hydraulic pressure from a hand pump.

9.3.3.4 Lubricator test sub: quick test sub

A lubricator test sub is used to reduce the time taken to confirm pressure integrity of the wireline pressure control equipment each time the lubricator is broken. Using a test sub also significantly reduces the amount of fluid that drops into the well after each pressure test. This can be particularly advantageous when operating in wells where there is a risk of hydrates, where the well is sub hydrostatic, or where the formation can be easily damaged.

A test sub is normally positioned in the pressure control equipment at the point where it will be disconnected/connected between wireline runs (usually immediately above the BOP). An initial (whole body) pressure test is carried out to confirm pressure integrity of all the pressure control

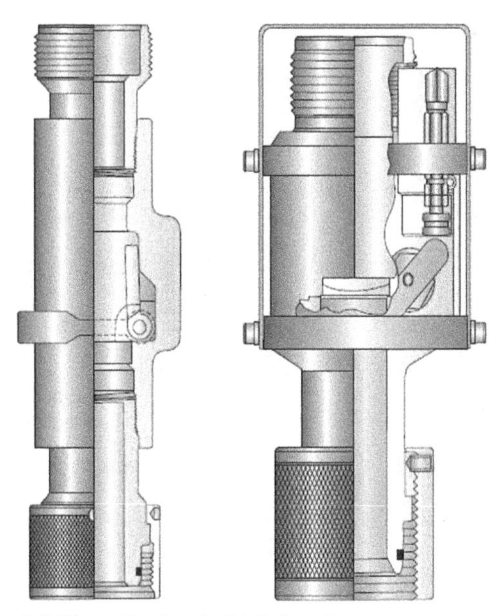

Figure 9.30 Manual (left) and hydraulic (right) tool trap.

components. Following this initial test, the lubricator is disconnected/connected at the quick test sub (if any other connection is used, the full body test must be repeated). A quick test sub has a stepped seal bore with two O-ring seals. A test port is located between the two O-rings. Each time the lubricator is made up, the void between the two O-rings is pressure tested via the test port, confirming the integrity of the connection. As the volume between the O-rings is small, the test can be carried out quickly and no fluid is lost to the well. The quick test sub has several advantages:

- Saves time.
- Reduces the amount of contaminating fluid dropped into the well.
- Reduces the chance of hydrate formation.
- Reduces crew exposure to high pressure testing.
- Eliminates the need to pressure test with explosive devices in the lubricator, i.e., perforating guns, explosive setting kits (Fig. 9.31).

9.3.3.5 Side entry sub (Y spool)

The side entry sub, or "Y" spool, is used during fishing operations. It is normally positioned at the bottom of the lubricator and above the BOP. If wire becomes stuck, the "Y" tool enables a second toolstring to be run alongside the stuck wire, whilst maintaining the ability to manipulate the stuck wire (Fig. 9.32).

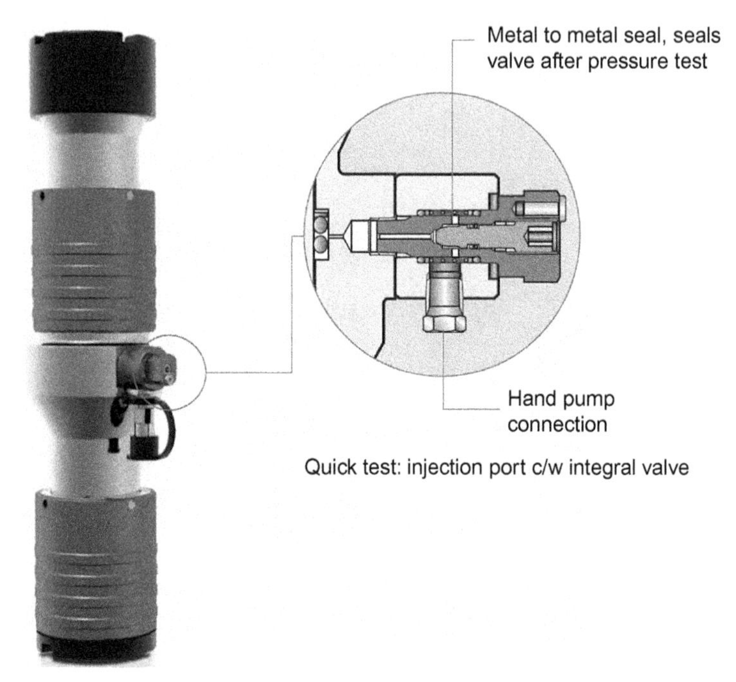

Figure 9.31 Lubricator test sub. *Image courtesy NOV Elmar.*

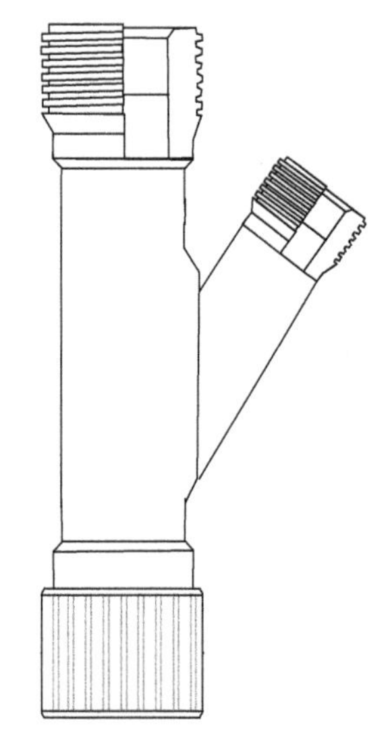

Figure 9.32 Side entry sub or "Y" spool.

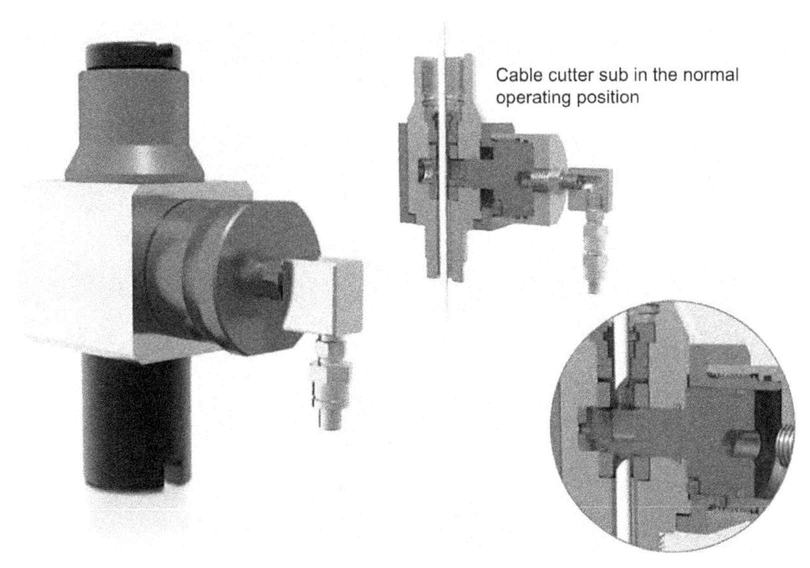

Figure 9.33 Cable cutting sub. *Image courtesy NOV Elmar.*

9.3.3.6 Cable cutting sub

A cable cutting sub is designed to be positioned immediately below the grease head. If the wire gets jammed in the flow tubes (for example the wire is stranded forming a bird's nest) the wireline can be cleanly cut well above the wireline valve (BOP), allowing an easier recovery (Fig. 9.33).

9.3.4 Single well control panels

At many locations, wells are linked into an automated emergency shutdown (ESD) system that is an integral part of the production facility. ESD systems are operated from a central panel that controls any fail-safe valves on the Christmas tree and in the well. In most cases this will mean the upper master valve (UMV), and may also include the wing valve, the SCSSSV, and annulus safety valve (ASV) (if run). If hydraulic (or pneumatic[b]) pressure is lost, the valves close and remain closed until panel pressure is restored (Fig. 9.34).

The conditions required to vent panel pressure and initiate a shutdown vary, but normally include:

- Over pressure in the production system.
- Under pressure in the production system.
- Hydrocarbon escape.

[b] Some tree valves rely on pneumatic pressure, but the fail-safe principle is the same.

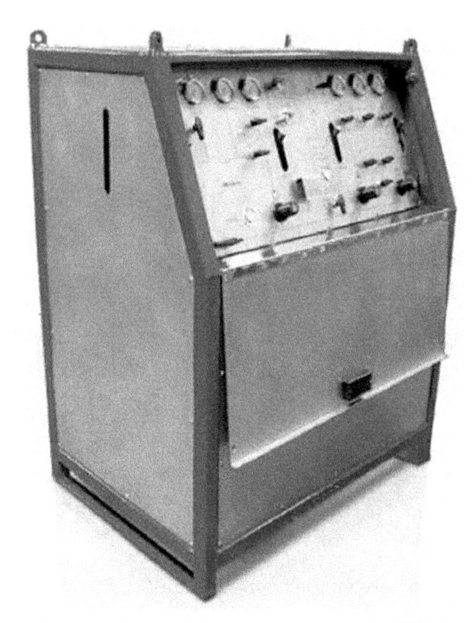

Figure 9.34 Single well control panel. *Image courtesy of NOV Elmar.*

- High temperature.
- Liquid overflow in the separator.
- Activation of fire protection systems.
- Manual initiation.

During an unplanned shut-down of the host facility, tree valves and downhole safety valves will close without warning. A closing tree valve will cut slickline and will either cut or severely damage braised cable. The outcome will be very serious:

- Potential to damage the tree valve gate and seat.
- Hydrocarbon release if broken end of wire is ejected from the stuffing box and the check valve fails.
- Downhole safety valve damage. Worst case scenario—falling wireline toolstring (above the SCSSSV when the wire parts) impacts on closing SCSSSV destroying the flapper.
- Prolonged and difficult operation to recover the broken wire and lost toolstring.

Whilst genuine emergencies are thankfully rare, spurious alarms are depressingly common. Unplanned and unwanted valve closure (and cut wire) can be prevented by overriding the main ESD system using a dedicated well intervention well control panel. This is a single well control panel that is monitored and operated by the well intervention crew.

Remote panels vary considerably in design and functionality. Some are designed only to operate tree and downhole valves, whilst others will incorporate additional functions, for example, operation of the wireline BOP and stuffing box pack-off (Fig. 9.35).

Well service crews and their supervisors must have an in-depth understanding of the panel and how it operates.

The main pump delivering hydraulic pressure to each function may be air, electric, or diesel engine driven. A well-designed pump will have a hand operated back-up pump. It will also have audible alarms that will sound if air pressure (air driven pump) or hydraulic pressure drops below pre-determined values.

Many pumps have an accumulator system that will allow limited operation of the valves if power (pneumatic or electric) is lost to the pump.

Irrespective of design, there are some common practices that must be implemented when using a well control panel:

- Proper consideration must be given to the location of the pump. In the event of a genuine emergency, the crew may need to shut in the well by manual manipulation of the panel, i.e., dumping the hydraulic pressure. The panel must be clear of hazardous areas and easily accessible. Ideally, the pump should have line of sight with the well, although this is not always practical.
- If it is not possible to locate the panel away from hazardous areas, it is good practice to have a secondary shut-down button or dump valve remote from the main panel. Each member of the crew must be made aware of the location of any secondary shut-down stations.
- Most panels have a low melting point fusible link that will vent the panel pressure should a fire prevent the crew from performing a manual shut-down.
- If the panel is equipped with an accumulator system, ensure the accumulator bottles are fully charged.
- When using air operated panels, consider having a back-up "air supply" in case the facility air fails. Compressed nitrogen is commonly used.
- Never leave the panel unattended when the tree valves are open.
- Have contingency plans in place to deal with panel malfunction, hydraulic hose leaks, loss of power pack, or rig air.

Each member of the wireline crew, as well as operating company supervisory staff, should know how the remote panel operates, and importantly how to initiate an emergency (total) shut-down. Single well

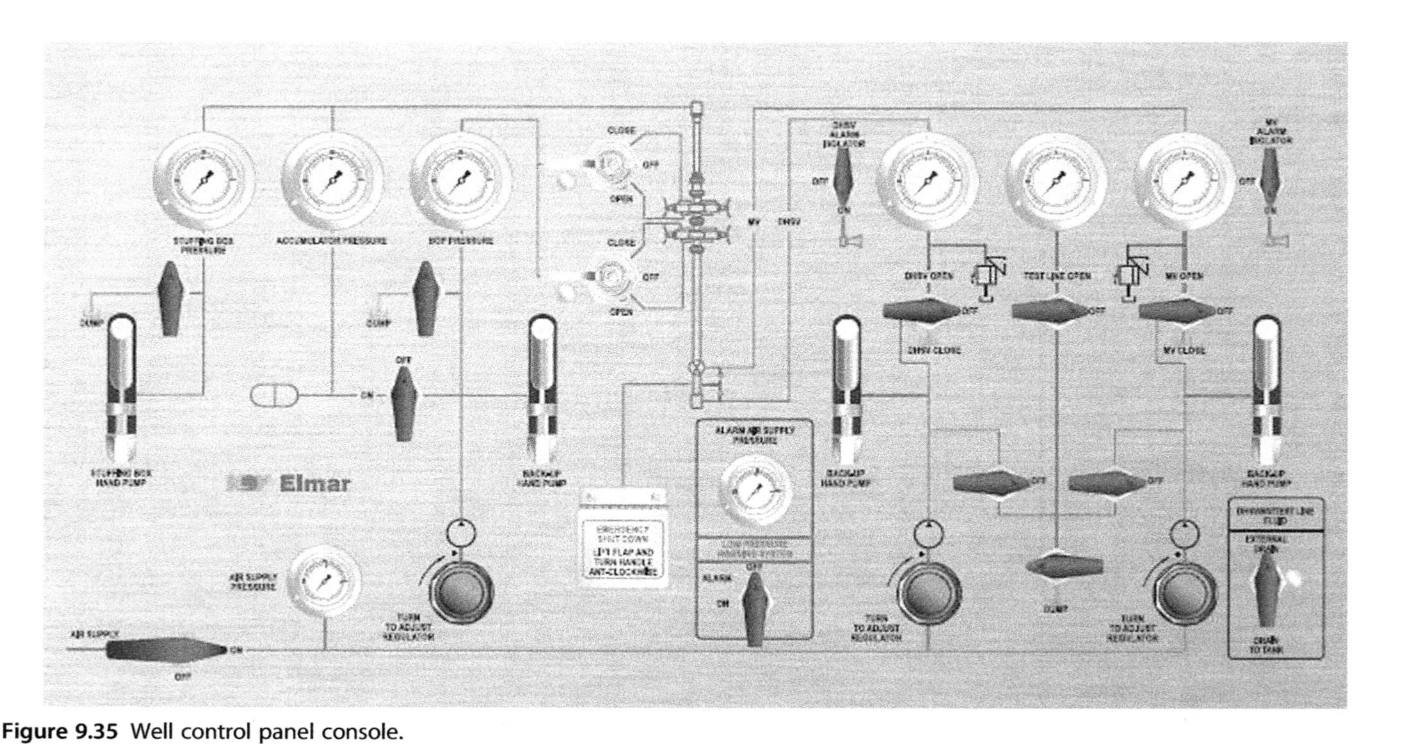

Figure 9.35 Well control panel console.

This panel can control the tree master valve, and the SCSSSV, as well as providing hydraulic power to the BOP and the stuffing box pack-off. *Image courtesy NOV Elmar.*

control panels allow operation of individual functions, for example the tree valve can be opened or closed independent of the SCSSSV and vice versa. It should also be possible to initiate ESD of the well in a single action. For example, striking a button or opening a dump valve closing all valves. The shut-down button or valve is normally prominently located on the control console and positioned below a protective cover, reducing the risk of accidental activation.

Single well control panels are not confined to wireline operations. Their use is recommended for all most interventions including coiled tubing, hydraulic workover, and pumping operations.

Finally, to emphasize the importance of all crew members knowing how to use the panel, not just senior and supervisory staff; consider the following scenario.

Memory pressure/temperature gauges are being run on slickline. Part of the data acquisition program calls for a 30-min stop at the top reservoir depth. The slickline crew chief puts the gauges on depth, pulls on the break. What next? The crew chief will almost certainly use the opportunity to obtain some refreshment, leaving the slickline unit and the associated pressure control equipment in the care of one of the junior members of the crew. Does that person have the necessary experience to know what to do if there is an unexpected loss of containment; a leak from the pressure control equipment? Would they know what to do if there was an emergency that required the well to be shut in with wire still in the hole?

9.3.5 Well control barriers during wireline operations

During live well interventions where the pressure control equipment is rigged up on top of a Christmas tree, the primary, secondary, and tertiary barriers are:

Primary pressure control (when the tree valves are open and wire is in the well):

- Stuffing box (slickline) or grease head (e-line and braided cable). The pressure containment envelope above the swab valve (SV) includes the tree connection, riser, wireline valve, and lubricator. It would include any bleed/gauge manifold.

  Primary pressure control; well shut in, tree valves closed:
- Tree valves. To establish double barrier isolation both the UMV and SV are normally closed. The UMV is the primary barrier.

- If a well is equipped with a downhole lubricator valve (see Chapter 3: Completion Equipment) the primary barrier during toolstring deployment is a closed, and inflow tested, SCSSSV. Since the lubricator valve is normally positioned above the SCSSV it becomes the secondary barrier.

Secondary pressure control (tree valves open—wire in the hole):

With wire in the well the location of the secondary well control barrier is dependent on where and how the primary barrier fails.

- If wire breaks and is ejected from the stuffing box or grease head, the internal BOP gland or ball check should seat and prevent any loss of containment through the hole left by the missing wire. The ball check or BOP are therefore a secondary barrier. It should be noted that these secondary barriers do not always (rarely) work. In any event, the tree valves would be closed following a wire break.
- If a leak develops in the pressure control equipment above the BOP (stuffing box or lubricator connection) then the wireline valve (BOP) rams would be closed. The wireline valve is therefore a secondary barrier.
- If a leak develops in the primary pressure containment envelope below the BOP, the only safe course of action would be to close the tree valves (breaking the wire). As a consequence the tree valves are the secondary barrier.

Tertiary barrier:

- Wire cutting valve or BOP.
- Christmas tree valve.

9.3.6 Rigging up on wells equipped with a conventional (vertical) Christmas tree

Almost all wireline interventions are carried out on live wells equipped with a conventional (vertical) Christmas tree. The pressure control equipment is rigged up directly onto the tree with the closed tree valves providing isolation. A typical sequence of events for rigging up the pressure control equipment and entering the well might be as follows:

- Close in production (or injection) at the choke and wing valve. In many locations closing in of the well will be carried out by the production crew before the well is formally handed over to the wireline crew.
- Close in the UMV and SV.
- Vent pressure from above the SV and remove swab cap.

- Rig up riser (if required).
- Rig up the wireline BOP.
- Make up the wireline toolstring.
- Thread the wire through the stuffing box (or grease head).
- Tie the rope socket or cable head.
- Make up the rope socket to the toolstring (or part toolstring).
- Insert the toolstring into the top of the lubricator.
- Make up the stuffing box or grease head to the lubricator.
- Pick up the lubricator.
- Use the winch to lower the toolstring out of the lubricator—connect any additional tools to the toolstring.
- Pick up the toolstring to place it fully inside the lubricator.
- Make up the lubricator on top of the BOP and carry out pressure testing as required.
- Open the UMV then SV.
- Run into the well.

At the end of each wireline run, the tree valves are closed and the lubricator depressurized. With the lubricator fully vented, it can be disconnected and lifted to allow toolstring reconfiguration before the next run takes place. At the end of the operation, the pressure control equipment is removed and the tree cap reinstated. After confirming the integrity of the tree cap, the well can be put back into production (Fig. 9.36).

9.3.7 Rigging up on wells equipped with horizontal (spool) trees

Live well interventions on wells equipped with "horizontal" or "spool" trees require additional pressure control equipment.

Horizontal (spool) trees differ from conventional (vertical) trees in that the horizontal tree has no gate valves across the bore of the tree. Instead the valves are mounted externally on the tree block.

Hydrocarbons are retained within the tree using two wireline set mechanical plugs; a crown plug (upper) and a tubing hanger plug (lower). To gain intervention access to the tubing string both plugs must first be removed. Clearly, with both plugs removed there are no pressure barriers in the bore of the tree. It is necessary to provide temporary mechanical barriers in the form of gate valves as part of the pressure control rig up.

Pressure control equipment is rigged up as shown in Fig. 9.37. The gate valves provide the required degree of isolation (two mechanical barriers). Slickline is used to recover first the crown plug, then the tubing

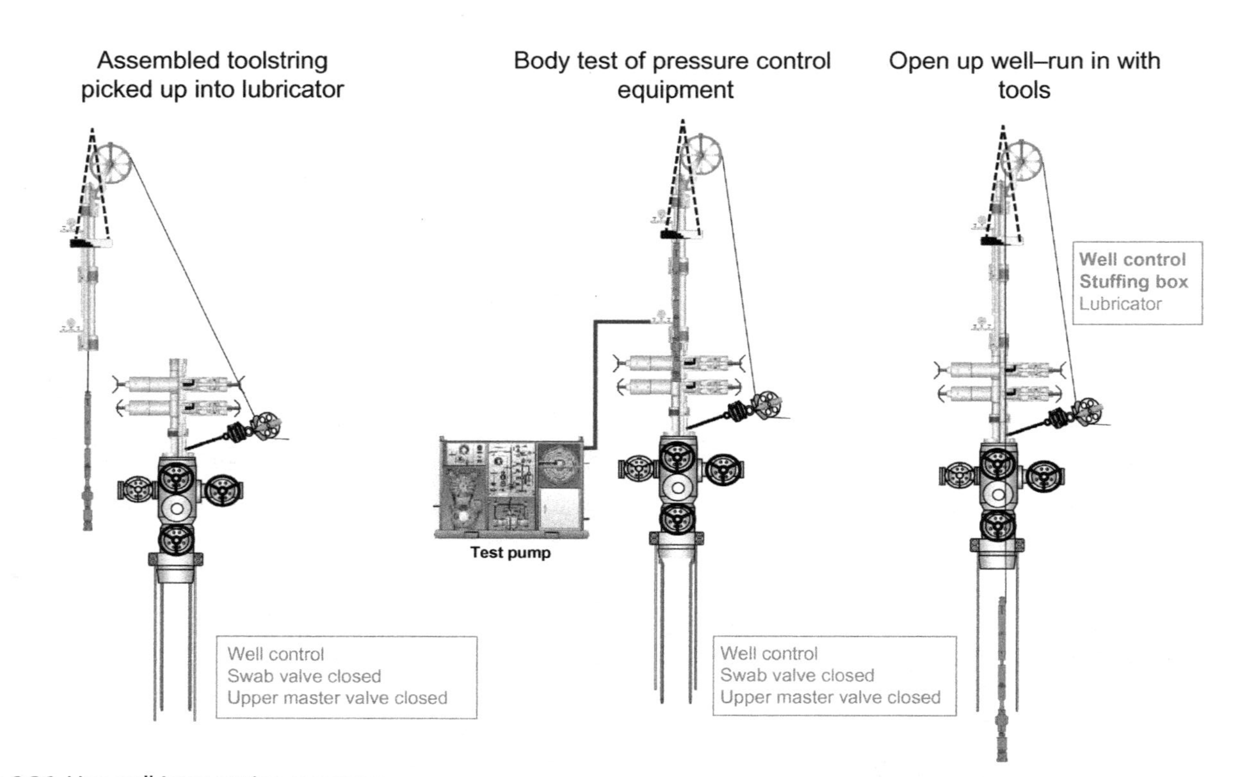

Figure 9.36 Live well intervention sequence.

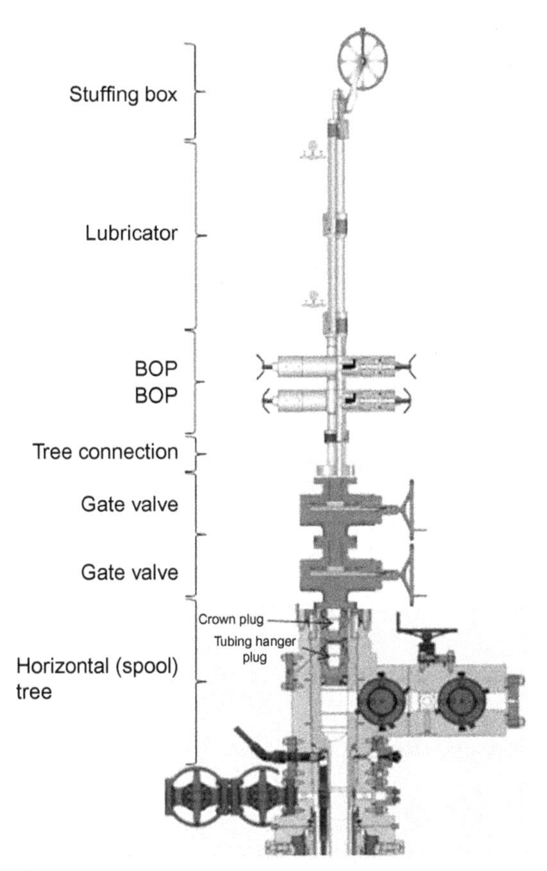

Figure 9.37 Wireline pressure control for a horizontal tree.

hanger plug. With the crown and tubing hanger plug removed, normal intervention operations can take place with the gate valves allowing the well to be safely closed-in between runs.

At the end of the intervention, both plugs must be set and tested before the pressure control equipment can be removed from the tree.

9.3.8 Drilling blow out preventer still in place: production tubing or landing string in the rotary table

Wireline operations are frequently required during the installation of a completion, for example, running a plug prior to hydraulically setting a packer. Pressure control equipment is routinely rigged up on the completion tubing (or landing string) in the rotary table. A cross-over from the tubing thread type to the ACME thread of the pressure control equipment is required and may need to be manufactured specifically for the completion tubulars.

In exceptional circumstances, some operators will allow wireline to be used "open hole", i.e., without pressure control equipment. For example, this might be permitted where cemented casing has been inflow tested, kill weight fluid is in the hole, and the casing has not been perforated.

When rigging up pressure control equipment during a completion consider the following:

- A gate valve(s) (manual or actuated) should be used to provide isolation for the installation and removal of tools from the lubricator. Normally two valves will be used to ensure double valve isolation when breaking the lubricator.
- Most rig ups will include a pump-in tee for circulation/well kill/stimulation. This should be equipped with a double barrier isolation (normally 2 × 1502 lo-torque valves).
- If a tee is included in the rig up, it should be located above the gate valve (Fig. 9.38).

9.3.9 Drilling blow out preventer in place: no tubing in the well

Many completion and workover operations require wireline to be run into the wellbore before the completion is installed. Some of these interventions require the use of pressure control equipment. For example, cement bond logs are often run with pressure applied at the surface to reduce micro-annulus effects. Pre-completion perforating on wireline also requires the use of pressure control equipment.

Different systems are commonly used when rigging up pressure control equipment above the drilling BOP, including the Dutch lockdown system and the shooting nipple.

The Dutch lockdown system is a two-part assembly. The outer shroud is bolted onto the top of the annular BOP. An inner mandrel is then lowered through the outer shroud and spaced out across the annulus preventer. Hold down bolts secure the mandrel in place and the annular bag is closed around the inner mandrel to obtain a seal. Pressure control equipment is rigged up on the threaded connection on the top of the inner mandrel (Fig. 9.39).

A shooting nipple is a length of pipe with a connection for the pressure control equipment at the top. It is positioned across the drilling BOP with the pipe rams closed to effect a seal. The nipple is prevented from

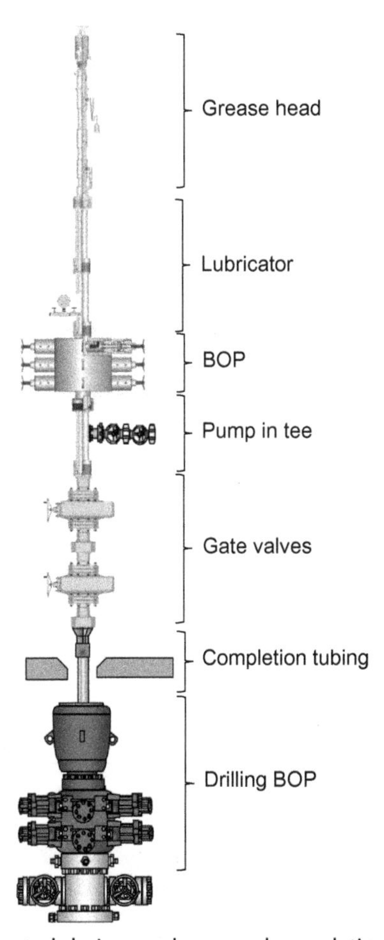

Figure 9.38 Pressure control during workover and completion operations.

being blown out through the top of the BOP by a collar stop or "donut" ring positioned below the closed BOP ram (Fig. 9.40).

There have been several serious incidents when using shooting nipples; mostly whilst perforating. Before using a shooting nipple, the risks must be properly understood and assessed. Some shooting nipple failures have been a direct result of poor manufacturing standards, such as shooting nipples being manufactured in local machine ships and fabrication yards with little or no quality control. Other problems have been a direct result of poor practice and incorrect use. The IADC website provides a stark reminder of how easily the incorrect use of a shooting nipple can degenerate into a major incident.

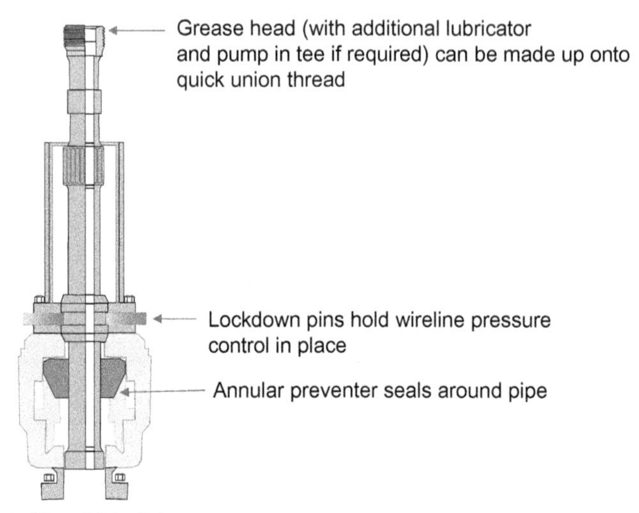

Figure 9.39 "Dutch" lockdown system.

Figure 9.40 A shooting nipple.

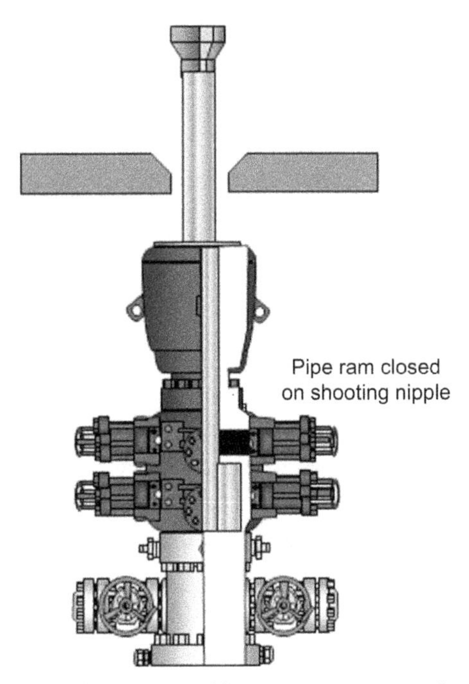

Figure 9.41 Shooting nipple set across blow out preventer with pipe rams closed to effect seal. Incorrect space-out prevents closure of the blind ram.

A well was being perforated using e-line. Pressure control equipment had been rigged up above a shooting nipple positioned across the drilling BOP. The well was filled with what was thought to be an overbalanced brine. Three perforating runs were carried out without incident. On the fourth perforating run, pressure was observed at the surface. An attempt to kill the well by bullheading kill fluid was thought to have succeeded, although only a limited volume of kill fluid had been pumped because of set limits on applied surface pressure. When pulling back to the surface on a fifth perforating run, pressure once again was seen at the surface. As an attempt was made to bleed off pressure, the long thread connection (LTC), on the top of the shooting nipple, jumped a thread and began leaking to the atmosphere. Within 3 min, flow from the well was reaching the crown. It was not possible to close the BOP blind ram, because the shooting nipple was incorrectly spaced out inside the BOP and prevented closure (Fig. 9.41). The well continued to flow and after 4 days caught fire and was destroyed.[6]

Before using a shooting nipple the following concerns must be addressed:

- Has the equipment been properly tested and certified to ensure it is capable of withstanding maximum closed-in wellhead pressure?
- Check the shooting nipple dimensions are compatible with the drilling BOP with regards to length and diameter. Pipe rams or the annular BOP must form a pressure tight seal around the body of the shooting nipple. The shooting nipple must not prevent closure of the blind ram.
- If pressure is observed at the surface, how easy will it be to regain control of the well? If gas reaches the surface the bullhead velocity will need to be higher than the gas migration velocity. In a clear brine, migration velocity can be significant and it might not be possible to pump down large diameter casing at the necessary rate.
- Bullheading risks fracturing or damaging the formation.

If a shooting nipple is used, it is advisable to include a pump-in tee. This would enable the well to be killed with the pressure control equipment still in place.

9.3.10 Pressure control: sub-surface lubricator valve

Some completions are equipped with a lubricator valve, a surface controlled full opening ball valve, run as an integral part of the completion tubing. Unlike a downhole safety valve (SCSSSV) the lubricator valve holds pressure from above and below. It is not, however, "fail-safe". A failed lubricator valve stays in the last known position.

Lubricator valves are used where there is a height restriction on the length of wireline lubricator that can be used, a problem on some offshore platforms. They are also used where there is an operational requirement to run a very long toolstring; long perforating guns for example.

A lubricator valve, used in combination with the SCSSSV, enables intervention tools to be lowered into a live well whilst maintaining a two-barrier isolation. Positioning the lubricator valve above the SCSSSV means the safety valve is protected by the lubricator valve (much more robust) if part or all of the toolstring is accidentally dropped in to the well (Fig. 9.42).

When using a lubricator valve, steps must be taken to ensure the integrity of both downhole valves before the tree valves are open. Similarly, when deploying tools out of the well, steps must be taken to

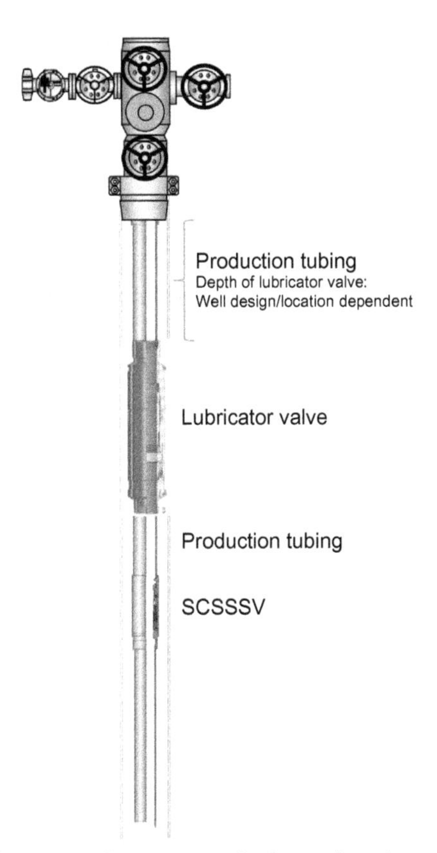

Figure 9.42 The lubricator valve is normally located a short distance above the SCSSSV.

confirm the SCSSSV and lubricator valve integrity before the lubricator can be broken. Normally, the sequence of operations will be as follows:

- Close in production.
- Close in the SCSSSV and carry out an inflow (bleed off) test. The bleed off will normally be to the production facilities through the flow line. *It is good practice to leave some residual pressure in the tubing so any leakage becomes immediately apparent.*
- Once SCSSSV integrity is ensured, apply pressure to the hydraulic control line of the lubricator valve to move the ball in to the closed position.
- Pressure the tubing above the closed lubricator valve to confirm closure/integrity. Test pressure must be more than the wellhead closed-in pressure (WHCIP). If not, a leak past the lubricator valve would not be identified as pressure would act against, but not leak past the closed SCSSSV.

- Having confirmed the integrity of the lubricator valve, bleed back the test pressure.
- Remove the tree cap.
- Rig up shear seal (if used) and wireline BOP.
- Open the tree valves (SV and UMV).
- Deploy the intervention toolstring into the well. Once it is fully made up it will be clamped and supported on top of the wireline BOP.
- Make up the wireline lubricator (if used) and stuffing box or grease head.
- Connect the rope socket or cable head to the toolstring resting in the BOP and take the weight of the tools.
- Remove the support clamp from the toolstring before connecting the wireline pressure control equipment.
- Test the pressure control equipment against the closed lubricator valve to more than closed-in wellhead pressure.
- Adjust the test pressure until it is 100–200 psi (20–30 bar) above closed-in pressure and open the lubricator valve. As the valve opens, tubing pressure should drop to match the previously recorded WHCIP as equalization cross the SCSSSV takes place.
- Open the SCSSSV. The tool is now ready to be run in to the well.

To remove the toolstring from the well:

- After reaching surface and tagging the stuffing box or grease head, bleed down the control line to close the SCSSSV.
- Bleed off above the closed SCSSSV—normally via the flow-line to the production facilities. This may not be sufficient to fully depressurize the tubing, and any residual pressure might have to be bled to the closed drain (if available) or to the atmosphere.
- Once valve integrity is ensured, close the lubricator valve via the hydraulic control line.
- Pressure test above the valve to confirm closure and valve integrity. Bleed off test pressure.
- The wireline lubricator can now be broken. Once broken the lubricator can be lifted.
- Clamp the toolstring and lower it to rest on top of the BOP.
- Rig down the lubricator.
- Remove the toolstring from the well.
- Lay down the BOP, close the tree valves, and replace and integrity test the tree cap.
- The lubricator valve and SCSSSV can now be opened and the well put back into production.

9.4 WIRELINE DOWNHOLE EQUIPMENT

It is necessary for wireline professionals have detailed knowledge of a wide range of downhole equipment. Operating company supervisory staff are not expected to have the same detailed knowledge. However, when they are responsible for well integrity and control during a live well wireline intervention, an appreciation of the more commonly used wireline tools and toolstring configuration is extremely useful.

9.4.1 Basic slickline toolstring

Many slickline operations rely on the mechanical manipulation of downhole tools. For example, high impact force is needed when setting a plug in a nipple profile; V-packing has to be forced into a seal bore, steel or brass pins in the running tool must be sheared to enable the lock mandrel to locate in the nipple, and more pins sheared to release the running tool from the lock mandrel.

The impact force essential for pushing the packing into the seal bore and shearing pins is generated by opening or closing the toolstring jar. The amount of impact force is a function of velocity and the weight of toolstring above the jar. Friction, deviation, and line drag all act to reduce the impact force.

Velocity (line speed) is manipulated using the hydraulic controls on the wireline winch. When operating at very shallow depths, the brake can be applied and the toolstring jars manipulated by hand.

A simple wireline toolstring (Fig. 9.43) will consist of three basic components:

- A rope socket: To attach the wire to the toolstring (Fig. 9.44).
- Wireline stem: Weight bar; necessary to overcome the piston effect of pressure acting against the cross-sectional area of the wire as it passes through the stuffing box. Weight is also needed to generate impact force.
- Jar: Usually a mechanical link jar (commonly called a spang jar). Tubular jars can be used, although these are normally reserved for wire recovery fishing operations (Fig. 9.45).

In addition to the basic toolstring components many wireline specialists will include a set of power jars; oil jars, or spring jars. These give a much greater impact when jarring up. The disadvantage, particularly if oil jars are used, is that the jar acts as a shock absorber, reducing impact force when jarring down. Some operating companies mandate the use of

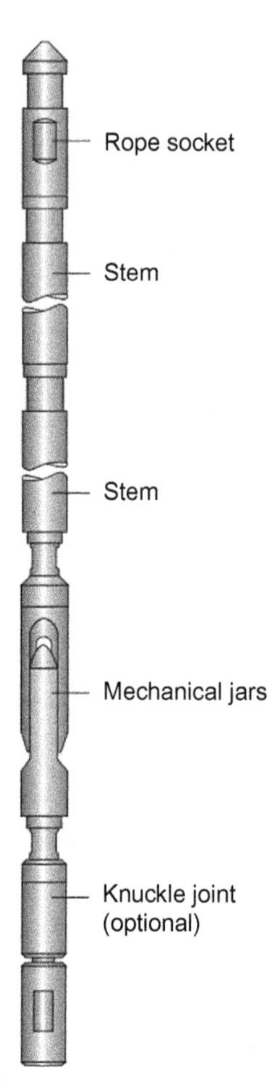

Figure 9.43 Basic slickline toolstring.

a power jar in every slickline toolstring. However, there are occasions when the power jar is a hindrance, for example, when trying to perform delicate operations, such as trying to locate a valve in a side pocket mandrel. Unwanted firing of the jar can result in a misrun, even lost or damaged tools (Fig. 9.46).

Other commonly used slickline toolstring components:

- Rollers, to help overcome friction in high angle wells.
- Centralizers.
- Knuckle joints to make the toolstring more flexible.
- Swivels to reduce wire fatigue.
- Accelerators increase the impact delivered by power jars—particularly where the toolstring is at a shallow depth.

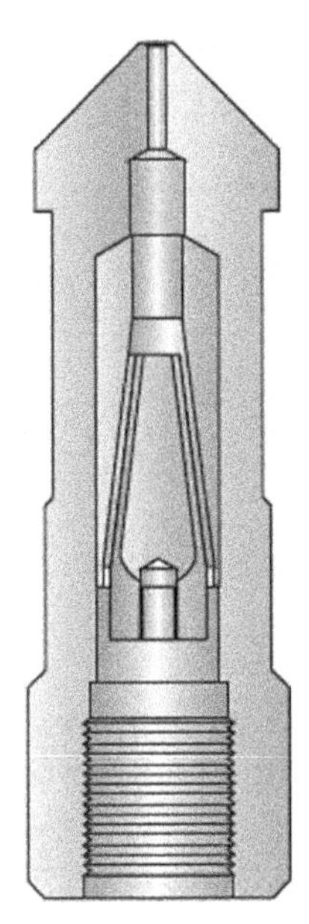

Figure 9.44 Slickline rope socket.

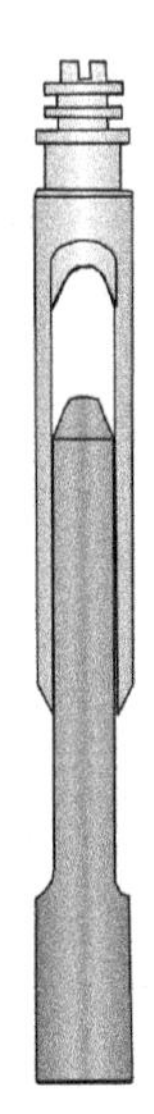

Figure 9.45 Link jar or "spang" jar.

9.4.2 Toolstring configuration

Many different toolstring components and different configurations are used during slickline and e-line use. From the well control and well integrity perspective, the main concern is toolstring weight.

Figure 9.46 Oil jars.

During wireline interventions, well pressure acts against the cross-sectional area of the wire. The amount of (upwards) force acting on the wire is the cross-sectional area of the wire multiplied by wellhead pressure; toolstring weight must be sufficient to overcome this force. This is easily calculated if the wellhead pressure and wire outer diameter is known (Table 9.7).

$$\text{Force} = \text{Pressure (wellhead)} \times \text{area (wire)}.$$

For example, using $^{7}/_{32}''$ cable in a well with a surface pressure of 3250 psi, the toolstring weight needed to balance well pressure will be $3250 \times 0.037 = 120.25$ lbs. Additional weight is needed to overcome friction in the grease head, and the calculation makes no allowance for buoyancy or well angle.

During many wireline interventions, the toolstring will be lighter when pulling out of the well, for example after setting a bridge plug, or after firing a perforating gun. Pressure can increase whilst the toolstring is in the well; for example, after equalizing across a plug. Loss of toolstring weight, or increased well pressure can have consequences for maintaining the integrity of the pressure control equipment. If a toolstring is

Table 9.7 Cross-sectional area for standard diameter wirelines $\frac{\pi}{4} \times ID^2$

Slickline size and cross-sectional area

Slickline diameter (in.)	Cross-sectional area (in.)
0.072″	0.0037″
0.082″	0.0053″
0.092″	0.0066″
0.108″	0.0092″
0.125″	0.0123″

Braided cable and e-line cross-sectional area

Cable diameter (in.)	Cross-sectional area (in.)
$^{3}/_{16}$″	0.027″
$^{7}/_{32}$″	0.037″
$^{1}/_{4}$″	0.049″
$^{5}/_{16}$″	0.077″

underweight, wire will be blown out of the well as the toolstring nears the surface. In some cases, the winch will not be able to keep up and the rope socket or cable head will impact the base of the stuffing box or grease head, almost certainly breaking the wire and leading to an escape of hydrocarbons. The dropped toolstring has the potential to damage completion components and might be difficult to fish. When calculating the minimum toolstring weight required to operate in a live well the following considerations should be made:

- Calculated toolstring weight must not include any components that will be released and left in the well, for example wireline plugs.
- Many perforating guns lose weight once they are fired. Make allowances for this.
- Make allowances for any change in wellhead pressure that may occur because of the operation being performed, for example opening a sliding sleeve to access a production zone, adding new perforations or equalizing across barriers.

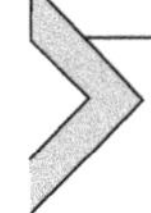

9.5 WELL CONTROL DURING WIRELINE INTERVENTIONS

As so many wireline interventions are carried out on live wells, well control will normally focus on monitoring and maintaining the integrity of the pressure control equipment. Should integrity be lost, then well site supervisors must know what immediate actions need to be taken to regain control of the well. Good planning and enforcing some basic precautions reduces the risk of an incident.

9.5.1 Pre-intervention preparations

Before beginning any well intervention, determine the well status and research the well history. Take note of:

- Wellhead closed-in pressure.
- Well fluids (Water cut and water chemistry, CO_2, and H_2S).
- Recent production history.
- Condition of the completion tubing and completion components (if known).
- Annulus pressure history.
- Annulus fluid and fluid level (if known).
- Most recent SCSSSV (if fitted) integrity test records.
- Recent intervention history.
- Tree and wellhead maintenance records. When was the tree last serviced and integrity tested—are the SV and master valve functioning and holding pressure?
- Tree connection (for rigging up the pressure control equipment).

Having established that it is safe to proceed, the next step will be to rig up and test the wireline pressure control equipment:

9.5.2 Rigging up: equipment location and layout

Wireline equipment configuration is both location- and operation-dependent. Equipment used on offshore platforms differs from that used at a land location. Irrespective of location or operation, there are some basic considerations when preparing to intervene in a live well:

- The power pack must be positioned in accordance with zone classification. A hot work (spark potential) permit may be a requirement for operating the power pack.
- Identify access and exit routes from the worksite to a safe area.
- Location of well control panel and secondary shut-down button (in a safe area).
- Location of emergency equipment.
- Proximity of essential services.

  - Fuel for diesel power pack.
  - Power source for electric power pack.
  - Rig (compressed) air for operating well control panel and pressure test pump.
  - Location of closed drains for bleeding down hydrocarbons from pressure control equipment.
- Proximity of gas alarms. Could a leaking grease head or stuffing box or cold venting of the lubricator cause a facilities shut-down? Do gas alarms need to be temporarily isolated?
- Impact of facilities shut-down:
  - Will electrical power, rig air, and other services be lost?
  - What contingencies are in place?
  - How much autonomy is built in the wireline equipment—for example does the well control panel and grease injection pump need a continual air supply, and is a back-up available.
- Impact of simultaneous operations.
- Height restrictions. Overhead deck on offshore platforms. Crane limitations on land.
- What temporary barriers need to be in place to cordon off the work area.
- Line of sight—wireline unit to pressure control equipment. The operator should have a clear view of the lubricator and the lower hay pulley.
- Lifting equipment location and limitations.
- Wireline unit security. Even heavy wireline skids can move when tension is applied to the wire. In addition, lightweight wireline skids with a high center of gravity (i.e., some heli-portable designs) need to be guyed to prevent them from toppling when tension is applied to the wire. Truck mounted wireline units will move if the wire is under tension and the parking brake not set.
- Wireline masts and lubricators should be guyed for stability and safety.

9.5.3 Pressure testing surface equipment: preparation

There are several checks of the pressure control equipment that need to be carried out before and during use. Properly implemented, these checks significantly reduce the potential for leaks whilst the equipment is exposed to well pressure.

- Before rigging up, all necessary pressure control conformance certification should be inspected. Equipment supplied without the relevant certification must not be used.
- Lubricator and BOP lifting bridals and lifting subs must have the relevant lifting certification.
- Visually inspect quick union threads, seal face, and O-ring seals.
- Using the pump that will be used during the intervention, function test (close and open) the BOPs.
- If a test stump is available, pressure test below the BOP rams before rigging up the BOP. For a braided cable BOP a properly designed test rod will need to be positioned across the closed rams. Do not use home-made rods (i.e., bits of bent shear stock or knotted wire). They risk being blown out of the BOP when pressure is applied. This can injure people and damage equipment.
- If the production Christmas tree is equipped with hydraulic or pneumatically operated valves, confirm the valve position (by measuring the length of exposed stem, or by visual inspection of indicator pins).
- If using a remote well control panel, a hydraulic jack should be fitted to the tree valve actuator and the valve functioned open/close. Note the time taken to close the valve. Some valves will have long closure times if wellhead pressure is low.

  Note: Lock out caps (solid and fusible) are not recommended.
- Confirm the number of turns required to open/close each manual valve on the Christmas tree.
- If using a long lubricator, it should be properly supported to reduce bending when it is picked up into the vertical.
- Once made up on the tree, the lubricator should be supported to keep it in the vertical position.

9.5.4 Pressure testing surface equipment

Where possible the integrity of the pressure control equipment will be confirmed by pressure testing before it is exposed to well pressure. Testing is normally carried out using water. If there is a risk of hydrates a water/glycol mix is used. Test pressure will normally be WHCIP plus an agreed margin. In some jurisdictions, the margin is defined. For example, in Canada, Enform recommend 1.3 × the maximum WHCIP.[7] Where no recommendation exists, 10% above maximum WHCIP is commonly applied. Some well intervention specialists require the equipment to be tested to its certified working pressure. This exposes the crew to unnecessary risk, especially where high pressure equipment is used on a well with low surface pressure.

The duration of each high-pressure test is determined by operating company policy and local regulations, but will not normally be less than 10 or 15 min. Longer tests are sometimes stipulated for high pressure wells, or wells with a dangerous level of H_2S. It is good practice to carry out a low-pressure test (typically $\pm$ 100 psi) before increasing to the final test pressure.

9.5.5 Testing the blow out preventer (wireline valve)

If the BOP has been successfully tested on a stump before it is rigged up over the well, then there is no need to test the ram integrity again. A body test will still be required to confirm connection integrity. Ram testing can also be carried out once the BOP has been rigged up.

9.5.5.1 Testing the shear seal blow out preventer

If a shear seal BOP is used, it will almost always be rigged up directly above the tree, or as close to the tree as possible. When operating on high pressure/high temperature (HPHT) wells, shear seal BOPs are often connected to the tree using a flange instead of a quick union. The shear seal BOPs should be tested from below by applying a differential pressure across the closed rams. It can be tested as soon as it is in place, before the rest of the pressure control equipment is rigged up.

With the BOP in place, the normal sequence of events for testing the valve is:

- Connect the test pump to the kill wing of the Christmas tree.
- Confirm the UMV is closed.
- Open the kill wing valve and SV.
- Pump fluid until returns are seen overflowing the top of the BOP.
- Close the BOP shear/seal ram and apply test pressure.
- Pressure can be blend back to the pump, or the BOP ram equalizing port can be opened.
- Open the shear/seal ram.

9.5.6 Slickline blow out preventer test

Slickline BOP blind rams are tested from below by applying a differential pressure across the closed rams. They will not hold pressure from above.

If desired, the BOP can be tested before the lubricator and stuffing box is rigged up. The BOP can be tested via the kill wing or via a test port on any riser below the BOP. The test sequence is similar to that used for the shear seal BOP.

Test using the Christmas tree kill wing:

- Connect the test pump to the kill wing of the Christmas tree.
- Confirm the UMV is closed.
- Open the kill wing valve and SV.
- Pump fluid until returns are seen overflowing the top of the BOP.
- Close the BOP blind ram and apply test pressure.
- Pressure can be blend back to the pump, or the BOP ram equalizing port can be opened.
- Open the blind ram.

Using a test port or pump-in tee:

- Confirm the tree valves are closed.
- Connect the test pump to tee or test port.
- Pump fluid until returns are seen overflowing the top of the BOP.
- Close the BOP blind ram and apply test pressure.
- Pressure can be blend back to the pump, or the BOP ram equalizing port can be opened.
- Open the blind ram.

9.5.7 Braided cable (e-line) blow out preventer test

Braided cable (and e-line) BOP dual rams should be differentially tested from below and with a test rod across the rams. Test pressure can be via the tree, or through a test port or pump-in tee on any riser below the BOP.

Pressure test using a test rod:

- Ensure both BOP rams are fully open.
- Place the test rod inside the BOP straddling both rams.
- Fill the riser/BOP with test fluid.
- Close the both BOP rams.
- Apply test pressure.

9.5.8 Pressure control equipment: full body test

Before exposing the pressure control equipment to well pressure, a body test of the assembled components is carried out. Location and equipment configuration will determine the best test method.

- Fluid can be pumped into the pressure control equipment through the kill wing, a lubricator tee-piece or one of the bleed ports. The body test is normally carried out through the tee-piece or test port.
  - Grease head: Leave the grease head unpressurized until fluid is seen spilling over the top. Stop pumping into the lubricator, then apply pressure to the grease head (grease pressure should be higher than the lubricator test pressure). Apply test pressure to the lubricator.
  - Stuffing box: Bleeding all the air from the lubricator will be difficult unless there is a bleed port close to the top of the lubricator (recommended). A bleed port enables trapped air to be removed.
- Fluid used for this test will drop into the well. If there is any possibility of hydrate formation, the fluid must be mixed with a suitable hydrate inhibitor. In most cases this will be a 50:50 mix of glycol and water.
- Once the pressure control equipment has been successfully tested, it is good practice to drop the pressure to a value slightly higher (± 100 psi) than WHCIP. It is easier to open tree valves at lower differential pressure.
- The tree valves can now be opened.
  - The master valve should be opened first, particularly if the valve is actuated, as the speed of opening cannot be easily controlled.
  - The SV should be opened slowly until any differential pressure across the valve is equalized. If higher than wellhead pressure was left in the lubricator, it will drop as the valve opens.
  - Always count the number of turns needed to open a valve.
- Any time the lubricator is broken and re-stabbed (to remove and install tools) integrity must be confirmed before the well is opened once more. The amount of fluid dropped into the well can be reduced if a lubricator test sub is used (Fig. 9.47).

In most circumstances, well pressure is not used to integrity test pressure control equipment. In remote locations where there are limited facilities (i.e., some land locations or unmanned offshore satellite towers), opening the tree valves and exposing the equipment to well pressure is often the only way of proving integrity. This is not good practice, but it still occurs.

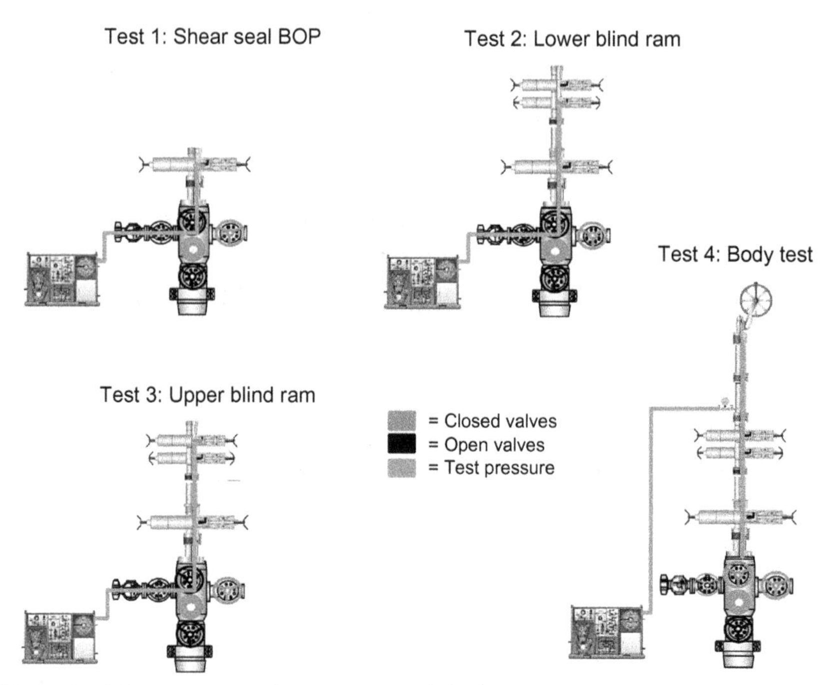

Figure 9.47 Pressure control equipment. Full body test.

9.5.9 Pressure control equipment: bleed down after wireline intervention

Wireline operations on live wells will leave the lubricator filled with pressurized hydrocarbons and possibly other noxious gases, such as H_2S and CO_2. Cold venting of the lubricator to the atmosphere is common, but is only acceptable where it is safe to do so. If the well has liquid hydrocarbons at the surface, if H_2S is present in dangerous concentrations, of if there are ignition sources in the proximity of the lubricator, then venting must be contained in a closed system. On many offshore production platforms a closed drain system is available and is designed to take pressurized hydrocarbon. Facilities operators should be informed when closed drains are used. Most closed drain systems will not fully depressurize pressure control equipment. Any residual pressure will have to be cold vented to the atmosphere. Certain precautions need to be taken when cold venting gas:

- Ensure the area is adequately ventilated. If the wellhead is in an enclosed area, a hose should be used to divert the gas to a safe, well ventilated, area.

- Stay upwind of the lubricator during venting—particularly if there is H_2S present.
- Gas alarms may need to be temporarily isolated whilst the pressure control equipment is being depressurized—know the location of the gas alarms and inform the facilities personnel when venting is taking place. The facilities operators must also be informed when venting is complete so gas alarms can be reinstated.
- If there is liquid at the surface, vented fluid must be contained for proper disposal.
- When (wet) gas is cold vented through a needle valve, Jules Thompson cooling often leads to ice blockage across the needle valve. Cycling the needle valve open and closed will normally free any blockage.

Bleeding down pressure control equipment, especially long, large diameter risers and lubricators, takes time. Be patient. Once fully depressurized, the quick union can be backed off and the lubricator lifted. If the quick union will not turn it may be because there is still pressure in the lubricator. Never try to force the quick union.

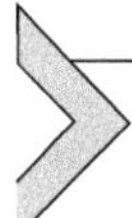

9.6 WELL CONTROL DURING LIVE WELL WIRELINE INTERVENTIONS

The main concern during any live well intervention is maintaining the pressure integrity of the surface equipment. Personnel must know how to respond to barrier failure, and how to immediately reinstate a failed barrier. The following pages outline the most common causes of barrier/integrity failure, and provide the immediate actions needed to reinstate the failed barrier.

9.6.1 Leaking Stuffing Box (Slickline)

Leaks from the stuffing box are common, especially where there is gas at surface. They are caused by excessive wear of the packing rubbers. Leaks will happen less often if:

- The seal elements (packing rubbers) are replaced frequently.
- The wire is properly lubricated.
- The correct pressure is applied to the pack-off. Too high a pressure accelerates wear.

- There is no roughness (rust/corrosion) on the surface of the wire. Stuffing box leaks must be dealt with immediately.
- Stop the winch and apply the brake.
- For hydraulic stuffing boxes, apply pressure to the hydraulic packing piston until the leak stops. Do not over-pressure the piston, as this can prevent free movement of the wire and accelerate wear of the packing rubbers.
- If a manual stuffing box is used, a member of the crew will need to gain access to the stuffing box and tighten the packing nut by hand.

If the leak cannot be contained by energizing the stuffing box packing, then a temporary repair will enable to wire to be pulled out of the well:

- Stop the winch and apply the brake.
- Close and lock the BOP blind ram(s). Depressurize the lubricator above the closed rams.
- Confirm ram closure/integrity by observing for pressure build up (PBU) in the lubricator.
- Once satisfied that the rams are holding, back out the stuffing box hydraulic piston to expose the uppermost stuffing box packing.
- Using a new piece of stuffing box packing, make a lengthways cut through half the diameter of the packing.
- Fit the cut packing onto the wire (Fig. 9.48).
- Slide the packing down the wire into the packing barrel recess and replace the hydraulic packing piston.
- Apply hydraulic pressure to the packing nut.
- Equalize well pressure across the BOP ram to confirm the repair is working.
- Pull out of the well.
- Replace the stuffing box packing before continuing with operations.

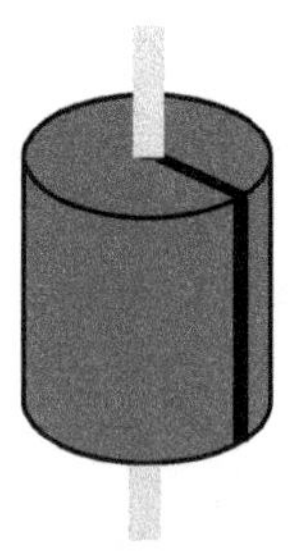

Figure 9.48 Packing is cut along its length through half the diameter and pushed onto the wire.

It is worth noting that some equipment vendors (e.g., NOV Elmar) supply spiral cut stuffing box packing that is specifically designed to be replaced whilst the wire is in the hole. This significantly reduces the risks that some operators associate with using cut packing to temporarily fix a stuffing box leak, and is an excellent idea (Fig. 9.49).

If the temporary stuffing box repair is unsuccessful, then the following course of action can be taken to restore stuffing box integrity and enable the wire to be pulled from the well.

- Stop winch movement and apply the brake.
- Close the wireline valve and bleed off above the rams.
- Observe for PBU above the rams.
- Once satisfied that the wireline valve is holding pressure, back off the quick union (normally the connection immediately above the BOP) and lift the lubricator to expose the wire.
- Firmly clamp the wire at the top of the BOP.
- Slack off on the wire and confirm the clamp is holding.

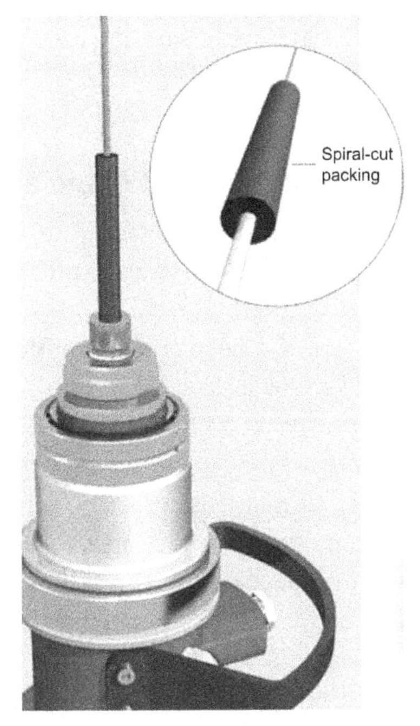

Figure 9.49 NOV Elmar spiral cut packing is a safer alternative—allowing packing to be replaced whilst wire is still in the well. *Image courtesy of NOV Elmar.*

- At the unit, pull a generous amount of slack wire off the drum.
- Cut the wire close to the wireline unit.
- Pull the cut wire down out of the lubricator.
- Lay down the lubricator and back out the stuffing box.
- Re-pack the stuffing box with new packing.
- Reverse feed the tail of the clamped wire back through the lubricator.
- Reverse feed the wire through the stuffing box then connect the stuffing box to the lubricator.
- Tie the two wire ends together (the wire in the well to the wire on the winch).
- Pick up the lubricator and stuffing box, taking care not to kink the wire between the BOP and the bottom of the lubricator.
- With the lubricator hanging a short distance above the BOP, pick up slack until there is enough tension on the wire to enable the clamp to be removed.
- Lower and stab the lubricator whilst simultaneously picking up the wire. Make up the quick union connection.
- Equalize across the wireline valve and check for leaks. Open the BOP rams and pull out of the hole.

If the repair was carried out with wire deep in the well, the wire on the unit will have to be replaced.

9.6.2 Leaking Grease Head (Braided Cable and E-Line)

A properly maintained and monitored grease head should not leak. Leaks can be prevented by implementing some basic precautions and following best practice:

- Ensure the flow tubes are the correct size for the wire—oversized or worn tubes are more likely to leak.
- New cable should be pre-packed with grease before or when spooling onto the winch.
- It is usually much easier to maintain a grease seal when using seasoned (used) cable, but there are exceptions. Cable coming off the top of the drum and therefore going deeper into the well will be used more than cable on the bottom of the drum. The cable at the top will consequently have more wear. Flow tubes sized for the worn cable may impede the passage of the thicker wire lower down the drum. If flow tubes are sized for the less worn cable, it will be difficult to maintain a seal when the worn cable is across the flow tubes—study the cable

history. If a cable has been used excessively in shallow wells, be cautious about running into a deep well for the first time.

- Ensure there are enough flow tubes and grease injection points for the pressure and fluid type (liquids or gas)—see Table 9.4.
- Keep grease pump pressure above wellhead pressure. Ensure there is enough margin between the working pressure of the grease head and grease injection equipment and wellhead pressure. If the grease seal is lost, it may be necessary to pump at significantly above wellhead pressure to regain the seal.
- Have an adequate supply of grease on location. Include a contingency for unplanned events such as stuck cable.
- Only use waste (used and contaminated) grease as a last resort. In some locations using waste grease is forbidden, but it may be all that is available in the short-term, and will probably be adequate to get the wire out of the well. The only alternative is to cut the wire and secure the well until new supplies reach the location.
- If rig supplied compressed air is used to operate the grease pump, have contingency plans in place in case of loss of rig air, i.e., bottled compressed nitrogen.
- Run at a line speed commensurate with grease supply, number of flow tubes, and wellhead pressure.
- Anticipate increased well pressure, for example when perforating or pumping into the well. It will be necessary to increase grease pressure to compensate.

Even with these measures in place, the potential for a leak remains. Stopping a leak will depend on the cause, well pressure, and the type of fluid in the well. For example, a leak caused by having incorrectly sized (too large) flow tubes will be difficult to stop. Similarly, regaining control will be more difficult when there is gas to the surface. The difficulty will be intensified as well pressure increases. Any leak has the potential to escalate and become serious, so a quick response is needed.

If a leak occurs:

- Stop wire movement and immediately increase grease injection pressure.
- If the leak occurred when pulling out of the well, try slowly running in. Running in can assist in regaining the seal, as a pre-packed section of the wire is run across the flow tubes, and gas percolating up thorough the cable is pushed back into the well.

For oil wells and low-pressure gas wells this will normally be enough to regain control of the well. In higher pressure systems, additional measures are usually necessary:

- If the grease head is fitted with a pack-off or line wiper, these should be energized. As the pack-off and line wiper are energized, some of the hydrocarbons leaking from the grease head will be diverted down the grease return line.
- Closing the return line from the grease head whilst increasing grease supply pressure will sometimes slow the leak, but this is only a temporary fix.
- Once the leak has been contained, open the grease return line and allow grease to circulate.
- Ease the pressure on the stuffing box and resume operations.

If the leak has still not been contained the next step is to:

- Try to slow the leak rate by maintaining a high grease pressure. If the leak continues unabated, there is a risk of freezing (Jules Thomson effect) and the formation of hydrates.
- Close the BOP (upper and lower rams) and pack the void space between the rams with high pressure grease. For two-barrier isolation, two sets of rams or a triple ram BOP configuration are required.
- If the BOP rams are holding pressure the leak from the grease head will diminish and stop. Bleed down any residual pressure in the lubricator.
- If freezing has taken place in the grease head, time will be needed for the system to thaw. Where ambient temperatures are very low, some heating (i.e., from a steam hose or hot water) can be used to speed up the process.
- Circulate grease to flush out any contaminated grease. (Where ambient temperatures are low, a thinner oil may be required to flush congealed grease out of the flow tubes before fresh grease circulation can be established.)
- With the head packed with fresh grease and pressure applied (above wellhead pressure) equalize across the BOP and open the rams.
- Confirm the grease seal remains intact whilst picking up on the cable and pull out of the hole. Speed may have to be moderated to retain the seal.

If grease head pressure integrity cannot be restored, but BOP integrity has been established, it will be necessary to cut the cable then repair or modify the grease head. Cutting the cable enables flow tubes to be replaced or more flow tubes (and grease injection points) added.

- With BOP integrity established, back off the quick union above the BOP and lift the lubricator.
- Securely clamp the wire above the BOP.
- Pull a generous length of cable off the wireline unit drum—enough to place the wire splice at least three or four wraps down the drum once the repairs have been made.
- Cut the wire immediately in front of the unit and then pull the slack (cut) wire down through the lubricator/grease head.
- Lay down the lubricator and repair the grease head—replacing and adding flow tubes/grease injection points as required.
- Reverse feed the tail of the clamped cable back through the lubricator.
- Reverse feed the cable through the grease head then connect the grease head to the lubricator.
- Splice the two cable ends together (the cable in the well to the cable on the winch).
- Pick up the lubricator and grease, taking care not to kink the cable between the BOP and the bottom of the lubricator.
- With the lubricator hanging a short distance above the BOP, pick up slack until there is enough tension on the cable to enable the clamp to be removed.
- Lower and stab the lubricator whilst simultaneously picking up the cable. Make up the quick union connection.
- Equalize across the BOP and check for leaks. Open all BOP rams and pull out of the hole.

It may be necessary to replace the cable—this will depend on how far down the drum the cut had to be made.

9.6.3 Braided Cable and E-Line. Broken Wire Strand (Bird's Nest)

When using multi-strand braided cable (or e-line), there is a risk of parting a single strand. A broken strand is more likely to occur after BOPs have been closed or the wire has been clamped. Braided cable is also more vulnerable after prolonged working, particularly if prolonged downwards jarring was necessary. When a strand of wire breaks, the winch operator may notice a dark spiral on the cable because of the missing strand. More commonly, the operator will experience erratic fluctuations in weight and an overall weight increase as the broken strand strips off the cable, bunching and tangling below the grease head. The tangled wire has the appearance of a bird's nest—hence the name.

If the problem is not noticed quickly, the loose strand can interfere with, and prevent sealing of, the BOP. If BOP integrity cannot be achieved, there is no alternative but to cut the wire using the shear seal BOP (if fitted) or with a tree valve, normally the swab. In extreme cases, a bird's nest can make it impossible to close in the well securely. It would have to be killed before the lubricator could be safely removed.

If a stranded cable is identified early, before a bird's nest has fully formed, it may be possible to cut away the broken strand and bed the end in to the cable, such that it can be pulled through the grease head and the wire recovered to the surface. The ease with which this seemingly simple task can be accomplished will depend not just on the extent of the damage, but also on the location of the rig up. For example, if the lubricator has been rigged up in a derrick there will be sufficient height and additional hoists (tuggers) to enable the lubricator to be lifted high above the BOP and, if necessary, broken in more than one place. The procedure that follows is intended as a guide only. On-site personnel should decide how best to proceed based on how the equipment is configured and the facilities available.

- As soon as a "bird's nest" is identified or suspected, stop the winch and apply the brake.
- Inspect the wire to confirm a missing strand.
- Close the BOP rams and inject grease in the void space between the rams. Grease will have to be injected at above wellhead pressure.
- Attempt to bleed off above the BOP. If damaged cable is sitting across the rams it will be difficult, probably impossible to bleed down.
- Once BOP integrity has been confirmed, the quick union above the BOP can be backed off.
- The lubricator can now be lifted. Take care when lifting the lubricator. Make sure the lubricator/grease head is sliding over the cable, not pulling the cable up through the BOP ram.

At this point the options will depend on how high the lubricator can be lifted—working in a derrick it may be possible to lift the lubricator clear of the bird's nest to fully expose it—in other locations this may not be possible. Both options are discussed here:

- If possible, lift the lubricator until undamaged wire is observed—this assumes the grease head is sliding freely up the cable as it is raised. Pull away the loose strand and cut it close to the point where there is no damage. Bed the cut end into the lay of the wire. If the cut end is sitting proud it will need to be filed down and dressed off to prevent it jamming in the flow tubes.

- If additional hoists are available it is worth breaking off the grease head and raising it above the upper lubricator section. A visual inspection can be carried out to ensure there are no loose strands of wire remaining below the grease head.

  Note: This method will involve the use of "man-riding" winches to enable personnel to access the damaged cable and the grease head quick union connection.
- Make up the grease head and lubricator and lower them down on to the BOP.
- Make up the quick union and equalize across the BOP.
- Slowly pick up on the wire—keep a close eye on cable tension until undamaged wire is through the grease head. Continue to pull out of the hole.

Where a mast is being used, or where rig up height is limited a different approach is needed:

- After lifting the lubricator, clamp the wire above the BOP.
- If the wire is moving freely through the grease head pull it down through the lubricator until the damaged section is found—this assumes that the wire immediately above the BOP is undamaged.
- Pull away the loose strand and cut it close to the point where there is no damage. Bed the cut end into the lay of the wire. If the cut end is sitting proud it will need to be filed down and dressed off to prevent it jamming in the flow tubes.
- Taking care not to further damage the wire, pull through additional slack and lay down the lubricator.
- Break off the grease head and clear any broken strands of wire from inside the lubricator.
- Make up the grease head and pick up the lubricator.
- Carefully spool the slack wire back on to the lubricator until tension is seen.
- Remove the fishing clamp
- Lower the lubricator and re-stab.
- Make up the quick union and equalize across the BOP.
- Slowly pick up on the wire—keep a close eye on cable tension until undamaged wire is through the grease head. Continue to pull out of the hole.

A severe bird's nest can be large enough to obstruct the BOP. If this happens it will not be possible to obtain a seal across the rams. There will be no alternative but to cut the wire (using the shear seal BOP if fitted or the tree valve). The wire will then have to be fished.

9.6.4 Leaking pressure control equipment: above the blow out preventer

Leaks from lubricator connections and manifolds sometimes develop whilst wire is in the well. In most cases leaks are caused by a quick union O-ring failure. Some measures can be implemented to help reduce the chance of this type of leak:

- Regularly replace the O-ring seal in the pressure control equipment quick union connections—particularly at any connections that are frequently broken and made up.
- Ensure that the O-rings used are of the correct size and material.
- Visually inspect the O-ring for damage before rigging up.
- Properly support the lubricator using the hoist or guy lines, so as to limit the amount of bending and flexing.
- Ensure that all pressure control equipment is properly inspected, tested, and certificated before it arrives at the wellsite.

If a leak from the lubricator does develop, it should be managed as follows. Some wellsite supervisory staff will allow the wire to be recovered from the well if the leak is small, shows no immediate sign of worsening, and the wire is not too deep into the well. This is always a contentious issue, and many operating companies will insist on repairing a leak, no matter how small. The acceptability of living with the leak will depend on several factors:

- Leak volume.
- The nature of the leaking well-bore products, gas, liquids, and especially the presence of H_2S.
- Potential consequences—a leak on a manned offshore platform will be viewed more seriously than, for example, a remote land location far away from people or infrastructure.
- Time needed to recover wire to the surface and close in the well.
- Crew experience.
- Company policy.

The immediate actions necessary to stop a leak are:

- Stop wire movement and apply the winch brake.

- Close in the BOPs. The leak should diminish and stop once the BOP is closed.
- Bleed off any residual pressure above the BOP. It will not be possible to carry out a PBU test, since the equipment above the BOP is leaking. Observer to confirm the BOP is holding pressure.
- Break the lubricator at the leaking connection, and lift to expose the leaking O-ring.
- Remove and discard the O-ring.
- Select a new O-ring, and using a sharp blade (box cutter/Stanley knife) cut the O-ring. A 45° cut works best, as it provides a larger surface area for the glue but avoids too big a taper on the cut.
- Place the O-ring around the exposed wire. Use a cyanoacrylate adhesive (super glue) to bond the ends of the O-ring. Once the adhesive is set, slot the O-ring into the groove on the quick union.
- Re-stab the lubricator and make up the quick union.
- Equalize across the BOP and observe for leaks from the repaired connection.
- Recover the toolstring to the surface and close in the well.
- Replace the temporarily repaired (cut) O-ring.

If the cut and glued O-ring fails to stop the leak, the wire will have to be cut and an new (uncut) O-ring used:

- Stop winch movement and apply the brake.
- Close the wireline valve and bleed off above the rams.
- Once satisfied that the wireline valve is holding pressure, back off the quick union (normally the connection immediately above the BOP) and lift the lubricator to expose the wire.
- Firmly clamp the wire at the top of the BOP.
- Slack off on the wire and confirm the clamp is holding.
- At the unit, pull a generous amount of slack wire off the drum.
- Cut the wire close to the wireline unit.
- Pull the cut wire down out of the lubricator.
- Lay down the lubricator. Remove the grease head (or stuffing box). Replace the O-ring in the leaking connection.
- Reverse feed the tail of the clamped wire back through the lubricator.
- Reverse feed the wire through the stuffing box or grease head and connect to the lubricator.
- Tie the two wire ends together (the wire in the well to the wire on the winch).

- Pick up the lubricator taking care not to kink the wire between the BOP and the bottom of the lubricator.
- With the lubricator hanging a short distance above the BOP, pick up slack until there is enough tension on the wire to enable the clamp to be removed.
- Lower and stab the lubricator whilst simultaneously picking up the wire. Make up the quick union connection.
- Equalize across the wireline valve and check for leaks. Open the BOP rams and pull out of the hole.

9.6.5 Leaking Pressure Control Equipment: Leak Below the Blow Out Preventer

If a leak develops below the BOP, the only way of stopping the leak is cutting the wire, shutting the tree valves and depressurizing the pressure control equipment.

If a shear seal BOP is part of the rig-up, this will be used to cut the wire. Wire cutting tree valves can also be used. Cutting wire with standard tree valves risks damaging the valve and may necessitate the replacement of the gate and seat.

Thankfully, leaks below the BOP are rare, and the risk can be reduced by using flanged connections below the BOP.

9.6.6 Pressure Control During Wireline Fishing Operations

If a toolstring is stuck, or if wire parts and falls back in to the well, it must be fished. Fishing operations on live wells need to be conducted in a systematic manner if pressure integrity is to be maintained. Summarized here are the most commonly used fishing techniques and the main areas of concern in fishing operations.

By following a few simple guidelines, the potential for parted wire or cable can be reduced:

- Keep records of wire use.
- Limit time spent jarring (down as well as up for braided cables).
- Limit maximum line tension—most wire manufacturers recommend 75% of minimum breaking load.
- Ensure weight indicator reading is accurate (calibrate load cell—account for fleet angle variation at the bottom sheave).
- Ensure correct pulley diameter is used.
- Slip and cut wire frequently.

- Ensure wire is compatible with well fluids (sour service requirements).
- Spool wire with adequate tension.
- Use of friction reducers.
- Operator skill.

Even if all the suggested precautions are adhered to, tools can still become stuck, and wire can part.

9.6.7 Dealing with a stuck toolstring

A toolstring can become stuck for a variety of reasons, some beyond the control of the wireline crew:

- Debris (wax, scale, asphaltene, or junk).
- Sand fill.
- Tubing collapse or partial collapse.
- Gun swell (perforating guns).
- Torn or swollen elastomeric packing.
- Operator error.
- Tool failure.
- Differential pressure (blocked equalizing ports).
- Blow up (or down) hole.

If the toolstring is stuck, the wire can be worked to try to free the string. However, if wire is worked for too long it will fatigue and part. Wire will generally part at the surface, and can drop back in to the well. When this happens, hydrocarbons can escape through the stuffing box or grease head if the BOP gland or ball check fail to seal. Immediate closure of the tree valves and bleeding the lubricator is the fastest method of regaining control. Fishing broken wire can be difficult; if the toolstring does not free up relatively soon after becoming stuck it is far better to make a controlled cut or release from the rope socket/cable head before the wire parts.

When non-conductive wireline (slickline and braided cable) is in use, the decision to drop a cutter will depend on whether or not it is still possible to operate the jars. If there is no jar action, there is little to be gained by working the wire. However, if jar action is present, the length of time that jarring should be allowed to continue is very much down to operator skill and experience. Factors that can influence the decision will include:

- The age and condition of the wire.
- Type of wire (carbon steel is more fatigue resistant than stainless steel alloys).

- Operational location. Are daylight only operations enforced?
- Is it possible to move the fatigue points on the wire—moving the top sheave, hay pulley, and the unit (or taking the wire out of the counter head wheel).

If jar action is lost, or continued jarring risks breaking the wire, the wire will be deliberately cut in a controlled manner. The aim is to cut the wire at the rope socket or cable head, enabling the wire to be recovered from the well.

9.6.8 Pulling the weak point (e-line operations)

If logging tools are stuck and cannot be freed, pulling the "weak point" releases the wire from the cable head and allows it to be recovered to the surface:

- Pick up at surface to the tension calculated to pull the wire out of the cable-head.
- As the cable head releases, tension should drop to cable weight + drag. The weight of the toolstring should be gone.
- Start to pull out of the hole. As the end of the cable approaches the surface, it will begin to blow out of the grease head.
  - Keep people away from the wire as it starts to blow out.
  - The point at which the cable begins to blow out can be calculated from the cable diameter, well pressure, and cable weight. Cable weight is normally given in lbs/1000 ft.
- Be ready to shut in at the tree as soon as the cable is ejected from the grease head.
- Do not rely on the secondary (ball and seat) seal in the grease head stopping the escape of hydrocarbons—it might not work.
- Where gas alarms are likely to be affected by the unplanned and unwanted release of gas, warn the facilities operators. Gas alarms may have to be isolated until the pressure control equipment has been depressurized. Failure to do so could cause an unplanned shut-down of the facilities.
- In deep wells or wells with high friction (tortuous trajectory), it may not be possible to transmit enough tension from the surface to the cable head. Wireline drag modeling using software such as Cerberus can be used to determine if the weak point can be used. An alternative

is to use an "addressable cable head". This can be detached from the toolstring by electrical signal from the surface.

- If pulling out of the hole after activating the addressable cable head, the wire will begin to blow out of the well as the cable head nears surface. The cable head should stop the wire exiting the grease head. However, if wellhead pressure is high, the cable velocity might be enough to pull the cable-head from the cable. A tool-catcher or tool trap reduces the risk of the cable head falling back down the well.

9.6.9 Dropping a cutter bar: well control considerations

Cutter bars are designed to cut the wire at, or close to, the rope socket. There are several different types of cutter bar available, and the model chosen will usually depend on what is available at the well site. Irrespective of type, a cutter bar will only work if it impacts on a solid surface—usually the rope socket. If the toolstring is buried below debris, or if the toolstring has been blown up the hole and is tangled up in loose wire, the cutter bar is unlikely to work. In that case the procedure is to first drop a "go-devil", a flat-bottomed bar that gives a solid surface for the cutter bar to work against. Whilst the intricacies of fishing operations are beyond the scope of this book, there are well control and pressure integrity concerns when installing a cutter bar or go-devil into the wireline lubricator during a live well intervention. The procedure for installing a cutter bar or "go-devil" is as follows:

Step 1:

- Close the wireline BOP rams on the wire.
- For dual ram BOPs (e-line and braided cable) grease must be packed between the rams at above WHCIP.
- Bleed off well pressure above the BOP using the lubricator needle valve.
- Close the needle valve and observe for PBU.
- If no build up is observed, it is safe to lift the BOP (Fig. 9.50).

Step 2:

- Back off the lubricator quick union nut and carefully lift the lubricator.
- The lubricator needs to be high enough to allow the cutter bar (or go-devil) to be installed.

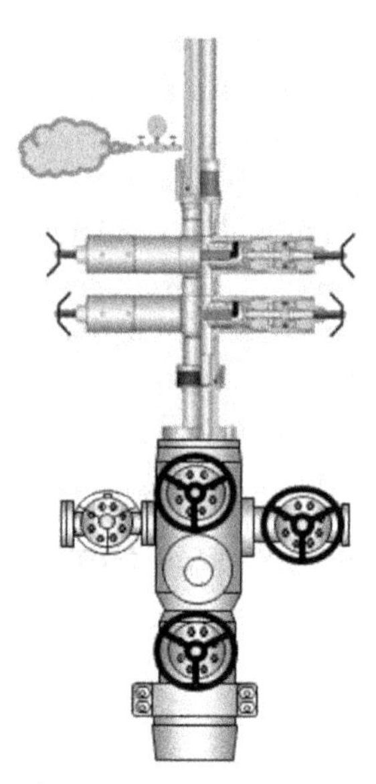

Figure 9.50 Dropping a cutter bar. Step 1.

- Make the cutter bar (or go devil) up on to the wire—the method of attachment depends on the type of cutter bar used. Ensure the cutter bar slides easily up and down the wire.
- Rest the cutter bar on top of the BOP ram.
- Lower the lubricator and make up the quick union (Fig. 9.51).
  Step 3:
- Equalize across the BOP rams until full WHCIP registers on the lubricator pressure gauge.
- Check for lubricator leaks before opening the BOP ram.
- Open the BOP rams, allowing the cutter bar to drop in to the well (Fig. 9.52).
  Step 4:
- Adjust the tension on the wire so the cutter bar drops easily. Too much tension will pin the cutter bar against the tubing wall, or cause a premature cut at upsets in the tubing (cross-over, side pocket, etc.). Too little tension (slack wire) can cause premature cut or impede cutter bar progress as it drops down the well.
- Allow time for the cutter bar to reach the fish. More time is needed if the well is fluid filled and where there is deviation.
- Once the bar has had time to reach the fish, work the wire. Sometimes the bar will only make a partial cut. Tension will sometimes be enough to finish the cut.

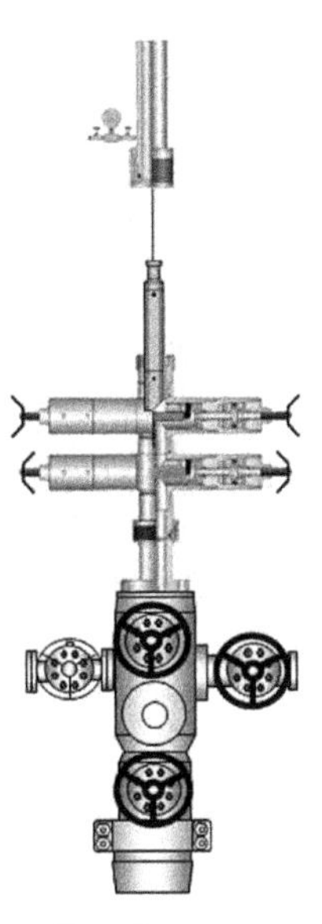

Figure 9.51 Dropping a cutter bar. Step 2.

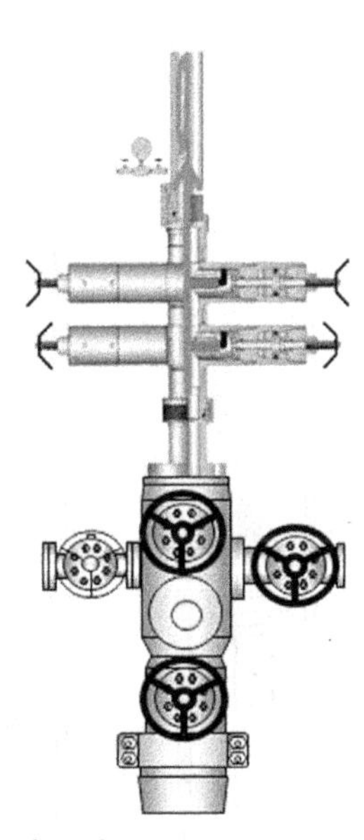

Figure 9.52 Dropping a cutter bar. Step 3.

- If no cut is made, consider dropping a second bar (or go devil).
- Once the cut is made, the wire can be spooled from the well.
- As the wire leaves the stuffing box or grease head, be ready to close in the well (Fig. 9.53).

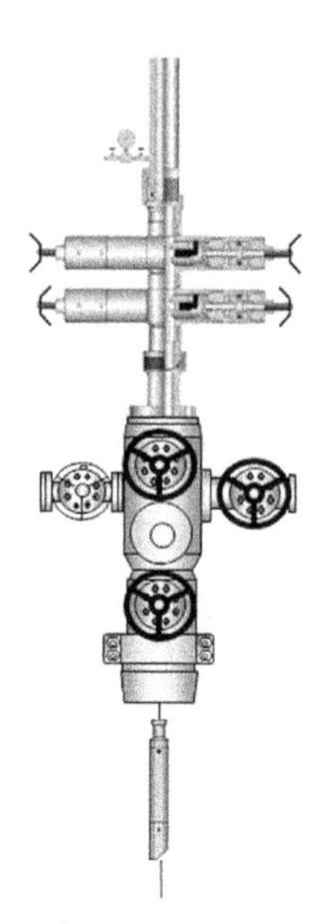

Figure 9.53 Dropping a cutter bar. Step 4.

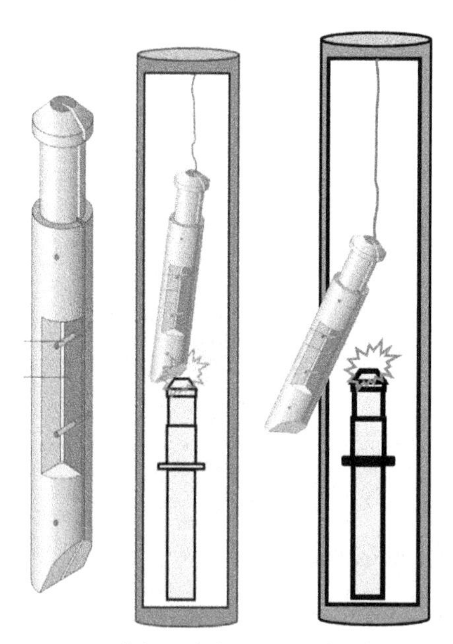

Figure 9.54 The consequence of thoughtless cutter bar use.

There is a type of cutter bar that, if not used correctly, can punch a hole in the production tubing. This is a simple cutter bar with a beveled base. It cuts the wire by deflecting off the rope socket as it impacts. It is cheap, reliable, simple to use, and is still available in many locations. It should never be used in a dry gas well, particularly if the well is vertical, or where tubing corrosion/erosion is suspected. The cutter bar, traveling at speed, can punch a hole through the tubing after deflecting off the rope socket, which is not the desired outcome (Fig. 9.54).

9.6.10 Using a lubricator "Y" sub (slickline)

An alternative, favored by some wireline specialists, is to rig up with a "Y" sub. This allows a second toolstring to be run alongside the stuck wire. A blind box (flat-bottomed steel bar) is run down to the stuck toolstring and the wire at the rope socket is "boxed off"—jarring on the wire until it parts (Fig. 9.55).

The advantage of using a "Y" spool lies in the ability to jar at the rope socket to make the cut. This improves the chance of making a clean cut at the rope socket. However, there are risks associated with running through the stuck wire, particularly when pulling back after the cut is made. In addition, by cutting the stuck wire at the surface, a commitment is made to re-spool the unit with new wire. Dropping a cutter bar can free the wire without having to cut at the surface.

9.6.11 Wire parts at the surface

Normally, when wire breaks at the surface, the tail of the wire will fall back into the well. However, occasionally the tail of the wire will remain at the surface. If this is the case it might be possible to recover the wire in the hole. Before trying to recover the wire, consider the following:

- It the wire was moving freely before it parted, and the toolstring was free, it should be possible to splice the ends of the wire together and pull it out of the well.
- If the wire was stuck, it is probably best to try to cut the wire at the cable head (or rope socket) using a cutter bar.

If the wire breaks at the surface and falls back into the well, a release of hydrocarbon from the stuffing box or grease head will occur if the BOP gland or ball check fail to seat. The well must be secured by closing the tree valves and depressurizing the pressure control equipment.

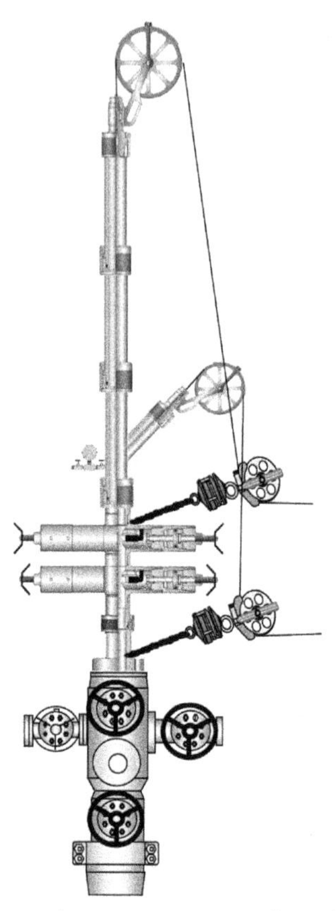

Figure 9.55 Using a "Y" spool to run a second toolstring and "box off" the stuck wire.

If working on a well with actuated tree valves, and if the well is being controlled from a remote well intervention panel, then closing the master valve by dumping the hydraulic pressure is normally the quickest and safest method. Once the master valve has closed, the SV can be manually closed. If the valve is equipped with manual valves, close the master valve and the SV.

Once the well has been made safe, recovery of the lost wire and toolstring can be planned. The complexities of fishing broken wire from a well are beyond the scope of this book, however, an overview of the procedure enables the critical well control and well integrity concerns to be summarized. If the toolstring was stuck at the time the wire parted, it will be necessary to first cut the wire from the cable head/rope socket. This is normally done by dropping a "blind box", allowing it to fall through the broken wire and impact on the rope socket. Under some circumstances the blind box can be run on wire, although running wire through broken wire or cable is not without risks.

Once the wire has been separated from the rope socket, the next operation would be fish the loose wire from the hole. The normal method is:

- Before starting to fish the wire, reconfigure the rig up.
  - Consider rigging up additional lubricator. This will probably be necessary for fishing the toolstring, but is also beneficial when fishing wire.
  - If not already in place, rig up a shear seal BOP immediately above the Christmas tree.
- Estimate the depth to the top of the broken wire.
- Run a wire finder. Locate the top of the wire and ball it up enough to be able to "grab" the wire with a wireline grab.
- Run a wireline grab and catch the ball of wire left behind by the wire finder.

  Note: Some operators will opt to use a baited wireline grab or a combination wireline finder/grab to ball and grab the wire in one run.
- Pull the grab and the fished wire back to the surface. The aim is to have a single strand of wire across the BOP once the grab is back inside the lubricator.
- Close the wireline BOP and bleed off above the rams—observe for PBU. If there are problems closing in or bleeding down, it will probably be because there are multiple strands of wire across the BOP rams.

  Note: If braided cable or logging cable are being fished, it will not be possible to close the wireline BOP if there is more than a single strand of cable across the rams.

  If slickline is being fished, ram closure might be possible where more than one strand is present. However, when the lubricator is lifted, there will be no way of knowing which strand is the "long strand".
- Rigging up additional lubricator will improve the chances of pulling back into the lubricator with only a single strand across the BOP; the ball of wire latched by the grab will be high enough above the BOP rams to ensure proper closure.

Multiple wire strands across the BOP will prevent it from sealing. There is no option other than cutting the wire. A shear ram or shear seal BOP located immediately above the Christmas tree is the best solution to this problem. If no shear ram has been rigged up, it will be necessary to cut the wire using the tree valves. Although some Christmas trees are equipped with wire cutting valves (SV and or master valve), many are not. Non wire cutting valves can be used to cut wire, but this risks damaging the valve gate and seat, and damage is more likely if multiple strands have to be cut. A valve repair (gate and seat replacement) will be necessary.

Having cut the wire, the tree valves can be closed-in and the lubricator depressurized, disconnected, and raised. Any wire in the lubricator can be removed before a further attempt to recover the wire is made. Subsequent runs are likely to be complicated, because of loose strands of cut wire (Fig. 9.56).

Once a single strand has been pulled into the lubricator and the BOP has been successfully closed, the next step is recovery of the broken wire:

- Having confirmed the integrity of the closed BOP, carefully lift the lubricator.
- Confirm that only one strand of wire is across the BOP.
- Place a fishing clamp on the wire, resting the clamp on top of the BOP.

Figure 9.56 What you hope not to see. Multiple wire strands compromise the blow out preventer and complicate the fishing operation.

- Slack off on the wire to lower the wireline grab out of the bottom of the lubricator.
- Cut the wire as close as possible to the grab, being careful to remove any kinked or damaged wire.
- Rig down the grab toolstring and lay down the lubricator.
- The wire tail above the clamp must be long enough to be able to be back-fed up through the lubricator and stuffing box. If there is not enough wire, it may be possible to strip some through the BOP. Great care needs to be taken if stripping is attempted:
  - Stop immediately if a leak develops. If the leak cannot be contained the well will have to be closed-in (at the tree valves or shear seal BOP if fitted).
  - If braided cable (or e-line) is being stripped, care will need to be taken to maintain the grease seal between the rams. Keep line speed to a minimum and ensure grease pressure is maintained. Using a tugger, crane, or block and tackle to strip is unadvisable. There is no way of monitoring tension. If wire must be stripped through the BOP, it is best to run it through the weight indicator and back to the unit. This enables line tension to be monitored whilst stripping.
  - Strip wire slowly and monitor the BOP for leaks.
- Stripping should not be attempted if there is a harmful concentration of H_2S, or where the closed-in wellhead pressure is high, particularly if a gas cap is present. Many wireline specialists will not allow wire to be stripped through BOP's under any circumstances. If, after closing the BOP and lifting the lubricator, the tail of wire above the rams is too short to enable it to be back-threaded through the stuffing box or grease head, proceed as follows.
- If conditions allow, rig up as much lubricator as possible.
- Reduce the length of the toolstring in the lubricator—typically to a rope-socket, and a short section of stem.
- Connect an inverted rope socket to the tail of wire above the BOP.
- Using a pin-pin cross-over, connect the inverted rope socket to the short toolstring.
- Pick up on the wire until tension is observed—remove the fishing clamp, lower and make up the lubricator.
- Equalize across the BOP. Confirm lubricator pressure integrity.
- Open the BOP rams and pick up until the short toolstring tags the stuffing box.

- Close the BOP rams and bleed down—again confirming integrity by observing PBU.
- Carefully lift the lubricator, whilst simultaneously easing off on the wire (remember the rope socket is hard up against the stuffing box).
- Place the fishing clamp on the wire above the BOP. Slack off and pull the tail of wire (and the short toolstring) down out of the lubricator.
- Remove some (or all) of the lubricator sections.
- The tail of wire above the fishing clamp should now be long enough to back thread the grease head or stuffing box. Depending on the length, the grease head/stuffing box may have to be made up directly onto the BOP.
- Tie the end of the wire back to the unit and pick up tension.
- Remove the fishing clamp and make up the stuffing box/grease head (and any lubricator used).
- Equalize across and open the BOP.

If the toolstring was free before the wire parted, and is being pulled out with the wire, make sure there is enough lubricator at the surface to accommodate the toolstring length. It may be necessary to cut the wire again and re-install the previously discarded lubricator.

If the wire was cut at the rope socket before fishing of the wire began, then the normal well integrity concerns associated with the end of the wire reaching surface apply, i.e., a release of hydrocarbon as the wire is ejected from the well.

9.6.12 Failure of wireline winch power pack or mechanical failure of the winch

Contingency procedures must be in place to cover the crew response should the power pack or the wireline winch have mechanical problems. For example, most wireline crews would recommend clamping the wire whilst repairs are carried out. The pressure control equipment should never be left unattended, and should always be monitored.

REFERENCES

1. American Petroleum Institute. *Specification for wire rope. API specification 9A*, 26th ed; May 2011.
2. International Standards Organisation. *ISO 2232:1990. Round drawn wire for general purpose non-alloy steel wire ropes and for large diameter steel wire ropes — specifications.*

3. Institute of Petroleum. *Model code of safe practices Part 1. The selection, installation, inspection, and maintenance of electrical and non-electrical apparatus in hazardous areas*, 8th ed.
4. API RP 53. *Recommended practices for Blowout Prevention Equipment Systems for Drilling Wells*.
5. NORSOK standard. D-SR-008. System requirements. Wireline equipment. Rev 1. October 1996.
6. http://www.iadc.org/wpcontent/uploads/Well_Control_Incident_Lateral_Learing_Document.pdf.
7. ENFORM. *Slickline operations. An Industry recommended practice for the canadian oil and gas industry*. Volume 13; 2007. Paragraph 13.4.4.

CHAPTER TEN

Coiled Tubing Well Control

10.1 INTRODUCTION

Coiled tubing has been used to service wells since the early 1960s. Thanks to continuous improvements in both technology and reliability, coiled tubing is now a common intervention technique. Coiled tubing is manufactured as a continuous length of pipe, with diameters ranging from 1″ up to 4½″. The pipe is spooled onto a large drum that enables it to be transported to and from the wellsite. At the wellsite the coiled tubing is spooled from the transport drum to an injector head. It is the injector head that grips and controls the pipe, enabling it to be run into, and pulled from, a well. Pressure control equipment located below the injector head enables coiled tubing to be used in live wells. It is the ability to operate in live wells, relatively quickly and easily, that makes coiled tubing an obvious choice for many interventions, especially where there is a requirement to pump fluids.

A modern coil tubing unit is capable of many well intervention applications. The most common are:

- Well unloading (getting the well to flow) using nitrogen lift.
- Wellbore cleanout (removal of sand, proppant, and other unwanted debris).
- Acid treatment (acid wash to remove carbonate scale; formation matrix treatment).
- Remedial cementing.
- Mechanical interventions in high angle wells (plug setting, opening, and closing SSD).
- Fishing operations.
- Underbalanced drilling.
- Velocity string installation.

Most coiled tubing interventions are performed on live wells using pressure control equipment. Anyone planning, supervising, or running a

Well Control for Completions and Interventions.
DOI: https://doi.org/10.1016/B978-0-08-100196-7.00010-5

coiled tubing intervention on a live well must understand how this equipment is assembled, how it works, and crucially, what actions need to be taken in the event of an emergency and what measures need to be in place and enforced to prevent a well control incident.

10.2 COILED TUBING EQUIPMENT

To carry out a coiled tubing intervention in a live well, the following equipment is needed:

- Control cabin and power-pack.
- Coiled tubing reel.
- Injector head.
- Pressure control equipment (pack-off or stripper, BOPs, and riser).

Additional equipment requirements will depend on the nature of the intervention. Most coiled tubing interventions involve pumping fluid through the coil, and in many cases fluid returns will have to be handled at surface (Fig. 10.1).

10.2.1 Coiled tubing: the pipe

Coiled tubing is manufactured from long strips of plate steel. At the coiled tubing mill, these strips are formed into a tube, and the edges are welded together with a continuous longitudinal seam. In most cases the internal seam is left in place, indeed, beading can be seen on the inside of the pipe. Nearly all coiled tubing is manufactured from low-alloy carbon steel with yields of between 55 and 120 Kpsi, and is commercially available in the size range 1″ to 4½″ OD. Most intervention work is carried out using coiled tubing at the smaller end of the range (1½″, 1¾″).

When in use, coiled tubing is under stress from internal and external pressure (burst and collapse), and multiple axial loads. In deep, highly tortuous wells, high axial tension loads are generated at the surface by a combination of weight and friction, with surface tension being highest when picking up from deep in the well. To better cope with high drag wells, many coiled tubing strings are internally tapered with heavy wall (thicker) pipe at the top of the string (on the bottom of the reel), and lighter thin wall tubing at the top of the reel—the bottom of the well. Tapered strings can work in deeper, more tortuous wells than a nontapered string of an

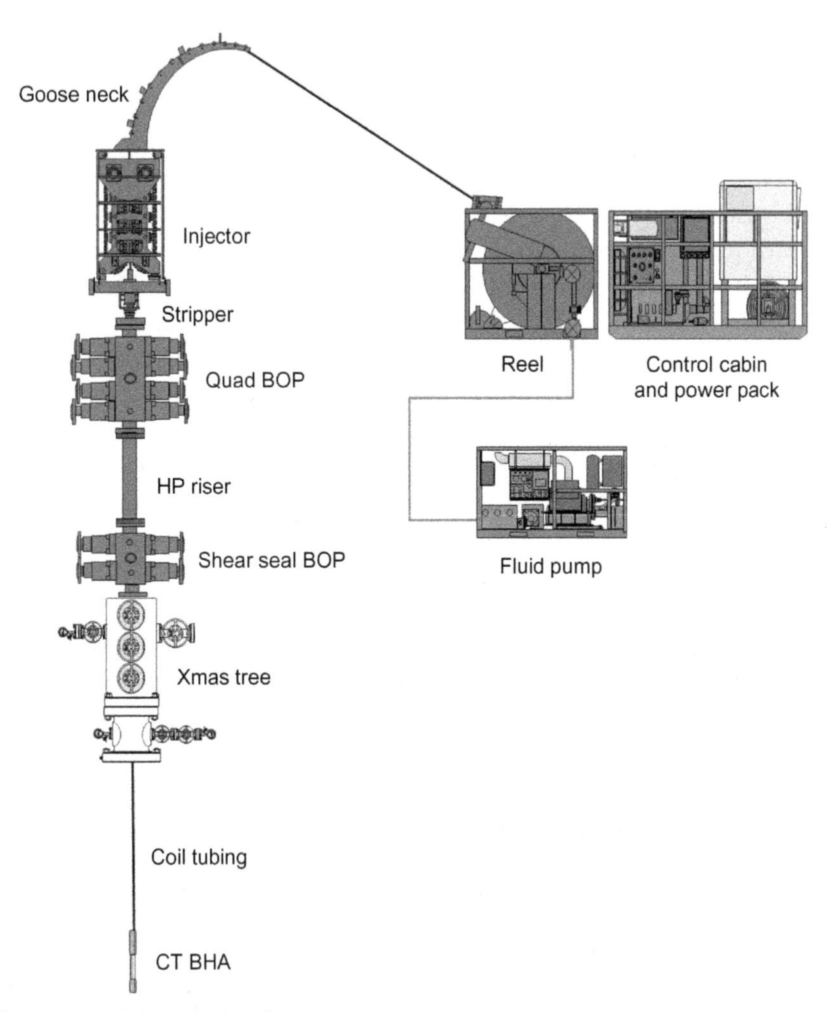

Figure 10.1 Main coiled tubing components.

equivalent length, since the heavier wall tubing, close to surface, has a greater cross-sectional area and therefore greater axial strength. Tapered strings have multiple sections, each of a different ID. String design may be job-specific, but is more commonly a general design that can be used in a range of wells.

10.2.1.1 Pipe performance

Metal has a degree of elasticity. The modulus of elasticity of low-alloy carbon steel, of the type usually used for the manufacture of coiled tubing, is approximately 30×10^6 psi. If a tension load (stress) is applied to a length of pipe, it will elongate. The fractional length change (strain) is proportional to the stress applied, and is dimensionless. If the stress applied to the pipe is less than the yield, then no permanent deformation results. However, if stress exceeds the yield of the material, permanent plastic deformation will occur. Usually, we go to great lengths to avoid bending jointed pipe beyond the elastic limits of the steel. Not so with coiled tubing. It is repeatedly plastically deformed during routine use. Deformation of the coiled tubing is caused by the bending and unbending of the steel as it is run into and pulled out of the well. Tripping the pipe into and out of the well involves a number of bending events and cycles.

- Running into the well:
  - The pipe straightens as it is pulled off the reel: Event 1.
  - The pipe bends over the guide arch (goose neck): Event 2.
  - The pipe straightens as it runs through the injector and into the well: Event 3.
- Pulling out of the well:
  - The pipe bends as it is pulled over the guide arch: Event 4.
  - The pipe straightens as it comes off the guide arch towards the reel: Event 5.
  - The pipe bends as it runs onto the reel: Event 6 (Fig. 10.2).

Each pair of bending events results in a bending cycle. For example, the straightening of the pipe as it comes off the reel, and the bending when it is pulled back onto the reel, is one cycle. During a trip into the well, some parts of the reel will be exposed to more cycles and events than others. For example, as the pipe is run into the well, it is good practice to stop and check pick-up weight/drag at regular intervals. Each pick-up on the pipe will add to the number of cycles and events.

During coiled tubing operations, the pipe is continually being loaded beyond the yield point, subsequently unloaded, then loaded in the opposite direction as it goes through the various cycles and events. Repeated bending and unbending produces yield stress in the coiled tubing. Internal reel pressure increases the effect.

An important part of reel integrity and therefore pressure containment when using coiled tubing is the ability to predict the deterioration in material yield caused by cycling and fatigue. Fatigue modeling for coiled tubing is firmly established.[1] Coil tubing vendors now routinely use software to monitor fatigue, enabling pipe to be withdrawn from use before failure occurs.

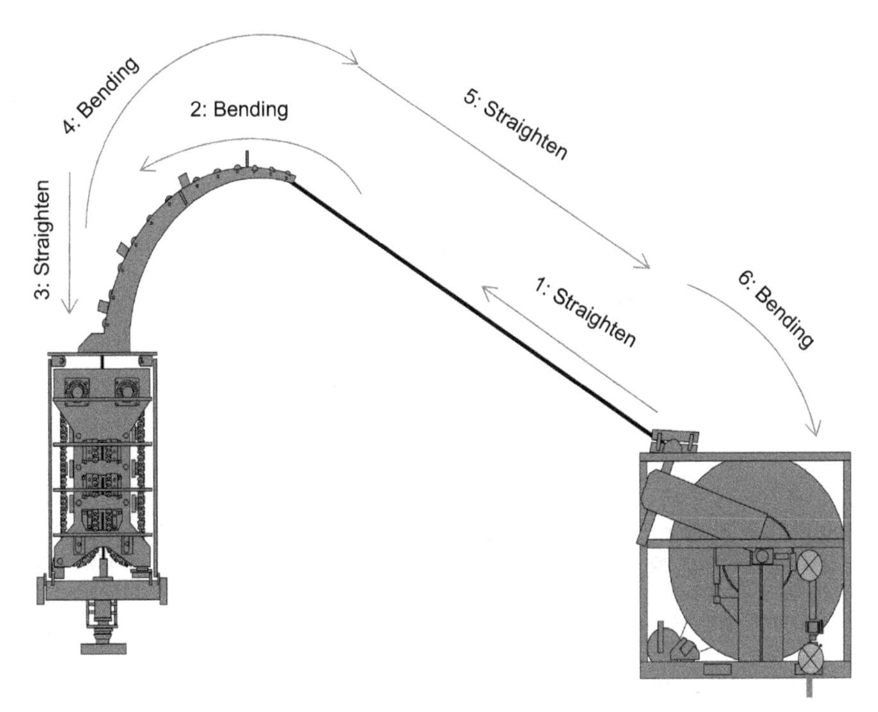

Figure 10.2 Bending cycles during coiled tubing use.

10.2.1.2 Corrosion resistance

Until quite recently, 2004,[2] many coiled tubing operations were limited by the corrosion resistance of the tubing, since low-alloy carbon steel is particularly susceptible to CO_2 corrosion. The introduction of 16% Cr coil alleviates some of these problems.[3] Erosion resistance (to abrasive slurry) has improved, as has resistance to corrosion in high CO_2 environments. 16% Cr coil is increasingly finding use as a velocity string where exposure to wellbore fluids is of a longer duration.

10.3 WELL CONTROL EQUIPMENT

Well control equipment enables interventions to be carried out with pressure at the surface. Equipment specification and configuration is determined primarily by well pressure, location, operational requirements, operating company policy, and government legislation. At minimum, pressure control equipment will consist of a set of coiled tubing BOP's and a stripper (pack-off).

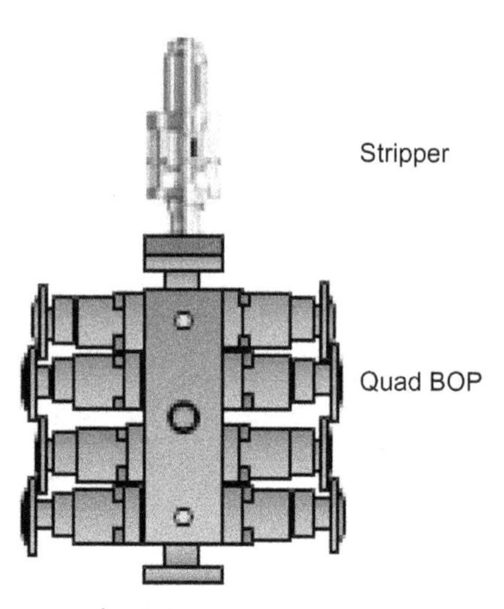

Figure 10.3 Pressure control minimum requirement. However, this configuration would not allow two barrier isolation if the stripper rubbers have to be changed when coiled tubing is in the well, and would therefore be in violation of many operating companies well control policy.

Whilst this may be deemed suitable for some low pressure or subhydrostatic operations, a more robust pressure control configuration would usually be required, particularly if operating a two barrier policy (Fig. 10.3).

Most operations would include—from bottom to top:

- Shear seal BOP (made up directly above the Christmas tree).
- High pressure riser (and pump in tee if required).
- Quad (or dual ram combi) BOP with a pump-in circulating port between the slip and shear rams.
- Stripper (pack-off) or more commonly dual (tandem) strippers.

The use of flanged connections is increasingly common, and is mandatory in some locations, particularly for connecting the shear seal BOP to the Christmas tree (Fig. 10.4).

10.3.1 The stripper or pack-off

A stripper (or pack-off) forms the primary well control barrier during coiled tubing interventions on live wells, and is positioned immediately below the injector head. The stripper contains two semicircular elastomer seals forming a dynamic seal when the pipe is in motion, and a static seal when the pipe is stationary. Pipe movement wears the elastomer seal, and as the seal wears, leaks are likely to occur. To stop stripper leaks, the

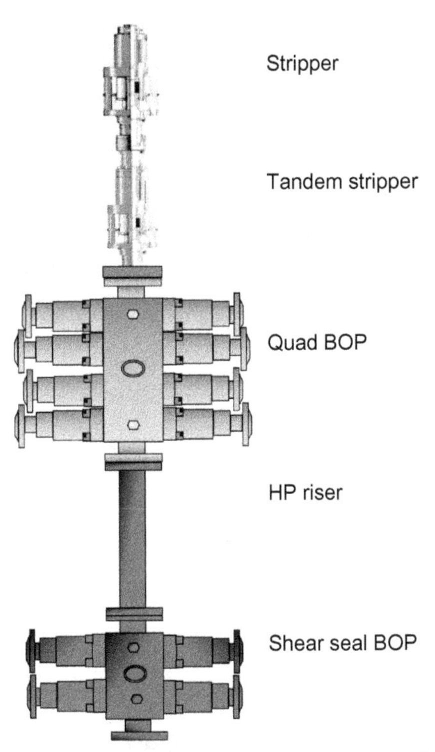

Figure 10.4 Coiled tubing pressure control equipment. This configuration would allow the upper stripper rubber to be replaced whilst maintaining two barrier isolation.

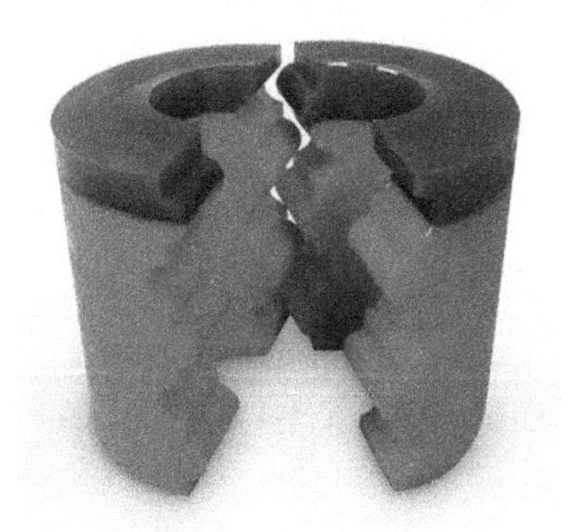

Figure 10.5 Stripper rubber packing elements. Source: *Photo courtesy of Supreme Manufacturing Company, Inc.*

assembly is equipped with a hydraulic piston. Pressure applied to the piston compresses the elastomer, thus enabling the seal to be maintained. Stripper packing elements should be resilient, with a low coefficient of friction, as well as good chemical and abrasion resistance (Fig. 10.5).

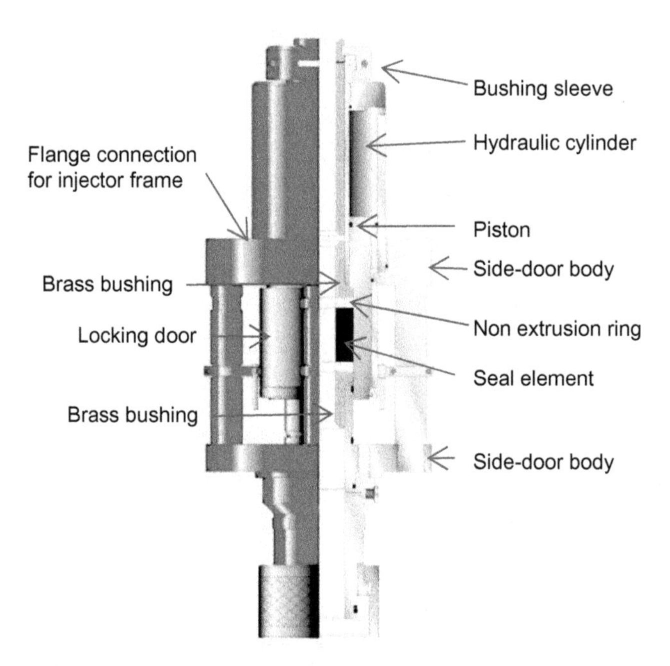

Figure 10.6 A side-door stripper.

The early design of an "in-line" stripper is rarely used today, as the design does not allow the seal element to be replaced when coiled tubing is in the well. In-line strippers have been replaced by "side-door" strippers (Fig. 10.6).

Side-door strippers allow the seal element to be replaced while the pipe is in the well. Wellhead pressure can be enough to energize the seal by forcing a piston against the bottom end of the elastomer, compressing it until it seals. The coiled tubing operator is able to apply an additional sealing force by applying hydraulic pressure to a piston. However, the application of too much pressure will result in increased frictional drag, and will alter weight indicator readings. In extreme cases, too high a stripper rubber pressure when running into the well can buckle the pipe between the bottom of the injector chains and the top of the stripper.

Many injector heads have the stripper permanently attached to a flange on the lower frame. Permanent fixing helps maintain correct alignment with the injector chains and simplifies rig-up procedures (Fig. 10.7). It is important that the gap between the injector chains and the stripper is minimized. This prevents buckling unsupported tubing where high force must be used when running in the hole (Fig. 10.8).

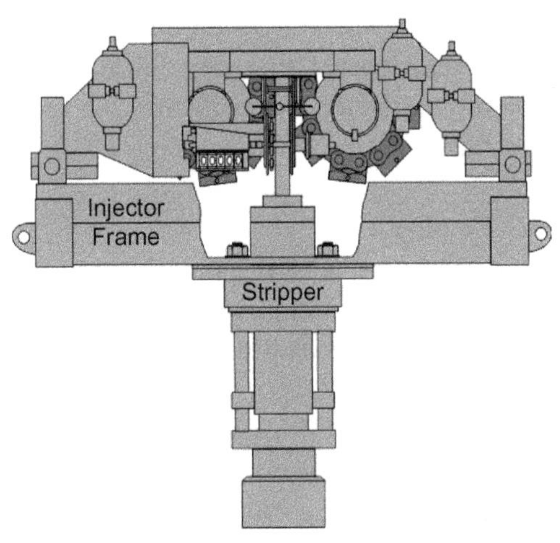

Figure 10.7 The upper stripper can be permanently attached to the injector frame.

Figure 10.8 The gap between the injector chain and the stripper. The pipe has been braced to reduce bending/buckling.

In most applications today, tandem strippers will be used, with well pressure being controlled by the upper stripper. If a leak develops, the lower stripper and the pipe rams will be closed (double barrier isolation) whilst the seal element is replaced.

10.3.2 Coiled tubing blow out preventers

Coiled tubing BOP's are designed to prevent the escape of wellbore pressure, and are configured specifically for use during coiled tubing operations. The BOP's will be selected based on the size of coil tubing in use, and the maximum wellhead closed in pressure (WHCIP). Pressure rating of the BOP should correspond to API 6A,[4] and to API 16A.[5]

The number and type of rams in a coiled tubing BOP varies, depending on operational requirements, with a four ram (quad) BOP being the most widely used. Quad (four ram) BOP's are configured (top to bottom) as follows:

- Blind ram. Seals the wellbore when the pipe has been removed from the BOP.
- Shear ram. Cuts the pipe.
- Slip ram. Supports the weight of the pipe hanging in the well below the BOP. Some slip rams are designed to prevent upwards movement of the pipe in a "pipe-light" situation.
- Pipe ram. Seals around the pipe.

The rams on coiled tubing BOP's are hydraulically operated from the console in the coiled tubing control cabin.

In addition to the rams, a quad BOP should have a circulating port positioned between the shear ram and slip ram. Finally, the BOP must have a method of equalizing across sealing rams (pipe and blind) when a differential pressure exists (Fig. 10.9).

The BOP's are available in a range of sizes that usually conform to API flange sizing. Table 10.1 is a guide to the pipe diameter that can be used with a range of BOP sizes.

10.3.2.1 Blind ram assembly

Blind rams are sealing rams that keep well pressure and fluids below the BOP when there is no coil tubing across the BOP. Blind rams consist of

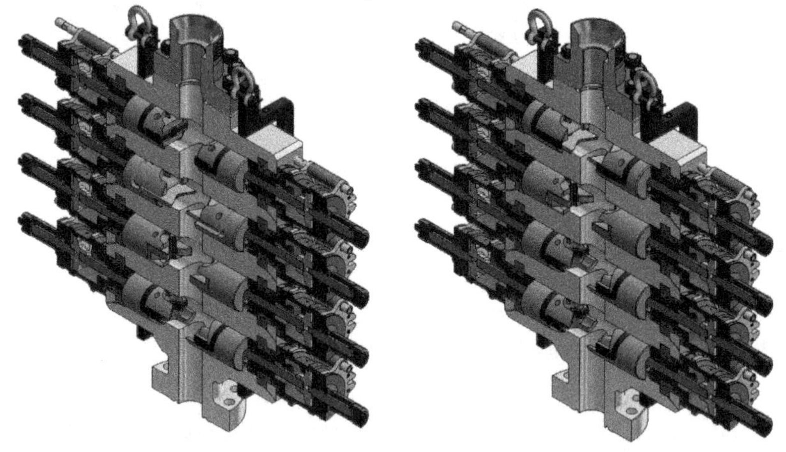

Figure 10.9 The left-hand drawing shows a quad BOP with the standard ram configuration (top to bottom) of blind – shear – slip – pipe. An alternative better configuration is shown in the right-hand drawing (top to bottom) and is shear/blind – slip – pipe – pipe. *Image courtesy of Lee Specialties.*

Table 10.1 Coiled tubing blow out preventer range

Blow out preventer size	Coiled tubing OD
2.56″	0.75″–2.00″
3.06″	
4.06″	1.00″–2.875″
5.12″	1.25″–3.5″
6.375″	
7.06″	

identical opposing ram bodies, each equipped with a front seal and a rear seal. When closed, the front seals of the two opposing rams are in contact, and the well is sealed. The rear seal prevents pressure from passing behind the rams.

10.3.2.2 Shear rams

Shear rams are equipped with blades that cut coiled tubing. These blades should be capable of cutting any coil run into the well, irrespective of wall thickness, yield, or material type. If the coil tubing is carrying internal wires or cables, these will need to be cut as well. Ideally, the sheared portion of the coiled tubing should be left with a "fishable profile that has a residual outside diameter equal to or less than the original outside diameter."[6] It is also important that, after cutting, a clear flow-path remains (Fig. 10.10), since there may be a necessity to kill the well through the cut coil.

10.3.2.3 Slip rams

When closed, slip rams grip the pipe, preventing it from dropping into the well. Most slip rams are bi-directional, and will prevent the pipe from being blown out of the well in a snubbing (pipe-light) situation. Slips should be capable of supporting a hanging load equivalent to the yield of the pipe. Where bi-directional slips are used, the slip is usually designed to hold 50% of the pipe yield in an upwards direction. Closing slip rams can mark the tubing, and those bite marks can reduce the fatigue life of the coiled tubing, in some cases by as much as 70%. Testing has established that an interrupted tooth design is best for reduced damage and improved fatigue life (Fig. 10.11).[6]

Figure 10.10 After the cut. This reel was run with internal e-line (smart coil).

Figure 10.11 Slip rams.

10.3.2.4 Pipe rams

Pipe rams close around the outside of the coiled tubing to create a seal. Each opposing ram has a semicircular seal element that is sized for the coiled tubing in use. When closed, the seal element deforms to seal around the coiled tubing. Once closed and inflow tested, the pipe ram forms a single well control barrier.

10.3.2.5 Coiled tubing blow out preventer operating sequence

The sequence for operating the rams dictates their configuration. In the event of a requirement to cut the pipe, e.g., an uncontrollable leak at the surface, the following steps would be taken.

- The slip ram and pipe ram are closed.
- Pressure above the pipe ram is bled off via the side outlet.
- Shear rams are closed, cutting the pipe.
- The cut pipe, above the shear ram, is picked up above the blind rams and the blind rams are closed (Fig. 10.12).

Using the ram configuration closing sequence as described above, the cut coil tubing is left suspended in the slip rams with the pipe rams sealed around the pipe. The open end of the cut pipe sits below the closed blind ram. Fluids can be circulated through the coiled tubing via the port located between the shear ram and the slip ram. Returns can be taken from a pump-in tee on the riser (below the BOP), or through the wing valve or kill wing on the Christmas tree.

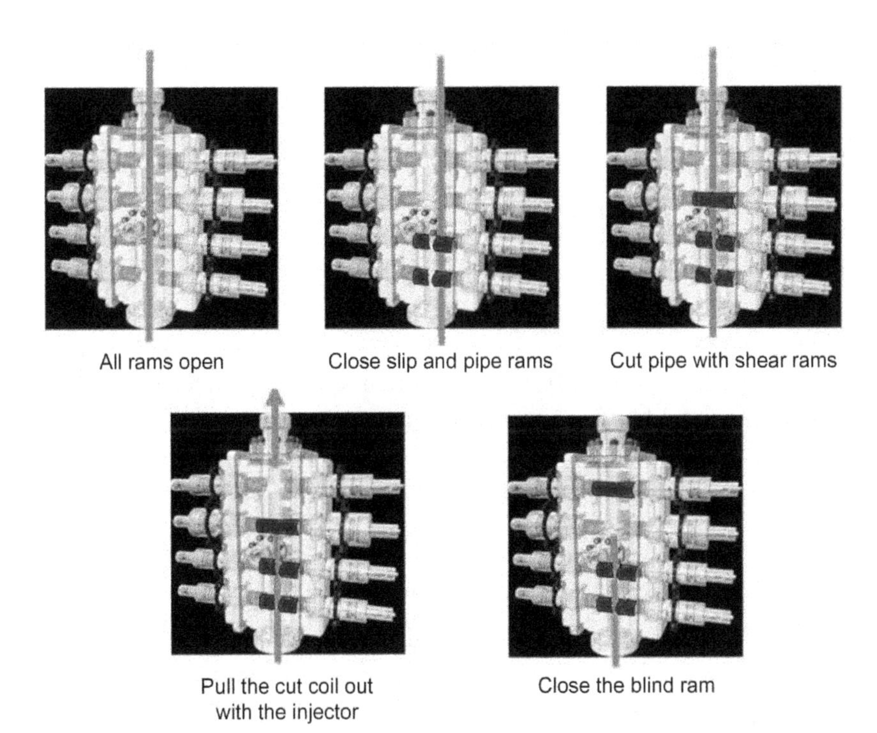

Figure 10.12 Ram closure sequence for sealing the well, cutting and removing the coiled tubing.

10.3.2.6 Changing the ram configuration in quad blow out preventers

It is becoming increasingly common for coil tubing vendors to switch the pipe and slip rams, resulting in the following configuration:

- Blind
- Shear
- Pipe
- Slip.

The reason for swapping the pipe and the slip ram is to aid recovery of the cut pipe from within the body of the BOP. Think of a situation where, for well control purposes, it has been necessary to cut the pipe using the shear rams. The coil above the closed shear rams is pulled clear of the BOP and the blind ram closed, securing the well. The cut pipe remains in the well, hung from the slip rams. Pipe rams seal the pipe/tubing annulus. After killing the well, the cut coiled tubing can be recovered. This would usually be accomplished by running an overshot, and grabbing the stump protruding above the closed slip ram. In BOP's where the slip ram is uppermost, the stump of exposed pipe above the slip ram is short, therefore grip of the overshot on the pipe can be tenuous. By switching the slip and pipe ram, the stump, protruding above the slip ram, is longer, giving a more secure grip.

Other vendors are completely re-configuring the quad BOP in what is arguably a much safer configuration that is (top to bottom):

- Shear/Blind
- Slip
- Pipe
- Pipe.

Having a shear seal BOP in the top ram is arguably safer than having separate shear and blind ram functions (as described in 10.3.3). The shear seal function frees up one of the four rams for an additional pipe ram. This enables double barrier isolation to be maintained when working on the stripper (Fig. 10.9).

10.3.3 Combination blow out preventers

Combination BOP's combine the blind ram and shear ram functions in the upper ram, and the slip and pipe ram functions into the lower ram. Combi BOP's have two notable advantages over the quad.

- They are more compact, thus reducing the height of the rig up.
- Many coiled tubing specialists regard the combi BOP as being safer than the quad in one important respect. A situation could arise where

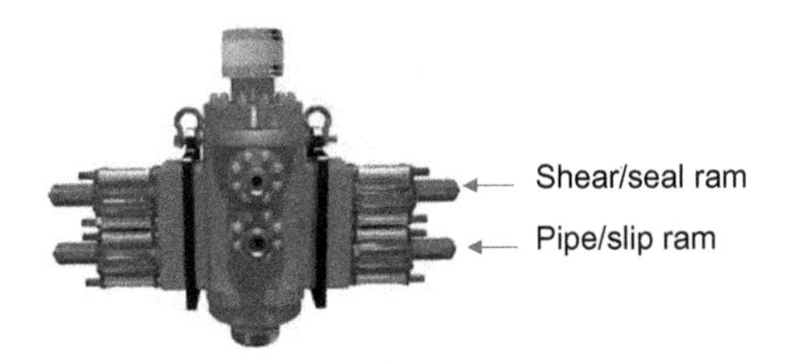

Figure 10.13 Combi BOP. Source: *Image courtesy of Lee Specialties.*

the shear rams have been functioned to cut the coiled tubing, but it has not been possible to raise the cut coil above the blind ram. This could be caused by a power pack failure, or by an inability to pull collapsed coiled tubing through the stripper. Irrespective of the cause, the cut coil would prevent the blind ram being closed, and therefore compromise the ability to make the well safe. A shear/seal ram eliminates this potential problem (Fig. 10.13).

10.3.4 Shear seal blow out preventer

A shear seal BOP, as the name suggests, uses a single ram that both shears the coiled tubing and the seal in the well. They are frequently used in addition to the standard quad or combi BOP, and are usually rigged up immediately above the Christmas tree.

In the event of an emergency, it may be necessary to shear and drop the coiled tubing before closing in the valves on the tree. It is vital that the cut coiled tubing drops below the Christmas tree, since coiled tubing remaining across the body of the Christmas tree would prevent the valves from being closed. This is a serious safety concern, but a condition that could easily arise if a long riser is in use and the quad (or combi) BOP is positioned high above the Christmas tree. By positioning a cutting ram immediately above the tree, the distance the cut coiled tubing needs to drop to clear the tree is minimized (Fig. 10.14).

10.3.5 Blow out preventer operations: opening and closing the rams

A BOP usually operates with approximately 1500 psi (10,350 kPa) hydraulic pressure. Two hoses must be connected to each cylinder for operation of a BOP. One hose is used to close the ram, the other to open it. Correct hook up of the hoses should confirmed by functioning all of the rams (Fig. 10.15).

Figure 10.14 Shear seal blow out preventer.

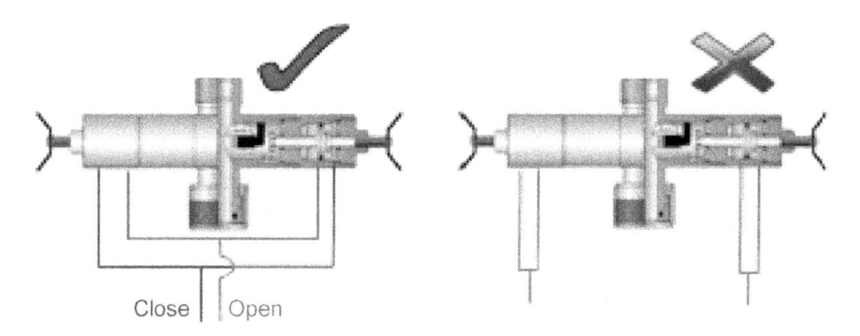

Figure 10.15 The right and wrong way to hook up the hydraulic hoses. The one on the right does happen!

10.3.5.1 Closing and locking and opening the rams

Note: It is not advisable to use the slip rams during normal operations. If they are used, the pipe should be inspected for signs of surface damage.

Rams are usually operated using control levers on the console in the control cabin, each ram having its own control leaver. After hydraulically closing a ram, the manual lock should be closed. Closing the manual lock will retain each ram in the closed position in the event of loss of hydraulic pressure.

If the hydraulic system fails prior to closing each ram, it may be possible to close the rams manually. Manual closure requires the hydraulic fluid on the wellbore side of the piston to be released, otherwise hydraulic lock occurs. This can be accomplished by placing the control lever in the closed position. If that does not work, the hydraulic fittings will need to be removed to allow the hydraulic fluid to drain.

The ability to close the valves manually will depend to some degree on the size of the BOP and well pressure.

Before the rams are opened, any differential pressure across the blind ram and pipe ram must be equalized. Failure to do so can cause damage to the ram seals, and there is a risk that they will no longer work.

Fully back out the manual lock (anticlockwise rotation). Apply hydraulic pressure to fully open the rams. It is not possible to manually open the rams.

The rams must be fully open so they do not interfere with the bottom hole assembly (BHA).

10.3.6 Coiled tubing barriers

When coiled tubing is used in a well where there is pressure at surface (a live well), the barriers are generally defined as follows.

10.3.6.1 Well pressure retaining barriers (barriers external to the coiled tubing)

Primary Barrier.

- Stripper rubbers, BOP body and riser.
- Christmas tree valves when deploying tools in and out of the riser.
- If a downhole lubricator valve is fitted (and used), this would be the primary barrier when deploying a BHA into or out of the well (lubricator valve description can be found in the chapter on completion equipment).
- Gate valves if a coiled tubing tool-sting deployment system is being used.

Secondary Barriers.

- BOP pipe rams.
- SCSSSV or Lubricator valve if the coil tubing BHA is above.

Tertiary Barrier.

- Shear and seal capability in BOP. For many operations, a shear/seal combi BOP will be mounted immediately above the tree.
- Some trees have been fitted with gate valves able to shear coil tubing.

10.3.6.2 Coiled tubing reel: internal pressure control

Primary Barrier.

- Check valves (nonreturn valves) in the BHA.

Secondary Barriers.

- Fluid pumped through the coil to prevent hydrocarbon ingress.

Tertiary Barrier.

- Shear and seal capability in BOP. For many operations, a shear/seal combi BOP will be mounted immediately above the tree.
- Some trees have been fitted with gate valves able to shear coil tubing.

Note: Some coil tubing operations use reverse circulation to clean out the wellbore. This means no check valves are included in the BHA. In these circumstances, primary internal provided by overbalance pressure prevents hydrocarbons from entering the coiled tubing.

10.4 THE INJECTOR HEAD

Coiled tubing injector heads come in a range of sizes, and it is the prime mover for the coiled tubing. Performance capability is typically categorized by pulling power, and is usually given in pounds of force. In addition to supporting tension loads, injector heads must be able to push (snub) the pipe into the well against well pressure. Snubbing creates axial compression. Most units have a snubbing capability of approximately half of the pulling limit. The injector will be selected based on the coiled tubing size (OD) and the yield of the strongest (heaviest wall) coiled tubing to be run into the well. Most operating companies and coiled tubing vendors will select an injector head that has a maximum load equal to or greater than 80% of the yield of the coiled tubing. The highest tension loads will occur when picking up off the bottom, where the pipe is at maximum depth. The highest compression (snubbing) load will occur as the pipe leaves surface, where the well is closed in (at maximum wellhead pressure) (Fig. 10.16).

Figure 10.16 Coiled tubing injector heads (with guide arch). Source: *Image courtesy of CT Logics.*

An injector should be designed to:

- Apply a dynamic axial force to the coiled tubing in such a way as to control movement of the coiled tubing in and out of the well.
- Maintain enough traction to prevent slippage of the coiled tubing.
- Maintain enough grip to prevent pipe movement when the unit is stopped.
- Enable mounting of depth measurement and weight measurement equipment.

A conventional injector head feeds the coiled tubing between two sets of contra rotating endless chains. Short "gripper block segments," mounted along the entire length of each chain, are designed to grip the pipe as it feeds between the chains. The gripper block is sized according to coiled tubing diameter. A variable speed hydraulic motor drives the chains, converting rotation into linear (up and down) movement. The coiled tubing, gripped between the chains, moves at the same speed as the chain. Most injector heads are equipped with a failsafe hydraulic break. Loss of hydraulic pressure stops the sprockets, locking the chain in a stationary position (Fig. 10.17).

Fig. 10.18 illustrates the chain configuration of a typical injector head. A series of pads called "skates" (2) apply force to the chain and thus create sufficient grip/traction. The skates are connected to hydraulic pistons. Pressure is used to adjust force (grip). Outside chain tension can be adjusted (3), since a properly tensioned chain is more efficient and will prevent damage to the coiled tubing.

Figure 10.17 Injector chain and gripper blocks.

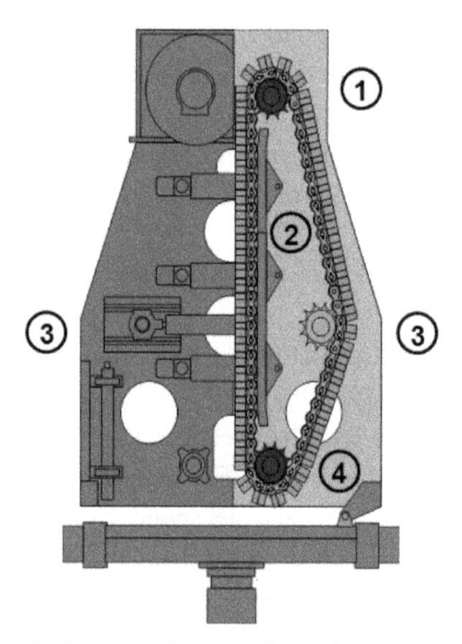

Figure 10.18 Chain tensioning and skate configuration.

10.4.1 Guide arch (gooseneck)

Most operations require the use of a guide arch, colloquially known as the gooseneck. This guides the pipe from the reel and into the injector head. Rollers are built in at intervals to reduce friction. The guide arch controls the radius of the bend, and therefore must be correctly sized for the pipe diameter in use. The guide arch must also be configured to accommodate the fleet angle between the reel and the injector as the pipe is spooled on and off. Finally, the guide arch must have sufficient strength to support loads induced by any reel back-pressure.

10.4.2 Weight indicator

Coiled tubing weight indicators are usually built into the injector frame. Various types (electronic and hydraulic) are used. A popular system uses an injector head frame hinged on one side, with a hydraulic load cell on the opposite side. The load cell sends a calibrated signal to the operator console with a reading of tension or compression in lbs or kg. Forces exerted by the action of the drive system and the tubing weight are applied along the centerline of the tubing, and cause the subframe to pivot; the resulting deflection measures direct force or load acting on the hydraulic load cell bladder. Individual hydraulic load cells will measure

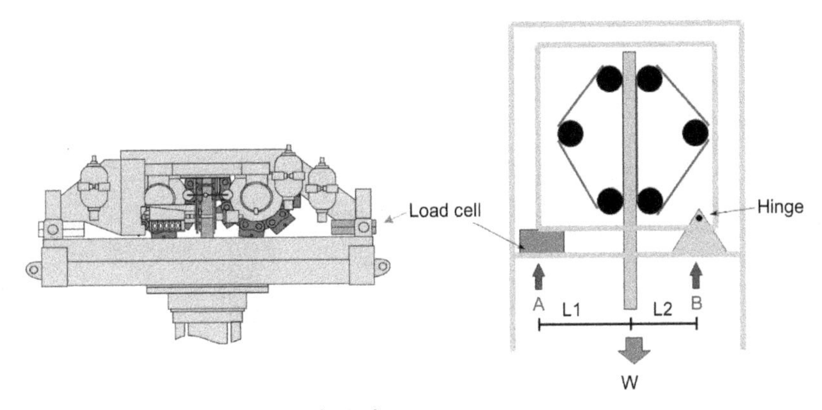

Figure 10.19 Coiled tubing weight indicator.

either pipe-heavy (positive value) or pipe-light (negative value). Proper calibration and maintenance of the load cell is essential. An incorrect weight reading could result in damage to the coiled tubing, and consequently a well control incident (Fig. 10.19).

10.4.3 Depth measurement equipment

Coil tubing depth is measured electronically, mechanically, or more commonly using both methods. Depth counters are usually mounted on the injector head—an electronic encoder assembly on the injector head drives chain and friction wheel counters between the chain and the stripper. Friction wheel counters are also be mounted by the level wind on the reel.

Note: Accurate depth determination with coil tubing can be difficult. Friction wheel slippage, encoder calibration errors, and pipe stretch can combine to give significant errors (Fig. 10.20).

10.4.4 The reel

The coiled tubing reel is primarily a storage device. It allows the pipe to be transported to and from the well location with relative ease. Offshore reels are a stand-alone skid. For land operations, the reel is usually permanently mounted on a truck or trailer chassis.

Most reels use a hydraulic motor-operated chain drive to rotate the drum. When pulling out of the hole, the reel motor uses slightly more torque than would be required to match the speed of the injector head. This ensures that the pipe between the reel and the injector is continually

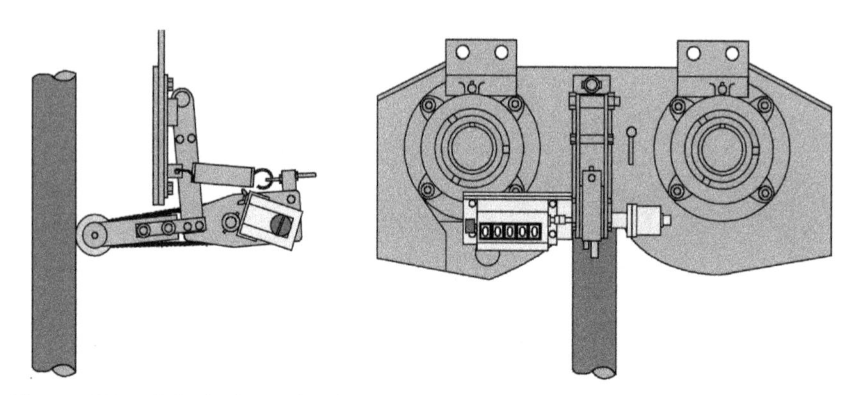

Figure 10.20 Coiled tubing depth measuring equipment.

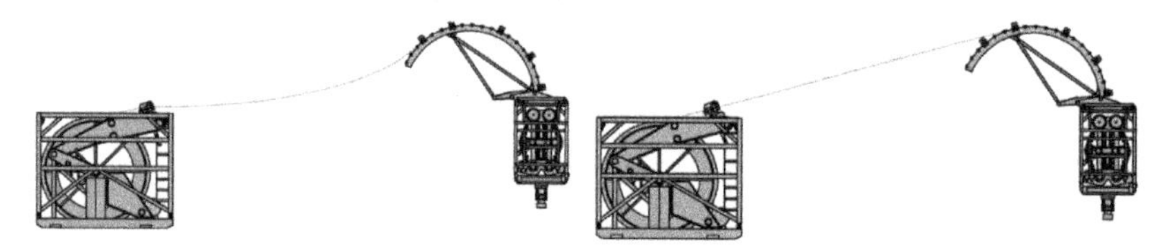

Figure 10.21 Tension must be maintained between the reel and the guide arch. Left: incorrect. Right: correct.

in tension. Maintaining tension ensures that no slack pipe is spooled onto the reel, and that there are no additional and unnecessary bending cycles.

When the pipe is run into the well, the hydraulic motor acts as a dynamic break—the injector is pulling the pipe off the drum against the reel motor hydraulic pressure. This keeps the pipe between the reel and the injector taut, again minimizing bending, and preventing the damage that would occur if the pipe "jumped" the guide arch (Fig. 10.21).

In addition to storing the coiled tubing, the reel is also equipped with fluid and (where necessary) electrical connections. Fluid is pumped into the center of the reel via a high pressure swivel—usually equipped with a 2″ 1502 hammer union and lo-torque valves. This enables fluid to be pumped whilst the coiled tubing is in motion. Most reels will also have a method for launching balls, wiper plugs, and darts. Some coiled tubing reels have an electrical cable running through the pipe, allowing the deployment of logging tools. A slip ring on the side of the reel allows the electrical current to pass from the moving pipe to surface readout equipment (Fig. 10.22).

Figure 10.22 Pipe connections at the center of the reel.

10.4.5 The power pack

The power pack supplies hydraulic power to the various systems that make up the coiled tubing unit. Most coiled tubing units use diesel engines to power hydraulic pumps. These in turn supply pressurized fluid that runs the hydraulic motors needed to operate the injector, reel motor pressure control equipment, and other ancillary functions. Different functions require differing pressure and flow rates. As a consequence, most power packs have multiple pumps. Typically, six hydraulic circuits are used:

- Main injector head motor circuit.
- Reel motor circuit power.
- Levelwind circuit.
- Blow out preventer circuit.
- Priority circuit running various control and ancillary features (e.g., skates and chain tensioning).
- Ancillary circuit for any additional hydraulic functions.

Most coiled tubing units have nitrogen-charged accumulators built into the power pack. These allow well control equipment (BOP and strippers) to be functioned in should the power pack fail. Accumulators are the standard bladder type, designed to retain the precharge, even under extreme conditions.

10.4.6 Control cabin

Ideally, the control cabin will be positioned to allow the operator to have a good view of the injector and pressure control equipment. If the

Christmas tree and wellhead can be seen from the control cabin, so much the better. Inside the cabin are the controls needed to operate the coiled tubing unit and associated well control equipment. These would include, but need not be limited to:

- Injector head. Direction control and hydraulic pressure.
- Traction cylinder pressure.
- Blow out preventer rams function and hydraulic supply pressure.
- Pack-off/stripper function and hydraulic supply pressure.
- Weight indicator.
- Wellhead pressure (coil tubing/completion tubing annulus).
- Pressure in the reel (at surface).
- Depth measurement.
- Power pack engine monitoring and control.

Most coiled tubing units have electronic data monitoring and recording. Recorded functions can include:

- Pump discharge pressure.
- Wellhead pressure (coiled tubing annulus).
- Depth.
- Reel speed.
- Weight.
- Stripper hydraulic pressure.
- Hydraulic power supply pressure (Fig. 10.23).

Figure 10.23 Control layout in the coiled tubing cabin.

10.4.7 Rigging up coiled tubing on floating rigs and platforms

Vessel motion, principally heave, is an obvious problem during interventions on floating rigs. During coiled tubing operations the intervention riser, attached to the Christmas tree on the sea floor, remains motionless whilst the rig heaves up and down due to wave motion. The relative motion between the moving rig and stationary riser would be a major problem without an additional item of equipment; the coiled tubing lift frame.

With the intervention riser in place, the lift frame is picked up using the rig traveling block. Bails and an elevator attached to the bottom of the lift frame are then latched to the top of the intervention riser. Since the rig traveling block is connected to a heave compensating system, the riser is supported and kept under near constant tension throughout the coiled tubing intervention (Fig. 10.24).

Attached to the top of the lift frame is the winch that is used to lift and support the injector head. As the winch is integral to the lift frame, and because the lift frame is connected to the riser, there is no heave induced movement on the injector when it is connected and disconnected from the BOP and riser. Trying to stab the injector in an uncompensated system would be extremely hazardous for the crew. Lift frames are also widely used for wireline interventions.[a]

10.4.8 Calculating stripping and snubbing forces

Well pressure at surface opposes movement of coiled tubing into a well. The opposing force can easily be calculated. Simply multiply the wellhead pressure by the cross-sectional area of the coiled tubing OD.

Example:

Wellhead closed-in pressure (WHCIP) = 5000 psi

Coiled tubing OD = 1.5″

Area of tubing OD = $\frac{\pi}{4} \times 1.5^2 = 1.767$ in.^{2}

Force = Pressure × Area. 5000 psi × 1.767 = 8836 lbs

In the context of coiled tubing operations, the injector head must "push" the coiled tubing into the well against 8836 lbs of upwards (negative) force. The action of pushing the pipe into the well against well pressure is known as "snubbing." The term "pipe-light" is also used (Fig. 10.25).

[a] Additional information about lift frames is in Chapter 14.

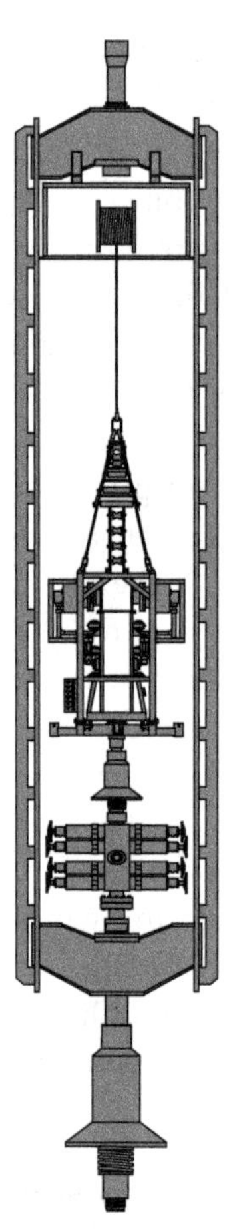

Figure 10.24 Coiled tubing lift frame.

In addition to a piston force, friction created by the stripper seal element must be accounted for. Stripper friction acts against the pipe as it is run into the well. When running in, friction increases axial compression at the stripper, and is therefore a negative (−) value.

Example:

Piston force: −8836 lbs

Stripper (pack-off) friction: −2000 lbs

Total force to run into the hole: −10,836 lbs

As more coiled tubing is run into the well, the axial load at the surface increases until the weight of pipe in the well is equal to the upwards force created by the well pressure acting against the pipe. The point at which

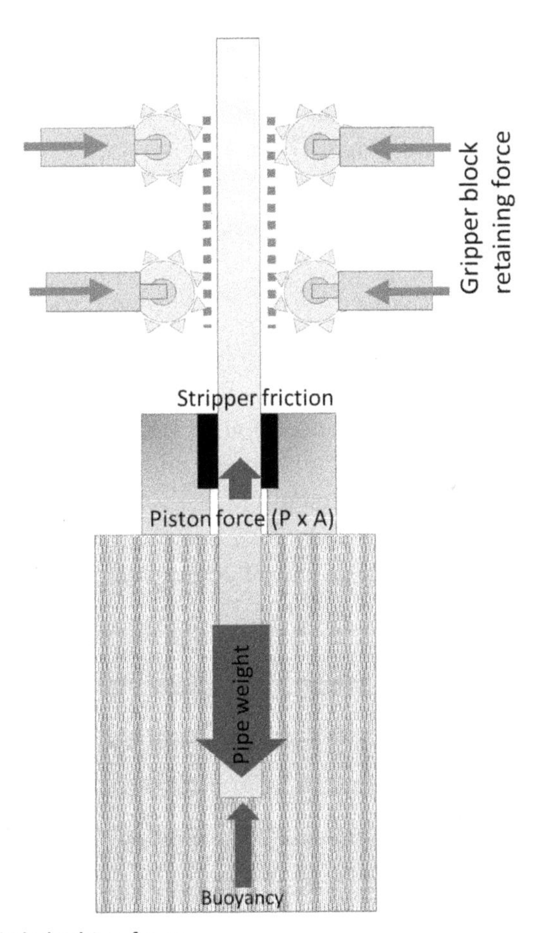

Figure 10.25 Coiled tubing forces.

the two forces are equal is called the balance point. Below the balance point, "snubbing" transitions to "stripping"; also known as "pipe-heavy."

To calculate the balance point, the point at which the tubing changes from being pipe-light to pipe-heavy, it is necessary to first calculate the buoyed weight of the coiled tubing. This can be calculated from the equation:

$$W = W_{\text{air}} + \left(\frac{\rho_{\text{i}}A_{\text{i}} - \rho_{\text{o}}A_{\text{o}}}{19.25}\right) \tag{10.1}$$

where:

W_{air} is the tubing weight in air (lbs/ft)

ρ_i is the density of the fluid in the coiled tubing in lb/gall
A_i is the area of the tubing ID (in.^{2})
ρ_o is the density of the fluid in the production tubing (outside the coil)
A_o is the area of the tubing OD (in.^{2})
To continue the example:

Pipe diameter:	1.5 in. OD
Pipe weight:	1.619 lbs/ft
Wall thickness	0.109 in.
Fluid in coiled tubing	9.5 ppg
Fluid in wellbore	9.5 ppg

When calculating the balance point, stripper friction should be ignored. This is because it is unknown, and can only be estimated. Moreover, it is continually changing.

The area of the OD has already been calculated as 1.767 in.^{2}

The (calculated) ID = OD−2 × wall thickness. 1.5 − (2 × 0.109) = 1.282

$$= \frac{\pi}{4} \times 1.282^2 = 1.290 \text{ in.}^2$$

$$W = W_{air} + \left(\frac{\rho_i A_i - \rho_o A_o}{19.25}\right)$$

$$\text{Becomes } W = 1.619 + \left(\frac{(9.5 \times 1.290) - (9.5 \times 1.767)}{19.25}\right) = 1.383$$

The buoyed weight of the coiled tubing in the well is therefore 1.383 lbs/ft, putting the balance point at:

8836/1.383 = 6389 ft below the stripper.

It should be noted that at this point stripper friction remains and still influences the weight at the surface. In theory, if the stripper friction value of 2000 lbs remains unchanged, then additional coiled tubing needs to be run to reach the condition where the injector is no longer "pushing" pipe into the well, but has transitioned to supporting the pipe; snubbing changes to stripping. This would now occur:

(8836 + 2000)/1.383 = 7835 ft below the stripper.

In the example used, the fluid in the coiled tubing and the coiled tubing annulus (production tubing) is the same. In reality they are likely to be different. This will change the buoyancy, and thus the depth of the balance point.

To continue the example:

Pipe diameter:	1.5 in. OD
Pipe weight:	1.619 lbs/ft
Wall thickness:	0.109 in.
Fluid in coiled tubing:	Nitrogen under pressure with an average density of 1.1 ppg
Fluid in wellbore:	9.5 ppg

$$W = W_{\text{air}} + \left(\rho_i A_i - \rho_o A_o / 19.25\right)$$

$$\text{Becomes } W = 1.619 + \left(\frac{(1.1 \times 1.290) - (9.5 \times 1.767)}{19.25}\right)$$

$$= 0.820 \text{ lbs/ft}$$

Clearly much more buoyant; the balance point is much deeper at 8836/0.820 = 10,776 ft.

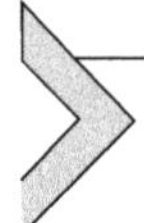

10.5 DOWNHOLE TOOLS AND THE COILED TUBING BOTTOM HOLE ASSEMBLY

Coiled tubing can be used for a wide range of interventions, many of which require specialist tools. Bottom hole assemblies vary considerably, from the simple straight bar and nozzle used when gas lifting a well, to more complex logging assemblies, mud motors, and milling tools. However, the intricacies of most of this downhole equipment are beyond the scope of this book. Described here are only the BHA components that are common to most operations, and that have a direct bearing on pressure control and well integrity.

10.5.1 Coiled tubing connectors

Coiled tubing connectors provide a means of attaching the BHA to the end of the coiled tubing. Different methods are available. The most widely used are the roll on connector, the dimple connector, and the grapple connector (Fig. 10.26).

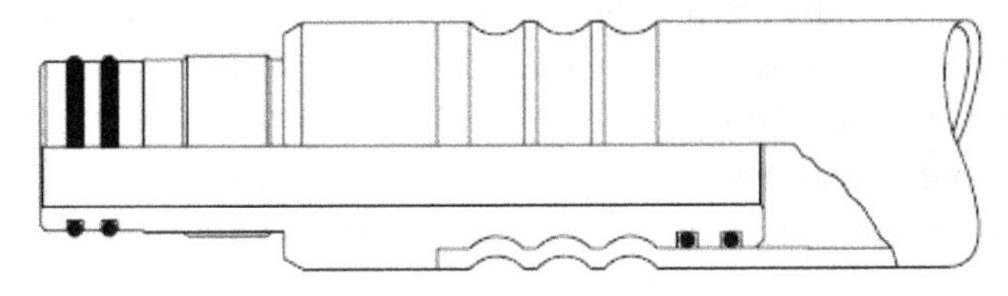

Figure 10.26 A roll on connector.

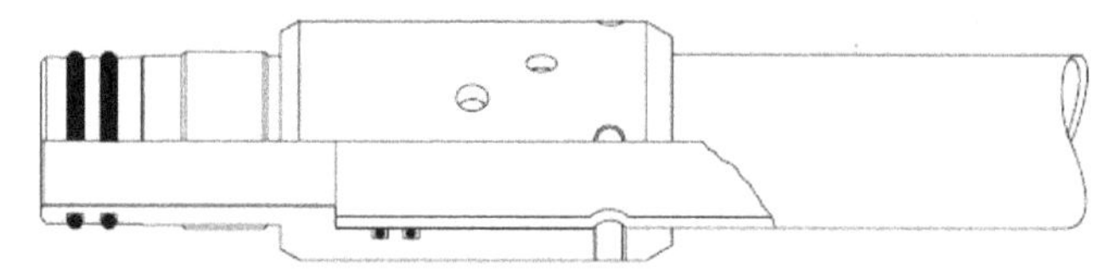

Figure 10.27 Dimple connector.

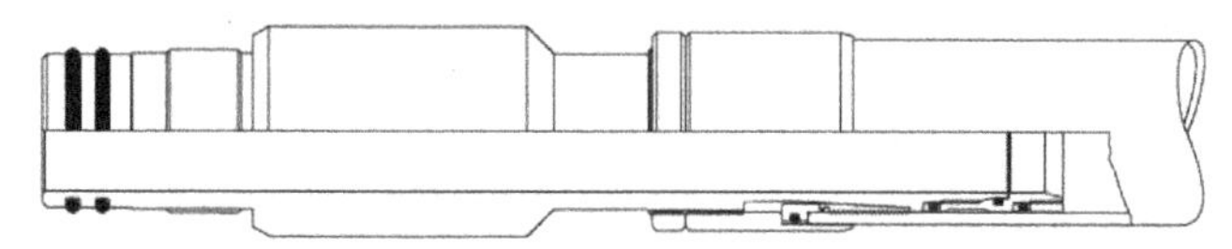

Figure 10.28 Grapple slip on connector.

After the roll on connector has been pushed on the end of the coil, a crimping tool is used to indent the coil into grooves in the connector, thus securing the connector to the coil.

The dimple connector uses a hydraulic jig to indent the coil onto the connector. It takes about 10 minutes to make the connection, and is acknowledged to be an improvement on the roll on connector (Fig. 10.27).

Grapple connectors use a slip arrangement to grip the pipe. Unlike the dimple and roll on connector, there is no internal deformation or restriction in the pipe, leaving a clear path for drop ball operated equipment (Fig. 10.28).

10.5.2 Check valves

Nearly all coiled tubing operations require the use of a check (nonreturn) valve, or more commonly dual check valves in the BHA. Check valves prevent well pressure and hydrocarbons from entering the coiled tubing, and are the primary internal well control barrier when coiled tubing is in the well. The only time check valves are not used is during reverse circulating operations (see Section 10.6.3).

There are several types of check valve in use (Figs. 10.29 and 10.30).

Since check valve stop well pressure (and fluids) from entering the coiled tubing, it is possible for pressure in the coil to be lower than the production tubing. Too great a differential pressure will collapse the coiled tubing. Measures must be taken to keep differential pressure to below the collapse pressure of the coil. Operations where external pressure exceeds internal pressure include:

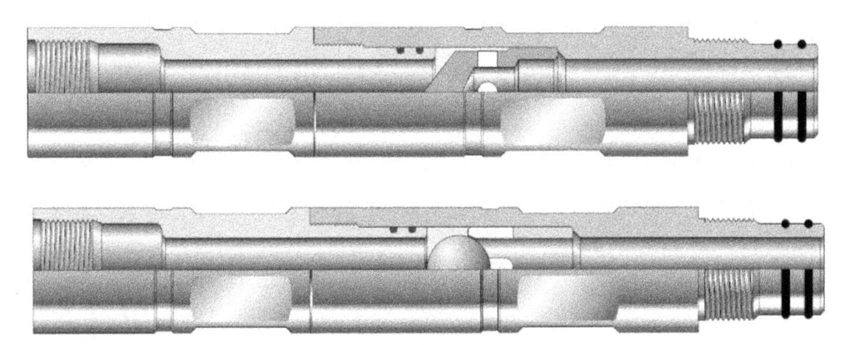

Figure 10.29 Dart check valve (top) and ball check valve (bottom): single valve.

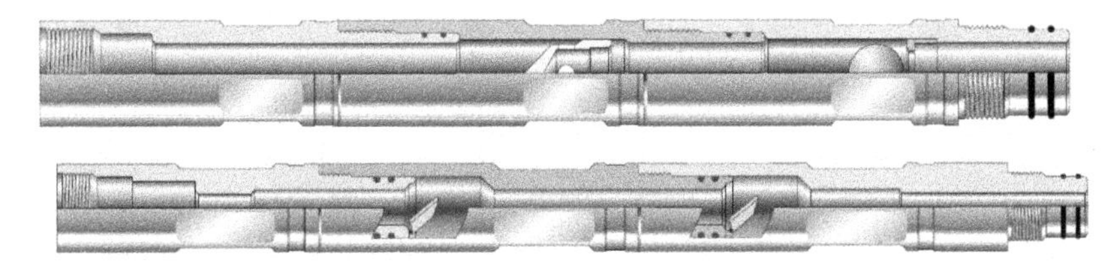

Figure 10.30 Double check valves: dart/ball (top) double flapper (bottom).

- Gas lifting operations where (lower density) nitrogen is in the coil and higher density liquids (water and hydrocarbon) are in the tubing.
- Running coiled tubing into a closed system, e.g., a well with unperforated liner or casing, will cause a rise in pressure due to pipe displacement. If the well is filled with a water-based fluid, the pressure increase will be significant, as water is not very compressible. Unless measures are taken to monitor and bleed off pressure build up caused by pipe being run into the well, it will collapse.
- Shutting in a well that is flowing will result in a rise in pressure. Internal reel pressure might have to be increased to compensate.
- Pressure testing of the production tubing when coil is in the well can lead to a collapse, unless the coil internal pressure is increased to compensate. The safest method is to pressure up through the coil. This ensures that coil pressure and wellbore tubing pressure are balanced (fluid density difference excepted).

10.5.3 Back pressure valve

There will be occasions when the hydrostatic pressure of a column of fluid in the coil tubing is greater than the pressure in the wellbore at the coiled tubing depth. This problem is more likely to arise where a dense

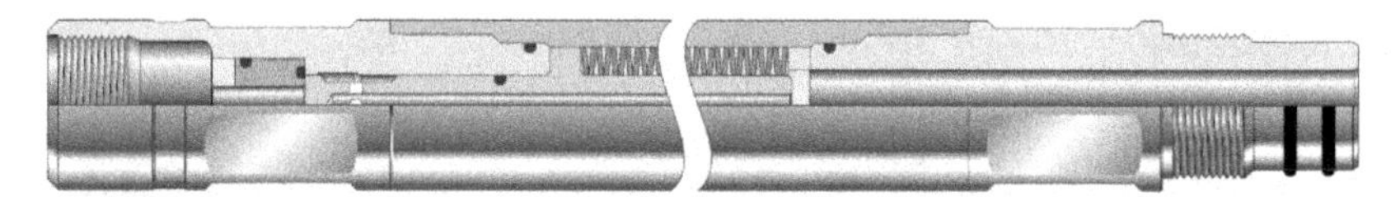

Figure 10.31 Back pressure valve.

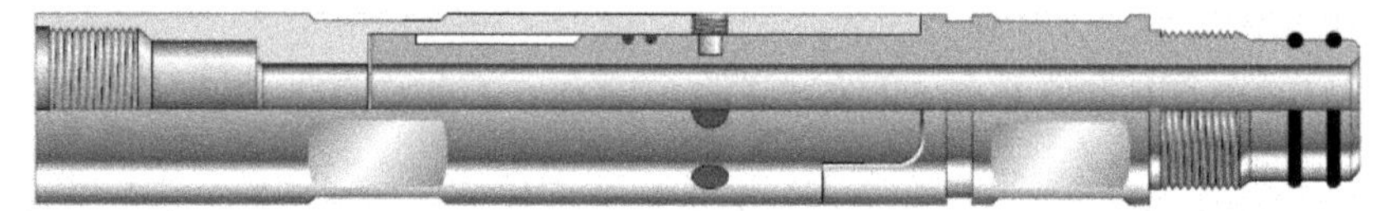

Figure 10.32 Shear disconnect sub.

fluid is in use, e.g., during a through coil cementing operation. A back pressure valve (BPV) is designed to keep fluid in the coiled tubing until it is ready to be displaced. Once on depth, increasing internal reel pressure overcomes a pretensioned spring, allowing the content of the coiled tubing to exit into the well. The BPV can be adjusted at the wellsite, and will be set in accordance with operational requirements (Fig. 10.31).

10.5.4 Disconnect sub

A disconnect sub is run below the coiled tubing connector and check valves. If the BHA gets stuck in the well, activation of the release sub allows the coiled tubing to be pulled from the well leaving the BHA behind. Two types of disconnect sub are in widespread use.

- Shear release. This is activated by applying tension (overpull). Tension shears pins in the release sub, freeing the coiled tubing from the BHA.
- Hydraulic disconnect. The hydraulic disconnect is activated by pumping a ball through the reel. When the ball seats at the disconnect sub, pressure applied at surface shears pins (or a disk) located in the sub, thus releasing the BHA (Fig. 10.32).

Placing the disconnect sub below the check valves ensures the primary barrier remains intact if the BHA needs to be released. In addition, the check valve must be of a design that allows the passage of the ball; i.e., a flapper, rather than dart or ball and seat.

10.5.5 Circulating sub

Circulation subs are used to open a circulation path above the BHA, or to regain circulation if it is lost due to a blockage in the BHA. There are

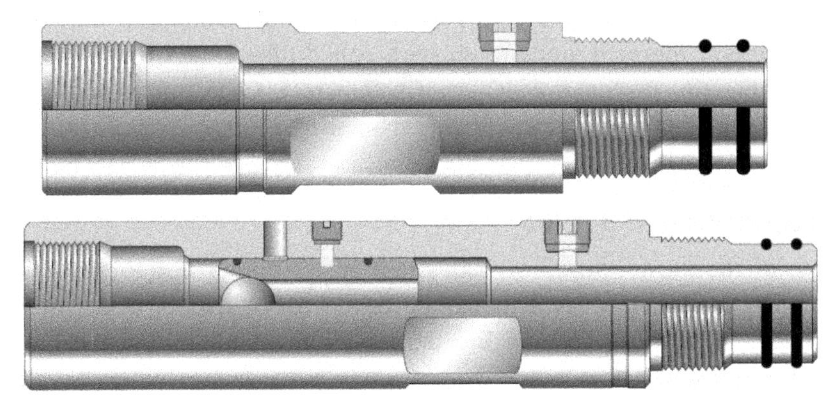

Figure 10.33 Circulation sub: Burst disk (top) and dual circulating and burst disk sub (bottom).

three options; a simple burst disk, a shear out ball drop, or a combination of the two. In the latter case, if circulation is lost, pressure is increased to rupture the burst disk. This gives enough circulation rate to enable a ball to be dropped, this shears the ball seat, uncovering ports that allow circulation at the maximum possible rate (Fig. 10.33).

10.5.6 Straight bar

Since coiled tubing is stored on a drum, it has residual curvature when being run into a well. This natural tendency to curve can impede progress when running into a well, as the ends of the pipe will push against one side of the tubing and naturally "hang-up" on any ledges. Gas lift mandrels can be particularly problematic. A straight bar is a short length of rigid pipe. It may be sized to match the coiled tubing OD, or it may be a larger diameter. Length varies, with most bars being in the region of 3–5 ft. Using a straight bar offsets the worst effects of bending in the coil, and helps limit the tendency to hang up. Centralizers are also used in conjunction with the BHA.

10.6 COILED TUBING OPERATIONS

Any response to a well control incident when using coiled tubing will, in part, be determined by the type of operation being performed.

Personnel at the wellsite should have a broad understanding of the types of operation routinely undertaken with coiled tubing, and the nature of any difficulties that can arise. Described here are the most commonly applied coiled tubing operations and techniques.

10.6.1 Nitrogen gas lift

One of the most commonly performed coiled tubing operations is the nitrogen gas lift. It is used to initiate flow from a dead or "lazy" well. Circulating nitrogen reduces the density of the fluid in the wellbore. Lighter, less dense fluid will flow more easily, as the hydrostatic pressure in the fluid column is reduced. Positioning the end of the coiled tubing deep in the well is more effective, as the greatest possible hydrostatic reduction is obtained. Gas lift operations need some additional equipment:

- Transport tanks with liquid nitrogen.
- Nitrogen convertor to convert liquid nitrogen to gas.

Although nitrogen lifting is generally thought of as a relatively straightforward operation, there are risks that can create pressure control problems. If internal reel pressure is too low, the pipe will collapse. Collapsed coil gives rise to a number of well control complications. Firstly, collapsed coil cannot be pulled through the stripper, and the collapsed section of pipe will create a significant leak path when it comes into contact with the base of the stripper. Pipe rams will not seal on the collapsed section, nor can they be held in the slips. The collapse also prevents or severely restricts the ability to circulate through the coil. In most cases where the pipe has collapsed, there is little choice but bullhead to kill the well and then recover the coil. To reduce the possibility of coiled tubing collapse during gas lifting operations, these precautions are used:

- Ensure the coiled tubing is in good condition and within ovality limits. Oval pipe has a much lower collapse pressure.
- Begin to pump nitrogen as soon as the coil begins to run into the well. This elevates the pressure in the reel and reduces the density, and thus the hydrostatic pressure, of the fluid in the well.
- At the start of the gas lift, ensure the choke is fully open.
- Route production through a test separator (if available). The separator should be set at the lowest possible pressure (as close to atmospheric as possible).
- Expect tubing head pressure to increase as the heavier fluids are removed from the well. Reducing the choke to control flow, or

closing in the well once flow has stabilized will bring about further increases in wellhead pressure. Ensure that internal coil pressure is maintained to compensate.

Other concerns have more to do with operational efficiency. For example, frictional losses can be significant when using large diameter coil in small diameter production tubing. Pumping at too high a rate can result in well fluid being forced back into the formation. Similarly, if small diameter coil is used in large diameter tubing, slippage prevents the nitrogen from lifting efficiently.

Liquid nitrogen has a very low boiling point, −196°C or −321°F. A spill will have serious consequences, especially on the metal deck of offshore platforms; they can crack. Measures must be in place to deal with any nitrogen spill, and the crew properly trained to respond. The first response is to hose away the nitrogen with copious amounts of water. On offshore platforms, precharged fire hoses are laid out as a contingency.

10.6.2 Wellbore clean out operations

Removing sand, proppant, or other types of fill from a wellbore is the most common type of intervention carried out using coiled tubing.[7] Sand will restrict production, and prevent access to the reservoir for logging tools (Fig. 10.34).

Conventional coil tubing clean out operations involve running coil to the top of the fill and circulating fluid that will lift any debris back to surface. For clean out operations to succeed, the annular velocity of the circulated fluid must be higher than the settling (slip) velocity of any solids being removed. Cleaning large diameter production tubing using small diameter coil is problematic. In some instances it will be difficult, if not impossible to achieve the circulation rates required to effectively remove the solids, as frictional pressure loss in the reel will limit circulation rate. Friction can be reduced and circulating rate increased by:

- Selecting the largest ID string available that can safely be transported to the wellsite.
- Selecting a reel that has the minimum length of coil necessary to reach the bottom of the well. Pipe left on the drum merely adds to frictional pressure drop.
- Using friction reducers.

Flowing the well whilst circulating increases annular velocity, and can be useful where pump rate through the coil is limited. The solids lifting

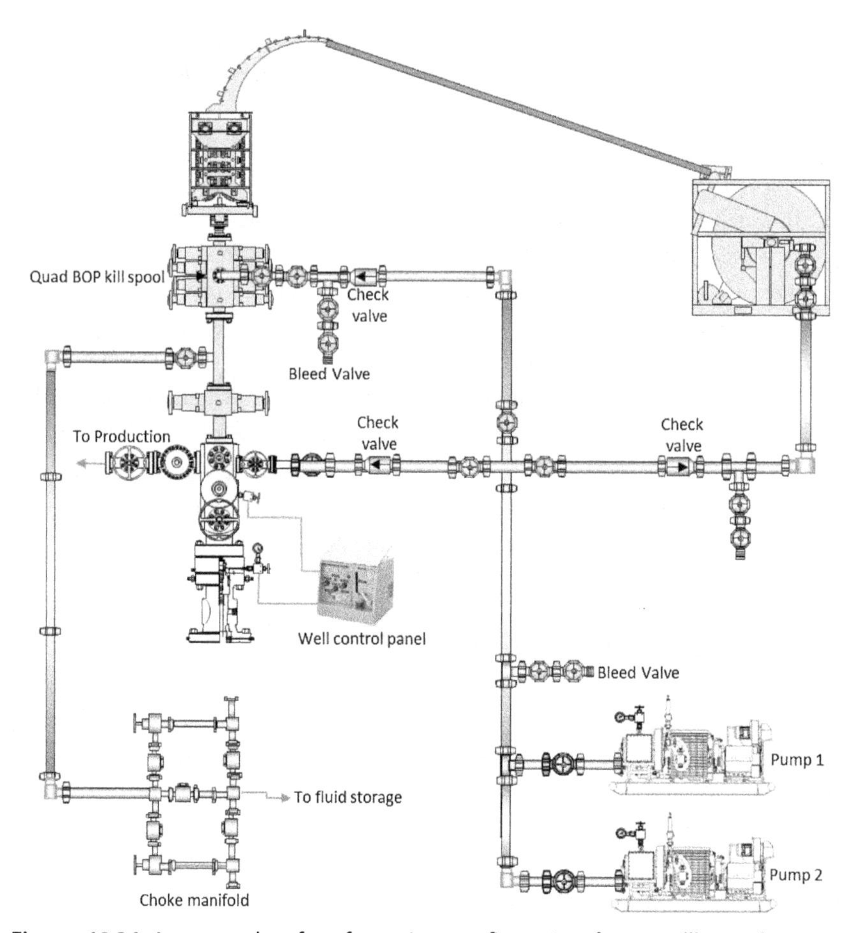

Figure 10.34 An example of surface pipe configuration for a wellbore clean out operation. Note the use of two fluid pumps and the single well control panel.

capability of circulated fluids can also be improved by using viscous gel sweeps or nitrified foam.

If the methods described are not sufficient, and annular velocity is still less than slip velocity, solids will begin to accumulate in the annulus above the nozzle and pipe can easily become stuck. Stuck pipe has the potential to lead to well control complications. If it cannot be pulled free, it will have to be cut to allow the free pipe to be pulled from the well. Cutting the pipe removes the primary (internal) barrier in the coil—the check valves. Recovery of the stuck pipe will almost certainly require the well to be killed.

The critical velocity for lifting solids from the well is a function of the particle size, particle density and shape, and the viscosity of the lifting fluid. Slip velocity can be calculated using Eq. (10.2).[8]

$$Vs = 0.45\left(PV/(MW)(Dp)\right)\left[\sqrt{\frac{36,800}{\left(PV/(MW)(Dp)\right)^{2}} \times (Dp)\left(\frac{\text{Den P}}{MW} - 1\right) + 1^{-1}}\right] \quad (10.2)$$

where:

Vs = Slip velocity in ft/min
PV = plastic viscosity in cps (centipoise)
MW = is fluid density in ppg
Dp = diameter of particle in inches
Den P = Density of particle in ppg

Note: Errors have occurred in the past because people have used bulk density instead of the material density. For example, 20/40 frac sand has a bulk density of 12.3–13.3 ppg (1.48–1.60 sg), whereas the sand grains (93%–95% silica) have a density of 21.99–22.15 ppg (2.64–2.66 sg).

To put this into context, the table below illustrates the slip velocity for two materials (sand and bauxite) of different size in fluids of different viscosity, which were calculated from Eq. (10.2) (Table 10.2).

If the slip velocity is known, then a simple calculation can be made to determine the circulation rate required to remove solids from the well.

Table 10.2 Single particle settling rates

	10/20 Mesh sand		20/40 Mesh sand	
Fluid	**ft/min**	**m/min**	**ft/min**	**m/min**
Water	30	9.1	19	5.7
Diesel (3 cp)	30	9.1	17	5.2
Polymer gel (50 cp)	10.9	3.3	2.23	0.7
Polymer gel (100 cp)	6	1.8	1.12	0.3
Polymer gel (500 cp)	1.3	0.4	0.22	0.07
	10/20 Bauxite		**20/40 Bauxite**	
Fluid	**ft/min**	**m/min**	**ft/min**	**m/min**
Water	34	10.3	21	6.4
Diesel (3 cp)	34	10.3	19	5.7
Polymer gel (50 cp)	13	4	3	0.9
Polymer gel (100 cp)	7.4	2.2	1.4	0.4
Polymer gel (500 cp)	1.6	.5	0.3	0.1

$$PO = \frac{AV(ID^2 - OD^2)}{1029.4} \quad (10.3)$$

where:

PO = Pump output required in bbls/min
AV = Annulus velocity in ft/min
ID = Internal diameter of production tubing, casing or well bore
OD = outer diameter of coiled tubing
Example:
Coiled tubing OD = 1.5″
Production tubing ID = 4.892″
Annular velocity (slip velocity) = 30 ft min

$$PO = \frac{30(4.892^2 - 1.5^2)}{1029.4}$$

$$PO = \frac{30 \times 21.68}{1029.4} = PO = \frac{650.4}{1029.4} = 0.631 \ \text{bbls/min.}$$

Pump output can also be calculated in gals/min, using the equation:

$$PO = \frac{AV(ID^2 - OD^2)}{24.5} \quad (10.4)$$

To calculate the answer in pump strokes:

$$SPM = \frac{\text{Annulus velocity (ft/min)} \times \text{annulus capcity (bbl/ft)}}{\text{pump output bbl/stk}} \quad (10.5)$$

Whist it may be useful for wellsite personnel to be able to estimate slip velocities and pump output corresponding to slip velocity, in reality the hydraulics modeling for coiled tubing sand clean out operations is more complex. A circulating velocity of approximately twice the slip velocity is typically recommended for vertical wells. However, as well angle increases, much higher velocities are required to prevent solids accumulating above and behind the nozzle. Further complications arise when non-Newtonian fluids (gels) are used, since viscosity will vary with temperature and flow rate (shear thinning). In most instances, hydraulics modeling software is used to calculate the pump rate and fluid viscosity requirements. Output from hydraulics models is also used to generate operational programs, where other parameters such as solids loading in the annulus and rate of penetration into the fill are defined. Understanding how coiled tubing can get stuck is an important part of prevention. Stuck pipe produces well control complications.

10.6.3 Reverse circulation wellbore clean out

There are circumstances where it is not possible to effectively clean a wellbore using conventional forward circulation. Conventional circulation is unlikely to work if:

- The coiled tubing outer diameter is small, relative to the internal diameter of the production tubing.
- The well cannot be flowed to assist lifting of solids, or where flow rate is limited.
- High density material, such as barium sulfate scale, needs to be removed.

Operating at, or close to, the limit of a conventional forward circulation is a high risk activity, and the risk of becoming stuck is significant. For example, it is difficult to achieve the high liquid rates required to clean sand from 7″ casing using 1.5″ coil tubing unless complex and expensive foam systems are used. In high angle and horizontal wells, the problem of achieving a high enough liquid rate is amplified to the point where wellbore cleaning cannot be accomplished, and the risk of getting stuck is too great. Reverse circulation dramatically reduces that risk, since some of the factors that increase risk during conventional operations diminish, or are eliminated. For example:

- Sufficient velocity in the coil to lift solids—significantly reduced chance of sticking pipe.
- No solids on the back-side of the coil, so nothing to fall back if the pumps fail.
- No (significant) change in flowpath ID within the coil. By comparison, the wellbore may contain dramatic and troublesome dimensional changes, e.g., at a liner top or tubing taper.
- The size of the "cuttings" is controlled by the nozzle port size.

A report from 2004 records not one single pipe sticking incident during more than 1600 reverse circulation coiled tubing clean out operations. Operational problems were limited to three cases of coiled tubing collapse and two cases where pin-hole leaks developed. There were no well control incidents.[9]

In spite of the success of many reverse circulation operations, there remains a reluctance to use the technique. The main concern to reversing through coil is the well control concern of having to remove the check valves (primary barrier) from the end of the coil, and worries about collapsing the coil.

10.6.3.1 Reverse circulation clean out: an overview

During reverse circulation, pumped fluid is routed through the four-way manifold to the pump-in tee on the riser (or Christmas tree kill wing valve). The fluid (along with any cuttings) returns through the coiled tubing, and is routed back through the four-way manifold for disposal, usually to a gauge or surge tank. The four-way manifold allows the direction of circulation (forward or reverse) to be switched quickly and easily (Fig. 10.35).

Since the coil tubing has a much smaller capacity than the surrounding production tubing, achieving the critical velocity required to lift fill off the bottom and out of the well can be accomplished at relatively low circulation rates. For example, at 1 bbl/min, velocity inside a 1½″ coil (0.109 in. wall) is 627 ft/min. If the same 1½″ coil were run inside 5½″ 17 lb/ft tubing (ID 4.892″), a forward circulation rate of 19.8 bbls/min would be necessary to achieve the same velocity in the coil tubing/production tubing annulus.

Since fluids return to the surface through the coiled tubing, hydrocarbons must be prevented from entering the coil. This is accomplished by keeping the bottom hole circulating pressure above reservoir pressure at all times.

The four-way manifold (Fig. 10.36) allows for quick and easy switching between reverse and forward circulation. This is usually only necessary when hard fill is encountered. Hard fill can be broken up by jetting and

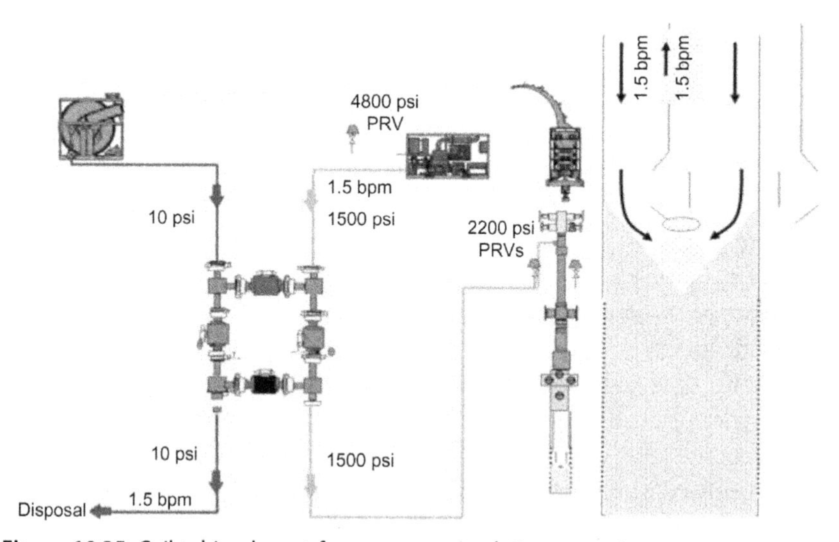

Figure 10.35 Coil tubing layout for a reverse circulation operation.

Figure 10.36 Four-way manifold.

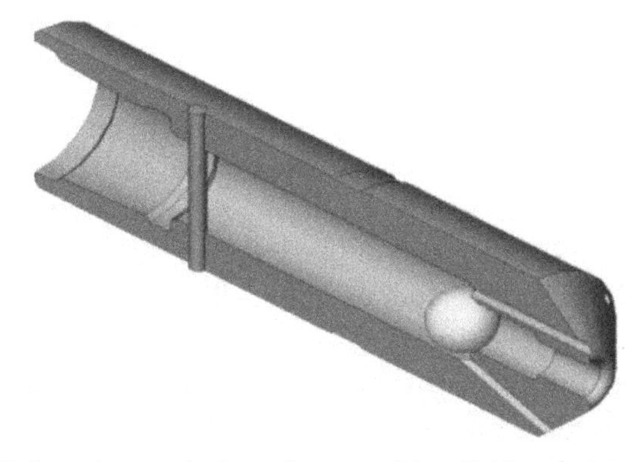

Figure 10.37 Ball and seat design allows rapid switching between forward and reverse circulation. Forward circulation results in a jetting action. During reverse circulation, the large diameter orifice allows sizable chunks of debris to enter the coil and be lifted back to the surface.

mechanical impact. The nozzle used in a reverse circulation BHA differs from that used for conventional circulation, and is designed to provide a jetting action when forward circulating whilst providing an opening large enough to accommodate larger chunks of fill when reversing (Fig. 10.37).

10.6.3.2 Reverse circulation clean out: candidate wells

Because reverse circulation operations must be conducted with the well in an overbalance condition, not all wells are suitable, and some types of well carry higher risks than others, such as:

- Any well that is not overbalanced by a column of the fluid selected for the clean out operation (additional circulating pressure has to be held to prevent hydrocarbon entering the well). Clean out fluid used for reverse circulation is almost always water-based.
- Wells that produce but have low or no injectivity.
- Multizone wells that are known to cross-flow.
- Deep deviated wells where the reel would be at maximum yield on pick-up (collapse resistance is lower when the pipe is in tension).
- Wells where losses are likely to result in excessive formation damage.

Because pressurized fluid is pumped into the annular space between the coiled tubing and the production tubing or casing in the well, the line between the four-way manifold and the well must be equipped with a pressure relief valve (PRV). This protects the coiled tubing from excessive collapse pressure.

10.6.3.3 Bottom hole assembly selection

The BHA nozzle should:

- Provide a powerful jetting action during forward circulation.
- Have a single large hole in the center is useful for lifting larger chunks of fill. However, the orifice in the nozzle must be smaller than the reel swivel joint ID, so as not to block the reel.
- Have a ball and seat or flapper design to allow rapid switching between lifting and jetting (Fig. 10.37).
- Have a pointed shape that will help break up compacted and hard fill.
- Have a nozzle that will fit through restrictions in the completion. A circumferential clearance of at least ½″ is recommended, to ensure minimum pressure drop around the nozzle.

10.6.3.4 Collapse prevention

The highest risks come from collapsing the reel. The collapse pressure is related to coil wall thickness, material yield, string tension, temperature, and ovality. Operators need a thorough understanding of the history and limits of the reel in use. In the North Sea (UK sector) there is a general policy of not exceeding 1500 psi differential from annulus to coil.

Precautions to reduce the risk of reel collapse include:

- The coil should be in good overall condition.
- Fatigue life of the coil should be less than 40% of the normal limit.
- High conformance to ovality limits; >2% is recommended (API limit is 4%).

- A PRV should be installed on the back-side (annulus) inlet to prevent coil collapse.
- Kill weight brine should be available.

10.6.3.5 Operational guidelines

Operational experience has shown that these guidelines work well during reversing operations:

- If hydrocarbons are in the wellbore, bullhead 2–3 full displacement volumes, or circulate conventionally to remove them.
- Carry out the circulation with the well overbalance to prevent hydrocarbon influx.
- Pump down the backside—up to 1500 psi. Return rate depends on reel size and length, but 1–1.5 bbls/min is effective in standard intervention coil sizes—1¼″ to 2″.
- It is important to know the loss rate in the well. High pump rates may be necessary to get the required rate of return in the coil to lift fill. For example, if losses to the formation are 3 bbls/min and the rate required to lift solids is 1.5 bpm, then a surface discharge rate of 4.5 bbls/min would be required.
- Monitor returns for any indication of gas or hydrocarbon in the reel at all times (increase in return pressure).
- Configure the surface pipework to enable pumping down the coil at any time.
- Crews should be familiar with the surface pipe configuration, and what to do to switch direction of circulation.
- Limit the volume of solids in the reel to 10% of its volume. This means regulating the rate of penetration in to fill. (Tubing and coil size dependent.)
- Pump down the coil to clear the nozzle of any debris (3500–4800 psi depending on reel type).
- Penetrate the fill to give about 5% solids fill in the reel—then gradually build to 10%. Do not exceed 10%. ROP = (capacity of wellbore at nozzle in ft/bbl)x(bbl/min return rate)x10%.
- Expect losses as perforations are approached.
- Friction reducers can be used to lower pump pressure.

10.6.4 Acid treatments

Acid is widely used during coiled tubing interventions. The most commonly applied applications are:

- Acid washing production tubing/casing to remove carbonate scale deposits using hydrochloric acid.
- Matrix (below formation fracture pressure) acid treatment of carbonate reservoirs to remove near-wellbore formation damage, and therefore improve well productivity.
- Spotting of mud acid (hydrochloric/hydrofluoric) to reduce formation damage.
- Acid fracture stimulation operations, e.g., Halliburton's "SurgiFrac" service.

There are clear advantages when using coiled tubing to carry out acid pumping operations.

- The CT pressure control equipment configuration allows the treatment to be performed on a live well. This avoids additional formation damage associated with a well kill.
- Associated operations can be performed as part of an integrated service, e.g., wellbore fill can be removed prior to the matrix treatment and nitrogen or artificial lift services may be applied to restore production following the treatment if required.
- Performing the treatment through CT avoids exposing the wellhead or completion tubular to direct contact with corrosive treatment fluids.
- Spotting the treatment fluid with CT will help ensure complete coverage of the interval. Using an appropriate diversion technique will help ensure uniform injection of fluid into the target zone. Spotting the treatment fluid also avoids the need to bullhead wellbore fluids into the formation ahead of the treatment.
- Long intervals can be more effectively treated using techniques and tools that have been developed for use with CT, e.g., a selective treatment system using straddle-pack isolation tools. This is particularly important in horizontal wellbores.

Pumping acid through coiled tubing is not without risks, and these need to be mitigated to avoid possible well control complications.

- All acid used must be properly inhibited to prevent corroding the coil tubing (and the completion tubulars). Uninhibited or poorly inhibited acid could damage coiled tubing to the point of failure.
- Only use H_2S service equipment, since H_2S is a possible by-product of some acid treatments.
- If acid is simultaneously being pumped down the coil and down the production tubing, (as happens during "SurgiFrac" operations), the point of injection into the coiled tubing/production tubing annulus

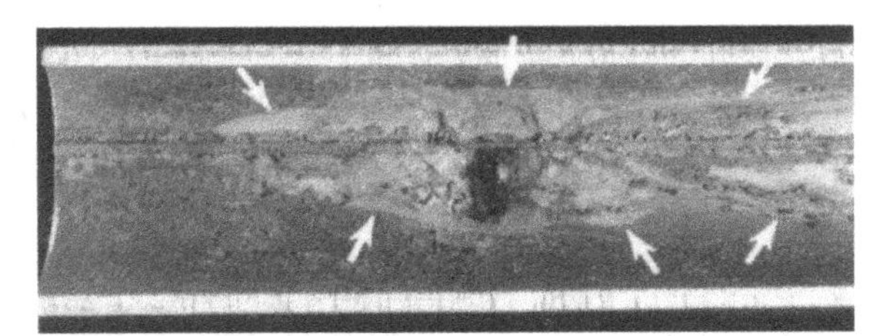

Figure 10.38 Corrosion caused by an acid puddle in the bottom of a wrap. *Photo courtesy of Precision Tubes.*

should be below any coiled tubing pressure control equipment. In other words, do not inject through a tee on the riser. It is better to pump in through the kill wing on the tree.

- When pumping through the coiled tubing and production tubing simultaneously, there is a risk of coil collapse unless reel pressure is maintained.
- At the end of the treatment, ensure no acid is left in the reel. Acid accumulating on the low side of a reel and can corrode the pipe (Fig. 10.38).

10.6.5 Removal of hard scale

Sulfate scale can deposit in wells as gypsum ($CaSO_4$, $2H_2O$), barium sulfate ($BaSO_4$), or strontium sulfate ($SrSO_4$) Gypsum can easily be dissolved using ethylenediaminetetraacetic acid; thankfully usually abbreviated to EDTA. Although barium sulfate and strontium can be dissolved with ETDA, it requires both high temperature and a long contact time to be effective. Because of the slow reaction rate, direct mechanical removal is commonly used for barium and sulfate scale removal.

Hydraulic jetting is a direct and effective method of scale removal. Carried out at low pressure (below 5000 psi), no specialist equipment is needed. Jetting operations carried out at above 5000 psi require more specialized equipment, and more detailed planning. Irrespective of pressure, scale removal will usually involve repeated reciprocation of the BHA across the scaled up section of the well, and as such will result in fatigue across a relatively short length of the coiled tubing. Careful monitoring and recording of fatigue cycles is an essential part of the operation.

The use of scale mills and under-reamers is an alternative to hydraulic jetting, and is commonly used.

Hole cleaning during a sulfate scale removal is often problematic, since scale is denser than sand or bauxite. Barium sulfate is 4.5 g/cc (37.5 ppg) and strontium sulfate is 3.96 g/cc (33 ppg). This, coupled with a generally larger particle size, means high annulus velocity is required to sweep solids from the well. In some cases, the required velocity might not be achievable.

10.6.6 Mechanical interventions with coiled tubing

Coiled tubing is frequently used for the setting and retrieving of mechanical intervention tools such as packers, bridge plugs, and straddles. It is also used to manipulate sliding sleeves and inflow control valves in intelligent completions where the hydraulic controls have failed. To minimize the risk of getting the reel stuck:

- Include a release mechanism (shear or hydraulic) in the BHA.
- Pulling tools for mechanical plugs must have a secondary release mechanism. This is usually operated by setting off weight and increasing circulating pressure.
- Before running large OD mechanical tools (bridge plugs, straddles, and packers) it is advisable to run a full bore drift. The drift run can be carried out on wireline before rigging up the CT unit.

10.6.7 Running perforating guns on coiled tubing

Running perforating guns on coiled tubing has a number of advantages:

- Live well gun running is simple and safe.
- A higher underbalance pressure can be used (by comparison with wireline).
- The coil can be used to introduce an underbalance cushion prior to gun detonation.
- Coil enables guns to be run into high angle and horizontal wells.
- Gun deployment systems enable long guns to be run into live wells—the potential exists to perforate long intervals in a single run, and with underbalance.
- Coil tubing trip times are faster that using jointed pipe.

Operational and well control considerations when perforating with coiled tubing include:

- Pressure control equipment—for most perforating operations the standard pressure control equipment is adequate. To operate in a live well, the riser must be long enough to accommodate the perforating guns. If there is a requirement to perforate an interval that is longer than the riser length available, multiple perforating runs can be carried out. Alternatively, a live well gun deployment system can be used.

- Gun deployment systems allow multiple gun sections to be installed in a well where the overall length of the gun is longer than the riser length available. Gun sections are joined using special connectors. Each connection is made inside the pressurized riser using specially designed deployment BOPs.

- Firing mechanism:
  - Electrical firing—to fire perforating guns electrically requires an electric cable inside the coiled tubing, and a firing head as part of the BHA. An electric cable has the additional advantage of enabling GR/CCL to be run with the coil for accurate depth correlation, overcoming the limitations of mechanical depth measurement.
  - Pressure activated firing—tubing conveyed guns can be fired using pressure activated firing heads. Most have an integral time delay with a predetermined duration between the application of pressure and gun detonation. Wellsite supervisors need to know how the firing head selected for any perforating operation works. Insufficient knowledge could be fatal.
- Depth control:
  - Coil tubing depth measurement is often rudimentary. The counters used on the injector head and level wind arm are prone to slippage, and they are not able to place a perforating gun on depth with enough accuracy. There will often be something in the well that can be used to confirm depth—e.g., a hard tag against a bridge plug.
  - If an electrical firing head is used, i.e., the coil is equipped with an internal electrical cable, and then GR/CCL depth correlation enables accurate depth determination.
- Perforating loose, unconsolidated sandstone formations risks sand entering the wellbore and sticking the gun. This risk increases with increasing underbalance. The use of high shot density "big-hole" charges also increases risk.

Coiled tubing perforating incident: Case history.[10]

In January 2010, coiled tubing perforating operations were taking place on a jack-up rig in the Dutch sector of the North Sea. A pressure activated firing head with a 30 minute time delay was in use. As the guns were run to depth, nitrogen was circulated through the coil tubing via a port above the firing head. Once the guns had been positioned at the correct firing depth, a ball was pumped/dropped through the coiled tubing, to isolate the circulating ports and allow reel pressure to be increased, activating the firing head. Activation of the firing head should have been confirmed by reestablishing circulation through the coiled tubing. This was not done. At the surface, the crew observed a pressure spike. Mistakenly thinking the guns had been activated, they stopped pumping and bled off pressure. Crucially, they did not attempt to circulate, nor did they confirm the volume of nitrogen pumped was sufficient to seat the dropped ball, i.e., the approximate volume of the coil. The crew, assuming the gun had fired, recovered it to surface.

After closing in the well, the crew began to depressurize the riser. As the riser was being depressurized, the coiled tubing operator in the control cabin noticed that pressure in the coil had not dropped along with the riser pressure as expected, and as should have been the case, had the firing head activated and the circulation path reestablished. Moreover, the differential pressure between that trapped in the coil and the rapidly depressurizing riser created enough differential pressure to initiate the time delay. Realizing that the time delay sequence had begun, the area around the pressure control equipment was evacuated. A short time later the guns fired, destroying the riser and BOP. Remarkably, no one was harmed.

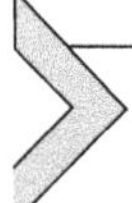

10.7 GENERAL COILED TUBING OPERATING GUIDELINES

Most operating companies will have operational and safety guidelines for coiled tubing interventions in live wells. Coiled tubing service suppliers will also have their own procedures and guidelines for safe operation. Operating company and service company procedures must include responses to well control. Whilst these responses will be adapted for the location and the operation in hand, there are some basic fundamentals that are universal.

10.7.1 Prejob checks

There are several prejob checks that, if implemented, will reduce the likelihood of accidents and incidents, including well control incidents. The list of prejob checks provided here is by no means exhaustive, but represents what most responsible operating companies view as a minimum requirement:

Crew competence:

- Operating and vendor company personnel should have the appropriate level of well control training.
- The vendor crew must be competent to operate all of the coiled tubing equipment, and know how to respond to a loss of (pressure) containment during all phases of the operation.
- If a (single well) well control panel is used, each and every member of the coiled tubing crew must be briefed on how the panel works, and what actions are required to close in the well at the tree, i.e., how to close hydraulically actuated tree valves.
- The crew must also be fully conversant with the operation of ancillary equipment; nitrogen converters, tanks and transfer hoses, fluid pumps, and mixing equipment.
- Where relevant, the crew must be trained in the use of H_2S breathing apparatus—see Chapter 2 "Well Construction and Completion Design."

Briefing and communications:

- Before any operation begins, good lines of communication must be established between the coiled tubing control cabin, other key locations, and key personnel. This includes the operating company supervisor or representative, and those operating ancillary equipment, e.g., a fluid pump or nitrogen convertor. Good lines of communication should be established with facilities operators, especially if the well needs to be flowed during the coiled tubing operation.

Nonoperational personnel are to be kept away from potential hazards. This is best accomplished by erecting temporary barriers. Hazardous areas include:

- Temporary high pressure lines.
- Wellhead, riser, and BOP.
- Coiled tubing Injector head.
- Mixing and pumping equipment—particularly if hazardous chemicals (e.g., acid) are being mixed and pumped.

Precautions should be in place for the safe handling of hazardous chemicals, including:

- Hazard data sheets available for all chemicals.
- Availability of acid neutralizing agents, usually sodium carbonate—Na_2CO_3 (soda ash).

- If nitrogen is being used on offshore platforms, fire hoses should be positioned to wash away spills—liquid nitrogen can crack metal decks.
- Personnel need to be made aware that nitrogen is an asphyxiant.

All pressure control and lifting equipment must have the required certification:

- Pressure control equipment should be accompanied by in-date test certificates, and the pressure rating should be suitable for the planned operation. Where relevant, the pressure control equipment should be suitable for sour service.
- PRVs should be function tested to confirm they operate (relieve) at the specified pressure.
- Confirm that all lifting equipment is accompanied by up-to-date load certification.

Additional precautions:

- Ensure access and escape routes are properly marked, and the crew knows where they are.
- Only operating company supervisory staff or a competent vendor company person (nominated by the operating company supervisor), should open and close tree and wellhead valves.
- If the well is connected to the production facilities, ensure the integrity of all isolating valves is confirmed. Ensure that isolations in place have been formally secured and recorded as part of the operational permit to work.

The use of prejob briefings is increasingly common and to be encouraged. For coiled tubing operations, briefings should take place before work commences, and before any safety critical phase of the operation such as pressure testing, arming perforation guns or flowing the well. Properly conducted, a prejob briefing ensures:

- Clarity about the operational objectives.
- Clarity about roles and responsibilities.
- Awareness of any potential risks and actions during unplanned events.

10.7.2 Prejob testing of pressure control equipment

Routine function and hydraulic testing of all essential pressure control equipment is a fundamental requirement for any coiled tubing intervention on a live well. Some of the function tests are usually carried out in the workshop before the equipment is sent to the wellsite. Pressure testing is usually carried out on location, and immediately before the equipment is exposed to well pressure and well fluids.

10.7.2.1 Testing shear/seal blow out preventers

The shearing/sealing capability of the BOP is usually tested in the workshop. To properly test the BOP, a sample of coil of the same maximum wall thickness and yield to be used in the operation will be placed across the BOP and cut. After cutting the coiled tubing sample, the BOP should by hydro tested from below to a pressure equivalent to the maximum working pressure of the BOP.

At the wellsite, the sealing capability of the BOP will be tested in situ (after it has been rigged up). Shear seal BOP's are usually rigged up directly above the Christmas tree, and tested via the kill wing valve. Where there is a risk of hydrates, the test fluid must be suitably inhibited, normally with glycol. The following procedure is representative of how the test is performed at the wellsite:

- Close the upper master valve, swab valve, and production wing valve.
- Depressurize above the closed swab valve, and remove the swab cap and swab cap flange connection.
- Make up the shear seal BOP to the flange connection on top of the Christmas tree.
- Hook up high pressure temporary lines to the kill wing valve on the side arm of the Christmas tree.
- Confirm that the production wing valve and master valve remain closed.
- Open the kill wing valve and the swab valve.
- Pump test fluid through the temporary high pressure line until it overflows the top of the shear seal BOP.
- Close the BOP (shear seal) rams.
- Apply test pressure for the required duration. Most operating companies require the test to be recorded on a chart and witnessed by the wellsite supervisor.
- Bleed off pressure (Fig. 10.39).

Riser, if required, would normally be rigged up above any shear seal BOP's with the quad (or combi) BOP's rigged up on top of the riser.

10.7.2.2 Testing the blind rams, shear rams, and riser

Testing of the riser and BOP blind ram is usually performed through a high pressure line connected to the Christmas tree kill wing. Alternatively, the test can be carried out through a tee piece on the riser. This procedure applies to both quad (four ram) and combi (two ram) BOPs.

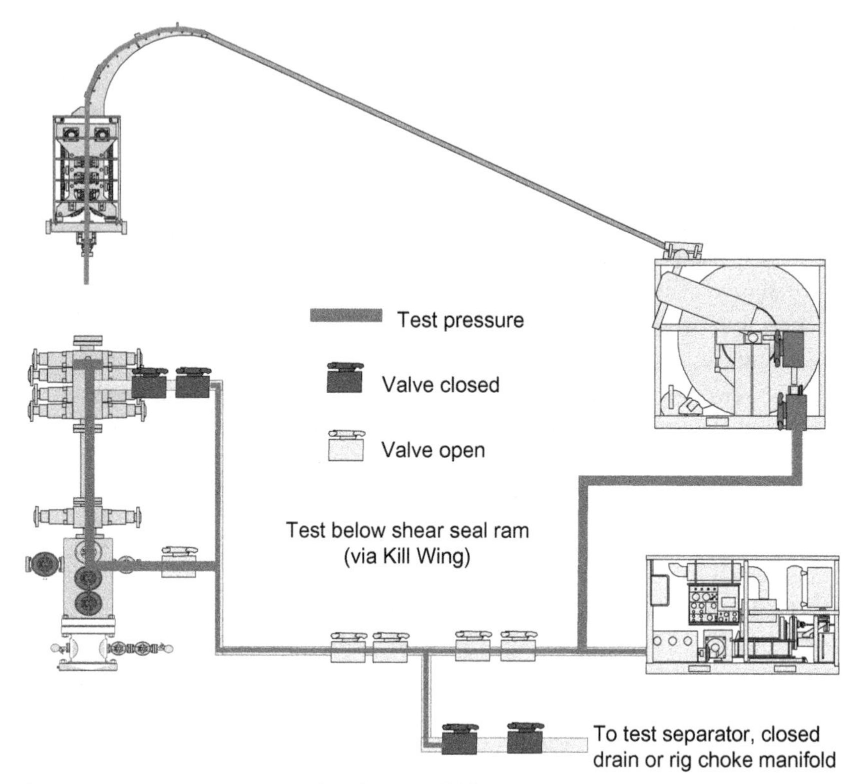

Figure 10.39 Pressure testing the shear/seal blow out preventer.

- Line up to pump into the riser via the kill wing (master valve closed, wing valve closed, swab valve and kill wing valve both open); or,
- Line up to pump via the riser tee (swab valve and master valve closed).
- Fill the riser and BOP with test fluid.
- Close and lock the BOP blind ram (or shear seal ram if using a combi BOP).
- Apply test pressure for the required duration (recorded and witnessed).
- Bleed off pressure.
- Open the blind (or shear seal) rams.

Shear rams (or shear seal rams) are usually function tested in the workshop before they are sent to the wellsite. As with the shear seal BOP's, the sample of coiled tubing that is cut must be representative of what will be used for the operation (Fig. 10.40).

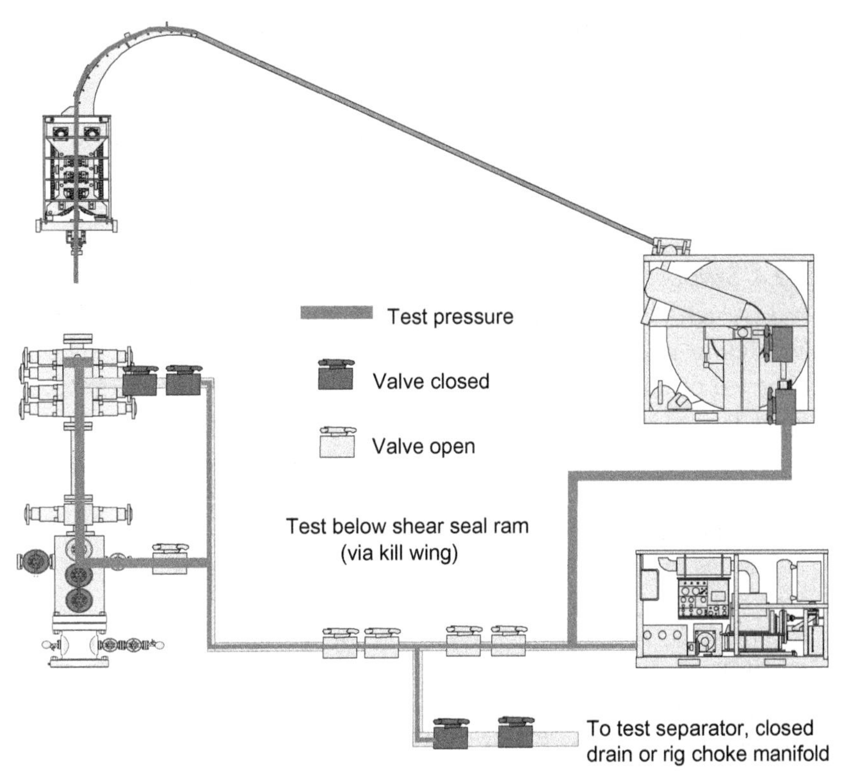

Figure 10.40 Pressure testing the blind ram.

10.7.2.3 Coiled tubing reel

The coiled tubing reel should be flushed through (usually with water) before testing. Pump at least twice the pipe capacity. This ensures that any residual debris and any potentially contaminating fluids left behind from previous operations are removed. Most coiled tubing specialists recommend pumping a "pig" or steel ball. This is especially important if a ball or pig have to be pumped through the coil during the operation. Flushing and pumping of a ball or pig is recommended, even if the pipe is new and has never been used. On at least one occasion in the past, ball drop functions in the BHA have been prevented from working by foam plugs left in the reel at the time of manufacture.

The reel can be pressure tested before or after the pipe is threaded through the injector chains, and this is usually the case for offshore operations where a skid mounted reel is used. For land operations where the reel and injector head are mounted on a truck chassis, it is common to leave the pipe threaded through the injector between operations.

- Fill the coiled tubing reel with test water from a fluid pump, displacing at least twice the reel capacity.
- Make up a test cap on the end of the pipe.
- Apply test pressure for the required duration. (Recorded and witnessed.)
- Bleed back pressure and remove the test cap.

10.7.2.4 Stripper

In most cases, the stripper is permanently attached to the base of the coiled tubing injector head. Before the stripper can be tested, the pipe must be fed through the injector chains, threaded through the stripper (or strippers if used in tandem), and the whole assembly made up on top of the BOP. The following procedure assumes that a tandem stripper is being used.

- Ensure no stripper pressure has been applied.
- Fill up via the reel until water overflows from the stripper.
- Stop the pump.
- Confirm the swab valve is closed. If there are lines running to a riser tee piece, these need to be isolated as well.
- Energize the stripper packer. If tandem strippers are used, the lower stripper is normally tested first.
- Apply test pressure through the reel and hold for the required duration. (Recorded and witnessed.)
- Bleed off pressure via the tree or pump-in tee on the riser. Bleeding back through the coil risks collapsing the pipe if the check valves are connected.
- If tandem strippers are in use, unpack the lower stripper.
- Pump through the reel until fluid is seen overflowing the upper stripper. This is to confirm the element in the lower stripper has relaxed.
- Pack-off the upper stripper seal element, apply test pressure through the reel, and hold for the required duration (recorded and witnessed).
- If the pipe rams are to be tested next, leave the test pressure applied. If not, bleed off pressure via the tree or pump-in tee on the riser. Bleeding back through the coil risks collapsing the pipe if the check valves are connected (Fig. 10.41).

10.7.2.5 Pipe rams

The pipe rams are usually tested as soon as the test on the stripper has been successfully concluded.

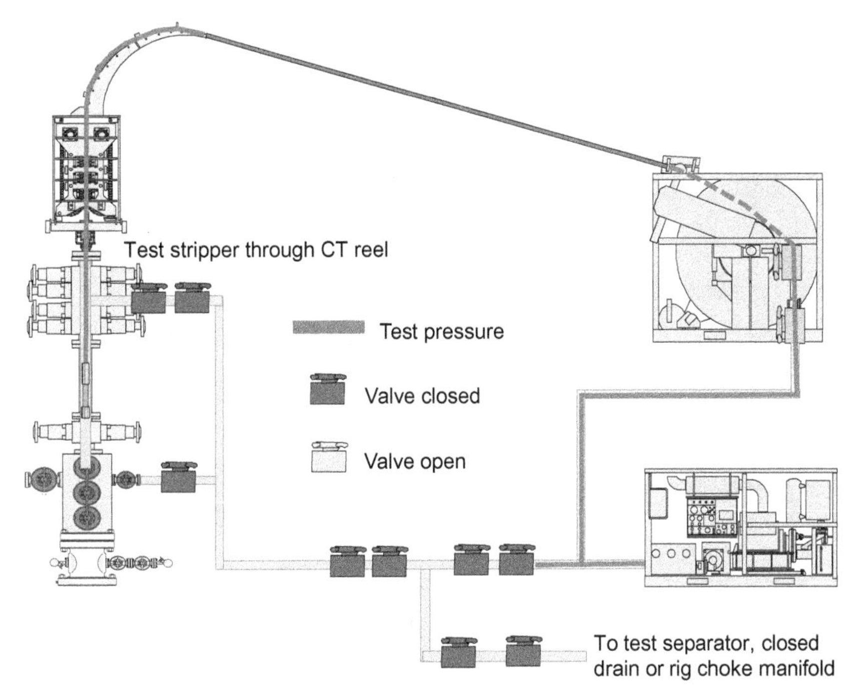

Figure 10.41 Pressure testing the stripper.

- Ensure coiled tubing pipe (not the straight bar or BHA) is across the BOP.
- With pressure in the riser still applied (following a successful stripper test), close the BOP pipe ram.
- Bleed off above the ram through the BOP circulating port. Close in the circulating port.
- Monitor pressure below the closed ram for the required duration. No pressure loss should be observed.
- On completion of a successful test, equalize across the rams and open the pipe rams (Fig. 10.42).

10.7.2.6 Bottom hole assembly internal check valves

The easiest way to confirm check valve integrity is to leave pressure in the riser and bleed back through the coil. Great care must be taken not to exceed coiled tubing collapse pressure. It will usually be necessary to bleed some of the riser pressure off through the riser tee or Christmas tree wing valve first.

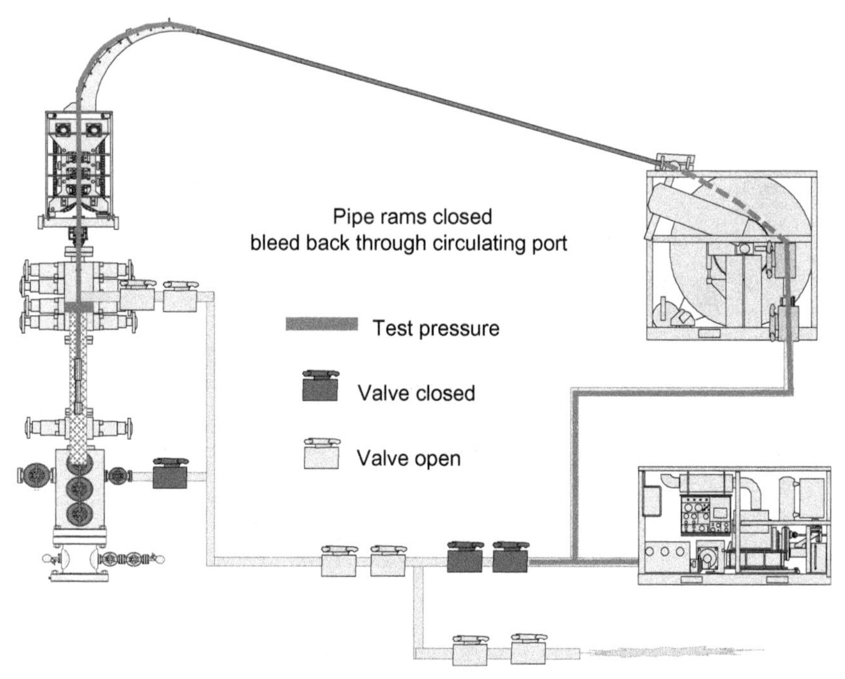

Figure 10.42 Pressure testing the pipe ram.

- If pressure remains in the riser following the testing of the pipe rams, bleed off through the riser pump-in tee or back through the Christmas tree wing valve until riser pressure is equal to the required check valve test pressure, then lock in the pressure.
- Bleed off through the reel. If the check valves are holding, the riser pressure should remain constant as the reel pressure is bled off.
- Once the reel has been depressurized, monitor riser pressure for the required duration.
- Bleed off through the tree or riser tee (Fig. 10.43).

10.7.3 Running coil in the well

During a coiled tubing intervention, pressure control equipment is exposed to pressurized hydrocarbon and corrosive chemicals, sometimes for a long time. Measures to reduce the risk of loss of containment include:

- If the injector head is supported by a crane, run the crane engine continuously while the load is suspended.
- Ensure the injector is properly supported at all times (Fig. 10.44).

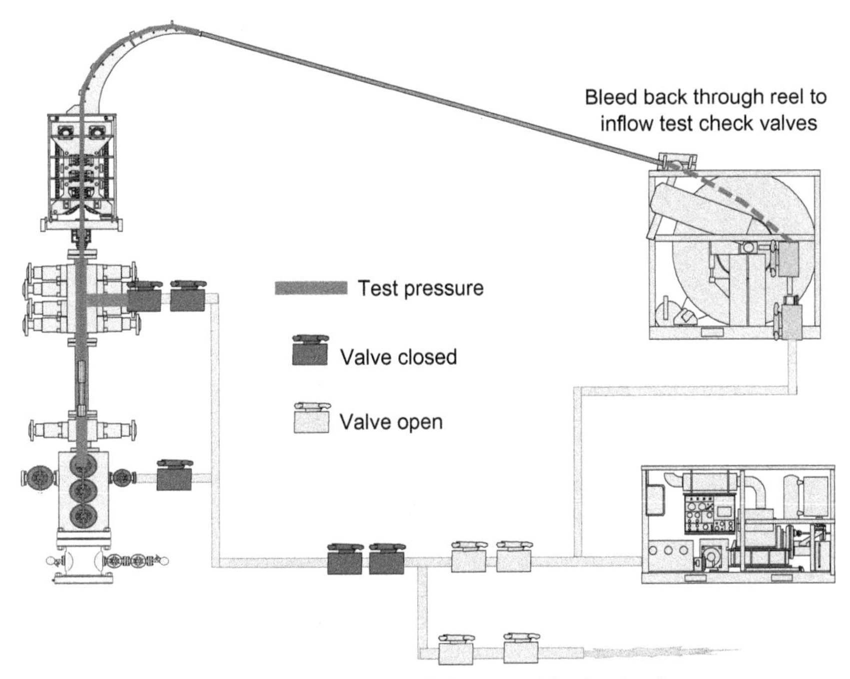

Figure 10.43 Pressure testing the bottom hole assembly check valve.

- Lubricate the coil while running in to protect the stripper elements.
- Zero the depth counters and weight indicator(s) before opening the well.
- Use the lowest possible stripper hydraulic pressure necessary to create a seal around the coiled tubing. Keeping pressure low reduces seal element wear.
- Avoid buckling the pipe on unexpected obstructions, using the minimum hydraulic pressure necessary to move the pipe at the required speed.
- Maintain positive pressure in the coiled tubing reel at all times. Agree a maximum allowable collapse pressure for the operation (typically ± 1500 psi).
- Do not use the slip rams during normal operations. If they have to be used, pick up and inspect the surface of the pipe contacted by the rams for any sign of damage before continuing. Do not run damaged pipe below the BOPs.
- Mark the pipe with paint at known depths for visual reference. Do not scratch, dent, or otherwise mark the surface of the pipe.

Figure 10.44 Properly support and secure the injector head throughout the operation!

- Agree a maximum allowable sit down weight before running into the well (usually >2000 lbs). The agreed value should be recorded in the work program.
- If in use, data acquisition should run continuously while the pipe is in the hole.
- Monitor the tri-axial (Von Mises Equivalent) stress in the pipe wall relative to the working limit prescribed by the work program.
- Monitor pick-up and RIH weight for comparison with predicted (modeled) values.
- Agree a maximum pick up weight before running into the well.
- If the expected parameters are incorrect, stop. Consult the coil tubing vendor and the onshore support group before continuing.

10.7.3.1 Recommended running speeds

The running speeds listed here are widely regarded as being reasonable for most operations and can be used if nothing is specified in the work instructions (Table 10.3).

Table 10.3 Recommended running speeds

Operation		Maximum speed	
		m/min	ft/min
Running in	Passing restrictions in the well	5	15
	When the BHA is leaving the riser and passing through the tree	5	15
	Within 500 ft (150 m) of tagging bottom or a known obstruction	8	25
	Uninterrupted running in	20	75
Pulling out	Uncertainly about depth—close to surface	5	15
	Passing restrictions in the well	5	15
	Bottom hole assembly passing the wellhead—entering riser	5	15
	Uninterrupted pulling out	30	100

10.7.3.2 Pull tests

Pull tests must be carried out to ensure that pick-up weight does not exceeding agreed tension limits. The results should be compared with modeled predictions. If pull test frequency or location is not specified, carry out pull tests:

- After running in to about 500 ft (150 m) to verify that all of the equipment is working properly.
- Every 300 m/1000 ft where the wellbore inclination is less than 60 degrees.
- Every 150 m/500 ft if the inclination is in excess of 60 degrees.
- After passing any restriction.
- During clean out operations.

More frequent checks are made if pick up weight is getting close to the maximum allowable tension, or if well conditions are giving cause for concern.

Note: Do not perform the pull test at the same point on every run, i.e., at exactly 1000 ft, 2000 ft, and so on. Change the point at which the pull test is taken to reduce bending fatigue in the reel.

At each pull test record:

- Depth.
- RIH, hanging, and POOH weight.
- Stripper pressure.
- Pump rate and pressure.
- Wellhead pressure.

10.8 WELL CONTROL AND EMERGENCY PROCEDURES

Things can, and do, go wrong during coiled tubing interventions. Contingency plans must be in place for a range of known problems. Moreover, all members of the crew should be aware of these contingencies, and how they are to be implemented. When an unplanned event or hazardous situation arises the crew should, in the first instance:

- Have the necessary knowledge and training to quickly identify the problem and immediately implement the most appropriate contingency plan.
- Clear nonessential personnel away from the area. If facilities allow, make a public announcement advising personnel to keep clear of the area.
- Inform the operating company supervisory staff.
- If acid or other hazardous chemicals were being pumped through the reel, begin to flush with water as soon as possible.

10.8.1 Power pack failure

Power pack failure is serious. There will be a complete loss of hydraulic power to the injector, reel motor, and ancillary equipment, including the BOP controls. Losing hydraulic pressure should activate the motor break, thus leaving the pipe supported. However, the risk of a coiled tubing runaway remains. Contingencies for dealing with a (pipe heavy) runaway into the well are described in Section 10.8.2.1. Section 10.8.2.2 describes a runaway out of the well (pipe-light). Following a power-pack failure:

- Close and lock the slip rams followed by the pipe rams (pipe/slip ram if combi BOP is used).
- Set the reel break.
- If, when the power pack failed, fluids were being circulated to remove solids such as scale, sand, or drill cuttings, it is important that circulation continues at the same rate (annular velocity) until all of the solids have been removed from the wellbore. Reducing rate, or stopping pumping, risks getting the coiled tubing stuck.
- Repair (or replace) the power pack.
- Equalize across the BOP.
- Unlock the pipe then slip rams.
- Release the reel break and set reel hydraulic pressure.

- Release the injector break.
- Pick up on the pipe to expose the area that had been supported in the slips. Inspect the surface of the pipe for damage.
  - If the pipe is damaged, pull out of the well. Carry out repairs or replace the coiled tubing reel.
  - If no damage has occurred, continue with operations.

10.8.2 Coiled tubing runaway

A coiled tubing "runaway" can happen when running into, or pulling out of the well. It is caused by a lack of grip between the contra-rotating drive chain and the pipe. Poor grip on the pipe is usually caused by insufficient hydraulic pressure to the gripper rams, or by under gauge pipe. With the introduction of 16% chrome pipe, coiled tubing, runaways are likely to become more common. Field trials carried out by BP noted that when slippage of the coil through the injector occurred it was "quieter and smoother" than with a carbon steel 80 ksi coil. Equipment wear (chain grippers) was more pronounced, because of the increased hardness of the 16% material.[11]

10.8.2.1 Pipe heavy coiled tubing runaway

When pipe-heavy (stripping), a runaway is most likely to occur when the coiled tubing is deep, well pressure is low, and the wellbore filled with gas. Snubbing force at the stripper will be low, there is little buoyancy to offset the weight of pipe in the well, and the axial load (tension) supported by the injector is high. Once a runaway has started it will be difficult to control. These actions may enable the pipe to be brought back under control:

- Attempt to speed up the injector to match the speed of the pipe, then:
- Increase inside chain tension to grip the pipe. If chains begin to grip, the injector can be slowed and brought to a stop. If grip is not improving, continue with the procedure.
- Increase stripper pressure to a maximum in an attempt to slow down the rate of runaway.
- If the runaway is hydraulic, manually set the injector break or reduce the hydraulic pressure to zero, to automatically set the break.

Note: Under certain circumstances, if the runaway tubing is at a speed above the critical speed, the back pressure created by the circulating hydraulic fluid may prevent the injector motor brakes from actuating. If this happens, select the pull mode for the injector and increase system hydraulic pressure until the tubing comes to a standstill.

- If none of this works:
- Close the slip rams. Closing the slips is a measure of last resort, and will almost certainly result in damage to the coiled tubing and to the slip rams. There is a very real possibility of the pipe parting as the slip ram is closed.

If the coiled tubing is not too far off the bottom when a runaway occurs, it may be practical (and safer) to simply let it fall to bottom, and then investigate the causes and carry out repairs. Clearly, this can only be done if there is sufficient tubing on the reel to reach the bottom. Allowing the pipe to impact the bottom of the well can damage the BHS and "helix" the pipe above the point of impact, making for a difficult recovery.

Once pipe movement has been arrested, either by regaining control with the injector, by closing the slips, or because it has reached the bottom of the well, it will be necessary to carry out injector repairs before the coiled tubing can be recovered from the well:

- Shut down the power pack and tag the control valves before any person climbs onto or works on the injector head.
- Open the injector chains and clean or repair the gripper blocks.
- Ensure that all personnel are safely clear of the injector.
- Close the injector chains, remove the tags from the injector controls, and reinstate hydraulic power to the injector head.
- Adjust the chain gripper pressure.
- Energize the stripper.
- Equalize pressure across the pipe rams. Open the pipe and then the slip rams, and inspect the surface of the CT contacted by the rams for signs of damage.
- If the CT is undamaged, continue CT operations.
- If the CT is damaged, POOH and make permanent repairs or change out the reel.

10.8.2.2 Pipe-light coiled tubing runaway

A runaway when pulling out of the well is most likely where wellhead pressure is high and pipe weight is low (BHA close to surface). The injector chains might completely lose their grip on the pipe (traction runaway) or, less common, the injector motors spin freely (hydraulic runaway). This mode of runaway is unusual, as the counter balance valves on the injector motor hydraulic would usually stop it happening.

If pipe is being blown out of the well, it needs to be stopped quickly. If not it can buckle below the stripper with catastrophic results. If the BHA reaches the stripper it is likely to shear and drop into the well.

If pipe is being blown out through the stuffing box, it might be possible to regain control by:

- Increasing injector speed to match that of the pipe being blown out of the well.
- Reel rotational speed should to be increased to stop loose wraps forming, or the pipe dropping off the side of the drum.
- Increasing inside chain tension to increase grip on the pipe, then begin to slow down the injector speed.
- Increase stripper pressure to increase friction on the pipe—this might help slow it down, but the rate of wear on the stripper will be high.
- If the runaway is hydraulic, try manually setting the injector brake. Alternatively, reduce the injector hydraulic pressure to zero to set the injector brake.
- Be ready to close the Christmas tree master valve if the pipe breaks and comes out of the stripper.
- Flowing the well will reduce snubbing force, but it is unlikely there would be time to start to produce the well. If the well is already flowing, and if it is possible to do so, increasing the choke might further reduce snubbing forces.
- If none of this works, and the slip rams are bi-directional, they can be closed. As with the pipe-heavy runaway, this risks pipe damaging or parting the pipe.
- If closing the rams does stop the pipe, engage the manual locks then close and manually lock the pipe rams.

10.8.3 Failure of reel hydraulic motor

If the coiled tubing reel motor loses power, it will not be possible to maintain tension between the injector head and reel. This can cause pipe on the drum to slacken off, and there is a risk of loops of pipe falling off the side of the drum and striking anyone alongside and damaging the pipe. If the reel motor fails:

- Immediately stop the injector.
- Inform all relevant personnel of the situation and warn them to keep away from the reel until the situation has been assessed.

- Slowly pick up a few feet to put some pipe slack between the guide arch and the reel.
- Close the BOP slip and then the pipe rams, and apply the manual locks.
- Apply the reel break (if fitted).
- Repair the reel drive and inspect the pipe for damage.
- If the pipe is damaged between the reel and the injector, make a temporary splice.
- Open the pipe and slip rams and POOH.
- If the pipe is undamaged, open the pipe and slip rams and run down as far as needed to correct the slack pipe on the reel. Avoid running any kinked or damaged tubing through the injector.
- Once all the slack tubing has been removed from the reel, pull out of the hole, spooling the pipe correctly.

If the reel cannot be repaired in situ, the coiled tubing above the injector head will have to be cut and the reel removed before a new reel is brought to the location.

- Confirm check valve integrity by bleeding off reel pressure if it is safe to do so—do not risk collapsing the reel if there is a high external differential pressure.
- Pumping water into the reel might be necessary. Flowing the well can also help reduce external reel pressure.
- If it is safe to do so, cut the pipe above the injector and remove the damaged reel.
- Replace damaged the reel and connect the end of the pipe protruding from the injector to the pipe on the new reel—make sure there is enough reel capacity before continuing.
- Open the pipe and slip rams and POOH.

10.8.4 Injector replacement

If the injector fails whilst coiled tubing is in the well, and it cannot be repaired in situ, it will have to be replaced. This will involve having to cut the coil tubing. Proceed as follows:

- Close the pipe rams and slip rams. Apply both manual locks.
- Ensure that the slips are holding the tubing.
- Back off the power reel pressure enough to get a little slack in the tubing at the gooseneck.
- Install a safety clamp on the coiled tubing approximately 4 ft (1.2 m) above the injector chains. Using suitable restraints, secure the tubing above the gooseneck on either side of the area of tubing to be cut.

- Ensure that the pressure is bled off the coiled tubing, and that the double check valve is holding pressure downhole. If there are concerns about collapsing the coil, a heavy fluid might have to be pumped into the reel to reduce collapse risk.
- Cut the tubing above the injector. Use a hacksaw initially, until a small hole has been cut. This will release any trapped pressure without the tubing breaking in two. Once all pressure has been removed, complete the cut with tubing cutters.
- With the tubing cut, spool the end back onto the reel.
- Depressurize any stripper pack-off pressure, and then depressurize the riser between the stripper and the (closed) pipe ram.
- Slack off inside chain tension, and very carefully and very slowly rotate the chains in the direction of the run in hole.
- In most coiled tubing pressure control configurations, the stripper is connected to the base of the injector head. To lift the injector, it will first be necessary break the connection between the base of the stripper and the BOP (quick union or flange).
- Carefully lift the injector with the hoist (i.e., crane or rig blocks) until it is possible to install a safety clamp on the tubing below the injector and above the BOP.
- Remove the safety clamp from above the injector, and lift the injector off the tubing protruding above the BOP.
- Lift the replacement injector into position above the free standing pipe.
- Reduce inside chain tension and rotate the chains out-hole slowly, while carefully lowering the injector to just above the safety clamp. Tubing should be picked up through the injector as it is slowly lowered.
- Place a safety clamp above the injector, then remove the safety clamp from below the injector.
- Continue to lower the injector until able to connect the union below the injector stripper to the BOP.
- Increase inside chain tension and apply tension equal to the buoyed weight of the tubing suspended in the slips.
- Open the manual locks on the pipe and slip rams.
- Equalize pressure across the pipe rams.
- Open the tubing and slip rams.
- Pull out of the hole until there is enough slack pipe above the injector to allow the pipe in the well to be spliced to the pipe on the reel using an in-line connector.

  Note: Do not run the inline connector back into well.

10.8.5 Hoist failure with pipe in the hole

If the hoist (crane, rig block, hydraulic lift) supporting the injector and BOP fails with the pipe in the hole:

- Check the injector, pressure control equipment, tree, and wellhead for damage.
- Close and manually lock the BOP slip and pipe rams whilst repairs are made.
- Equalize across the pipe rams. Unlock and open the pipe and slip rams, and continue operations.

10.8.6 Leaking coil tubing at surface (above the stripper)

A visible leak in the coiled tubing usually means a weakness that could easily lead to pipe failure (parted tubing) if cycling of the pipe over the guide arch were allowed to continue. Pinhole leaks are most commonly observed as the pipe passes over the guide arch, and are usually the result of internal corrosion or fatigue cracking. Action must be taken to secure the damaged pipe before it parts. The initial response will depend on the nature of the fluids being pumped, and system pressure.

- If acid or other hazardous chemicals are leaking from the reel, immediately begin to flush with water.
- If it is safe to do so, stop pumping and reduce the pressure in the reel. If the check valves in the BHA hold pressure the leak should diminish and pipe pressure reduce.
- If well pressure is high and internal reel pressure is low, then bleeding back risks collapsing the pipe.
  - Avoid exceeding pipe collapse pressure.
  - Consider pumping a heavier fluid and/or flowing the well to reduce pressure differential and therefore collapse risk. If there is a hydrate risk, the fluid pumped should be inhibited using glycol.

The next step is dependent on the integrity of the check valves. If they are holding and the leak rate is diminishing or stopped:

- Pull out of the hole to position the leak on the lower part of the reel (be prepared to deal with containment of any hazardous fluid).
- If hazardous chemicals were in the reel, e.g., acid, flush through the reel with water. Wash down and neutralize any spillage.
- If there are concerns about reel collapse, flowing the well and/or pumping through the reel will reduce the risk.
- Pull out of the hole and replace coiled tubing reel.

Note: In some locations, the practice when faced with a pinhole leak is to immediately run the leak through the stripper to contain it within the wellbore. Whilst this may appear attractive, it can make matters worse. The pipe will be weakened at the leak. If it is run into the well, the already weakened pipe will be subjected to additional loads and this might be enough to cause it to fail. In a pipe-light situation, the compression load to force the pipe through the stripper could cause a buckling failure between the bottom of the injector chain and the stripper. In a pipe-heavy (stripping) situation, running below the injector subjects the damaged pipe to maximum axial tension. Failure is a real possibility.

If the check valves are not holding, and the leak is not diminishing or getting worse:

- Observe the severity of the leak, and decide whether it is safe to pull out of the hole. Factors such as fluid type and area of dispersion will influence the decision. It may be necessary to pump through the coil to prevent an influx of wellbore fluid.
- If the leak is too severe to continue pulling out of the hole, the slip rams and pipe rams should be closed and manual lock set.
- Operate shear rams to cut the pipe and remove the cut pipe from the BOP before closing the blind rams.
- Circulate kill fluid through coiled tubing left in the well via the side outlet port between the shear ram and slip ram.
- Recover the remainder of the pipe from the well.

10.8.7 Leak in the coiled tubing below the stripper

If reel pressure is higher than well pressure, a leak from the coiled tubing will be seen as a drop in circulation pressure. When this is observed proceed as follows:

- Stop reel movement.
- Stop pumping for long enough for reel pressure to stabilize. It may be possible to estimate the depth of the leak from hydrostatic calculations, but beware of reel collapse.
- If acid, nitrogen, cement, or abrasive slurry is being pumped it is advisable to displace the reel to water or brine. Pump at a rate/pressure high enough to prevent hydrocarbons from entering the reel.

If difficulties are experienced calculating the depth of the leak, or if the leak is severe and getting worse, killing the well is advisable. Continued pumping, necessary to keep hydrocarbon out of the reel, risks washing out the pipe—this could cause it to part. However, if wellhead pressure is low, and the leak rate manageable, then it may be possible to recover the pipe from the well.

- Continue pulling out of the hole while circulating at a high enough rate to keep well bore fluid (hydrocarbons) out of the coil.
- Watch for signs of the leak passing the stuffing box. As long as the fluid escaping is the same as that being pumped (i.e., water or brine) it should be safe to pull the leak back on to the reel.
- Once the leak is seen at the surface, pull it back on to the reel. It is best to put the leak on the low side of the reel so it drains when pumping stops.
- If the well pressure is low (below the collapse rating of the pipe), it may be possible to stop pumping at this point and confirm the integrity of the check valves. If the check valves are not holding, circulation must be resumed to keep hydrocarbons from escaping at the surface.

If internal reel pressure is lower than well pressure, then the first indication of a failed check valve or pipe leak will be increasing circulating pressure. If this occurs:

- Stop pipe movement.
- Stop circulation for long enough to allow reel pressure to stabilize. Hydrostatic calculations can be used to try to establish where the leak is. The leak could be anywhere in the tubing, or in the BHA (check valve failure).
- Circulate water or brine into the reel to flush out hydrocarbons—if gas has entered the reel, the pump rate will need to be higher than the gas migration rate to remove gas from the pipe.
- If the leak is in the tubing, too high a pump rate risks washout and a parting of the pipe.
- With a safe (nonhydrocarbon) fluid in the coil, pull out slowly until the leak is above the stripper. It may be necessary to continue pumping to prevent collapsing the reel once the leak is at surface.
- If the leak is too severe, then run back into the well, set the slips, close the pipe rams, shear the pipe, and close the blind rams.

When pulling back to the surface with leaking tubing, there is always a risk of the tubing parting and pulling out of the stuffing box—there is no BHA to stop it. If this happens, try to stop the pipe before it leaves the top of the injector chains. Close the BOP.

Note: Reel pressure will bleed rapidly as soon as the fractured end exits the stripper. In high pressure wells, the loose end can be blown up and out of the injector and will flail about. It is important to ensure everyone is clear of the area before the pipe exits the well.

10.8.8 Riser leak

Major riser leaks are rare, and much less likely if flanged connections are used to make up the pressure control equipment. Leaks are more likely if the rig up is not properly supported. If the riser between the BOP and the Christmas tree begins to leak, take the following action:

When only a short section of pipe is in the hole:

- If it is thought safe to do so (as judged by the senior operating company supervisor on location), pull out of the hole to above the Christmas tree (and shear/seal BOP if used).
- Close the swab valve on the Christmas tree.
- Close hydraulic or manual master valve on the Christmas tree.
- Bleed off pressure in riser and repair leak.
- Pressure test all broken connections and recommence operations.

When the pipe is deep in the well and pulling back to surface would expose people and plant to too much risk:

- Close shear rams (or shear/seal BOPs) to cut coiled tubing.

  Note: If the coiled tubing was on the bottom, or very close to the bottom of the well, it should be picked up off the bottom a sufficient distance to ensure it drops below the tree when cut.
- Allow a few seconds for the coiled tubing to drop below the Christmas tree, then close in the master valve and swab valve.
- Repair leak in riser and pressure test all broken connections.
- Commence fishing operations.

If the coiled tubing is deep in the well, but the riser leak is small, it will be tempting to try to pull out of the well, thus sparing a lengthy well kill and fishing operation. The decision to attempt to recover the coil will be at the discretion of the operating company supervisory staff on site and operating company policy and procedure. Pulling out of the hole should only be attempted if:

- There is no H_2S in the leaking hydrocarbon.
- There is a shear seal BOP in the rig up, and the leak is above the shear seal BOP.

- The facility to pump into the tubing (coiled tubing/production tubing annulus) is already in place and can be used without delay.

If these conditions are met then proceed as follows:

- Pump water (or water glycol mix if there is a risk of hydrates) slowly down the production tubing, ensuring collapse pressure is not reached.
- Pull out of the hole. Monitor the leak continuously. Be prepared to cut the pipe and close in the well if it worsens.
- As soon as the pipe is back in the riser, close in the well and bleed off.
- Repair the leak.

10.8.9 Leaking stripper rubber

Stripper rubber wear is worse if packing pressure is too high, the surface of the pipe is rough (rusty), and corrosive fluids are present. That said, even during routine interventions, the seal element in the stripper can wear enough to leak.

If the stripper rubber begins to leak:

- Stop pipe movement.
- If the coil is being used to remove scale or fill from the well, continue pumping (circulating) fluid at the planned maximum rate to reduce the possibility of the pipe sticking.
- If a small volume of nonhazardous fluid (water) or gas (nitrogen) is leaking past the stripper, the coil should be pulled off the bottom, cleared of scale or fill, before movement is stopped—this will be at the discretion of the operating company supervisor.
- Increase packing pressure to the stripper to stop the leak.

If increasing the packer pressure does not stop the leak, the stripper rubber will have to be changed. If a policy of two barrier isolation is required, then the pressure control equipment should be rigged up with two independent sets of pipe rams or a second stripper. It is common nowadays to use tandem strippers.

- Close and manually lock the BOP pipe rams. If more than one set of pipe rams is available, both should be closed to provide two barrier isolation.
- Ensure the injector is in neutral and the brake engaged.
- Bleed off any pressure trapped above the pipe rams.
- If tandem strippers are in use, pack off the lower stripper—this assumes the upper stripper was the operational stripper, and is the one leaking. This will provide two barrier isolation.

- Replace the sealing elements in the leaking stripper.
- Energize the stripper.
- Equalize pressure across the BOP pipe rams.
- Unlock and open the pipe rams.
- Unpack the lower stripper.
- Pick up a few feet and inspect the surface of the CT contacted by the rams.
- If the CT is undamaged, resume CT operations.
- If the CT is damaged, POOH and make permanent repairs or substitute a different reel of tubing.

10.8.10 Collapsed pipe

Collapse pressure values are more difficult to predict than burst. Both burst and collapse are a function of wall thickness, pipe diameter, and material yield. However, collapse pressure is complicated by ovality, and ovality occurs more readily in coiled tubing than jointed pipe, because it is repeatedly cycled over the guide arch and passed through the gripper blocks. Newman[12] and Luft[13] have developed accurate methods of calculating theoretical coiled tubing collapse values using "plastic hinge" theory. However, the application of theory relies on continual monitoring of pipe condition, since ovality changes with use. Actual collapse values are further complicated by tri-axial considerations. Although maximum axial tension occurs immediately below the injector head, peak tri-axial loading can occur deeper in the well. Pipe can collapse well below the stripper, and at well below the theoretical collapse value. Coil tubing collapse is less common than it used to be, because of improved monitoring and modeling, however, it can, and does, still occur.

If pipe does collapse, it will usually result in a sudden and unexpected rise in circulating pressure. A collapse close to surface will present particular problems:

- The stripper will not seal around collapsed pipe, and collapsed pipe will cause a leak if pulled up against the underside of the stripper.
- It is unlikely that flattened pipe will pull back through the bushings in the stripper.
- Blow out preventer pipe rams will not seal of flattened pipe.
- Blow out preventer slip rams are unlikely to be able to support collapsed pipe.
- The injector chains will not grip.

- Collapsed pipe is more likely to part.
- Circulation will be limited or impossible.

If tubing is being pulled out of the well, and collapsed pipe pulled up against the bottom of the stripper causes a leak:

- Immediately run back in the well. Run down a sufficient distance to place undamaged (round) pipe across the stuffing box and BOP. Placing undamaged pipe across the stripper should stop any further release of well fluids.
- Close the pipe and slip rams.

Once the well is secure, measures can be used to reduce collapse pressure on the coil:

- Flow the well, or if already flowing, flow a higher rate. Stop the annular fluid injection if reverse circulating.
- Increase the coiled tubing internal pressure by attempting to circulate.

It will be necessary to kill the well before the collapsed pipe can be removed. Since the circulation path through the coiled tubing will be severely restricted by the collapse, a bullhead or lubricate and bleed kill will probably be required. The method used to recover the pipe will depend on whether or not the collapsed coil can be pulled through the stripper.

Procedure if the collapsed coil can be pulled through the stripper assembly.

- Clamp the coiled tubing above the injector head. Attach the pipe to hoisting equipment.
- Release the pressure from the stuffing box and open the pipe rams (and slip rams if closed).
- Cut the tubing at the gooseneck and use the rig or a crane to pull the collapsed section of tubing through the stripper and injector head.
- Reclamp the tubing above the injector head and cut off in 30 ft (10 m) sections (or as appropriate to the crane or rig).
- Continue pulling and cutting tubing until all of the collapsed tubing has been removed from the well, and the remaining tubing can be pulled using the injector head.
- Once undamaged (not collapsed) tubing is above the injector chains (plus some excess) close the slips and pipe rams. Remove the clamps.
- Using a dual roll-on connector, join the end of the pipe protruding from above the injector to the end left on the reel. Take up any slack.
- Open the pipe and slip rams.

- Continue to pull out of the hole.

  If the collapsed coil cannot be pulled through the stripper:
- Clamp the pipe where it protrudes above the injector chains.
- Disconnect the stripper from the riser or BOP so that it can be lifted with the injector head.
- Using the traveling block (rig supported operations) or crane, raise the injector head to expose the pipe below the stripper.
- Clamp the pipe so it is supported when rested on the riser top or BOP (depending on the configuration of the rig up).
- Cut the pipe between the stripper and the clamp, leaving the cut coil suspended in the well (hanging from the clamp).
- Remove the damaged pipe from the injector head and stripper. If the stripper has been damaged, make any necessary repairs.
- Using the crane (or traveling block), pick up on the coil tubing that remains in the well. The aim is to remove all of the collapsed section and bring undamaged pipe to the surface. It may be necessary to cut and clamp a number of times before undamaged pipe is at the surface (above the BOP or riser).
- With undamaged pipe at the surface, clamp the pipe with enough "stick-up" above the clamp to pass through the stripper/injector.
- Lower the stripper and injector head over the end of the pipe protruding above the BOP (or riser).
- Connect the tubing in the injector to the end on the reel using a temporary connector.
- Pull out of the well with the remainder of the coiled tubing.

If it is not possible to run down to place undamaged pipe across the stripper and there is a leak at the stripper:

- Close the pipe rams. This might reduce or even stop the leak.
- Clear all nonessential personnel from the vicinity.
- Close the shear rams to cut and drop the coiled tubing—the cut must be made above the collapse.
- Close the blind rams.

10.8.11 Tubing kinked

If coiled tubing is run into the well using excessive speed, or if the hydraulic pressure is higher than that needed to move the coil at the required speed, there is a real risk of buckling the pipe. Where pipe in the well has lateral support, i.e., not much clearance between the pipe

OD and wellbore ID, buckling will usually occur between the stripper top and injector chain bottom. Where clearance is large (small coil inside large production tubing), buckling failure will occur below the stripper, usually where the wellbore ID is largest. The pipe will usually kink several times forming plastic hinges; a type of failure that is difficult to detect.

10.8.11.1 Tubing kinked below the stuffing box

A sudden weight loss associated with hitting an obstruction whilst running into the well does not necessarily mean the pipe is kinked. Other indicators are needed. These are:

- An increase in pump pressure as fluid or gas is being forced through an additional restriction created by a hinge.
- A decrease in pump pressure indicating the tubing has parted.
- A loss of string weight—parted tubing.
- An increase in string weight when pulling caused by deformed pipe dragging on the tubing wall. In some circumstances this is followed by a sudden weight loss as the buckled tubing fails.
- Difficulty pulling the (damaged) tubing back through the stripper.

10.8.11.2 Tubing kinked above the stuffing box and below the injector chain

This phenomena is easy to diagnose, since the damaged pipe is visible above the stuffing box. That said the pipe between the injector chain and stuffing box is not always visible from the control cabin. If buckling damage has occurred:

- Stop the injector as quickly as possible.
- Close the slip and tubing rams and manually lock them.
- If the downhole check valves are holding, bleed the pressure in the tubing down and cut the tubing. Use a hacksaw to start the cut until it can be confirmed that there is no trapped pressure in the tubing. Remove as much of the kinked tubing as possible, leaving any undamaged tubing showing above stuffing box intact so that it may be rejoined later.
- Bleed the pressure from above the tubing rams and undo the connection below the injector.
- Slowly raise the injector until it is clear of the damaged tubing.
- Cut away any damaged tubing, dress the tubing, and install an inline connector.

- Run undamaged tubing down through the injector until it protrudes below the stuffing box.
- Lower the injector down over the pipe sticking out of the BOP.
- Attach the pipe in the injector to the pipe held in the BOP using the inline connector.
- Slack off the inside chain tension. Rotate the chains slowly in the out of the hole (in a counter-clockwise direction), whilst simultaneously lowering the injector head until the quick union connection below the stuffing box can be made.
- Increase inside chain tension. Pick up until tension is equal to the weight of the tubing suspended below the slips +2000 lbs (900 daN) for friction.
- Equalize the pressure across the tubing rams.
- Unlock the tubing and slip rams.
- Open the slip and tubing rams and pull out of the hole, repairing or replacing the tubing on the power reel.

If the downhole check will not hold, then the tubing will have to be cut using the shear rams and the well killed.

10.8.12 Tubing parts at the surface

When tubing parts at the surface, the results can be dramatic and potentially lethal. If the end of the pipe is pulled out of the injector chain, it will flail about until reel pressure dissipates or can be contained. An excellent reason for keeping all nonessential people away from the reel and injector, especially when pumping high energy fluids; nitrogen being the best example.

In the event that the coiled tubing parts at the surface:

- Attempt to spool as much coiled tubing back on to the reel to avoid whiplash.
- If the broken pipe is still retained by the injector head chains, stop the injector then close slip and pipe rams. If the pipe has dropped into the well, then the blind ram (and/or Christmas tree valves) must be closed.
- If personnel are in danger from fluid release and/or the check valves are not holding, operate the shear/seal rams and commence well kill operations.
    - If the check valve holds and the leak is contained:
        - Monitor WHP while contingency plans are reviewed.

- Kill well and make necessary repairs to coiled tubing.
- Remove injector and feed coiled tubing back through injector chains.
- Install fishing spear. (Depending on the tubing stick-up, other methods of attachment may be more appropriate—e.g., a pipe-to-pipe connector.)
- Rig up injector and stab into top of fish, pull test spear then release slips and pull out of the hole.

10.8.13 Stuck pipe

The most likely causes of coiled tubing getting stuck in the well are:

- Solids settling and packing off around the pipe. This will happen if the fluid pump fails during a well clean out, or if fluid velocity in the annulus is lower than the settling velocity of the lifted solids.
- Unexpected increase in friction or drag where regular pull tests are not performed.
- Mechanical obstructions in the well.
- Inability to recover or set mechanical intervention tools coupled with a failure of the release mechanism.
- Differential sticking.
- Guns stuck during underbalanced perforating operation in unconsolidated formation.

In some situations, the only way of getting the pipe out of the well will be to cut it above the stuck point. However, before taking this drastic measured, it is worth trying to pull free from the obstruction.

- Work the pipe. Most operating companies have a policy that limits over-pull to 80% of the pipe yield. Keep a close record of fatigue cycles; do not risk parting the pipe. Although pumping whilst working the pipe will accelerate fatigue, it may be necessary if the nozzle is buried in fill.
- Circulate a friction reducer.
- Rapidly bleed off annulus pressure (if possible) while pulling on the pipe. This may cause sufficient backflow to dislodge debris.
- Try to increase buoyancy by pumping heavier fluid into the annulus and displacing the coiled tubing to nitrogen, but watch collapse pressure.
- Release the BHA by using ball operated shear sub if circulation is possible.

If none of this works, the next step will be to try to cut the pipe above the free point. In a vertical well and where the coiled tubing has uniform wall thickness (not a tapered reel), the free point can be determined by measuring stretch in the pipe as load (tension) is applied. *For a nonvertical well or where a tapered string is being used, a CT simulator should be used to determine the free point.*

- From neutral weight, slowly pick up to the agreed limit of coiled tubing tension. A usually accepted tension is 70% of the yield of the pipe. With 70% of pipe yield (F_1) applied, note the depth (L_1).

  WARNING: Check collapse pressure before picking up—tension reduces collapse resistance (tri-axial load).
- Slack off until the pipe is at 20% of yield (F_2) and record the depth (L_2).

Calculate the distance from the injector to the free point from the following equation:

$$L_{\text{free}} = \frac{(L_2 - L_1) \times A \times E}{F_1 - F_2} \quad (10.6)$$

where:

E is the modulus of elasticity of steel, approximately $30\text{x}10^6$ psi for carbon steel

A is the cross-sectional area of the tubing wall

L_1 is the pipe depth with a load equivalent to 70% of the pipe body yield applied

L_2 is the pipe depth with a load equivalent to 20% of the pipe body yield applied

F_1 is 70% of pipe body yield

F_2 is 20% of pipe body yield

Note: yield can be calculated by multiplying the pipe material yield by the cross-sectional area of the tubing.

Example stretch calculation.

Calculate the free point using the following figures:

Pipe properties:

Pipe OD = 2.375 in.

Wall thickness = 0.190 in. (no taper in reel)

Weight = 4.445 lb/ft

Material −80 ksi yield carbon steel

First, calculate the yield of the coiled tubing.

Area of tubing OD = $\frac{\pi}{4} \times 2.375^2 = 4.4301$ in.^{2}

Area of tubing ID $= \frac{\pi}{4} \times 1.1995^2 = 3.1259$ in.^{2}
Wall area $= 4.4301 - 3.1259 = 1.3042$ in.^{2}
Yield = pressure × Area. 1.3042 × 80,000 psi = 104,336 lbs. Maximum pull is 70% of 104,336 = 73,035 lbs (F_1)
20% of 104,336 = 20,867 lbs (F_2)
The following depths have been recorded:
Depth when pulling 73,035 lbs = 9322 ft (L_1)
Depth when pulling 20,867 lbs = 9327 ft (L_2)
Free point calculation.

$$L_{\text{free}} = \frac{(9327 - 9322) \times 1.3042 \times 30,000,000}{73,035 - 20,867} = 3750 \text{ ft (free point)}$$

Having established the free point, the next step is to cut the pipe above the free point (as estimated above). Cutting the pipe means losing the integrity provided by the check valves.

Releasing stuck coiled tubing using a chemical or explosive cutter.

- Pull the coiled tubing into tension so that it is as straight a conduit as possible. Tension in the string will also help when the pipe is cut.
- Close the pipe and slip rams hydraulically, and set the manual locks. Energize the stuffing box and increase the injector head inside chain tensions, ensuring that the coiled tubing is secure. Set the reel brake.
- Monitor for pressure build-up across the pipe rams.
- Bleed off the pressure from the coiled tubing.
- Monitor for pressure build-up at the reel to confirm downhole check valve and coiled tubing integrity.
- If the integrity of the barriers is proven, build a scaffold tower around the injector head. This will allow safe working above the injector head.

Note: If the check valves fail to hold, an attempt will have to be made to kill the well. Problematic if there is no circulation path.

- Remove the gooseneck latch down rollers, and cut the coiled tubing approximately 5 ft (1.5 m) above the injector head. Ensure that the coiled tubing is supported on both sides prior to cutting.
- Spool the reel end of the tubing back onto the reel and secure.
- If required for access, remove the gooseneck.
- Straighten the tubing above the gooseneck to allow the passage of wireline.
- Fit a compression (swage) connection with a cross-over to (normally) a 1502 hammer union.
- Install two plug (lo-torque) valves, with a large enough ID to allow a cutter to pass through.

- Rig up a pump-in tee and wireline BOP.
- Rig up wireline lubricator long enough to accommodate the cutter (Fig. 10.45).
- Run in the hole with the cutter and a free point tool on the wireline. When at depth with the wireline, cycle the wire a few times to release all torque, then cut tubing.
- Pull out of well with the wireline and observe the well. If required, circulate kill fluid through the coiled tubing via the temporary pump-in connection.
- Rig down the wireline BOP and lubricator, remove the "T" and plug valves. Cut off the swage connection. Dress the pipe.
- Open the BOP's and attempt to pick up on the pipe.

Note: Sometimes it is necessary to work the coiled tubing a few times to get the explosive or chemically weakened tubing to break and come free.

- If successful, pull out of the hole until there is enough tubing to join the end to the reel. Connect the two ends using a dual roll on connector.

Figure 10.45 Rigging up above cut stuck coiled tubing. A perfect illustration of why it is better not to get stuck in the first place—think well control!

10.8.14 Tubing parted downhole

A sudden loss of weight and a decrease in circulating pressure usually means the pipe has parted downhole. Care must be taken when pulling out of the hole after the tubing has parted, since there is no BHA to prevent the broken pipe being pulled right out of the stripper and a consequent loss of containment. To prevent this happening:

- Circulate fluid to keep hydrocarbon out of the coil (when the reel parts the check valves will be lost).
- Estimate the length of the remaining pipe left in the hole from pick up weight.
- Slowly pull the broken end of the pipe back towards surface. When it is calculated that the end of the pipe is getting close to the stripper, try to close the swab valve; count the number of turns. If it will not close, reopen the valve.
- Pick up on the pipe a measured distance. The distance picked up must be less than the measured distance between the swab valve on the Christmas tree and the bottom of the stripper assembly.
- After picking up the measured amount, try again to close the swab valve. If it will not close, pick up the measured distance once more.
- Repeat until the swab valve can be fully closed.

Note: The aim here is to pull the end of the pipe back inside the riser, but not out of the stripper. There is no BHA, so there is nothing to stop the pipe exiting the stripper and allowing the release of hydrocarbons—except care and patience.

Once the well has been closed in at the swab valve:

- Close the hydraulic master valve.
- Prepare to fish the CT left in the well.

It may be that the first indication of parted tubing is the release of well fluids as the tubing stub is pulled out of the stripper. In this event, the well should be closed in by the fastest and safest possible means.

- If the Christmas tree is equipped with an actuated valve and it has been connected to a single well control panel, the fastest and safest method of securing the well might be to vent panel pressure allowing the master valve to close. However, if there is a large length of riser between the tree and the stripper, the best course of action will be closure of the quad BOP blind ram or combi BOP blind/shear ram. This assumes the BOP is placed at the top of the riser. Depending on the configuration (access to controls), it may be possible to close the blind ram and tree actuated valve simultaneously.
- If there is no tree in place, or if the tree is only equipped with manually operated valves, the safest and quickest method of closure is to

activate the blind ram. Once flow from the well has been stopped, the tree valves can be manually closed.

10.8.15 Tubing pulls out of stripper

If the tubing is accidentally pulled clear of the stripper:

- Immediately stop the injector to stop the end of the coil leaving the top of the chains.
- Close the blind rams. If you are sure there is no pipe across the tree, the UMV should be closed as well. Closing the UMV may be a quicker means of securing the well if it is fitted with an actuated valve operated from a well control panel.
- Depressurize the reel and riser.
- Plan to fish the lost pipe from the well.

10.8.16 Unable to circulate

If a blockage downhole prevents continued circulation through the reel:

- If circulation is lost during a well cleaning operation, immediately pick up off the bottom or away from any point where debris might accumulate. In high angle or horizontal wells, it is advisable to keep picking up until the end of the coil is back in the vertical section of the well.
- Once clear of possible obstructions, reciprocate the pipe—do not reciprocate between the same two depths, change the fatigue points on the pipe.
- Activate the circulation sub or by-pass if fitted, or release the BHA to try to regain circulation.
- If it is not possible to restore circulation, POOH.
- Monitor pressure while pulling out to avoid bursting or collapsing the pipe.

10.8.17 Offshore platform operations

When coiled tubing is used on an offshore platform, all members of the crew must know how to react if there is an emergency. Whilst procedures will vary between instillations, action to make a well safe is universal.

There must be communication with the installation control room at all times, and a channel kept clear for the ongoing operation.

10.8.17.1 Production shut-down

This procedure assumes that the well is equipped with a SCSSSV, and that the master valve and SCSSSV are connected to a well control panel operated by the coiled tubing crew or operating company supervisor. This scenario will be discussed and agreed with platform, rig, or location supervisors before work begins. The controls agreed should be documented and signed off by all relevant supervisors. A copy of the agreement will accompany the permit to work.

- Stop fluid pump (or nitrogen converter).
- If the well has an actuated wing valve, this will usually close in response to the shut-down. Automated wing valves are not usually under the control of the well intervention crew, although there are exceptions.
  - If the valve has been isolated and is under the control of the Well Interventions Supervisor, it must be closed and the facilities' operators informed.
  - If the well is equipped with a manual wing valve, it should be closed and the facilities' operators informed.
- If scale milling or sand washing operations are underway, pick up to get the nozzle clear of open perforations and above any fill that could settle and stick the pipe when circulation stops.
- If there is time, pull back the distance equivalent to the distance (height) between the cutting rams in the coiled tubing pressure control equipment and the downhole safety valve. Should the situation deteriorate and a platform abandonment become necessary, then the coiled tubing can be cut and dropped. It should fall far enough to clear the SCSSSV, which can then be closed.
- If the reel end is above the SCSSSV, and circumstances dictate, close the SCSSSV. If the end of the reel is below the SCSSSV, stop the pipe and apply the injector break. Close and lock the pipe rams. If not already closed, close the wing valve.
- If the BOP is connected to an accumulator, and you are sure there is enough pressure in the accumulator bottles to shut the BOP, and then shut down the power pack. If the power pack is needed to function the BOP, leave it running. Inform the platform control room/management of the well status and await instruction.

10.8.17.2 Muster alarm

As per the procedure for a production shut down. Nonessential members of the crew report to their muster station. The operating company supervisor and essential members of the crew remain in the coiled tubing unit, ready to cut and drop the pipe if the situation deteriorates. The definition of who is essential must be clear from the offset of the job. Pass the details of those remaining at the wellsite to the muster point coordinator.

10.8.17.3 Platform abandonment

Follow the procedure for a production shut down (Section 10.9.17.1), then:

- Cut the coiled tubing and secure the well.
- If shear/seal rams have been rigged up immediately above the tree, cut using these then close in the tree valves. The coil may or may not clear the SCSSSV, depending on where in the well the reel end was when the pipe was stopped (during the initial production shut down). In all probability, well control panel pressure will be dumped irrespective of the position of the tubing in the well.
- If no shear/seal rams are fitted, the pipe will be cut with the shear rams in the quad BOP. If the pipe had been picked up clear of the bottom before the injector was stopped, then it should fall clear of the tree and the tree can be closed in.
- If the pipe was stuck on the bottom when the emergency began, or there was no time to pull back, then the pipe will have to be cut, pulled out of the BOP, and the blind ram closed.
- Some Christmas trees are fitted with hydraulically-assisted master valve actuators that are capable of cutting coil. These are the exception rather that the rule. If a well is equipped with such valves, they would activate during a platform abandonment.
- Shut down the power pack, and go to the muster station.

REFERENCES

1. Newmand KR, Brown PA. Development of a Standard Coiled Tubing Fatigue Test. SPE 26539, 1993.
2. Martin JR, van Arnam WD, Normoyle B. QT-16Cr Coiled Tubing: A Review of Field Applications and Laboratory Testing. SPE 99857, 2006.
3. Martin JR, van Arnam WD. The Development of Corrosion-Resistant Coiled Tubing Product. SPE 81721, 2003.
4. ANSI/API Specification 6A. Specification for Wellhead and Christmas Tree Equipment. Nineteenth edition, July 2004.
5. ANSI/API Specification 16A. Specification for Drill Through Equipment. Third Edition, June 2004.
6. Palmer R, Newman K, Reaper, A. Developments in Coiled Tubing BOP Ram Design. OTC-7876-MS, 1994.
7. International Coiled Tubing Association (ICoTA). An introduction to Coiled Tubing, history, applications and benefits. <www.icota.com>, 2005.
8. Lapeyrouse NJ. *Formulas and Calculations for Drilling, Production and Workover.* Amsterdam: Gulf Professional Publishing; 2002.
9. Michel C, Stephens R, Smith D, Crow W, King G. Reverse Circulation with Coiled Tubing – Results of 1600 Jobs. SPE 89505, 2004.
10. Report from the International Perforating Forum website. https://www.perforators.org/incident/incident-01012010-1.

11. Julian M, Martin B, McNerlin. Special Report: Coiled Tubing, BP field tests 16% as North Slope work string. *Oil and Gas Journal* 07/10/2006.
12. Newman KR, Collapse Pressure of Oval CT, SPE Paper 24988, Nov 1992.
13. Luft HB, Wright BJ, Bouroumeau-Fuseau P. Expanding the Envelope of CT Collapse Ratings in High Pressure/High Temperature Wells, SPE Paper 77611, SPE ATCE, San Antonio, TX, October 2002.

CHAPTER ELEVEN

Hydraulic Workover (Snubbing) Operation

11.1 INTRODUCTION

Of all the intervention methods described in this book, snubbing has the unenviable reputation of being the most dangerous. Most people who carry out interventions on live wells understand the risks of working with hydrocarbons under pressure, none more so than the crew of a snubbing unit. If mistakes are made, or equipment fails, the consequences for a snubbing operation are likely to be more severe than when performing a wireline or coiled tubing intervention on the same well.

A hydraulic workover (HWO) unit enables drill-pipe or completion tubing to be run and pulled from a well without the use of a drilling derrick. Moreover, it can do so whilst the well is still pressurized (live), allowing some workover and completion operations to take place with the well in an underbalanced condition. The HWO is the modern successor to early rig assist snubbing units. These date back to the early 1920s, and were used mainly in well control situations where, following a kick, pipe was forced into the well (against pressure) to establish a circulation path. Early snubbing units were rudimentary mechanical devices that needed the drilling derrick traveling block to operate. Snubbing cables were passed through pulleys on the rig floor, attached to the traveling block and secured to traveling slips. Picking up on the blocks closed the traveling slips on the drill pipe, and then forced the drill pipe into the well through a closed annular preventer. This act of pushing the pipe into the well against well pressure was, and still is, termed snubbing.

The 1950s saw the introduction of the modern HWO unit. Replacing the snubbing cables with hydraulic cylinders enabled more force to be applied, and at the same time eliminated the need to have a drilling derrick on location. Modern HWO units are easily transportable and have a relatively small footprint, making them ideal for all manner of workover and completion activities on both live and dead

Well Control for Completions and Interventions.
DOI: https://doi.org/10.1016/B978-0-08-100196-7.00011-7

wells. At present, most HWO units are used to perform interventions and workovers (recompletions) on dead wells, where putting a full-size rig on location is costly or logistically problematic. Although very little of the work carried out by HWO units is, strictly speaking, "snubbing" the term has stuck, and most people will refer to a HWO unit as a snubbing unit, irrespective of how it is being used. For the purposes of clarity, the following definitions will be used in this book:

HWO Unit: A hydraulically powered jack capable of pulling and running jointed pipe from a well. When used with the appropriate pressure control equipment, pipe can be run and pulled from a live well. A HWO unit can also be used to perform workover or completion operations where the well control barrier is a column of kill weight fluid.

Snubbing: The process of running pipe into a live well through the seal elements of pressure control equipment where the force, exerted by well pressure acting against the cross-sectional area of the pipe, is greater than the buoyant weight of the pipe hanging in the well. Loss of restraining force (grip) on the pipe would therefore cause it to be ejected (blown out) from the well. A condition where the pipe is ejected from the well if not restrained is also known as "pipe light."

Stripping: The term applied to "pipe heavy" conditions. A condition where the buoyant weight of pipe hanging in the well is greater than the force generated by surface pressure acting upon the pipe at surface. The transition between pipe light (snubbing) and pipe heavy (stripping) occurs at the *Balance Point.*

11.2 HYDRAULIC WORKOVER OPERATIONS

HWO units are used to perform operations that are beyond the scope of those normally performed using wireline and coiled tubing. They include:

- Recompletion (workover) on locations where the use of a drilling derrick is restricted or difficult. In some cases, the operation will be performed with the well "live," i.e., without the need to kill the well.
- Pressure control—snubbing pipe into live wells to provide a circulation path for well kill operations.
- Wellbore clean out operations. HWO units can deploy larger and more robust tubing than that used by coiled tubing. This enables higher

annular velocities to be achieved, and therefore more efficient hole cleaning can be realized, important for high angle and horizontal wells.
- Fishing and milling.
- Reservoir stimulation—acid and propped hydraulic fracturing.
- Sand consolidation treatments.
- Gravel packing operations.
- Installation of velocity strings.
- Through tubing drilling (side track, well deepening). Both conventional (overbalance) and underbalanced drilling is possible.
- Deployment of long perforating guns.

11.3 HYDRAULIC WORKOVER UNITS: THE ADVANTAGES

HWO units have several obvious advantages over other well intervention methods. There are, however, some drawbacks as well.

Advantages
- Greater snubbing/stripping force than coiled tubing units.
- Ability to deploy large diameter tubing, thus enabling higher circulation rates for wellbore clean out operations.
- Ability to run heavy wall high yield tubing. Heavy wall and high yield pipe can tolerate high axial loads, and has a higher burst/collapse rating than coil. Coiled tubing material yield is limited because of the bending cycles it must undergo.
- Normally able to operate at higher wellhead pressure than with coiled tubing.
- Pipe rotational capability—use of a rotary table enables more rotational torque than can be delivered using a mud motor.
- Normally more portable than coiled tubing—equipment can be broken down into lighter, smaller loads (although there will be more of them).

Disadvantages
- Cost: HWO is significantly more expensive than wireline, and generally more expensive than coiled tubing.

- Pipe tripping times are normally much longer than coiled tubing and conventional rig-based operations, especially when operating on a live well.
- Risk: tripping pipe in and out of a live well will always carry a higher risk than performing the same operation on a dead well.

11.4 RIG UP CONFIGURATION: AN OVERVIEW

For through tubing interventions on "live" production wells, the HWO, along with the associated pressure control equipment, is rigged up above the production Christmas tree. When conducting operations on a killed well, or a well with barriers in place, the HWO and associated pressure control equipment can be rigged up on the wellhead. For some live well interventions, the HWO is used in conjunction with a drilling derrick (rig assist). More usually, the HWO is operated as an independent system.

HWO equipment is modular and is rigged up using a crane. Individual components are generally not very heavy, e.g., a high powered four-cylinder jack unit weighs approximately 6 tonnes. The modular nature of HWO units and the modest weight of the components make them ideal for difficult-to-access locations such mountainous regions, jungle, or limited facility offshore satellite towers.

Basic equipment consists of:
- The basic HWO unit—the hydraulic jacking system.
- The workstring, tubing, and bottom hole assembly (BHA).
- Well control components—Blow out preventer (BOP) and associated control equipment.
- Auxiliary equipment—pipe handling, fluid storage and handling, emergency escape equipment (Fig. 11.1).

11.5 HYDRAULIC WORKOVER UNIT

A modern HWO unit is assembled from the following components:
- Jack assembly
- Control system
- Guide tube assembly

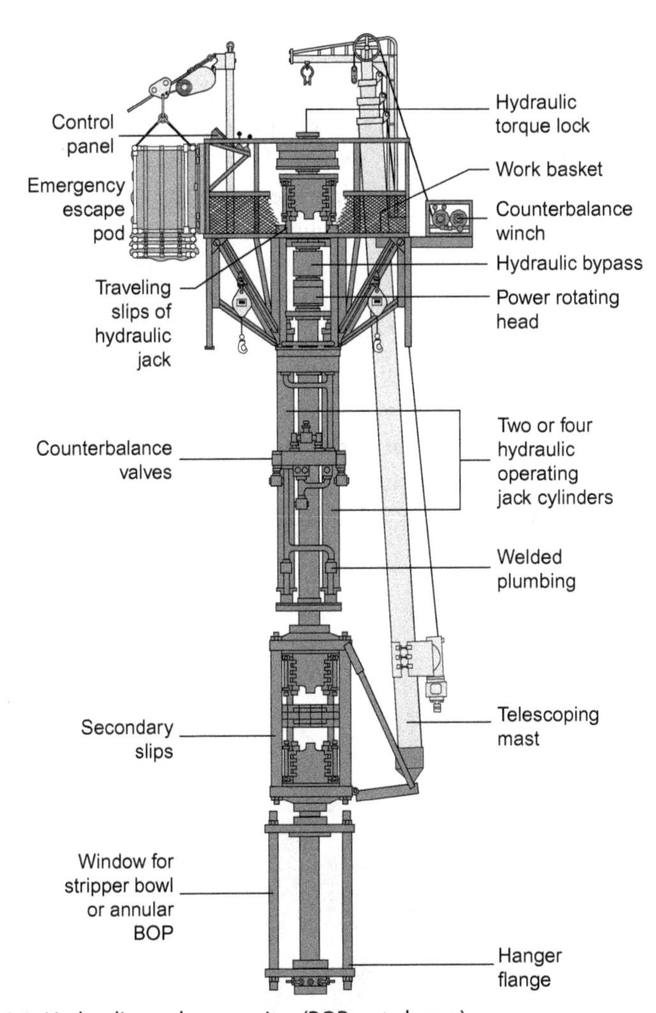

Figure 11.1 Hydraulic workover unit—(BOP not shown).

- Rotary table
- Traveling slips
- Stationary slips
- Guy wire system
- Work window
- Work basket
- Counterbalance winch
- Power pack.

11.5.1 The hydraulic jack

The hydraulic jack assembly consists of one or more hydraulic pistons that, when stroked up and down, move pipe in and out of the well. Most units are configured with either two or four cylinders. More cylinders mean more power, but slower tripping speeds. If the load is light, a four-cylinder unit can operate using two cylinders to increase trip speed. Most jacks use a regenerative hydraulic circuit to improve speed when tripping out. This circulates hydraulic fluid from the snub side of the piston to the lift side, reducing the volume of fluid required for each stroke, but, whilst speed is increased, lifting capacity is reduced (Fig. 11.2).

Lifting and snubbing capacity is a function of piston rod and cylinder size, the number of cylinders and the system pressure. Typically, lifting capability is normally about double the snubbing capability. The maximum lifting capacity of a unit is calculated from:

$$F_{\text{lift}} = \frac{\pi}{4} P_{\max} N C^2 \quad (11.1)$$

Figure 11.2 A four-cylinder hydraulic jack.

where:

F_{lift} = the maximum upwards force that can be applied (lbs)

P_{max} Maximum system pressure (psi)

N = the number of jack cylinders

C = cylinder ID (inches).

The maximum snubbing capacity is calculated from:

$$F_{snub} = \frac{\pi}{4} P_{max} N(C^2 - R^2) \tag{11.2}$$

where:

F_{snub} = the maximum snubbing force that can be applied (lbs)

R = piston rod OD (inches).

For example, a two-cylinder jack with a working pressure of 3000 psi, 9″ ID cylinders, and 6.5″ OD piston rods would be able to snub:

$$F_{snub} = \frac{\pi}{4} 3000 \times 2 \times (9^2 - 6.5^2) = 182,605 \text{ lbs}$$

and lift,

$$F_{lift} = \frac{\pi}{4} 3000 \times 2 \times 9^2 = 381,703 \text{ lbs.}$$

11.5.2 Slip window

The opening between the bottom of the inverted stationary slip and the top of any pressure control equipment is the slip window. This opening is used to make up equipment that is too large to pass through the guide tube. It is also used to allow hydraulic control lines, instrument cables, and ESP cables to be run during workover and completion operations (Fig. 11.3).

In live well snubbing operations a guide tube is installed in the slip window to prevent buckling of the tubing in the gap between the base of the inverted slips and the top of the annular of stripper rubber. The window guide must be installed and secured at any time the tubing is subjected to snubbing loads.

11.5.3 Traveling slips and stationary slips

Traveling slips are part of the traveling assembly which moves up and down as the jack cylinder piston rods extend and retract. The slips grip the pipe, transmitting piston force from the jack to the pipe, moving it in or out of the well. When snubbing (pipe light), the traveling slips are inverted,

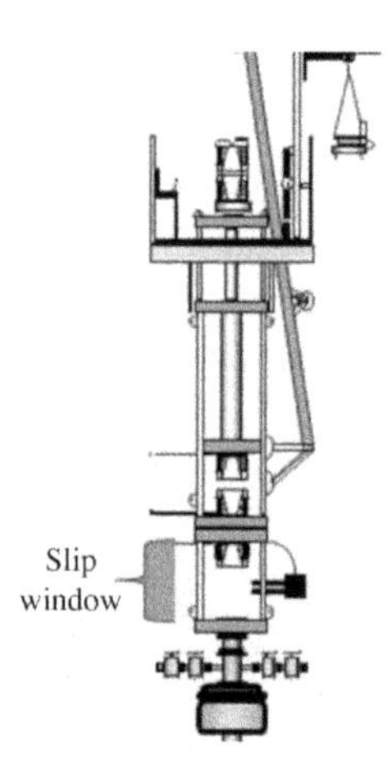

Figure 11.3 The slip window.

pushing the pipe through the primary pressure control barrier when running into the well. When pulling out, inverted slips restrain the pipe, preventing it from being blown out of the well. When stripping (pipe heavy) the slips are configured to support the weight of the pipe hanging in the well. Some HWO units use two sets of traveling slips. By having separate slips for pipe heavy and pipe light, a time advantage is gained; it is not necessary to invert the traveling set when the balance point is reached.

Stationary slips are attached to the base of the jack. They hold the pipe when the traveling slips are not engaged. The typical unit will have two opposed sets of stationary slips: a snubbing slip to prevent pipe upward movement of the tubing when pipe light; and a second pipe heavy slip to prevent downward movement when below the balance point and pipe heavy.

Where snubbing force is high (a combination of large OD tubing and high wellhead pressure) a second set of snubbing (pipe light) slips is sometimes used (Fig. 11.4).

11.5.4 The telescoping guide tube

During snubbing operations in live wells, force is required to push the tubing through the primary seal. As wellhead pressure increases, so the force required increases. Unless the tubing has good lateral support, it will buckle and fail. The telescoping guide tube provides the necessary lateral support, and should be used during all snubbing operations.

Figure 11.4 Inverted (top) and conventional (lower) stationary slips.

The guide tube, a subassembly of the jack, is attached to the bottom of the traveling slips. As the slips move up and down the tube telescopes, providing the necessary support for the tubing inside. The guide tube must be matched to the size of pipe being run (Figs. 11.5 and 11.6).

11.5.5 Rotary table

Some HWO units are equipped with a rotary table. The rotary table is attached to the traveling assembly and is hydraulically powered. Rotary speed (rpm) is normally controlled by fluid volume rather than hydraulic pressure. The rotary table is used mainly for milling and fishing operations when the well has been killed, although live well operations would be possible using a rotating annular of the type developed for underbalanced drilling.

When operating a rotary table, the traveling slips are mounted on top of the rotary table, to enable the pipe to be rotated and reciprocated simultaneously.

11.5.6 Power tongs

Power tongs, located in the work basket, are for making and breaking pipe connections. Tongs used in the work basket normally have integral back-up, and if premium connections are used will have computer analyzed make up.

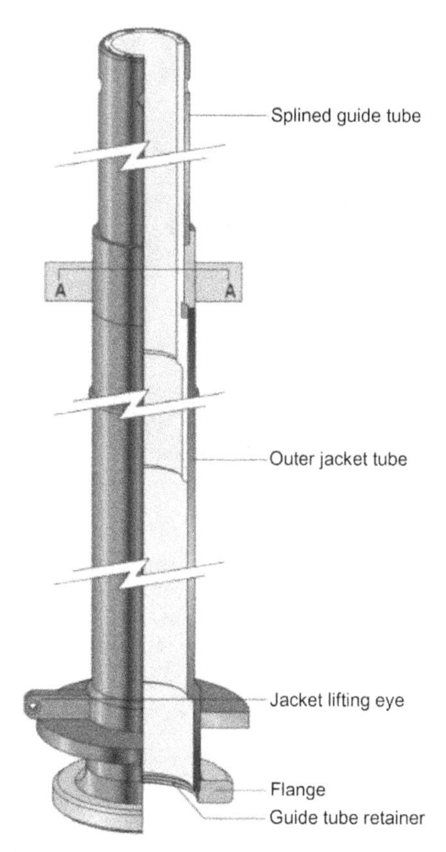

Figure 11.5 Guide tube.

Figure 11.6 Buckled tubing—the result of not using a guide tube. *Photo reproduced with permission of the SPE (from SPE paper 115534).*

11.5.7 Work basket and hydraulic workover control consoles

The work basket is positioned above the jack and contains all the controls needed to operate the HWO unit, including jack controls, slip controls, pressure control equipment, and the counterbalance winches.

Most HWO units will have several separate control consoles, both in the work basket and at remote (safe) locations. A typical arrangement is to have:

- Jack operator's console—controlling the jack, traveling slips and stationary slips.
- Assist operator's console—controlling the counterbalance winches and pressure control equipment.
- Remote BOP controls (located in a safe area away from the wellhead).
- Some units will have a second operator console, remote from the basket.

11.5.7.1 Operator's console

The jack operator's console (in the work basket):

- Jack direction and speed.
- Jack regenerative/full speed selector.
- Slip functions for both traveling and stationary slips.
- Metering of snubbing/lifting forces.
- Rotary table—if fitted.

Gauges usually included on the operator console are:

- Slips circuit pressure gauge.
- Lift pressure.
- Snub pressure.
- Weight indictor.
- Rotary table (if fitted) hydraulic pressure.

The snubbing operator monitors jack snubbing and lifting force on calibrated precision weight indicators. The gauges indicate output force by measuring hydraulic pressure directly from the active cylinder piston(s). Snubbing unit lifting and snubbing forces are a function of active hydraulic piston area and pressure. Because of the way the system works, on most units, weight (force) is only registered when the jack is in motion.

In addition to the primary hydraulic controls and weight monitoring, the console will normally have pneumatic control for the following functions:

- Power pack engine throttle.
- Power pack emergency shut down.

- Power pack start and stop (normal) function.
- Remote accumulator dump selection.
- Remote accumulator reset selection.

11.5.7.2 Assist operator's console (in the work basket)

A second operator's console, usually located on the opposite side of the work basket to the jack controls, is used to operate the pressure control equipment and counterbalance winch controls. In a standard configuration, the console will control:

- Upper and lower stripper rams, equalizing loop and vent.
- Annular BOP.
- BOP rams.

Most BOP consoles will have gauges to monitor the following functions:

- BOP system pressure.
- Auxiliary system pressure.
- Annular system pressure.
- Left and right counterbalance winch hydraulic pressure.

The auxiliary control valve is a three-way directional control that operates up to three hydraulic actuators on the equalizing and bleed off functions when stripping ram to ram.

The counterbalance winches are operated using two pressure adjustment valves, also mounted on or adjacent to the BOP control console.

11.5.7.3 Remote operator's console

Not all HWO units are equipped with a remote operator's console. A remote console is intended for emergency use only, and should be located as far from the wellhead as is practicable whilst remaining accessible. It normally has the controls necessary for emergency operation of the unit, e.g.:

- Jack direction and speed.
- Jack regenerative/speed selector.
- Slip operation.

Gauges on the remote panel include:

- Slip circuit pressure.
- Lift pressure.
- Snub pressure.
- Weight indicator.

11.5.7.4 Remote blow out preventer console

The remote BOP console is intended for emergency operation of the BOP if the work basket needs to be evacuated, or the crew in the basket is incapacitated. It is usually part of the accumulator package. A remote console should have the following functions:

- Control selector, to take control of the BOP at the remote station, overriding the basket control.
- Up to four ram BOP functions.

The remote console should have gauges to monitor:

- BOP system pressure.
- Bleed-off and equalization loop pressure.

11.5.8 Counterbalance winch

Two counterbalance winches are standard on any HWO unit. One winch supports the tubing as it is run or pulled through the jack, whilst at the same time the second winch lifts the next joint for running, or lowers a joint just pulled (Fig. 11.7).

11.5.9 Gin-pole (telescoping mast)

A gin-pole telescopes up above the work basket and supports sheaves for both counterbalance winches. This enables a full joint of tubing to be lifted above the basket for make up onto the pipe in the slips. Gin-poles are usually only used for stand-alone operations. Rig assist units will use

Figure 11.7 Counterbalance winch.

the rig derrick hoists to lift and lower pipe. Most gin-poles are extended and retracted using hydraulics.

11.5.10 Emergency evacuation of the work basket

Access to the work basket in most HWO units is by vertical ladders. Vertical ladders are not suitable for rapid evacuation of the basket in the event of an emergency, such as a blow out or fire. HWO units must therefore be equipped with a method of evacuating the crew quickly, easily, and safely. Many units are equipped with poles, in most cases two poles. They need to be:

- Sited to give easy egress from the basket.
- Located at either end or on either side of the work basket.
- Maintained in good working order.
- Have a landing area and route away from the well that is simple to follow and clear of all obstructions.

Escape poles are not suitable for all applications. Where the work basket is high above ground/deck level, alternative methods of egress such as zip lines or slides are used. Slides are widely used on land operations, and covered fireproof slides have recently been introduced to the industry (Fig. 11.8).

Figure 11.8 Fireproof escape chute—used where the work basket is high above ground level. *Image courtesy of Foxxhole™ Escape Sytems.*

For rig-assist operations, many operating companies follow the Canadian "ENFORM" recommendations and have a policy of not allowing anyone to access the derrick above the level of the HWO unit work basket.

All personnel who are required to work in the basket must be trained to use whatever emergency egress equipment is in use at the wellsite.

11.5.11 Hydraulic power pack and accessory equipment

Hydraulic power for the snubbing unit jack, rotary table, counterbalance winches, power tongs, and BOP functions is supplied by diesel powered hydraulic pumps. As well as the usual engine controls and (zone compliant) shut-down systems, the power-pack will have pressure gauges for each hydraulic circuit. Three main circuits are normally supplied by the power pack.

11.5.11.1 Main system pressure (jack pressure)

Main system pressure supplies the jack cylinders. The maximum jack pressure is set for snubbing in the hole, if the set value is exceeded, excess pressure will be relived, preventing buckling or tensile failure of the tubing.

11.5.11.2 Blow out preventer and slip operation pressure

This circuit is used to operate the BOP rams, including the stripper rams, the annular BOP, and all remaining ram preventers. It also supplies hydraulic pressure for the stationary and traveling slips.

11.5.11.3 Counterbalance winch pressure

Counterbalance winch pressure is regulated to operate at the minimum needed to lift the load. For example, when a joint of tubing is lifted into the basket, hydraulic pressure is adjusted to lift the load and no more. As the joint is run into the well, no further adjustment is needed, as the jack action will pull the winch cable off the drum whilst keeping tension on the joint to support it. Setting too high a pressure on the counterbalance winch can damage the gin-pole. There have been instances where the gin-pole has collapsed because winch back pressure has been set too high.

11.5.12 Hydraulic hoses

Hydraulic hoses are used to move hydraulic fluid to the jack, BOPs counterbalance winch, BOP, and rotary table. Return hoses route the hydraulic fluid back to the tanks of the power pack.

After rigging up, all hoses, fittings, and connections must be tested to the maximum rated working pressure of the unit. Both the snub and lift sides of the jack must be tested to the full test pressure. The test can be conducted with the jack in the fully extended and retracted positions. This exerts maximum hydraulic pressure on all lines and fittings in the jack loop.

11.5.13 Fluid (circulating) system

Fluid storage tanks, fluid pumps, high pressure hoses (or Chiksan®), and a circulating swivel make up the main components of the fluid system. If drilling operations are to be undertaken with the HWO unit, additional mud mixing and handling equipment would be added. In most respects, the fluid system is essentially the same as that used on a conventional drilling rig. Indeed, on rig-assist operations it is likely that the rig's own fluid system would be used.

Pumps need to be rated for the maximum anticipated wellhead pressure plus a margin, to allow for frictional pressure drop when circulating. Most pumps used for HWO operations on live wells will have to handle high pressure, and therefore are of limited output.

11.5.14 Guy wires and support system

For many stand-alone (not rig-assist) operations, guy wires are used to give lateral support. High rig ups will need different support wires at different levels on the structure.

Wellhead loading (and deck loading on offshore platforms) must also be calculated. A load supporting substructure might be needed to limit loads on the tree and wellhead.

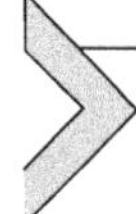

11.6 WELL CONTROL AND WELL CONTROL EQUIPMENT

All HWO unit operations are carried out through a BOP stack. For dead well operations, the criteria for configuring the BOP is the same as that used for any rig-based workover and completion operation (as described in Chapter 4: Well Control Surface Equipment). Operations on live wells require additional equipment that allows tubing to be run and pulled from a well under pressure. Where wellhead pressure is low, a stripper or annular BOP can be used as the primary external barrier. At higher pressures, pipe is stripped through ram type preventers. Ram preventers must be used where the pipe connections are not compatible with stripper bowls and annular preventers, irrespective of pressure. Ram type preventers are also used as secondary and tertiary barriers (pipe rams and blind rams). In addition to the BOP equipment, the tubing string BHA is equipped with non-return valves (NRV) to keep well pressure from entering the string. Before examining the equipment components and how they are configured, it is necessary to understand how barriers are defined for both live and dead well operations.

11.6.1 Barrier requirements and definitions (hydraulic workover)

Since HWO units have the capability to work on both live and dead wells, the configuration of well control and pressure control equipment varies enormously. Well control during HWO operations can be categorized as live well or dead well operations; either with or without a tree in place.

11.6.1.1 Dead well barrier definitions

Primary barrier.

- Kill weight fluid
  - If a tree is in place, closed tree valves would form the primary barrier during rig up, rig down, and when there is no tubing in the hole.
- Secondary barrier
  - BOP pipe rams or annular BOP.
- Tertiary barrier
  - BOP safety head—shear and blind rams.
  - If a tree is in place the tree valves would be closed after the tubing had been cut—providing the tubing dropped below the tree.

Since the well is dead, there is no requirement to have a check valve installed in the tubing string—the fluid barrier is the primary barrier. A full opening safety valve is the secondary barrier, and would be made up onto the top of the tubing and closed in if the well kicked. The tertiary barrier is the blind ram or blind shear ram.

11.6.1.2 Live well barrier definitions

Operations on live wells can be conducted through the existing completion with the tree in place. HWO units can also be used to run and pull completion on a live well where no tree has been installed.

11.6.1.3 External barriers during live well operations

Primary Barrier

- Where wellhead pressure is low, H_2S is not present or is at lower than harmful concentration, and where the tubing string has no external upset, or has tapered external upsets, a stripper bowl or annular preventer is generally used. The upper limit for an annular preventer or stripper bowl is normally 3000 psi (20,700 kPa).
- Where wellhead pressure is high, H_2S is present, and where the tubing connection design is such that strippers or an annular BOP would be damaged by connections being snubbed or stripped, the primary barrier is formed by either of two stripper rams.
- If a tree is in place, the tree valves would be the primary barrier during the rig up/rig down and whilst the tubing is out of the well.
- If no tree is in place, at least two independent (tested) mechanical barriers are required to enable the tree to be removed and the HWO unit to be rigged up safely.

  Secondary Barriers (pipe in the hole)
- If the primary barrier fails (a leak from the stripper bowl, annular BOP, or stripper rams) pipe ram preventers are closed to regain control of the well. Closed and tested pipe rams also allow the primary barrier to be repaired with pipe still in the hole. To conform with a policy of two barrier isolation during repairs, two pipe rams for each tubing size (tapered strings) must be included in the stack.

  Secondary Barrier (no pipe in the hole)
- Blind or blind shear ram preventer.

  Tertiary Barrier
- The tertiary barrier is reserved for use if both primary and secondary barriers fail. For snubbing operations on live wells, the tertiary barrier will be:
- A blind/shear ram or separate shear and blind ram.
- If a tree is in place the tree valves would be closed after the tubing had been cut—assuming the tubing drops below the tree.

11.6.1.4 Live well operation: pipe internal pressure control

During live well interventions, well pressure is prevented from entering the tubing string and flowing to surface using internal barriers.

Primary Barrier

- Check valve located in the BHA. In most instances, a double check valve will be used.

Secondary Barrier

- Full opening stab-in safety valve. This must always be kept in the work basket and be easily accessible. If the BHA check valve leaks the full opening valve is made up to the end of the tubing in the work basket and shut in.
- One or more wireline nipples are normally run as part of the work-string BHA. They are located a short distance above the check valves. If the check valves leak, wireline can be used to set a plug in the nipple profile. This restores the primary barrier and allows the tubing to be pulled to surface without having to kill the well. Depending on the design it may also be possible to pump a plug into the nipple.

Tertiary Barrier

- The internal tertiary barrier is the same as the external barrier—closure of the shear seal BOP will cut the pipe and shut in the well, providing both an internal and external barrier. If the tree is in place, and the cut tubing has dropped clear of the tree, the tree gate valves would also be closed to provide an additional barrier.

Several components are needed to ensure that the HWO pressure control equipment can fulfill barrier and well control requirements. These are described next.

11.6.2 Single well, well control panel

When HWO operations are conducted on a live well where the tree remains in place, a single well control panel (as described in Chapter 9: Wireline Operations) can be used to hold the tree valves and the surface controlled sub surface safety valve (SCSSSV), if fitted, in the open position. Unintentional closure of the master valve or SCSSSV whilst the pipe is in motion would be extremely damaging to the pipe, the tree valve, and the SCSSSV.

11.6.3 Stripper bowl

A stripper bowl assembly can be used as the primary barrier during live well operations, but only if wellhead pressure is low. It can be used with

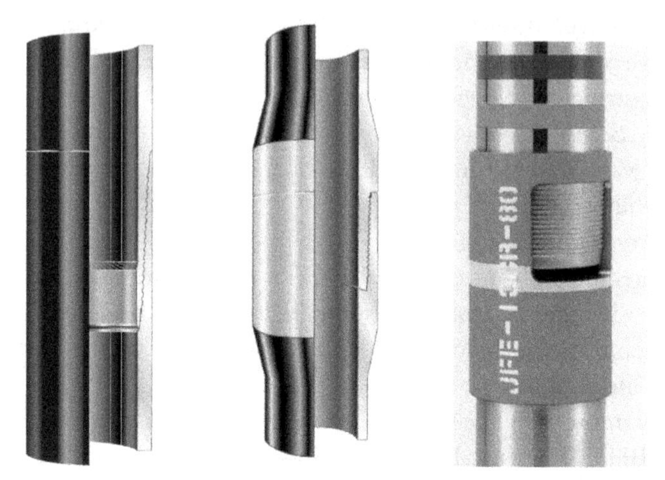

Figure 11.9 Flush joint (left). Tapered external upset (center). Threaded and coupled (right).

either a single or double stripper element. Pressure limits for the stripper bowl vary with pipe size and jurisdiction, but in general are not used when wellhead pressure exceeds 3000 psi. Some operating companies limit their use to below 2000 psi. A stripper bowl can also be used to:

- Wipe and clean pipe whilst pulling out
- Prevent debris from being dropped into the well.

A stripper bowl should only be used with tubing that has flush or externally tapered connections. Pipe with conventional threaded and coupling type connections would damage the seal element in the stripper bowl (Fig. 11.9).

Pressure control equipment should be configured to allow the stripper seal elements to be changed out whilst pipe is in the well. Unless the wellhead pressure is low or the well shallow, the stripper element will almost certainly need to be changed out at some time during the operation. The rate of wear and therefore the potential to a leak past the rubbers depends on:

- Wellhead pressure.
- Lubrication at the rubber.
- External pipe roughness.
- Connection type.
- Tripping speed.
- Seal element type (shore durometer of rubber).
- Wellbore fluid.
- Single or dual element.

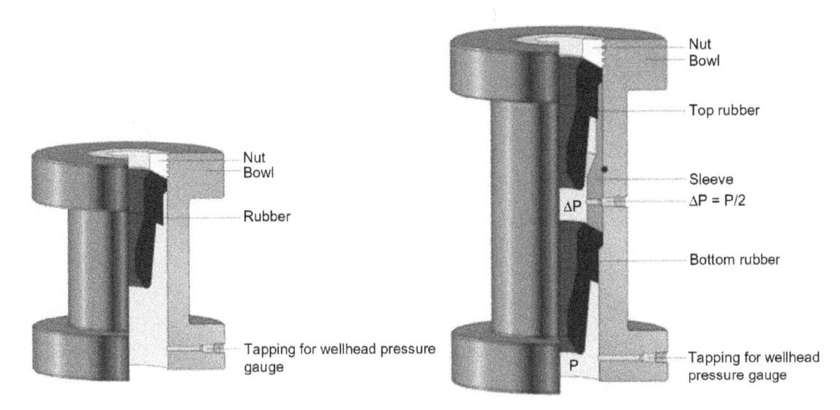

Figure 11.10 Single (left) and dual (right) stripper bowl.

In the dual element stripper, pressure equal to approximately half the closed in wellhead pressure is kept between the rubbers by a small pump and reservoir. This reduces the pressure differential across each of the elements. The single element stripper bowl only has one seal element, so differential pressure across the element is equivalent to the wellhead closed in pressure (Fig. 11.10).

11.6.4 Annular blow out preventer

Annular BOPs used for HWO operations are the same as those used in a conventional rig BOP stack. Annular BOPs can be used as a primary barrier in live well interventions. The limitations that are applied to the use of annular BOPs mirror those applied to the stripper bowl. Annular BOPs can only be used where:

- Wellhead pressure is low—normally 2000 psi or less.
- The tubing is equipped with flush connection or tapered external upset connections.

Annular BOPs have the advantage of being able to create a seal around equipment that is irregular in shape, or a different diameter from the tubing or workstring, e.g. a side pocket mandrel or flow couplings. Cameron, Hydril, and Shaffer annular BOPs are fully described in Chapter 4, Well Control Surface Equipment.

11.6.5 Stripping blow out preventer and associated equipment

Stripping BOPs are standard ram type BOPs with self-energizing elastomer elements that can seal on moving pipe. The seal inserts are designed to be robust, giving an improved life for stripping/snubbing.

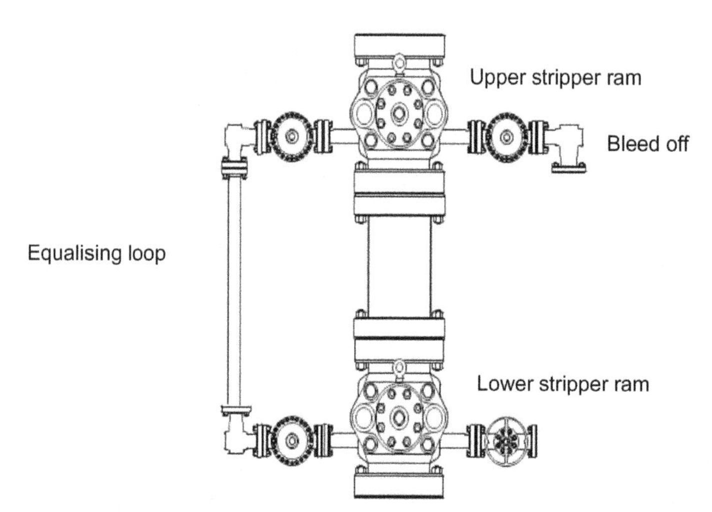

Figure 11.11 Stripping rams—equipment configuration.

Stripping rams are the primary well control barrier for live well HWO operations, and are used when the well pressure is too high for the annular BOP or stripper bowl. They are also used when the tubing connections are unsuitable for stripping through the BOP (threaded and coupled connections). Recovering and landing a tubing hanger (operations where the tree has been removed) are also carried out using stripper rams. Ram insert size is matched to the tubing OD, and the pressure rating of the rams must be more than the maximum closed in wellhead pressure (Fig. 11.11).

Stripping rams are used in pairs with a spacer spool between them. To enable tubing to be run past stripper rams (ram to ram stripping or snubbing) an equalizing loop and bleed-off are also needed.

11.6.5.1 Spacer spool

A spacer spool provides the clearance to allow tool joints (or other components, such as a tubing hanger) to be positioned between the two stripper rams.

11.6.5.2 Equalizing loop

The equalizing loop is a manifold connecting the two stripper rams. Running from below the lower ram to below the upper ram, it allows well pressure to be equalized across the lower ram before it is opened. Equalization is controlled by opening and closing a hydraulically actuated valve that is operated from the work basket. Most loops will also be equipped with manual gate valves or lo-torq valves in addition to the remotely operated valve. A positive choke located outboard of the low torque valves controls the pressure equalization rate and limits the effects of fluid hammer. The choke size depends on the wellbore fluid type. A larger choke is normally needed when liquid is equalized.

11.6.5.3 Bleed-off or vent line

The bleed-off or vent line runs from between the two stripper rams to a point away from the snubbing unit, and is used for bleeding off pressure between the rams. Vent lines are usually fitted with a hydraulically actuated valve that is operated from the work basket, and a second manual valve. Many vent lines are fitted with a variable choke positioned inside (upstream) the bleed-off valves. Using a choke reduces the tendency of the bleed-off valves to become flow cut. The route of the vent line is location and well condition dependent, and may be connected to the flare if one is available. Running or pulling tubing through stripper rams requires a "ram-to-ram" sequence to be followed, as described and illustrated in Fig. 11.12.

11.6.6 Blow out preventer ram preventers

Pipe rams close around the tubing to seal the well below the BOP. In both live and dead well operations, they function as a secondary barrier. In a live well operation, the pipe ram is positioned below the primary barrier and is closed in response to a leak from the primary barrier. A closed pipe ram provides the barrier necessary to enable the primary barrier to be repaired while pipe is in the hole. If the BOP stack has been configured to conform to API recommendations as laid out in API (RP) 53, the stack will have at least one blind ram. The inclusion of a shear ram or shear seal ram is largely dependent on well pressure and operational requirements.

11.6.7 Drilling spool

The drilling spool connects the kill and choke lines to the BOP. NORSOK recommend placing the spool "so that circulation for well control can be carried out with the workstring suspended in the BOP and the shear/blind ram closed."

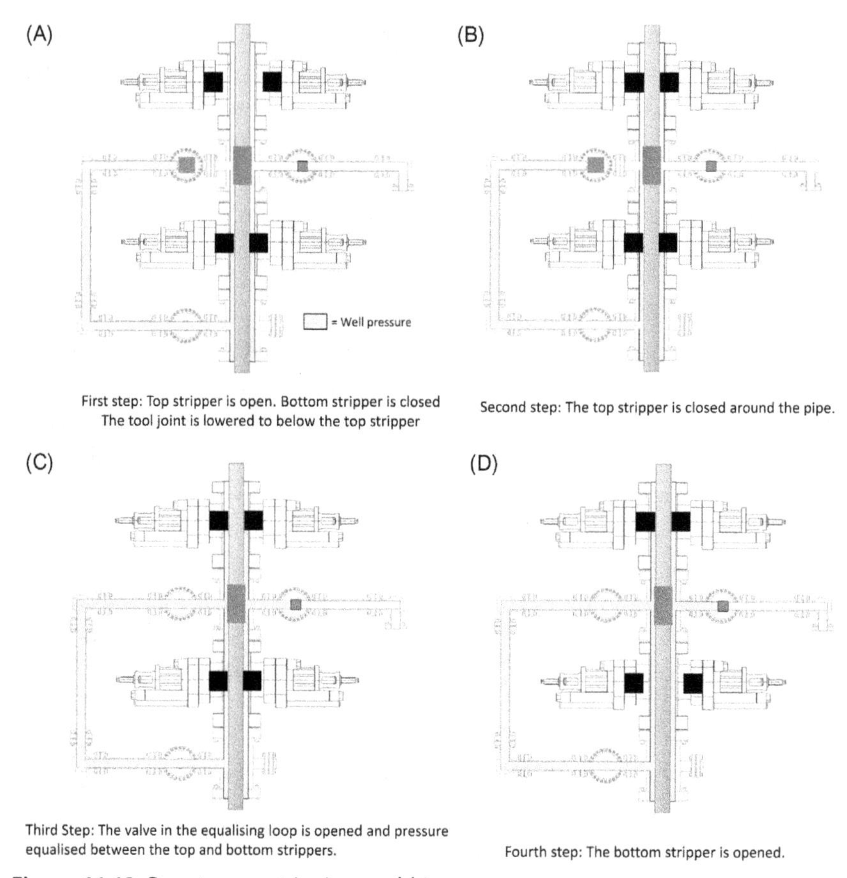

Figure 11.12 Ram-to-ram stripping snubbing sequence.

11.6.8 Lubrication

In the context of HWO, lubrication is the deployment into the well of BHA components that cannot be snubbed in because of their size or shape. The length of BHA that can be lubricated into the well is determined by how the BOP has been configured. The "Lubricating Space" is equivalent to the height between the base of the upper stripper ram and the lowest barrier in the stack able to isolate well pressure. Where applicable, Christmas tree valves form the lower barrier of the lubrication space. If no tree is in place, the lowermost blind ram is used. To comply with a policy of double valve isolation, two blind rams would be required (Fig. 11.13).

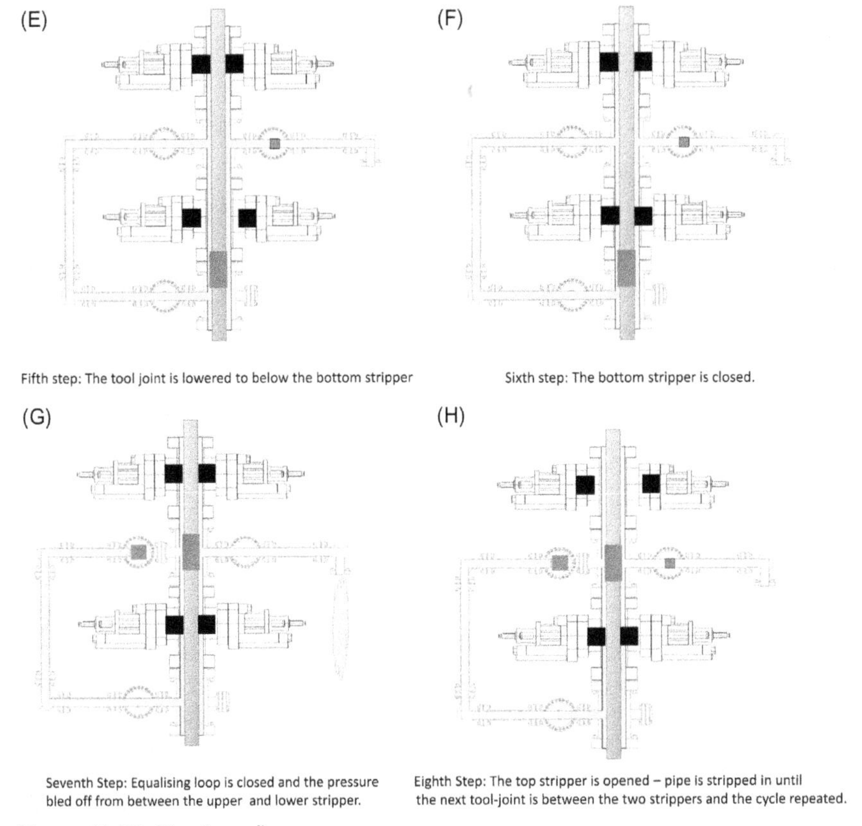

Figure 11.12 (Continued).

11.6.9 Blow out preventer control and operating system

Hydraulic pressure for the BOP and ancillary equipment is provided by the power pack. BOP functions are typically configured as follows:

- The operation of primary well control functions (stripper rams, stripper bowl, or annular BOP) is performed from the work basket control console.
- Secondary well control functions (pipe rams) should be accessible from at least two control consoles:
  - Work basket
  - Remote console, usually on the accumulator panel if it is in a safe area.
  - In some circumstances, a third remote panel may be located adjacent to an approved escape route.
- Tertiary well control (shear/seal ram) is usually operated from the remote accumulator panel, providing the panel is in a safe area. Alternatively, a third separate remote panel in a safe area can be used.

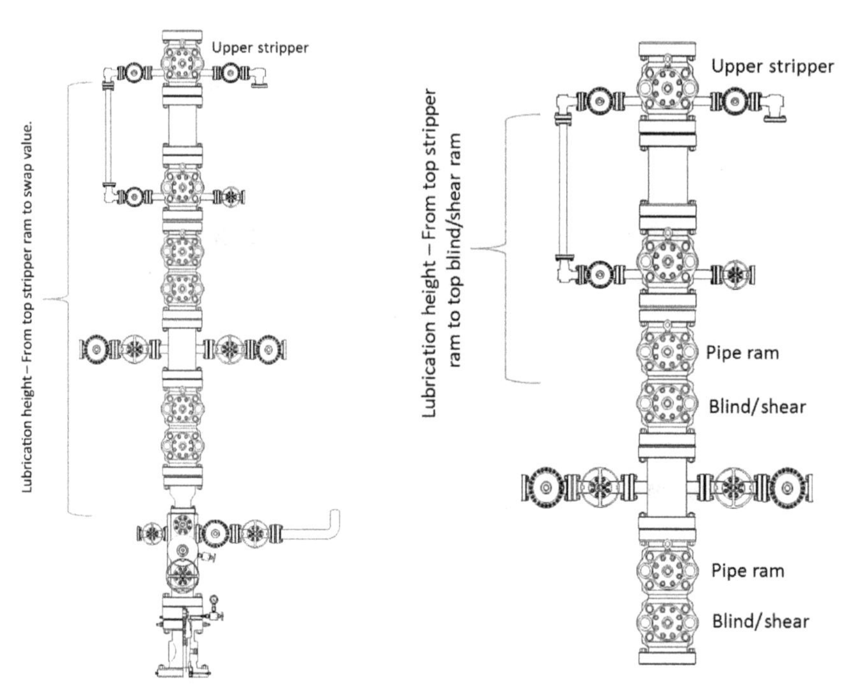

Figure 11.13 Lubrication space—through tubing (left) and operation with the tree removed (right).

Where possible, the accumulator unit console, used to operate the secondary and tertiary well control system, should be positioned in an area that would not be exposed in the event of an uncontrolled well control situation, i.e. away from the well and close to an escape route. For rig-assist HWO operations, the rig-floor is not usually considered as a safe area (Fig. 11.14).

The BOP control console and control system should have the following features:

- A panel that continually gives a visual indication of valve status (open or closed) for every function. Visual indication varies, but will usually be using selector leaver position, lights, or mechanical flags.
- Where appropriate, the control console should have a means to stop the accidental functioning of valves, particularly blind and shear rams.
- Audible alarms for all BOP control panels (basket and remote):
  - Loss of power (power pack)
  - Low accumulator pressure
  - Low level of control fluid.
- Regulators unaffected by a loss of power.

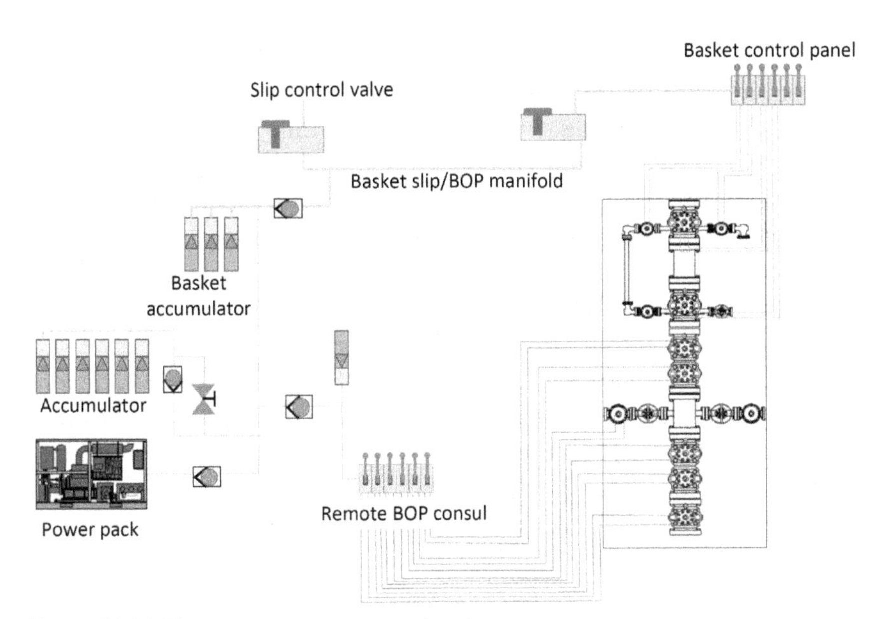

Figure 11.14 Blow out preventer control system.

- Closure time for ram BOP and annular BOP's must conform to recognized standards, and are the same as those detailed in Chapter 4, Well Control Surface Equipment:
  - BOP response time is the time between initiating closure at the control console, and the ram or valve completing its travel and sealing. Closing time should not exceed 30 seconds for annular preventers smaller than 18¾″ nominal bore and 45 seconds for annular preventers of 18¾″ and larger. Response time for choke and kill valves (either open or close) should not exceed the minimum observed ram close response time.
  - Measurement of closing response time begins at pushing the button or turning the control valve handle to operate the function, and ends when the BOP or valve is closed effecting a seal. A BOP may be considered closed when the regulated operating pressure has recovered to its nominal setting. If confirmation of seal off is required, pressure testing below the BOP or across the valve is necessary.
- Stripper rams are designed to respond more quickly (normally 5 seconds). This allows faster tripping speed when operating ram to ram.
- Equalizing loop, bleed-off line, and kill and choke valves have a rapid response time, usually approximately 1 second.

11.6.10 Well control system accumulator requirements

Accumulator capacity requirements for a BOP system used for HWO operations are basically the same as that used during conventional drilling operations. The API recommendation is that accumulators should have "sufficient usable hydraulic fluid volume (with pumps inoperative) to close one annular-type preventer, all ram-type preventers from a full-open position, and open one high closing ratio (HCR) valve against zero wellbore pressure."[1]

The NORSOK standard for BOP equipment used for interventions (including HWO) is more stringent, stating that:

> "*The accumulator capacity for operating a BOP stack with associated systems shall as a minimum have sufficient volumetric capacity to close, open and close all the installed BOP functions plus 25% of the volume for one closing operation for each one of the said BOP rams.*"
>
> *The following BOP's operated from the well control system shall be used as a basis when calculating accumulator capacity.*
>
> - *Annular preventer*
> - *Upper pipe ram*
> - *Shear/blind ram*
> - *Lower pipe ram*
> - *Equalizer loop valve*
> - *Bleed-off valve*
> - *Kill line valve*
> - *Choke line valve*
>
> *There shall be documentary evidence that the pressure capacity of the accumulators is capable of operating the rams according to the design requirements.*[2]

Irrespective of which standard is adhered to, operating in live wells will require more accumulator capacity because of the requirement to operate the ram-to-ram stripping equipment.

11.6.11 Blow out preventer equipment configuration live well interventions

The American Petroleum Institute make clear recommendations regarding the configuration of BOP equipment in API (RP) 53. Complexity is directly related to pressure rating. As pressure increases, more ram preventers are included in the stack. For HWO unit operations on dead wells, API (RP) 53 is valid and is used by most operating companies. However, the recommendations are only pertinent to dead well operations where fluid is used as a primary barrier. Operating in live wells requires a different approach, particularly regarding requirements for pipe ram preventers. If the primary barrier fails, pipe rams are closed to allow it to be repaired. Since most operating companies enforce a policy of two barrier isolation, it becomes necessary to have at least two sets of pipe rams for each pipe size (tapered strings) to comply. At present (2018), there are no industry-wide standards for live well snubbing/stripping pressure control requirements. However, regulatory authorities and advisory bodies for some jurisdictions have introduced recommendations and policy specifically for well operations with HWO units. Of these, NORSOK D-002 (Norway) and ENFORM IRP-15 (Canada), are the most comprehensive, and many aspects of their respective policies have been adopted by operating companies and vendors.

NORSOK HWO equipment standards are described fully in the NORSOK document, *System Requirements Well Intervention Equipment*. The main points relating to pressure control equipment are:

- The rams shall be able to close around the "workstring" and seal off the annular space in the BOP bore.
- Pipe and slip ram hang off capacity compatible with the workstring weight.
- Front packer on stripper rams shall be wear resistant for stripping.
- The annular preventer shall be able to close and seal off at maximum shut in or maximum expected operating pressure rating on open hole. The annular preventer shall operate (close) on both stationary and running objects. Stripping through the annular with the specified workstring shall be possible without damaging the tubular.
- The annular preventer shall be furnished with accumulator(s) as a surge dampener for emergency stripping of pipe and tubular.
- The shear/seal ram shall be capable of shearing the highest grade of workstring as well as sealing off the wellbore with lateral and face seals.
- The ram design needs to clear sheared workstring from the seal area as a part of the shearing and closing operation, to enable circulation rate through the workstring left in the well.
- The safety head shall be a combined shearing and sealing ram capable of shearing the workstring and sealing the bore with the maximum working pressure.
- Shear and seal capability shall be available when the "workstring" is unloaded, in tension, or in compression.

11.6.12 Blow out preventer equipment configuration: drawings

The drawings on the following pages are illustrative of the BOP equipment configuration for live well interventions. These drawings are intended for guidance only. Personnel planning a snubbing operation on a live well should ensure that the BOP configuration used conforms to local and company standards.

11.6.12.1 Example blow out preventer stack for live well interventions up to 5000 psi (34,500 kPa) working pressure: single pipe size

Fig. 11.15 shows a BOP stack for use on a live well where a single pipe size is to be used. It is compatible with NORSOK and ENFORM policy reccomendations. The main points to note are:

- Annular preventer in addition to the stripper rams.
- Two pipe rams (upper and lower) enable two valve isolation to be maintained when servicing the primary barrier (stripper bowl, stripper ram, or annular preventer).
- The upper pipe ram is positioned above the choke and kill line, enabling the well to be killed with the pipe ram (or blind ram) in the closed position.
- The positioning of the kill and choke line allows for circulation through the tubing after it has been cut, i.e., the pipe could be hung from the lower pipe ram, cut using the shear rams, and lifted above the blind ram. With the blind ram closed the well could be circulated or reverse circulated through the tubing suspended in the stack.
- Two tree valves act as the primary barrier when lubricating in the BHA. The distance between the swab valve and the lower stripper ram must be long enough to accommodate the BHA.
- A second shear/blind ram immediately above the tree (safety head) (Fig. 11.15).

11.6.12.2 Example blow out preventer stack for live well interventions up to 5000 psi (34,500 kPa) working pressure: tapered string with two pipe sizes

This configuration retains all the advantages of the single pipe diameter stack-up. Two additional pipe rams are required to enable double valve isolation for both pipe sizes. Variable bore rams are not recommended, as

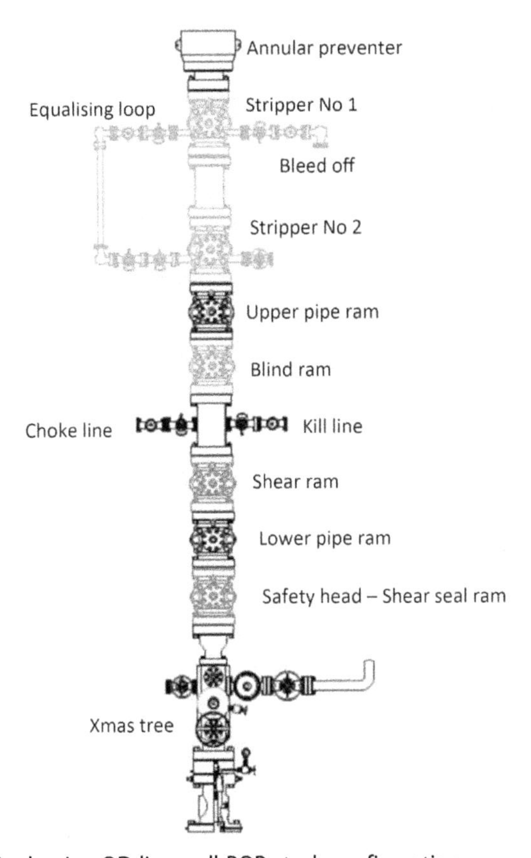

Figure 11.15 Single pipe OD live well BOP stack configuration.

they are unlikely to be able to support the weight of the string if it became necessary to hang off and cut (Fig. 11.16).

11.6.13 Tubing and workstring

There are two distinct classifications of pipe used with HWO units:

- Production/completion tubing. Although selection will be based primarily on the completion design requirements, the chosen tubing must be able to withstand any snubbing forces if the completion is being run into a live well.
- Workstring. Pipe used for through tubing interventions in live wells must be robust enough to withstand snubbing forces, compatible with well fluids, and allow repeated make up and break out.

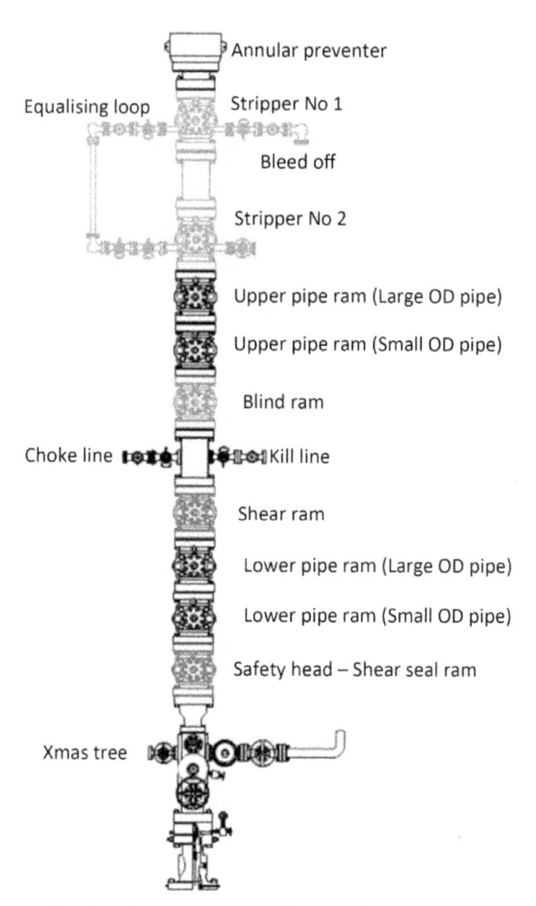

Figure 11.16 Tapered string (two pipe sizes) live well operation.

Snubbing and stripping into a live well will subject pipe to high stress. Of major concern are compression and collapse loads. Snubbing into high pressure wells places a large buckling and bending force on the tubing. Similarly, because the tubing must be run closed end to meet the requirement for an internal pressure barrier, there is an increased risk of collapse unless mitigating measures are taken.

Collapse data can be obtained from manufacturers tables. However, since collapse values are also a function of tension and compression, a more accurate prediction of collapse can only be obtained if a full triaxial analysis is performed. Tension limits will have to be adjusted to account for slip loading.

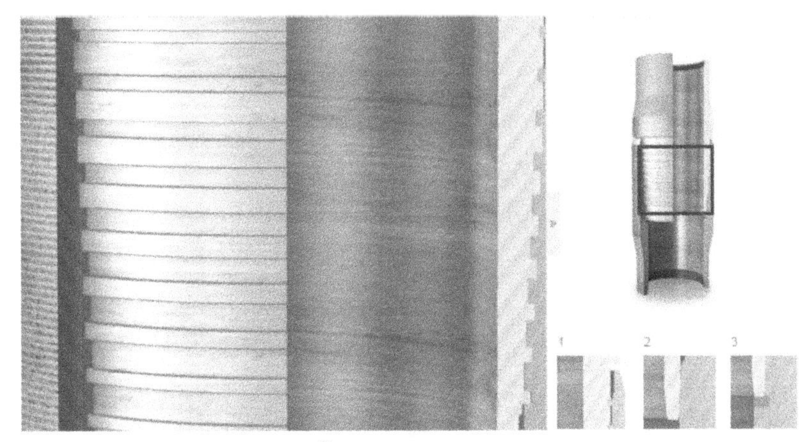

Figure 11.17 Tenaris Wedge 533® connection. *Courtesy of Tenaris Connections BV.*

11.6.13.1 Bending and buckling analysis

The workstring used for live well HWO operations needs resilient connections that can withstand multiple cycles of make-up and break-out. For example, a popular workstring tubular connection is the Tenaris/Hydril Wedge 533. This connection is designed for use with work string and drilling with tubing applications. It is robust enough to tolerate several make-up and break-out cycles. (Fig. 11.17).

11.6.14 Workstring well control barriers

During live well intervention, well pressure is prevented from entering the tubing using check valves located in the BHA. In most cases two are used, and these check valves form a primary barrier (element) and are part of the primary barrier pressure containment envelope, along with the tubing. If any part of the primary barrier fails; a leak past the check valves or a tubing leak, a full opening stabbing valve is used to shut off flow at surface.

11.6.14.1 The full opening stabbing valve

The stabbing valve needs to be available in the work basket at all times for immediate use, and is the secondary (internal) barrier. It should to be full opening, i.e., the bore through the valve should be at least as large as the tubing ID. Ideally, the stabbing valve should have the same thread connection as the workstring. If not, then the valve should be made up

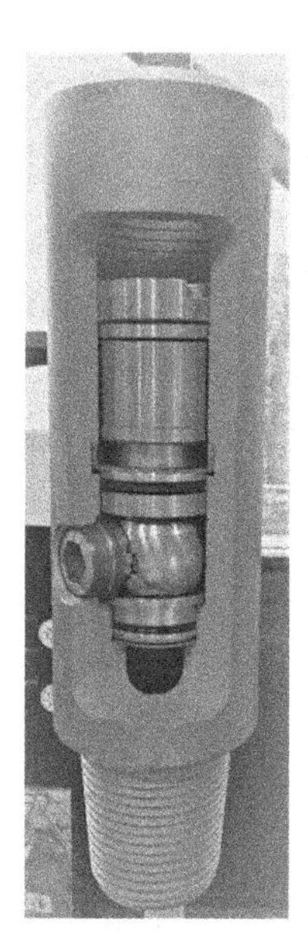

Figure 11.18 Full opening stabbing valve.

to a suitable cross-over (also refer to Chapter 4: Well Control Surface Equipment) (Fig. 11.18).

The stabbing valve should be left in the open position when not in use. If the well begins to flow through the tubing, the stabbing valve is screwed into the top of the pipe and the ball valve closed. Closure of the valve usually requires the ball to be moved one quarter turn using a hexagonal wrench. Ensure that the correct size wrench is kept in the work basket with the valve, and is easy to locate. The time to start looking is not when oil and gas are erupting from the pipe! The ball should move freely, and the crew must know how to use and operate the valve. NORSOK stipulate having at least two full opening stabbing valves (and cross-overs) at the worksite.

11.6.14.2 Back pressure valves

Back pressure valves (BPV), also called non return valve (NRV) or check valves, are the most commonly used type of internal barrier. They are normally a subassembly that is placed in the tubing string above the BHA. Most BPVs use either a ball and seat or flapper to prevent well pressure from entering at the bottom of the string. With a BPV in the string, forward circulation is possible; reverse circulation is not.

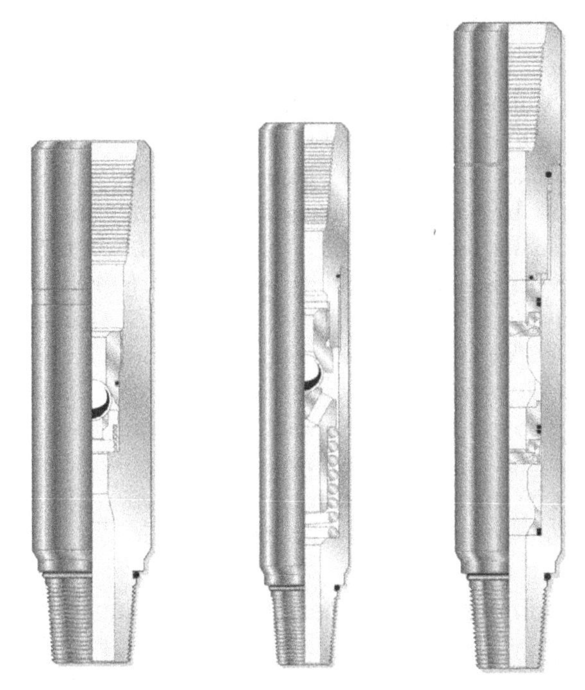

Figure 11.19 Ball check (left) ball and seat BPV (center) and dual flapper valve (right).

When a simple ball check is used, upward flow from the well forces a metal ball onto a seat, creating a seal. However, very low flow rate goes unchecked. A more reliable device is the ball and seat BPV; here a spring is used to hold the ball against the seat, meaning the valve seals even when flow from the well is minimal. Flapper valves are an alternative to the ball and seat. The advantage of using a flapper (Fig. 11.19 right) is that intervention tools can be run through it whilst the workstring is in the well. However, it needs flow and pressure differential to seat and seal the flapper.

NORSOK require that at least two BPV's are used in the tubing string, and further require that at least four valves are held at the work location. In addition, they also stipulate that the BPV allows balls and darts to pass through.

Measures should be in place to ensure that no pressure remains trapped between the check valves when dismantling the BHA. Pumping water through the BHA will usually remove any trapped pressure (Fig. 11.19).

11.6.15 Landing nipple

Some of the operations performed through the workstring are likely to lead to a loss of BPV integrity; prolonged high rate circulation for example. It is standard practice to run one or more landing nipples above the BPV in case of failure. A check valve or blanking plug can be locked in the nipple profile, restoring primary barrier pressure integrity in the workstring.

Slickline can be used to install a blanking plug or BPV in a nipple profile. However, rigging up slickline in the work basket can be difficult, and is certainly time consuming. It is quicker and easier to have a lock and plug (or BPV) that can be dropped and (or) pumped into place. However, not all nipple systems are compatible with locks that can be pumped into place, so part of the selection criteria for a workstring nipple will be compatibility with pump-down equipment. Wireline set locks are more likely to be used in completion strings (as opposed to intervention strings) where different design criteria will be applied.[i]

11.6.16 Plugging options where no nipple profile is available

Where a HWO unit is being used to pull an existing completion from a live well, it is probable that the well control barriers will have been installed before the unit moves onto location. Barriers are required to allow the Christmas tree to be removed, the BOP stack installed, and as an internal well control barrier during the recovery of the completion. Being able to install barriers and the associated type is entirely dependent on the design of the in situ completion. If any of the following conditions apply, an alternative to the conventional wireline set lock and plug cap will have to be found.

- If the completion is not equipped with two nipple profiles in (or close to) the tail-pipe.
- Nipple profiles are present but have become corroded or damaged to the point where plugs cannot be set.
- Where tubing integrity above the plugs cannot be ensured.

One option is to use a wireline set retrievable bridge plug with a plug cap or NRV. However, for a bridge plug to set correctly and maintain an effective seal against the tubing wall, the tubing must be in reasonable condition. If the tubing is in poor condition a live well workover is not an option, and the well must be killed.

[i] See Chapter 3, Completion Equipment for a Full Explanation of Wireline Nipple and Lock Systems.

11.6.17 Location of back pressure valves and nipple profiles in the bottom hole assembly

Most operating companies now insist on two BPVs, even when working in jurisdictions where they are not mandatory. Normally the first BPV is located immediately above the BHA or immediately above the tubing end if no BHA is required. The positioning of the second BPV in the string varies. It will usually be located a short distance above the lower valve with a pup joint or full tubing joint between the two. Dual flapper assemblies are available. These have the advantage of reducing length when lubrication space is limited. However, there is some doubt about whether they constitute two independent barriers (Fig. 11.20).

11.6.18 Downhole equipment

When preparing a snubbing programme, completion and intervention engineers should coordinate and plan downhole equipment selection and

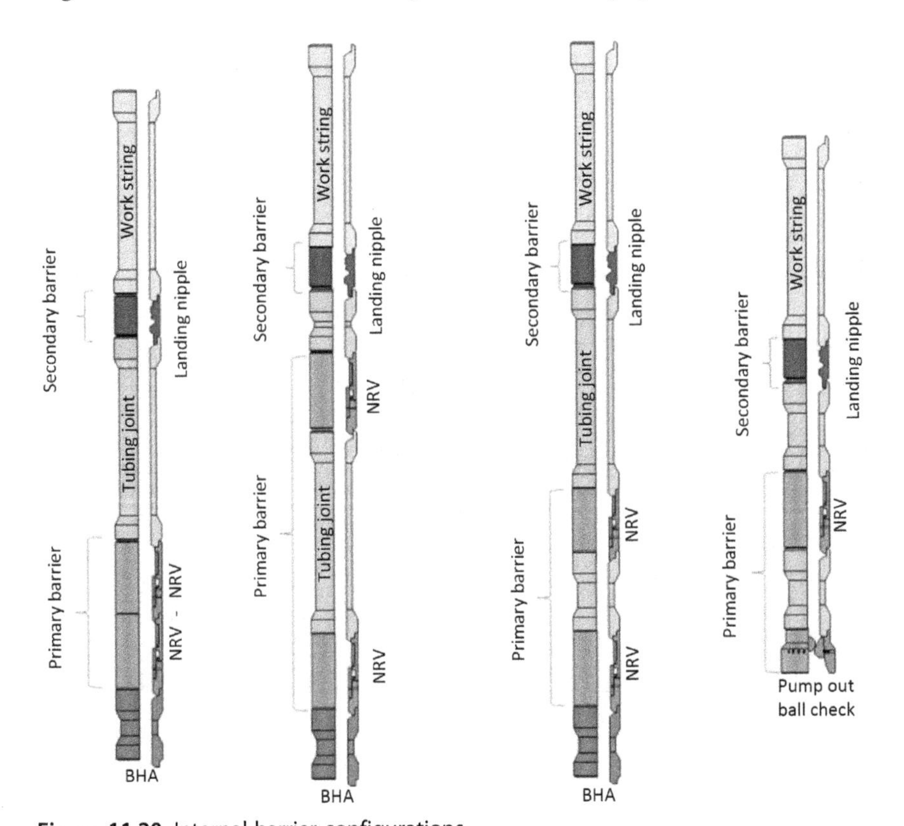

Figure 11.20 Internal barrier configurations.

deployment in close cooperation with the snubbing vendor. A well-designed BHA and tubing string will reduce operational problems, and reduce the probability of a well control incident.

- When equipment is being selected, string design must focus on maintaining lengths and configurations that are "snubbing friendly." This means that tools (such as packers, sliding sleeves, profile nipples, jars, collars, expansion joints, etc.) are short enough to stage through the snubbing stack when spaced out with pup joints that are long enough to allow the slips and rams to close.
- Minimum ID in the string must be large enough to allow the installation and retrieval of well control BPV and plugs.
- If the well is prone to issues such as sand production, scale deposition, corrosion, or hydrates, a simple BHA design and robust release mechanism is advised.

11.7 OPERATIONAL PLANNING AND PROCEDURES

Responsible operating companies and HWO vendors will have processes in place to ensure HWO operations are properly planned, as well as procedures for regularly performed HWO operation. In most cases, work programmes will be prepared by operating company engineers, with detailed input from the HWO unit vendor. In some jurisdictions, regulatory bodies provide guidance on planning requirements and operational procedures; an excellent example being the recommended practices produced by the Canadian Drilling and Completion Committee.[3]

11.7.1 Well parameters

Selection of a HWO unit and the associated pressure control equipment will be influenced by well conditions and the size (diameter) of pipe required for the operation to be performed. The main considerations are:

- For live well operations—maximum anticipated wellhead closed in pressure.
- The highest snubbing forces are generated where wellhead pressure is high and large diameter tubing is required for the operation.
- Additional pipe rams and/or shear and blind rams are usually required where wellhead pressure is high. This in turn will mean having more accumulator capacity.

- Well depth and trajectory.
- More pulling power will be required in deep wells. Deviation can add significant drag, especially where the well path is tortuous. Friction factors are likely to be much higher in live well operations, particularly in gas wells.
- Wellbore fluids, including the presence (and partial pressure) of H_2S and CO_2
- Additional barriers are normally required where H_2S is present, especially if present in harmful concentrations.
- Nature of the operation. Additional pulling power may be required for heavy duty fishing.

11.7.2 Prejob snubbing calculations

Prior to the start of any snubbing intervention, the anticipated forces necessary to snub/strip into the well must be calculated. The critical values that must be known beforehand are:

- Snub force calculations.
- Balance point calculation.
- Cylinder pressure hydraulic calculations.
- Pipe burst and collapse values.
- Pipe yield strength (including deration for slip load during overpull).
- Buckling loads.

11.7.2.1 Snubbing force

The method used to calculate snubbing force for coiled tubing can also be used for snubbing operations. The force needed to snub into a live well is calculated by multiplying wellhead pressure by the cross-sectional area of the tubing, plus seal element friction. If pipe is being snubbed through an annular BOP or stripper bowl, more snubbing force is needed to push the pipe connections through the barrier because of the larger surface area. For example, the force required to snub a 2⅞″ drill-pipe through an annular preventer against a wellhead pressure of 500 psi is:

Pipe OD = 2.875″. Connection OD = 3.219″

$$\left(\frac{\pi}{4}\times 2.875^2\right)\times 500 = 3246 \text{ lbs for the pipe body}$$

$$\left(\frac{\pi}{4}\times 3.219^2\right)\times 500 = 4069 \text{ lbs at each connection.}$$

When snubbing or stripping ram-to-ram, only the pipe body OD is relevant, as the connection is not snubbed through the closed pipe ram.

The buoyed string weight can be calculated using:

$$W = W_{air} + \left(\frac{\rho_i A_i - \rho_o A_o}{19.25} \right) \quad (11.3)$$

where:

W_{air} is the tubing weight in air (lbs/ft)
ρ_i is the density of the fluid in the coiled tubing in lb/gall
A_i is the area of the tubing ID (inches^2)
ρ_o is the density of the fluid in the production tubing (outside the coil)
A_o is the area of the tubing OD (inches^2).

If the pipe is being run into a well with a gas cap, buoyancy will change significantly when the liquid level is reached. To enable the balance point to be accurately determined, the depth of the liquid level in the well must be known.

The balance point is reached when the buoyed weight of the string is equal to the pressure force and can be calculated from:

$$\text{Snubbing force (lbs)/buoyed weight of pipe } \left(\text{lb/ft}\right).$$

If the pipe is being run dry, top filling with liquid will make it less buoyant. In some operations it will be possible to cross the balance point without having to move the pipe simply by filling the tubing with liquid.

When snubbing through an annular BOP of stripper bowl, the balance point should be calculated using pipe OD and not the connection OD, since it will be reached first.

11.7.2.2 Hydraulic cylinder pressure calculations

Recall that in the description of the hydraulic jack, Eqs. (11.1) and (11.2) are used to determine the maximum lifting and snubbing capacity of the HWO unit.

To find out how much hydraulic pressure is needed to snub or strip a load, both hydraulic jack piston area and the load need to be known. The piston area for snubbing is:

$$A_{snub} = \frac{\pi}{4}(C^2 - R^2)N \quad (11.4)$$

Piston area for stripping is:

$$A_{strip} = \frac{\pi}{4} N C^2 \tag{11.5}$$

For example, a hydraulic jack with two 9″ ID cylinders and 6.5″ OD piston rods would have a snubbing piston area of:

$$\frac{\pi}{4}(9^2 - 6.5^2)2 = 60.86 \text{ in}^2$$

And a stripping piston area of:

$$\frac{\pi}{4} 2 \times 9^2 = 127 \text{ in}^2$$

The hydraulic pressure needed to snub (pipe light) is:

$$P_{hy} = F/A_{snub}$$

and to strip (pipe heavy) is:

$$P_{hy} = F/A_{strip}$$

11.7.2.3 Pipe axial strength

The maximum axial force that can be applied to the tubing can be obtained from tables in reference books, i.e. "Tech Facts" or the "Halliburton Red Book." If no tables are available, it can be obtained by calculating the cross-sectional area of the tubing wall and multiplying by the yield stress of the pipe:

$$F_{a.max} = \frac{\pi}{4}(\text{OD}^2 - \text{ID}^2)\gamma_p$$

where:

$F_{a.max}$ is the maximum axial strength

Y_p is the yield stress (psi) of the pipe material.

For example, 4½″ 12.6 lb/ft (3.958″ ID) L-80 tubing has an axial limit of:

$$\frac{\pi}{4}(4.5^2 - 3.958^2) \times 80{,}000 = 288{,}036 \text{ lbs.}$$

11.7.2.4 Pipe burst

In live well interventions with a HWO unit, burst loadings should not be significant, as the pressure building in the pipe would equalize through the check valves to the annulus. However, if the check valves became blocked, or a positive plug was placed in the end of the tubing, burst loading could occur. As with pipe axial strength, and collapse, a burst value can be obtained from tables. American Petroleum Institute burst values obtained from tables are calculated using Barlow's formula for thin walled pipe.

$$P_b = \mathrm{Tol}\left(\frac{2Y_p t}{\mathrm{OD}}\right) \tag{11.6}$$

where:

Tol = API allowance for wall thickness

t = wall thickness.

The API wall thickness tolerance is 0.875 (12.5%) for API pipes. For cold worked corrosion resistant alloys, e.g., 22% and 25% chrome, the tolerance is 10%.

11.7.2.5 Pipe collapse

During HWO operations, closed ended tubing is run into a fluid filled high pressure environment. As a result, the tubing is exposed to collapse loading. Collapse rating is dependent on pipe diameter, wall thickness, and other harder to define properties, e.g., ovality. The API define four possible modes of collapse: elastic, transitional, plastic and yield, where the mode used to calculate collapse is determined by the slenderness ratio, i.e., the ratio of pipe OD to pipe wall thickness.

$$\text{Slenderness ratio} = D/t \tag{11.7}$$

where D is tubing outside diameter and t is wall thickness.

For example, 2⅞″ 8.60 lb/ft P110 tubing with 0.308″ wall thickness has a slenderness ratio of 2.875/0.308 = 9.33. American Petroleum Institute 5C3[4] contains empirically derived tables detailing which collapse mode should be used for a range of material yields and slenderness ratios. Having selected the mode, based on slenderness ratio, the corresponding formula is used to obtain a collapse value.

Continuing the example, Table 11.1 indicates that for tubing with a slenderness ratio of 9.33 and a yield of 110 ksi, the formula for yield collapse should be used.

Yield collapse is calculated from:

$$P_y = 2Y_p\left[\frac{(D/t) - 1}{(D/t)^2}\right] \tag{11.8}$$

Table 11.1 Collapse modes (as per API 5C3)

Grade (ksi)	Elastic collapse (*D/t*)	Transitional collapse (*D/t*)	Plastic collapse (*D/t*)	Yield collapse (*D/t*)
55	>37.21	25.01–37.21	14.81–25.01	<14.81
80	>31.02	22.47–31.02	13.38–22.47	<13.38
95	>28.36	21.33–28.36	12.85–21.33	<12.85
110	>26.22	20.41–26.22	12.44–20.41	<12.44

where

P_y = Yield collapse pressure (psi)

Y_p = pipe yield (psi).

$$P_y = 2 \times 110,000 \times \left[\frac{(2.875/0.304) - 1}{(2.875/0.304)^2}\right] = 21,043 \text{ psi.}$$

Heavy wall tubing, as typically used for snubbing workstrings, has low values for slenderness ratio and consequently collapse value is normally calculated using the yield collapse mode. However, when running completion tubulars other modes may apply. For example, 5½″ 17 lb/ft L-80 casing has an ID of 4.892, giving a wall thickness of 0.304. The slenderness ratio (D/t) is 5.5/0.304 = 18.09. From Table 11.1, the collapse mode is plastic. Plastic collapse is calculated from:

$$P_p = Y_p\left[\frac{A}{D/t} - B\right] - C \tag{11.9}$$

The values for A, B, and C are obtained from tables in API 5C3, and for 80,000 psi material the values are 3.071, 0.0667, and 1955 respectively. Placing these values into the equation, the collapse pressure of 5½″ 17 lb/ft L-80 tubing is:

$$P_p = 80,000\left[\frac{3.071}{18.09} - 0.0667\right] - 1955 = 6290 \text{ psi.}$$

Pipe having high slenderness values (large diameter/thin wall) is unlikely to be used during live well snubbing operation, since the risk of collapse and buckling would be too great. The formula for transitional collapse is:

$$P_t = Y_p\left(\frac{F}{D/t} - G\right) \tag{11.10}$$

The values for F and G are obtained from tables in API 5C3.

Elastic collapse:

$$P_e = \frac{46.95 \times 10^6}{(D/t)[(D/t) - 1]^2} \tag{11.11}$$

Collapse values published in various manuals and handbooks, such as "Baker Tech Facts" or the "Halliburton Red Book" are derived from equations and calculations in API bulletin 5C3 and the later technical report 5C3 (2008).[5] This was subsequently revised in 2015.[6]

The calculated collapse value assumes that there is no axial tension or internal pressure. However, during operations allowances must be made for internal pressure and axial load. Collapse resistance of pipe reduces with axial tension, and can be calculated by adjusting the pipe material yield stress to an axial stress equivalent grade using:

$$Y_{pa} = \left\{ \left[1 - 0.75\left(\frac{\sigma_a}{Y_p}\right)^2 \right]^{1/2} - 0.5\frac{\sigma_a}{Y_p} \right\} Y_p \tag{11.12}$$

where:

Y_{pa} = equivalent yield strength with axial load applied

σ_a = Axial stress

Y_p = minimum yield strength.

Collapse values are also affected when internal pressure is present. The effect of internal pressure is given by calculating an equivalent external pressure; P_c

$$P_c = P_o - \left(1 - \frac{2}{D/t}\right) P_i \tag{11.13}$$

where:

P_o = external pressure

P_i = pipe internal pressure.

The effect of equivalent external pressure is caused by external pressure acting on a larger surface than internal pressure.

For example, with the same 2⅞″ 8.60 lb/ft P110 tubing (0.304 wall) in the hole, a 7000 psi annulus test is performed, with 2000 psi applied to the workstring. Whilst the differential pressure is 5000 psi, the equivalent external pressure is:

$$P_c = 7000 - \left(1 - \frac{2}{2.875/0.304}\right) 5000 = 5928 \text{ psi}$$

11.7.2.6 Buckling calculations

Snubbing forces are highest when the first joint of tubing is run, since there is little balancing force from the weight of the pipe. Pushing the pipe through the pressure control seal creates significant compression loads that can lead to buckling failure of the pipe. Buckling takes place when the compressive load exceeds the compressive strength of the pipe. It will first occur in the maximum unsupported length of pipe, usually the window if no window support is used.

Two types of buckling can occur; elastic and inelastic. Elastic, or long column buckling does not exceed pipe yield, and no plastic deformation of the tubing takes place. Elastic buckling occurs in most completions where the string is in compression. Of more concern is inelastic buckling. Here, the pipe yield is exceeded and permanent plastic deformation occurs. Inelastic buckling takes two forms; intermediate and local (Fig. 11.21).

Severe buckling leads to pipe failure, loss of integrity, and a blow out. Knowing how much force is needed to buckle the pipe is essential.

The first step when determining buckling forces is to calculate the Column Slenderness Ratio (C_c) that divides elastic and inelastic buckling:

$$C_c = \pi\sqrt{\frac{2E}{Y_p}} \tag{11.14}$$

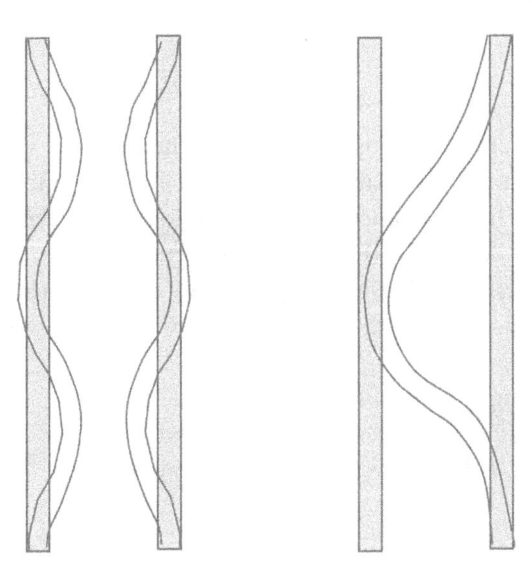

Figure 11.21 Buckling. Local inelastic buckling (left), and intermediate and long column (right).

where:

E = Modulus of Elasticity (approximately 30×10^6 for steel. Varies with grade).

The Radius of Gyration is defined as:

$$r = \sqrt{\frac{I_o}{A_x}} \tag{11.15}$$

where:

I_o = moment of inertia calculated from: $\frac{\pi}{64}(\mathrm{OD}^4 - \mathrm{ID}^4)$

A_x = cross-sectional area of tubing wall from $\frac{\pi}{4}(\mathrm{OD}^2 - \mathrm{ID}^2)$

The average radius is:

$$R = \frac{ID + t}{2} \tag{11.16}$$

where:

t = wall thickness (inches).

The Effective Slenderness Ratio is the larger value obtained from each of the two following equations:

$$S_{re} = \frac{L}{r} \tag{11.17}$$

where:

L = maximum unsupported pipe length (inches).

$$S_{re} = \sqrt{\frac{R}{t}}\left[4.8 + \frac{R}{225t}\right] \tag{11.18}$$

Inelastic column buckling can occur if the effective slenderness ratio (S_{re}) is less than the Column Slenderness Ratio (C_o) and is equal to, or less than, 250. Inelastic buckling is either local or intermediate. If the value obtained from S_{re} (Eq. 11.17) is less than the value from S_{re} (Eq. 11.18), local buckling occurs. If the S_{re} from Eq. (11.8) is less than the S_{re} from Eq. (11.17), and also less than C_o, then intermediate buckling occurs.

$S_{re} < C_o$ = Inelastic buckling.

S_{re} Eq. (11.17) $< S_{re}$ Eq. (11.18) = local buckling.

$C_o > S_{re}$ Eq. (11.17) $> S_{re}$ Eq. (11.18) = intermediate buckling.

The compressive load at which inelastic buckling can occur is calculated from:

$$F_{bkl} = Y_p \times \left(OD^2 - ID^2\right)\left[\frac{0.7854\ C_o^2 - 0.3927\ S_{re}^2}{C_o^2}\right] \tag{11.19}$$

where:

F_{bkl} = buckling load in lbs force.

Elastic (long column) buckling occurs if the effective slenderness ratio S_{re} is greater than the column slenderness ratio, and the effective slenderness ratio is equal to, or less than, 250.

Elastic buckling is calculated from:

$$F_{bkl} = \frac{225 \times 10^6 (OD^2 - ID^2)}{S_{re}^2} \tag{11.20}$$

Example: Calculate the buckling load for a 2⅞″ 8.6 lb/ft P110 work-string to be used with a HWO unit having a 27″ window (no support). Modulus of Elasticity (E) = 30×10^6

From tables, the ID of the 2.875″ 8.6 lb/ft tubing is 2.229″.

Wall thickness is (OD − ID)/2 = 0.308″.

Calculate the Column Slenderness Ratio (C_o).

$$C_o = \pi\sqrt{\frac{2E}{Y_p}} \text{ becomes } C_o = \pi\sqrt{\frac{2 \times 30,000,000}{110,000}} = 72.13$$

Moment of inertia: $\frac{\pi}{64}\left(2.875^4 - 2.259^4\right) = 2.075$

Calculate cross-sectional wall area $= \frac{\pi}{4}(2.875^2 - 2.259^2) = 1.388.$

Average radius:

$$R = \frac{2.259 + .308}{2} = 1.2835$$

Radius of gyration: $r = \sqrt{\frac{2.075}{1.388}} = 1.22268$

Calculate S_{re} (11.17)

$$S_{re} = \frac{L}{r} \qquad S_{re} = \frac{27}{1.22268} = 22.08$$

Calculate S_{re} (Eq. 11.18)

$$S_{re} = \sqrt{\frac{R}{t}}\left[4.8 + \frac{R}{225t}\right] \text{becomes } S_{re} = \sqrt{\frac{1.2835}{.308}}\left[4.8 + \frac{1.2835}{225 \times 0.308}\right] = 9.836$$

The correct Effective Slenderness Ratio is 22.08 since it is greater than 9.836. Since S_{re} 22.08 is less than C_o (72.13) and also less than 250, intermediate inelastic failure is predicted. Inelastic failure is determined using Eq. (11.16):

$$F_{bkl} = Y_p \times \left(OD^2 - ID^2\right)\left[\frac{0.7854\ C_o^2 - 0.3927\ S_{re}^2}{C_o^2}\right]$$

$$110{,}000 \times (2.875 - 2.259^2)\left[\frac{0.7854 \times 72.13^2 - 0.3927\ 22.08^2}{72.13^2}\right] = 260{,}423\ \text{lbs.}$$

11.7.3 Pipe design factors

It is useful to think of the stress applied to tubing during snubbing operations in terms of a safety factor (SF). The SF is the rating of the material divided by the load.

$$SF = \frac{\text{Rating}}{\text{Load}} \tag{11.21}$$

SFs can be calculated for all the components of stress (burst, collapse, axial, and tri-axial). For example, 2⅞″ 8.6 lb/ft P110 tubing has a burst rating of 20,620 psi. If a 15,000 psi pressure test were performed, the burst SF would be:

$$SF = \frac{20{,}620}{15{,}000} = 1.337.$$

Any prudent engineer will apply a design factor for burst, collapse, axial, and tri-axial loads. The design factor is the minimum acceptable SF. Different operating companies use different design factors, but those used by NORSOK are similar to the design factors used by many operators.

Failure mode	Design factor
Burst	1.1
Collapse	1.1
Axial	1.3
Tri-axial	1.25

11.7.4 Selection of the workstring

If production tubing is being run into the well as part of a completion, material selection is normally the responsibility of a completion engineer, often with assistance and advice from a materials specialist. In addition to any tubing stress analysis performed to confirm the selected tubing is suitable for all the anticipated service loads, the tubing must have sufficient strength to withstand any loads associated with the completion installation. The tubing must be corrosion resistant to any corrosive fluids associated with installation and possible future interventions, as well the corrosive effects of the nonhydrocarbon gases, H_2S and CO_2.

Workstring will be selected for its robust properties, but still must meet any necessary corrosion resistance requirements. In general, sweet (CO_2) corrosion is less of a concern, since the duration of exposure is much shorter than that of production tubing. Attack by H_2S is more of a concern, as hydrogen embrittlement can happen relatively quickly.

11.7.5 Mitigating explosive potential

In 2001, Petro-Canada experienced two downhole explosions, both in 2000 m deep natural gas wells. The lower two joints of tubing were split apart in each explosion, indicating where the explosions had occurred.[7]

Clearly there is a need for any snubbing programme to include an evaluation of the potential for fire or explosion. If air (containing oxygen) mixes with hydrocarbon gas and liquid at the right concentration and pressure, there is the potential for an explosion. This can occur in the workstring, or in the production tubing during live well interventions.

11.7.5.1 Explosive potential in the workstring/production tubing annulus

If swabbing is used to remove fluid from a well to create underbalanced conditions for perforating, air will be present. Following perforation, gas can enter the wellbore and mix with the air at a concentration that is potentially explosive. Snubbing into the well could cause an explosion. To remove the risk, the well should be flowed to remove the explosive mixture, i.e., flowed until hydrocarbons are at the surface and all the air has been removed.

11.7.5.2 Explosive potential in the workstring

When snubbing into a live well, the string will be air filled, unless it is top filled from the basket as it is run. Leaking check valves would result in hydrocarbon entering the tubing and mixing with the air present. However, by the time the leak had been identified and the full opening valve installed at the surface, most of the air will have been displaced.

Of more concern are operations where hydrocarbon gas is pumped into the workstring to equalize across the check valves. To reduce the risk of creating an explosive mixture, a spacer fluid should be pumped ahead of the gas. Water or completion brine is normally used, and should be mixed with glycol if there is a hydrate risk. An alternative is to equalize with nitrogen instead of hydrocarbon gas.

11.7.6 Specifying escape routes

Prior to rigging up, the plans for escape procedures must be in place. After rigging up, all personnel should be briefed on the location and use of escape routes, including the equipment used to evacuate the work basket. This should be done before operations commence, or when new personnel arrive at the work location.

11.7.7 Structural loads

Snubbing and stripping loads are transferred through the hydraulic jack to the Christmas tree and wellhead. Where the wellhead is not rated to the combined weight of the rig up and string weight (maximum expected pick up weight), then the load will have to be transferred from the tree and wellhead onto the surrounding ground, platform, or rig substructure. Some small offshore wellhead towers may also have limited deck loading capability, and beams will have to be used to spread the load.

11.7.8 Rigging up

Although some aspects of the rig up are site-specific, several functions are common to any HWO operation. Careful preparation of the equipment reduces the probability of a well control incident.

- When rigging up snubbing BOP's onto a Christmas tree, quick union connections or threaded connections should not be used. Replace the swab cap with a flanged connection.
- If not already in place, pressure gauges should be mounted on the casing strings.
- A remote well control panel should be used to control the subsurface safety valve (if fitted) and any hydraulically actuated tree valves.
- When operating through a Christmas tree, the number of turns to operate manual valves must be recorded.
- When rigging up the BOP, the ring gasket groove on top of the tree (or wellhead) must be thoroughly cleaned and inspected. Only new ring gaskets should be used.
- Confirm the ram configuration conforms to programme requirements.

- Ensure the lubricator spacing (swab valve to lower stripper rubber or annular BOP) is sufficient to accommodate the longest BHA or completion component.
- In most instances, hydraulic hoses will be attached to the jacking assembly before it is lifted into place.
- The rig up must be properly supported by guy-wires before the crane is released. The exact guy line configuration is rig up specific. Vendor guidance is advised. Guy wire anchors should be pull tested to API specification (land operations). For rig assist operations, the rig up should be braced inside the derrick.
- Confirm escape systems are secure and operational.
- Rig up the work basket, counterbalance winches, gin-pole, and BOP control systems.
- All high-pressure lines (circulating system) must be secured.

11.7.9 Pressure testing and function testing

The BOP stack should be pressure tested and function tested before it is shipped to location. There should be assurance that the shear rams are able to cut the weight and grade of tubing selected for the operation.

Once on location, a series of pressure tests will be performed to confirm the integrity of the primary, secondary, and tertiary well control barriers. Test pressure must be in excess of the maximum closed in wellhead pressure expected during the operation.

11.7.10 Setting the jack pressure

Whenever possible, the minimum force necessary to snub the pipe into the well should be used. It is vitally important that snubbing force does not exceed the buckling limits of the tubing as calculated during the planning stage. Setting the jack pressure so the pipe is snubbed in using the minimum required force reduces the possibility of pipe bending/buckling. The following procedure is a guide to the sequence that would normally be used to begin running tubing into the well. Where complex BHA or completion components are being introduced into the well, detailed sequential steps should be listed in the programme, along with tool dimensions.

- With the tree valves (or lower BOP blind rams closed for operations with no tree), lower the BHA into the snubbing stack using the counterbalance winch.

- Close both sets of snubbing slips (stationary and traveling). Use the traveling slips to pull tension to ensure the stationary snubbing slips are gripping the pipe. Maintain tension on the traveling snubbers.
- Close the annular (low wellhead pressure) or lower stripper ram (high pressure) around the tubing joint above the BHA.
- Equalize the stack-to-wellhead pressure, and confirm the annular or stripper ram is holding against well pressure. Confirm the NRV in the tubing are also holding pressure.
- Set the jack hydraulic pressure to 0. Increase the throttle to maximum.
- Select full "down" on the jack controller, then begin to increase hydraulic pressure until the begins to move (snub).

A slightly higher jack pressure is required to overcome static friction. Once the pipe is moving, the jack pressure to keep it moving is lower. As pipe is run into the well and weight increases, the jack pressure can be reduced. If snubbing through an annular BOP, jack pressure will have to be increased to snub pipe connections past the seal element.

Illustrated on the following pages is the slip sequence for running into a live well under pipe light (snubbing) conditions. When pulling out the sequence would be reversed.

Below the balance point the pipe heavy (traveling and stationary) slips would be used (Fig. 11.22).

11.7.11 Tripping pipe

Enform (Canada) use IRP 15 make some eminently sensible recommendations relating to routine tripping of pipe. These recommendations include the following:

- Intrinsically safe two-way radios can be used to improve communications between personnel in the work basket, at the BOP accumulator panel, pumping equipment, and supervisory personnel.
- Tubing movement should be stopped before:
  - Any worker climbs or descends the snubbing unit ladders
  - Any worker enters or exits the work floor
  - Any service rig/drilling rig workers climb the derrick ladders (rig assist operations).
- The driller must be able to see the snubbing unit heavy slips clearly. Tarps must be positioned so they do not interfere with the drillers line of sight.
- Picking up tubing over the snubbing operators panel must be avoided.

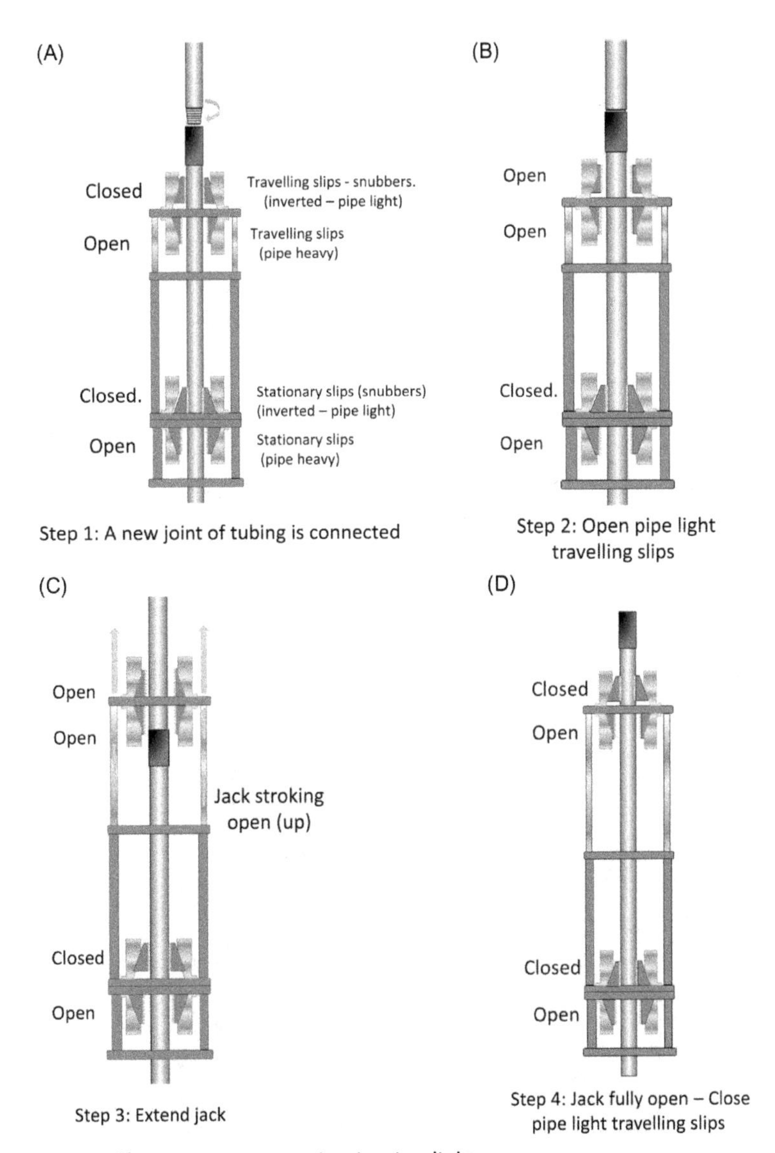

Figure 11.22 Slip sequence—running in pipe light.

11.7.12 Crossing the balance point

Of all the routine operations carried out during snubbing operations crossing the balance point, the transition between snubbing (pipe light) and stripping (pipe heavy) is the one that is most likely to create problems.

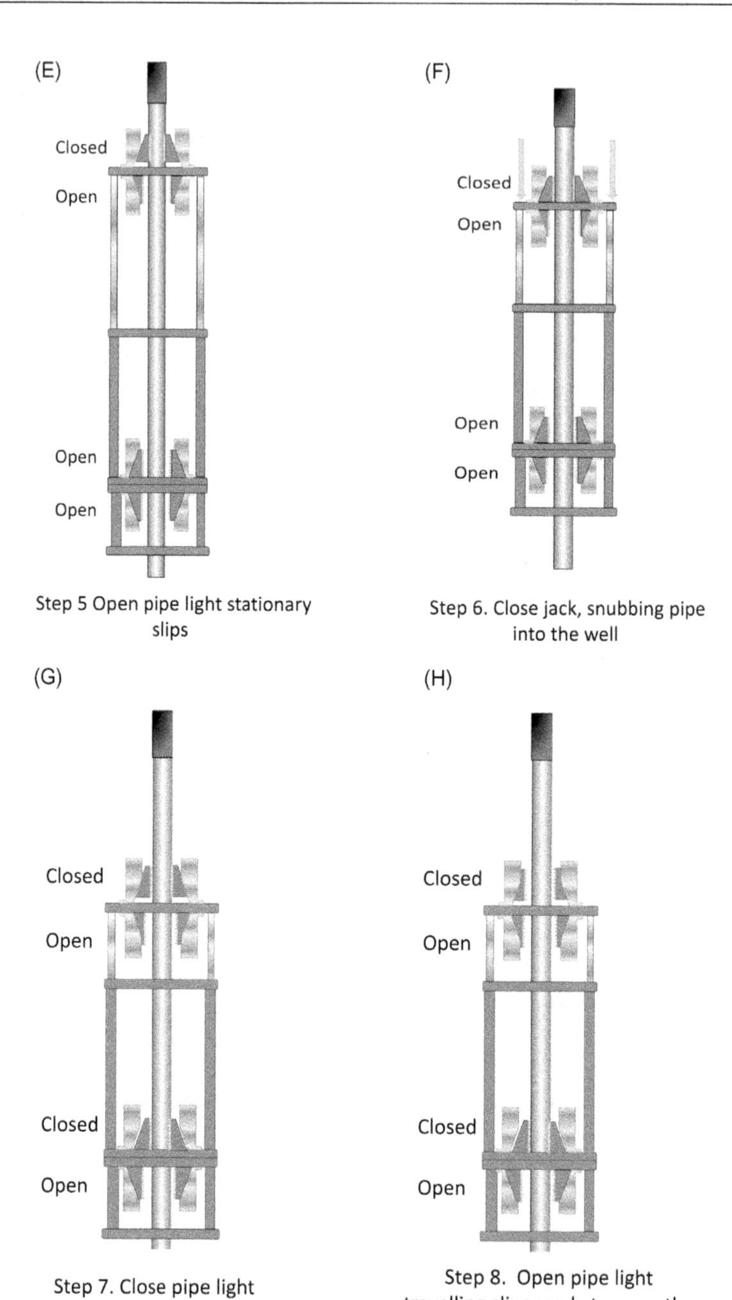

Figure 11.22 (Continued).

It is difficult to accurately determine when the balance point will be reached. Whist is relatively simple to calculate the force acting on the tubing, there are many variables affecting string weight:

- BOP hydraulic pressure.
- Stripper (or annular) ram friction.
- Weight of fluid and fluid level inside workstring.
- Fluid density and fluid level in casing (annulus around workstring).
- Differences in the coefficient of sliding friction and static friction.
- Difference between the OD of the tube and OD of tool joints (if snubbing through an annular BOP.

Manipulation of some of these variables can be used to assist passing the balance point, most commonly top filling the tubing to increase the string weight.

Often the biggest problem comes from the difference in the coefficient of static friction and the coefficient of sliding friction between the pipe and the BOP seal element—it takes more force to start the pipe moving than is required to keep it moving. If the traveling slips are turned over prematurely, it may be difficult to overcome the static friction to get the pipe moving. Operator training, experience, and technique are important factors in knowing when to "turn over." The following procedure is intended to make the transition from "pipe light" to "pipe heavy" successful.

Tripping in:

- An experienced and properly trained snubbing operator must be at the controls during the transition through the balance point.
- Calculate approximately how many joints will be needed to reach the balance point.
- Snub in hole to the calculated balance point.
- Filling the tubing with fluid at this point will ease the transition between pipe light and pipe heavy. If water is used it should be mixed with glycol if there is any risk of hydrates.
- Continue snubbing pipe, checking each joint for ability to pick up the pipe with traveling slips still in snub position.
- Continue snubbing in hole slowly until pipe starts to slip through the traveling slips. Use stationary pipe heavy slips as necessary to control pipe movement.
- Turn over traveling slips to heavy position.
- Continue to trip in hole.

Tripping out:

- Calculate the number of joints that must be pulled to reach the balance point.
- Strip pipe out of the well until close to the theoretical balance point.
- Continue pulling pipe with the traveling slips in the heavy position, checking each joint to see if pipe can be moved down.

  Note: Some slips will hold several hundred pounds of pipe light force when in the heavy position. Increased holding ability may be temporarily achieved by increasing the slip circuit hydraulic pressure. Once the balance point is crossed, hydraulic pressure should be regulated back to the standard setting.
- Continue pulling pipe until pipe starts to slip through the traveling slips, use the stationary snubbers as necessary to control pipe movement.
- Turn traveling slips over to snub position. Increased slip hydraulic pressure may still be required temporarily.
- Snub out of hole.

When tripping out of a live well, flowing the well will reduce pressure and enable more tubing to be pulled "pipe-heavy."

11.7.13 Well suspension: daylight only snubbing operations

At some land and limited facility offshore locations, operations are limited to daylight hours operations. Before the end of each day's work, the well must be secured and left in a safe condition.

- A supervisor with a valid well control certificate should be present at the wellsite to verify the closing and opening of the BOP when the well is secured (evening) and opened the following day.
- All appropriate primary rams must be locked after closure and remain locked until the next operation is ready to commence. The stack should be configured to allow pressure below any closed ram to be monitored and read before the ram is opened.
- Operations should be planned so that the tubing is either out of the well, or pipe heavy at the end of the day.
- When closing in the well, dual mechanical barriers must be used on both the tubing string and the annulus. For example:
  - Two close pipe rams (annulus) dual check valves and a closed, full opening safety valve on the tubing.
  - Two closed tree valves (tubing) production packer and tubing hanger (annulus) if no pipe is in the well.
- It is advisable to leave a responsible person to monitor the well overnight.

11.8 WELL CONTROL AND CONTINGENCY PROCEDURES

When operating on live wells, failure of either of the primary barriers can result in the sudden release of hydrocarbons. Equally, failure to maintain a proper restraining force on the pipe will result in it being dropped into the well or blown out of the well. The consequence of losing control of the pipe is likely to be a loss of pressure integrity. Well control during HWO operations will focus on:

- Maintaining a restraining force on the pipe to prevent uncontrolled movement.
- Preventing uncontrolled flow through the workstring (failure of primary internal barrier).
- Preventing uncontrolled flow from the annulus (loss of primary external barrier).

Maintaining effective well control at all times requires the crew to be properly trained in a range of contingencies. Training should be backed up by frequent well control drills. The procedures outlined here cover the most commonly occurring problems.

11.8.1 Power pack failure

If the power pack fails, hydraulic power will be lost and it will not be possible to raise the jack. Power to the slips and counterbalance winches will also be lost. There should be enough accumulator pressure for limited BOP functions. If power is lost the following actions should be carried out:

- If pipe light (snubbing) set both traveling and stationary slips.
  If pipe heavy (stripping)—bring the tubing down to basket level (if possible) and set slips as required.
- Close and lock both stripper rams and any pipe rams in the BOP stack.
- Make up the full opening stabbing valve and close it to secure the tubing.
- Secure any loads supported by the counterbalance winch with a mechanical clamp.
- Evacuate the work basket.

Once the well is properly secured, repair or replacement of the power pack can start.

11.8.2 Snubbing unit accumulator failure

Failure of the accumulator can result in failure of critical components of the pressure control equipment, including the annular preventer, stripping rams, bleed-off line, and equalizing loop. Accumulator pressure should provide enough reserve to enable the well to be made safe. Ideally, the snubbing unit should be equipped with nitrogen back-up systems designed so that the operator can maintain well control in the event of accumulator failure. If the accumulator fails, the following actions should be taken:

- If possible, position the string so tool joints do not prevent the closure of stripper rams and pipe rams.
- If possible, position the tubing connection at working level in the snubbing unit basket.
- Close and lock the lower stripper ram.
- Close and lock the open stripper, and lock the closed stripper (if annular preventer is used, consider increasing pack-off pressure).
- Install a stabbing valve into the tubing in the open position, tighten, and close.
- Set slips as required and pull tension into stationary snubbing slips. Use mechanical slip locks.
- Bleed off pressure in the snubbing stack above the primary BOP (including equalize line).
- Assess the situation with all onsite personnel to ensure well security before attempting any repairs.

11.8.3 Slip failure

Slip failure will have serious consequences. When pipe light the tubing can be (and has been) ejected from the well. When pipe heavy the tubing will drop into the well.

Before the operation begins, ensure that all of the slip dies are of the correct size and are compatible with the pipe material—higher grade (high yield) tubing and tubing with chrome content is harder and requires the correct dies to firmly grip the pipe. Even where the slips are in good condition at the start of a run into the well, they are prone to wear and tear. They are normally cycled four or five times per joint when running in with a short stroke unit. Coating or scale on the surface of the pipe accumulates on the die surface causing slips to lose grip. If slip failure occurs take the following actions:

- Close any available alternative slips.
- Close all available pipe rams.

- Install the full opening stabbing valve and close.
- Bleed-off stack pressure above the closed pipe rams.
- Carry out repairs to the faulty slips. The closed (pipe supporting) slips should be inspected.
- Confirm the repaired slips can support tubing loads.
- Inspect the tubing for damage then resume operations.

11.8.4 Stripper ram seal failure

Under normal operating conditions, failure of the stripper ram inner seal is gradual and progressive, starting with a slow leak. Increasing ram pressure can regain control, however, increasing regulated pressure will also increase the wear rate. If a leak cannot be stopped, the stripping ram inner seal will need to be changed. Changing out stripper ram inner seals is to be expected. Operations in high pressure wells can require multiple ram seal changes. In extreme cases, vendors have reported installing new stripper ram seals for every 6–8 joints of tubing run. To change the stripper ram seals, proceed as follows:

- Position the string so there are no tool-joints across the BOP pipe rams. If possible and spacing permits, position the top tool joint box as close as possible to the work basket.
- Close the upper pipe ram and bleed off pressure above the ram. Monitor and confirm the ram is holding pressure.
- Close the lower pipe ram and bleed off through the choke line (this assumes the stack is configured with a kill/choke spool between the two sets of pipe rams).
- Install a full opening safety valve (leave in the open position to monitor the well).
- Retract the stripper rams and replace the seal elements.
- Note: Consider replacing both sets of stripper ram seal inserts. If one ram is leaking, the other is likely to be similarly worn.
- Close upper stripper and test via the equalizing loop.
- Bleed off test pressure.
- Close the lower stripper.
- Equalize across the closed pipe rams. Observe for pressure build-up above the lower stripper.
- Open pipe rams and continue operations.

11.8.5 Annular seal failure

With pipe in the well, it is not normally possible to replace the seal in the annular preventer. This is because it is not possible to remove the bonnet with the jack rigged up.

To reduce the possibility of a leak from the annular preventer, a new seal element should be installed before running into the well. If the annular BOP does start to leak, the stripper rams should be used as the primary barrier (if in place). It will almost certainly be necessary to pull out of the well and rig down the jack to enable the annular BOP seal to be replaced.

11.8.6 Leaking pipe ram

Most BOP stacks are equipped with two sets of pipe rams (for each tubing size). This enables two barriers to be closed if the stripper rams require repair. A pipe ram leak compromises operational safety. Managing the situation will depend on which ram has failed, and why the rams needed to be closed in the first place. The following points should be considered when faced with a ram failure:

- If the upper stripper rubber requires repair and the lower stripper rubber can be closed and passes the integrity test, only having one pipe ram closed is sufficient to provide two barrier isolation. In these circumstances, the upper stripper can be replaced. However, it is recommended that the operation is terminated at this point, the tubing pulled from the hole and repairs made to the BOP.
- If the lower stripper ram is leaking and one of the two pipe rams are leaking then it will not be possible to repair the stripper ram whilst maintaining two barriers. Under these circumstances, a proper evaluation of the situation is required before proceeding. On a low pressure well where no H_2S is present, and where an escalation of the leak is unlikely to threaten surrounding infrastructure, e.g., a remote land location, then dispensation can be requested to allow repairs to proceed against a single barrier.
- In all other circumstances, the well must be killed before any repair against a single closed set of rams can be carried out.

11.8.7 External leak on the christmas tree or wellhead or the BOP below the lowermost pipe ram

An external leak from the surface pressure control equipment is potentially very serious; it allows the uncontrolled escape of hydrocarbons to the atmosphere, with the attendant risk of fire. Should a leak develop whilst the tubing string is in the well, the following must be considered:

- A rapid assessment of the leak rate is required. It could be anything from a minor weep at a flange connection to the catastrophic failure of a major component.
- Fluid and gas composition. Is H_2S present?
- Workstring weight and snubbing or stripping force.
- Impact on simultaneous operations.
- BHA position relative to SCSSSV (if relevant).
- Depth of the BHA relative to the bottom of the well, or any obstruction that would prevent the string from dropping if cut

There are two ways of stopping such a leak, pull out of the well and shut in, or cut and drop the pipe and shut in. Clearly, pulling out of the hole is going to take time and will only be attempted if the leak is minor and the onsite supervisory staff judge that an escalation is unlikely in the time it will take to recover the tubing to the surface. Possible options to contain the leak include:

- Killing the well
  - Bullhead from surface if string depth is shallow
  - If the string is deep enough into the well, forward circulation
  - Combination of circulation and bullhead.
- If there is a tubing retrievable SCSSSV in the well, and the workstring BHA is above the valve. Close the SCSSSV. Depressurize the tubing above the closed valve and recover the tubing to the surface. If the BHA is a short distance below the SCSSSV, the leak minor, and no H_2S is present, it might be possible to pick up above the SCSSSV. The on-site supervisors will have to judge if this action is acceptable.
- If it is a bad leak, is visibly getting worse, or if H_2S is present, then the string will have to be cut and dropped. Before dropping the string consider the following:
  - As soon as the tubing is cut, well pressure will enter the tubing string and will escape through the open top of the string unless a full opening safety valve is fitted.
  - If a full opening safety valve is fitted, the stump of pipe above the point of cutting will immediately go "pipe light," and will be ejected from the strippers unless restrained by the snubbing slips.
  - Rig up on wellhead (tree removed): As soon as the pipe has been cut, the blind rams should be closed and the stack above

depressurized (unless shear seal rams have been used). It will be necessary to pick up on the pipe to clear the blind ram.
- Rig up above tree. As soon as the pipe has been cut and dropped, the tree valves should be closed and the stack above depressurized. Fastest (and safest) closure will be by activating the hydraulic master valve from the single well control panel.

11.8.8 Bottom hole assembly check valve failure

It is not unusual for BHA check valves to fail, particularly after high rate, high volume circulation. If check valve(s) fail, well fluid will enter the tubing and migrate to the surface. Often the first indication of check valve failure is flow through the tubing at the surface. When this happens:

- Make up the full opening safety valve and close in.

  Note: An increase in string weight may be observed as well fluid leaks into the string. This is particularly noticeable if the closed end string was being run dry and the tubing is filling with a dense fluid. An example calculation is shown below.
- Hold the string in the appropriate slips (pipe light or pipe heavy).
- If the BHA is equipped with a nipple profile designed to accept a pump-down plug then:
  - Place the plug above (resting on) the full opening safety valve.
  - Make up a circulating head above the full opening safety valve.
  - Open the full opening valve and allow the plug to drop into place. The surface pump can be used to pump the plug down, or assist in seating the plug if necessary.
  - Inflow test the plug.
  - Pull out of the hole (wet trip).
- If the BHA is equipped with a nipple that can only accept wireline set plugs then:
  - Rig up slickline pressure control equipment above the full opening valve.
  - Integrity test the pressure control equipment.
  - Set a plug in the wireline nipple.
  - Rig down wireline.
- Pull out of hole (wet trip).

If the newly installed check valve fails to hold pressure the well will have to be killed.

If the BHA check valves suffer a catastrophic failure and the escape of fluids and gas through the top of the tubing prevents making up of the full opening safety valve, then the tubing must be sheared and the well closed in.

Example: String weight change due to check valve or connection leak.

Tubing size: 2⅞″ (2.875″ OD) 8.6 lb/ft. 2.259″ ID.

Wellbore fluid: 7.5 ppg oil/water.

WHCIP: 1750 psi.

Depth of tubing end when leak first observed: 1600 ft.

Tubing is being run dry—no liquid has been pumped during run in. How much will the string weight change if wellbore fluid fills the work-string due to a leak?

Dry string weight calculated from $W = W_{air} + \left(\frac{\rho_i A_i - \rho_o A_o}{19.25}\right)$

Area of tubing OD = 6.49.

Area of tubing ID = 4.00

Air density is negligible.

$$W = 8.6 + \left(\frac{(0.1 \times 4) - (7.5 \times 6.49)}{19.25}\right) = 6.09 \text{ lb/ft}$$

String weight at 1600 ft: 1600 × 6.09 = 9747 lbs. Snubbing force is 1750 × 6.49 = 11,357 lbs. At 1600 ft the tubing is pipe light (−1610 lbs) and will be held by the snubbing (pipe light) slips.

A leak from the production tubing past the check valves (or through a tubing connection) would change the string weight to:

$$W = 8.6 + \left(\frac{(7.5 \times 4) - (7.5 \times 6.49)}{19.25}\right) = 7.629 \text{ lb/ft}$$

String weight at 1600 ft becomes 1600 × 7.629 = 12,206. Slightly pipe heavy.

11.8.9 Tubing leak above the bottom hole assembly

If the workstring is properly maintained, and has correctly made up premium connections, the possibility of connection washouts and pinhole leaks occurring are reduced. However, leaks in the string above the check valves can and do occur. In the first instance, the crew will probably not know if the leak is coming from the check valve(s) or the tubing.

Continuation of a leak after a pump down or wireline plug has been run is an indication. If a tubing leak is suspected the following can be tried:

- It might be possible to estimate the depth and severity of the wash out by recording circulating pressure or circulating dye. Alternatively, wireline can be used to locate a leak (mares tail or noise logs).
- Pump kill weight fluid.
- If pumping of kill weight fluid is not possible or desired, then consideration can be given to a wireline intervention, the aim being to:
  - Set isolation straddle packers across the hole in the tubing, allowing resumption of circulation.
  - Set bridge plugs in tubing above the wash out. Once the hole is sealed off, the string can be pulled until the hole of the wash out is just below the BOP stack.

Exact knowledge of the location of the plug(s) and hole depth is required. Personnel in the work basket could be exposed to trapped pressure in the workstring if pipe tally errors are made. If two plugs are set, pressure trapped between the plugs may equal or exceed the hydrostatic force at the original setting depths of the plugs. When two plugs are used, the hot tap method may be needed to bleed-off the trapped pressure safely.

To remove the failed joint:

- Continue to pull out of hole until the wash out is located between stripper rams.
- Close lower stripper ram.
- Bleed internal string pressure to zero (if the BPVs in the string are holding pressure).
- Pull wash out to work basket area.
- Replace washed out joint(s) and resume running.

11.9 WHY WELL CONTROL MATTERS

In November 2004, 2⅜″ (60.3 mm) pipe was being pulled out of a live well using a service rig. The pipe was being stripped through the annular preventer against a wellhead pressure of approximately 1450 psi (10 MPa). The pipe was being pulled in double joint stands, and racked back by the derrickman who was working on the monkey board. After pulling 25 stands, the annular preventer began to leak. The Canadian Petroleum Safety council report details what happened next[8]:

In an effort to prevent gas leaking past the element/tubing interface, the Snubbing Operator increased hydraulic pressure to the annular preventer. This corrective action was not entirely successful. As a further step to obtain a better seal, prior to stripping out the 26th stand of pipe, the Snubbing Operator closed the stripping pipe rams onto the tubing so that he could flex the annular preventer rubber element by opening and closing the annular.

The evidence from the investigation indicates that the Rig Operator began hoisting the 26th stand of tubing without knowing the stripping pipe rams were closed. The tubing coupling, which has a larger outside diameter than the pipe, could not pass through the closed stripping rams. The Rig Operator was not able to stop the vertical motion of the tubing string before the uppermost joint of tubing failed in tension. When the top joint of tubing parted, the remainder of the tubing string dropped into the well. The wellhead, rig Blow Out Preventer (BOP) stack and snubbing stack were now without tubing. As a result, well gas flowed out of the top of the snubbing annular.

Gas from the well exited at velocity and ignited instantaneously. Flames extended out of the snubbing annular to 15 metres to 20 metres above the crown of the rig, engulfing the monkey board level where the Derrickman was working. The flames likely prevented the Derrickman from escaping from the monkey board by using the Geronimo escape buggy secured to the side of the monkey board. Both workers in the snubbing basket jumped to the rig floor and proceeded to ground level.

In the midst of the incident, the Rig Operator stopped the well fire by closing the manual master valve on the wellhead. Spot fires were extinguished with portable fire extinguishers and First Aid was rendered to the two Snubbing workers. Municipal responders responded within 30 minutes and the two injured workers were transported by ground ambulance to the Burn Unit at Foothills Hospital in Calgary, AB. The Derrickman was pronounced dead at the site.

REFERENCES

1. Recommended Practices for Blowout Prevention Equipment Systems for Drilling WellsAPI Recommended Practice 53 Third Edition, March 1997.
2. Norsok standard. System requirements well intervention equipment D-002 Rev. 1, October 2000.
3. Drilling and Completions Committee. IRP 15. An Industry Recommended Practice (IRP) for the Canadian Oil and Gas Industry. Snubbing Operations. Volume 15. Edition 3.1. May 2015.
4. API Bulletin 5C3. Bulletin on Formulas and Calculations for Casing, Tubing, Drill Pipe and Line Pipe Properties. Sith Edition. October 1 1994.
5. API TR 5C3, Technical Report on Equations and Calculations for Casing, Tubing, and Line Pipe Used as Casing or Tubing; and Performance Properties Tables for Casing and Tubing, First Edition, December 2008, Washington, DC, API.

6. Addendum. October 2015. ANSI/API Technical Report 5C3/ISO 10400:2007, Technical Report on Equations and Calculations for Casing, Tubing, and Line Pipe Used as Casing or Tubing; and Performance Properties Tables for Casing and Tubing, Washington, DC, API.
7. Avoid Downhole Explosions, Buckled Pipe and Parted Tubing During Snubbing Operations. Grant J. Duncan, SPE Paper 115534. 2008.
8. Canadian Petroleum Safety Council. Safety Alert - #26-2005 Gas Release and Fire During Snubbing Operations. Release date: April 25, 2005.

CHAPTER TWELVE

Well Control During Well Test Operations

12.1 INTRODUCTION

Well tests are routinely carried out throughout the life of many wells. Reservoir engineers use well test data to monitor well performance, validate the reservoir model, and as a diagnostic tool for detecting changes that might indicate a problem in the well.

In many oil fields a test separator is a permanent part of the production plant. Where such a facility exists, testing a well is a relatively straightforward operation. Flow is diverted away from the main bulk separation train and into the test separator. During the well test, produced fluids flow through permanent facilities that are protected by automated shut-down systems. Providing these facilities are correctly designed, installed, and maintained, a well test carries no more risk than routine day-to-day production. And yet, in the preface to the book "Operational Aspects of Oil and Gas Well Testing" the author states, "Well testing is recognised by many operating oil and gas companies to be the most hazardous operation that they routinely undertake. The potential for loss of life, loss of assets or environmental catastrophe are proportionately higher than at any other time in the drilling or operating of oil and gas wells."[1] Well testing in this context is not a routine test through permanent production facilities; rather, the author is referring to the use of a temporary production system. When an exploration or appraisal well is drilled, it will often be tested before the completion is installed. This will involve producing the well through a temporary completion (drill stem test (DST) string), temporary tree, temporary flowline, and temporary production facilities.

Initiating flow is analogous to deliberately allowing the well to "kick." At the end of the test, the well must be killed to enable the DST string to be recovered before the permanent completion is run. The potential

Well Control for Completions and Interventions.
DOI: https://doi.org/10.1016/B978-0-08-100196-7.00012-9

for loss of integrity is arguably far greater than at any other point during the life of a well.

In addition to flowing the well through a DST, wells are often flowed through the final (permanent) completion to a temporary production system; a well test package. Post-completion production through temporary systems is usually carried out to clean up completion fluid and detritus before the well is allowed to flow into the permanent production facility. Clean-up flow can be extended to allow the newly installed completion to be properly evaluated by gathering well test data. Because the well is flowing through a completion rather than a DST string, and because the Christmas tree is usually in place before flow begins, postcompletion well clean-up and well test operations are less potentially hazardous than an exploration well test.

Many people, from different disciplines and service companies, are needed to carry out a well test. Supervisors may have to coordinate and control the activities of a rig crew, well test crew, logging and slickline specialists, DST and tubing conveyed perforation (TCP) specialists, and perhaps the facilities operators. Ensuring a safe and efficient operation requires a good understanding of surface equipment, subsurface equipment, and a detailed knowledge about well test processes and procedures.

12.2 INDUSTRY STANDARDS

There are a number of industry standards that relate to well testing operations. These include, but are not limited to:

- Norwegian NORSOK D-007 Well testing system (Edition 2, September 2013).
- ISO 10407 (API RP 7G) Petroleum and natural gas industries—Drilling and production equipment—Drill stem design and operating limits.
- ISO 10418 Petroleum and natural gas industries—Offshore production Installations—Analysis, design, installation, and testing of basic surface process safety systems, equivalent to API RP 14C.
- ISO 13703 (API RP 14E) Petroleum and natural gas industries—Design and installation of piping systems on offshore production platforms.

12.2.1 Well test objectives

Well testing can reveal a great deal about well performance and reservoir behavior. Most (but not all) well tests follow a sequence of flow (drawdown) followed by shut-in (pressure build up). During the flowing period essential parameters are recorded:

- Oil flow rate.
- Gas flow rate.
- gas oil ratio.
- Water cut.
- Flowing pressure (surface).
- Flowing pressure (downhole).
- Flowing temperature (surface and downhole).
- Solids production.

Downhole gauges record bottom hole temperature and pressure throughout the test; downhole tools are used to obtain samples of the produced fluid under reservoir conditions.

In complex multilayered reservoirs, e-line logging will be used to determine the zonal contribution whilst the well is producing. It is standard practice to flow the well at different rates, a "step rate test." Each step will usually be of the same duration, with the rate increasing as the job progresses. At the end of the final flow period the well is closed in and reservoir pressure monitored. By plotting pressure build up against time, valuable information about reservoir properties is gathered.

A properly designed and conducted test provides valuable information about:

- Formation permeability.
- Reservoir pressure.
- Reservoir boundaries.
- Reservoir layering and zonal contributions—heterogeneity.
- Near wellbore damage (skin).
- Well productivity—productivity index and absolute open flow potential.

Data obtained during well tests can be used to "match" and fine tune inflow performance and tubing performance models such as Proper, Wellflo, and Olga.

Bottom hole and surface samples obtained during the well test are taken to obtain essential information about:

- Pressure volume temperature relationship of the hydrocarbons—essential information for completion and reservoir engineers, enabling them to select the optimum size of tubing for the development wells.
- Hydrocarbon chemistry. Wax and asphaltine potential.
- Water chemistry. Used for predicting scale potential.
- Presence of nonhydrocarbon gasses (CO_2 H_2S).

Note: Interpreting well test results is beyond the scope of this book.

Clarity about the objectives of a well test is important. A well test program must be explicit about the expected production rate, the number of steps and the duration of each, as well as the duration of the shut-in.

12.3 WELL OFFLOADING AND CLEAN-UP

Many production facilities are not designed for, and are poorly equipped to process, some of the fluids and solid detritus left behind at the end of a completion, workover, or well intervention. It is fairly common to produce a well through temporary facilities to remove (offload) some or all of the following:

- Non-reservoir fluids and gases:
  - Brine.
  - Mud.
  - Foams.
  - Chemicals (spent acid).
  - Base oil and diesel.
  - Nitrogen.
  - Viscous pills (polymer).
  - Emulsions.
- Solids
  - Mud particulates.
  - LCM material—calcium carbonate ($CaCO_3$).
  - Frac proppant.
  - Gravel pack sand.
  - Formation sand.
  - Perforating debris.
  - Metal cuttings (packer of casing window milling).

Once the by-products of the completion or intervention have been removed, the well is usually flowed until the produced fluids reach a level of cleanliness that is acceptable for the host facility.

Unlike a true well test, clean-up operations are normally conducted through the completion after the production tree has been installed. Where plant configuration allows, oil from the test separator can be routed to the host facilities. This reduces the amount of flaring required; an important consideration in these environmentally conscious times. A relatively simple, safe, and effective way of getting produced fluids into

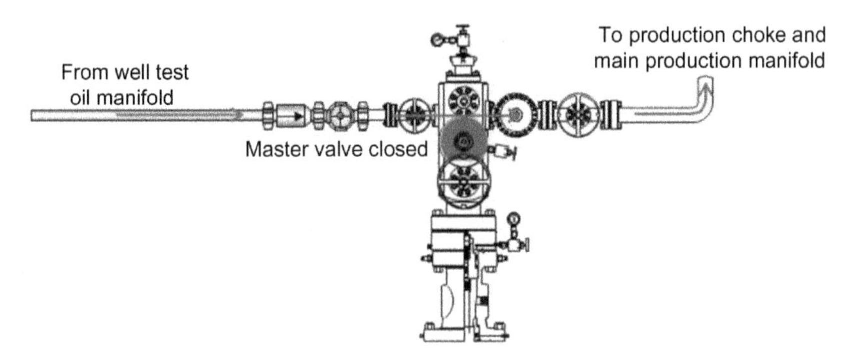

Figure 12.1 Using an existing well to route fluids from a well test package to the process facilities.

the process facilities is to route production across the top of another tree at the same facility.

Fig. 12.1 illustrates a simple but secure method of routing fluids from a well test oil header (downstream of the separator) across the top of a closed-in production well. The advantage of this method is that the produced fluids are routed into, and through, the permanent production facility, and these in turn are tied into the host facility shut-down system. This system has the added benefit of capturing the produced fluids for export, rather than burning them off through a flare.

Utilizing the host facility for hydrocarbon disposal requires close cooperation with production technicians.

Setting clear objectives for a well clean-up are important. Diverting flow away from the well test package and into the production facilities too early has the potential to damage the plant and will cause plant upset. A clean well is normally one where:

- Solids content is within acceptable limits (normally 0.5% by volume).
- Water cut is stable and within acceptable limits.
- Flowing wellhead pressure and temperature are nominally stable.

12.4 WELL TEST SURFACE EQUIPMENT

Well test surface equipment is used to process the produced fluids from a well independent of other wells in the field. Some items of equipment are common to all tests, whilst other equipment is well test specific. For example, a heat exchanger is only required when ambient temperature is low, or where high viscosity oil is being produced.

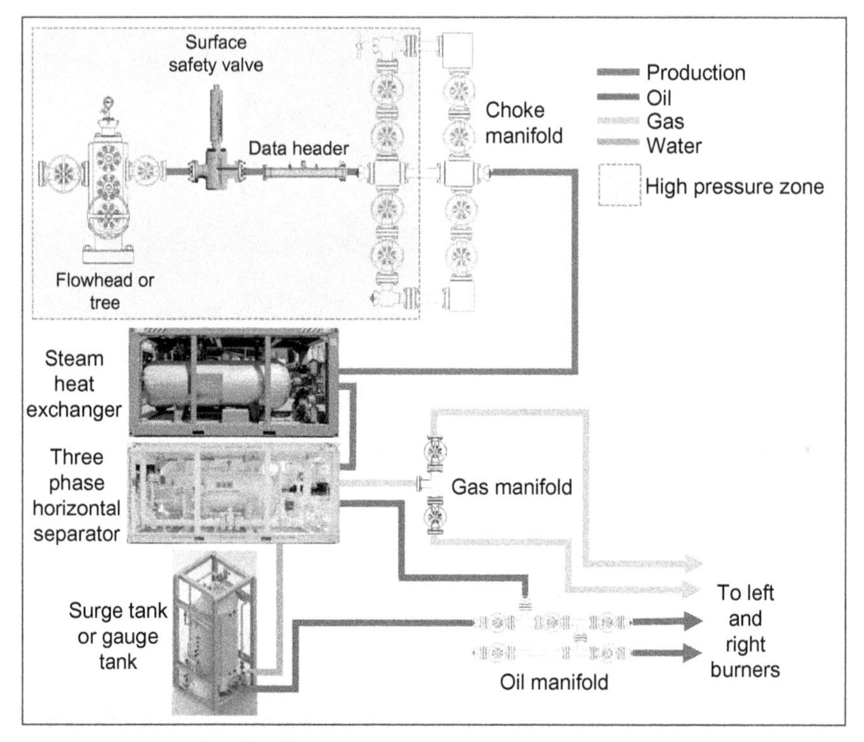

Figure 12.2 Example of well test equipment layout.

A well test package is made up from different items of portable equipment, each with a specific function. When connected together using temporary pipework, the individual components form a complete hydrocarbon process facility. The main process components are:

- Wellhead to choke manifold.
- Choke manifold to the separator inlet.
- Separator to gas, oil, and water phase outlets.
- Phase outlet metering devices to distribution manifolds.
- Distribution manifolds to storage tanks, flowline, or flare (Fig. 12.2).

12.4.1 The wellhead to the choke manifold

Equipment used between the wellhead and the choke manifold is classified as the high pressure side (end) of the rig up. All of the components used must be rated for and tested to a pressure in excess (typically 10%) of the maximum closed-in wellhead pressure. Maximum wellhead pressure is normally calculated by taking the reservoir pressure minus the hydrostatic pressure of a column of gas.

12.4.1.1 *Surface test tree or flowhead*

As noted, well clean-up and well test operations are sometimes carried out after the completion has been run and the production Christmas tree installed. If the drilling BOP is still in place, a surface test tree (STT) or flowhead is used to control production from the well. It is connected to the top of the tubing string and performs the same functions as the Christmas tree, isolating flow from the well through a series of gate valves. Blow out preventer pipe rams are closed around the drill pipe or workstring to provide annulus integrity. If a production tree is in place, the tubing hanger provides the mechanical barrier at the top of the production annulus.

Most modern flowheads are of a mono-block design and have at least one fail-safe (actuated) valve. Actuated valves are operated from a remote panel and ideally will be tied into an emergency shut-down (ESD) system. STTs (flowhead) are supplied by the well test vendors, and will have some or all of the following features:

- Kill wing valve on the kill wing outlet. The kill side of the STT is usually connected to a fluid pump. If the drilling rig is still on location, the cement unit or rig pumps are connected via the stand-pipe manifold.
- Flow wing valve on the flow outlet, connected to the choke manifold. If the wing valve is actuated, this will normally be connected to the ESD system.
- Swab valve for isolation of the vertical wireline or coil tubing access.
- Handling sub which is the lubricator connection for wireline or coiled tubing, and is also for lifting the tree.
- Pressure swivel which allows string rotation with the flow and kill lines connected. Some SST are configured with a gate valve below the swivel.

Treating iron (Chiksan) is used to connect the kill wing to a kill pump and the flow wing to the choke manifold. For land, platform, and jack-up rig operations, hard piping may be used. High pressure flexible lines (coflex) are used on floating rigs and drill-ships. These are necessary to allow for vessel movement (heave). Increasingly, vendors are using coflex hose for land, platform, and jack-up operations (Fig. 12.3).

12.4.1.2 *Temporary flowline*

High pressure temporary pipework is used to connect the flowhead (or Xmas tree) to the choke manifold. In general, the pipework upstream of the choke is classified as being "high pressure" irrespective of the

Figure 12.3 A surface test tree (flowhead) and coflex hose in use on a semisubmersible rig.

wellhead pressure or the actual rating of the lines. Everything on the downstream side of the choke manifold is nominally "low pressure."

Chiksan lines connected with hammer unions are still used for some low pressure applications. To comply with many operating company policies, Chiksan connections need to be manufactured with integral welded connections, not threaded line pipe connections. *Note: Guidance on the correct use of hammer connections can be found in Chapter 8, Pumping and Stimulation.*

Many operating companies do not permit the use of "WECO" style hammer connections on HPHT wells or high rate gas wells; the elastomeric seals are not reliable. Temporary lines for HPHT applications should be fitted with hub and clamp connectors of the "Graylok" or "Techlok" type (Fig. 12.4).

Hub and clamp connectors make pipe assembly more difficult and time consuming, but when correctly assembled, they rarely leak. Comprehensive assembly and disassembly procedures are available on the

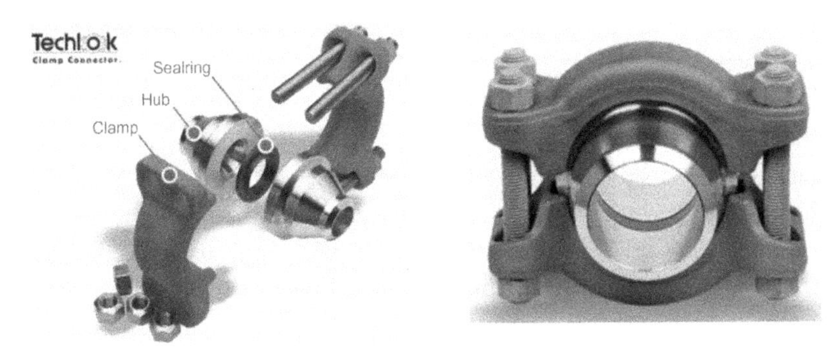

Figure 12.4 Hub and clamp connector. *Image courtesy of .VECTOR.*

Techlok website (www.vectorint.com). One point of note: do not allow crews to grease the gasket ring to hold it in place whilst assembling the coupling. This practice is contrary to the manufacturers recommendations—it causes leaks as the gasket is prevented from seating properly. Once the lines are assembled, the clamp fastening bolt torque should be checked regularly, especially where the lines are subject to vibration.

Hub and clamp connectors use a metal gasket. Providing the connection is correctly assembled, and the correct torque is used, a ring gasket cannot be crushed or overloaded. Initial make up "self energizes" the seal, and the seal is further energized as internal pressure increases. Seal integrity is designed to match or exceed the burst rating of the base pipe. Hub and clamp connectors are able to withstand high bending loads without leaking.

12.4.1.3 Coflex hoses

Coflex hoses must be connected so they hang vertically from the flowhead wing valve. They should never be hung across a windwall or from a horizontal connection, unless there is a preformed support to ensure they are not bent any tighter than their minimum radius of 5 ft. If available, flexible hoses are preferred to hard-pipe lines, since they are easier to hook up. On floating rigs they are essential for connecting the surface tree (attached to the seabed through the riser) to the moving (heaving) rig.

12.4.1.4 Rig or facility permanent pipework

Some of the pipework used during a well test might be a permanent part of the rig or production facility infrastructure. Before it is used for a well test it must be surveyed and pressure tested to confirm its suitability. If in doubt, it should be by-passed using temporary pipework supplied by the well test vendor.

12.4.1.5 Data header

A data header is short section of pipe that forms an integral part of the high pressure flowline which is normally placed immediately upstream of the choke manifold (Fig. 12.5). It is made with multiple threaded ports that enable a range of measuring and sampling functions. These would typically include:

- Chemical injection, normally methanol for hydrate prevention.
- Pressure gauge (upstream of the choke).
- Temperature (upstream of the choke).
- Pressure/temperature recording sensor.
- Deadweight tester connection.
- Fluid sample point.
- Sand erosion monitoring (coupons).
- Bubble hose.

12.4.1.6 Sand detection

A simple method of sand detection is to take frequent basic sediment and water (BS&W) samples at the data/injection point upstream of the choke. Correct positioning of the sample point is important. At low flow rates, samples taken from the high side of the flowline could incorrectly show little or no sand. Sample points should be available on both the low and high side of the manifold and samples should be collected from both

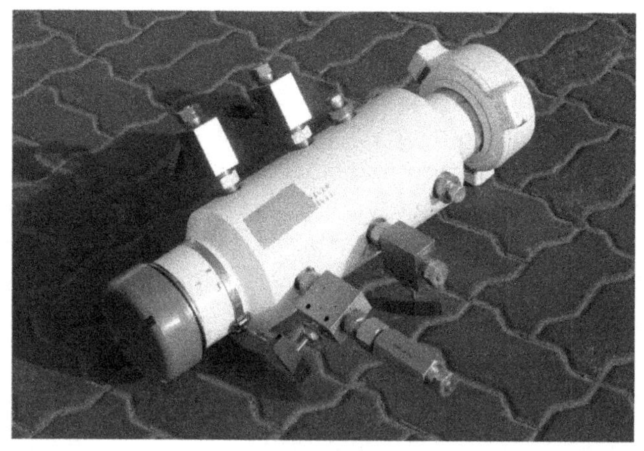

Figure 12.5 A data header. *Image courtesy of WOM. (www.womusa.com).*

points. If sand production is evident, an erosion coupon or probe should be installed at the data header to monitor erosion.

A more accurate method of monitoring sand production is using a sonic sand detector. Installed upstream of the choke, this device operates by detecting sand particles as they impact on a probe protruding into the flowstream. Accuracy in single phase gas flow is good, but deteriorates in multiphase flow.

12.4.1.7 Sand filters and knock-out pots

If a well test is performed on an unconsolidated sandstone formation before a sand control completion is installed, sand will be produced during the test. Similarly, sand production is inevitable during a clean-up flow or well test that follows a propped frac operation.

Sand filters, sand traps, or knock-out pots are positioned upstream of the choke to remove sand from the produced fluids (Fig. 12.6). Different vendors use different systems for sand removal. Whichever system is used, it must be sized to allow safe and efficient sand removal at maximum flow rates. Most vendors use more than one pot, enabling sand to be removed from one side of the system whilst production continues on the other.

12.4.1.8 Surface safety valve

Many operating companies recommend the use of a free standing hydraulically operated fail-safe valve placed upstream of the data header and choke. This is in addition to any actuated valves on the flowhead (or tree). The valve forms part of the ESD system, and limits hydrocarbon escape if the temporary flowline leaks or fails (Fig. 12.7).

12.4.2 The choke manifold to the separator inlet

The choke manifold is the where the transition from high pressure flow (upstream) to low pressure flow (downstream) takes place.

12.4.2.1 The choke manifold

Well test choke manifolds normally have two chokes, one variable and one of a fixed diameter. The variable choke is normally of a needle and seat design. During production start-up, or when changing rate, the variable choke is used. Flow is diverted through the fixed choke for periods of steady flow, e.g., during step rate tests. The orifice in the fixed choke is changed in line with the requirements of the test.

Figure 12.6 Sand filter. Source: *Image courtesy of eProcess Technologies.*

Figure 12.7 Hydraulically operated (fail-safe) SSV. *Image courtesy of WOM. (www.womusa.com).*

A correctly designed choke manifold will allow choke inserts to be removed and replaced with the well flowing, and whilst still maintaining two valve isolation. For two valve isolation to be maintained it is necessary to have two valves between the upstream (high pressure) inlet and the choke in question. Two valves are also required between the each choke and the downstream (low pressure) outlet (Fig. 12.8).

12.4.2.2 Steam heat exchanger and stem generator

Heaters are used to warm up produced fluid during some well tests. Produced fluid needs to be heated to:

- Prevent the formation of hydrates.
- Prevent wax deposition.
- Break foams or emulsions.
- Reduce the viscosity of heavy oil.

Heaters are classified as either direct or indirect. In a direct heater the fluid to be heated passes through tubes surrounded by a fire box; they are normally diesel fueled. Direct heaters should not be used for well tests. Instead, indirect heating is used. Steam is generated in a heater located a safe distance from the well test equipment. From there it is piped through insulated lines to a heat exchanger where it heats the oil. This system is much safer and is easily regulated. Fluids leaving the heat exchanger then flow to the test separator (Fig. 12.9).

Figure 12.8 Choke manifold in use during a well test. This manifold has two valves between the high pressure inlet and each choke and two valves between the low pressure outlet and each choke. Hammer unions allow easy access to the choke inserts. The data header is also visible on the upstream (high pressure) side of the manifold.

Figure 12.9 Well test equipment with the separator (left) and the heat exchanger (right).

12.4.2.3 The test separator

A separator is used to separate immiscible fluids—hence the name. Gas is separated from liquid and water from oil. Separation is achieved as the liquids are immiscible and have different densities. A test separator should:

- Be able to separate the different types of effluent (gas, oil, and water).
- Enable metering of the different phases.
- Enable sampling of the different phases.
- Allow the well to produce at a rate that is sufficient for effective clean-up.
- Tolerate effluent containing impurities such as mud solids or mud acid when the well is cleaning up.
- Be compact and easily transportable.

Test separators are often used to test exploration wells where the nature of the effluent is unknown. Consequently, they must be able to treat a wide range of hydrocarbons from a light dry gas through to heavy crude oil, along with produced water and well construction related contamination, mud, and mud solids. Versatility comes at a price and test separators rarely match permanent production separators in terms of separating efficiency.

The most commonly used test separator is a horizontal three phase, 1440 psi (9931 kPa) working pressure unit. This can handle up to 60 mmscf/day (1.699 m^3/day) of dry gas, or up to 10,000 bopd

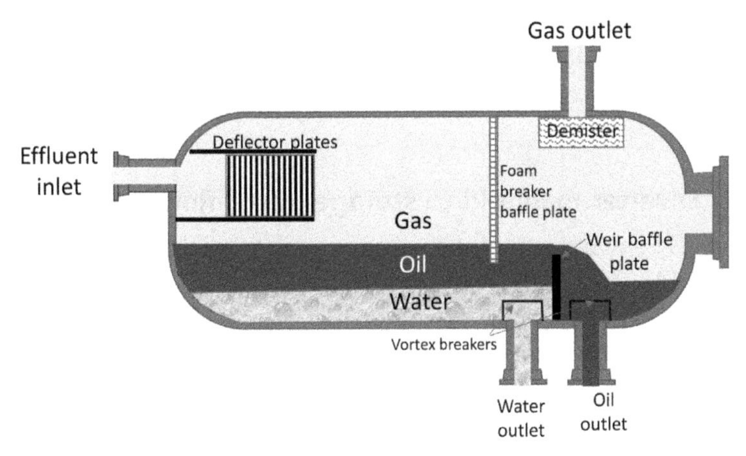

Figure 12.10 Test separator.

(1590 m^3/day) and associated gas at its working pressure. Other types of separator, such as vertical or spherical models and two-phase units are also used (Fig. 12.10).

Gas is metered using a Daniel's (or similar) orifice plate meter. Static pressure, pressure drop across the orifice plate, and temperature are all recorded, enabling gas rate to be calculated. Liquid flow rate is measured by positive displacement or vortex meters. Oil shrinkage (formation volume factor, Bo) is measured by allowing a known volume of oil to depressurize and cool to ambient conditions.

Separator pressure is controlled from a back pressure control valve located downstream of the gas meter (orifice plate). For practical reasons, the separator is normally run at a pressure lower than 50% of the wellhead flowing pressure (pressure upstream of the choke). If the separator pressure is high, relative to wellhead flowing pressure, fluctuations in the separator operating pressure will be transmitted down the well and show on the downhole gauges, making the pressure data from the gauge unreliable.

Overpressure could lead to a catastrophic failure of the separator vessel. A relief valve stops this happening. Most separators are fitted with two separate pressure relief systems: a pressure relief valve and a rupture disk. Valve relief pressure is normally set at 95% of the working pressure of the separator vessel, so for the "standard" 1440 psi separator relief is set at 1368 psi. Rupture disk pressure is normally set at the vessel working pressure. Rupture disks are normally replaced after each test.

Pressurized fluid and gas exhausted from the relief valve must be piped away to a safe area through vent lines. On land wells, the relief line runs to the relief and vent flare pit located a safe distance from the test package, but close enough to prevent back pressure in the relief line impeding its operation. On offshore rigs and platforms, the relief line is directed overboard.

12.4.3 Diverter manifold to storage tanks, flowline or flare

Immediately downstream of the separator outlets are the oil and gas diverter manifolds.

12.4.3.1 Oil and gas diverter manifold

A gas diverter manifold enables gas from the separator to be routed to the downwind burner boom (offshore) or downwind flare pit (onshore). Oil diverter manifolds serve the same purpose but have more valves, enabling flow to be directed to the surge tank, gauge tank, or either flare (Fig. 12.11).

12.4.3.2 Atmospheric gauge tank

The atmospheric gauge tank is a nonpressurized vessel used to measure low flow rates or to calibrate metering devices on separator oil lines. When flow rate is too low to properly drive oil to the flare, oil can be temporarily stored in the tank. Gauge tanks are useful during the

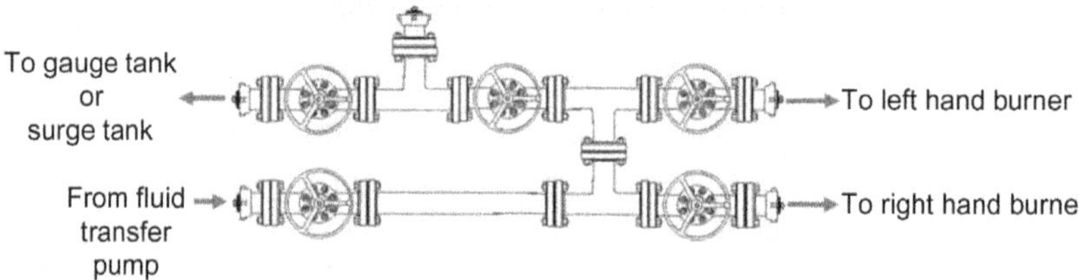

Figure 12.11 A diverter manifold in use.

Figure 12.12 Atmospheric gauge tank. *Image courtesy of WOM. (www.womusa.com).*

offloading of a well to store completion fluids and other noncombustible waste products.

Gauge tanks should not be used if there is H_2S. Many operators prefer to use surge tanks (Fig. 12.12).

12.4.3.3 Surge tank

A surge tank is an H_2S service pressurized vessel for the storage of hydrocarbons after separation. Most tanks have a working pressure of 50 psi. Surge tanks are used to measure liquid flow rate and the combined shrinkage and meter factor. They are also used as a second stage separator, and hold a constant back-pressure by using an automatic pressure control valve on the gas outlet. Surge tanks are also used for storage.

Some tanks are split vertically giving two 50 bbl compartments. Others are a single 100 bbl tank. The tank should be fitted with high and low level alarms to prevent spillage (Fig. 12.13).

12.4.3.4 Burners

Flaring during land based well tests generally involves piping oil and gas to a flare pit where well fluids are ignited.

On offshore rigs a more sophisticated arrangement is required. Burner heads are mounted on the end of a boom attached to the side of the rig. In almost all cases two booms are employed, one on each side of the rig,

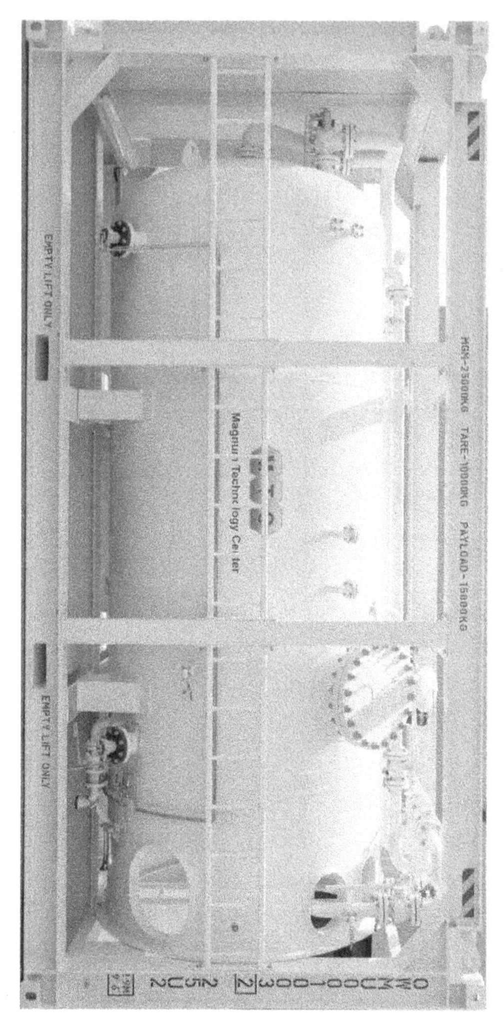

Figure 12.13 Vertical surge tank. Source: Photograph courtesy of WOM - www.womusa.com.

allowing burning to continue without being unduly affected by changes in wind direction (Fig. 12.14).

Burners have a ring of atomizers or nozzles to break up flow and improve combustion. Combustion is further assisted by introducing compressed air into the flow stream. The rig or facilities own compressed air supply should not be used, as there is a risk of hydrocarbons leaking back into the system. In addition, most rig and facilities compressors are not be able to supply air at anywhere close to the volume required. Compressed air for well test burners is supplied by heavy duty high volume portable compressors. The number of compressors required is flow rate dependent, and high rate

Figure 12.14 Offshore well test.

Figure 12.15 This well test required six air compressors.

wells will need multiple compressors. On some offshore rigs and platforms, deck space becomes a problem. It is advisable to manifold the air-lines together, and to have excess compressor capacity in case of break down.

Most well test companies now offer burners that are promoted as "green" or "clean" type burners. Whilst they are less polluting, having superior burning technology, they require significantly more compressed air to operate properly. Deck space can become an issue (Fig. 12.15).

12.4.4 The emergency shut-down system

An ESD system is fundamental to well control during well test operations. The ESD system controls the hydraulically actuated surface safety valve (SSV) valve, if used, and any actuated valves on the flowhead. A pump in the ESD panel supplies pressurized hydraulic fluid to actuated valves. Loss of hydraulic pressure closes the valve(s). Closure can be initiated manually, by activating any of the ESD stations positioned at strategic points across the location. An ESD can also be triggered automatically. Automatic shut-down is a response to high pressure, low pressure, or high liquid levels in the test separator or any of the surge tanks.

Striking any of the remote shut-down buttons causes signal air in the line between the button and the ESD panel to vent. Loss of line pressure in turn causes the ESD panel to vent the pressurized hydraulic fluid that keeps the actuated valves open. Supervisors and every member of the well test crew must know where each ESD station is located. Well test crews (who know their equipment) should agree the location of manual shut-down buttons with supervisory staff, who know the most suitable locations. These might include:

- By the well test spread (by the choke, separator or better still, both).
- On the rig floor (drillers dog house).
- Site office (company man or toolpusher).
- On escape routes (offshore by lifeboat station).

Automated shut-down is controlled by pressure pilots, level sensors, and gas detection equipment. It will normally be configured to respond to:

- High or low pressure upstream of the choke manifold.
- High or low pressure downstream of the choke manifold.
- High liquid level or high pressure in the separator.
- High liquid level or overpressure in the surge tank.
- High liquid level in stock tanks.
- Gas detection (local gas detectors).
- Local H_2S detection.
- High pressure in the steam exchanger jacket (if in use).

Pilot pressure is well specific, and therefore set on a job-by-job basis.

In the case of a well where the completion has been run and the tree installed, any actuated tree valves (master valve or wing valve) and the subsurface safety valve (if fitted) should be hooked up to the well test ESD system.

The safety relief system and the ESD system should be in compliance with the API standard API RP 520—Recommended Practice for the

Design and Installation of Pressure-Relieving Systems, Part 1—Design, and Part 2—Installation API STD 521 Guide for Pressure Relieving and Depressurizing Systems.

12.4.5 Surface equipment pressure rating

Surface equipment must be rated in excess of the maximum pressure to which it will be exposed. Different operating companies implement different standards, but in principle any surface well testing equipment should be designed with regard to the maximum closed in pressure (upstream of the choke) and maximum flowing pressure downstream of the choke.

- A working pressure minimum safety margin of at least 10% above the maximum anticipated operating pressure is usually recommend for surface testing jobs, plus any temperature derating for pressure.
- This means that the equipment should have a WP rating of 1.1 times the maximum potential wellhead pressure.

12.5 WELL TESTING: DOWNHOLE EQUIPMENT

On exploration or appraisal wells it is common to produce the well through a temporary production string, more commonly referred to as a DST string.

In the early days of well testing, DST strings were simple, but not exactly safe. Tests were normally carried out in open rather than cased holes. This saved the time and expense of running casing should the well prove to be dry, or noncommercial. Downhole tools and a packer were usually run on the same drill pipe that had been used to drill the well, hence the name. Using a drill string was an attractive option, in that it could be run quickly and at little cost. A closed pressure barrier in the form of a "tester valve" would be run on the drill pipe. The pipe above the tester valve was left empty, or only partially filled with kill weight fluid, thus providing the required degree of underbalance necessary to begin flow. Below the tester valve was a mechanically set, retrievable packer. Mechanical "Amerada" pressure gauges were positioned between the tester valve and the packer.

With the DST string at the correct depth, the packer would be set, sealing off the annulus above the packer from the reservoir. Setting additional weight down on the packer would open the tester valve. Because the tubing above the tester valve was empty or only partially filled with fluid, the well was able to flow. In these early tests, fluid was not brought to surface but was allowed to rise in the drill pipe until it was above the level of the tester valve. The string was then raised enough to close the tester valve—but not enough to unseat the packer. Simple mechanical Amerada pressure gauges recorded the pressure response to opening and closing the well, enabling some basic parameters to be estimated—providing of course someone had remembered to wind the clock!

At the end of the pressure build up, the string was raised to open the circulating ports in the tester valve. This enabled the hydrocarbons in the drill pipe to be reversed out and replaced with kill weight fluid. Once the well had been killed, the drill string could be recovered to the surface. It will come as no surprise to learn that many of these early well tests were problematic. Loss of containment and blowouts were sadly common, and DST operations were regarded with deep suspicion, justifiably so. Improvements in DST design and technology have reduced the risk—nevertheless many operating companies do not allow the use of drill pipes for well testing. Instead they specify the use of a workstring with premium (gas tight) connections.

A basic DST string combines a number of essential functions:

- Pressure isolation between:
  - Hydrostatic pressure of the kill weight fluid in the annulus P_h.
  - Formation pressure P_f.
  - Hydrostatic pressure of the underbalanced cushion P_c.

Isolation between the hydrostatic pressure of the kill weight fluid in the annulus and reservoir pressure is provided by the packer. In most DST string this will be a mechanical set pull to release device. Isolation between the underbalanced cushion and formation pressure is provided by the tester valve (Fig. 12.16).

To enable the well to flow, the cushion pressure (tubing pressure above the tester valve) needs to be lower than the reservoir pressure. This can be achieved by either running with the tester valve closed and the tubing above empty, or running with the tubing above the tester valve only partially filled. Alternatively, a lighter fluid can be circulated into the well through a multioperation circulating valve.

All DST strings will have a packer, tester valve downhole gauges, and a circulating device in common. Further tools will be added in

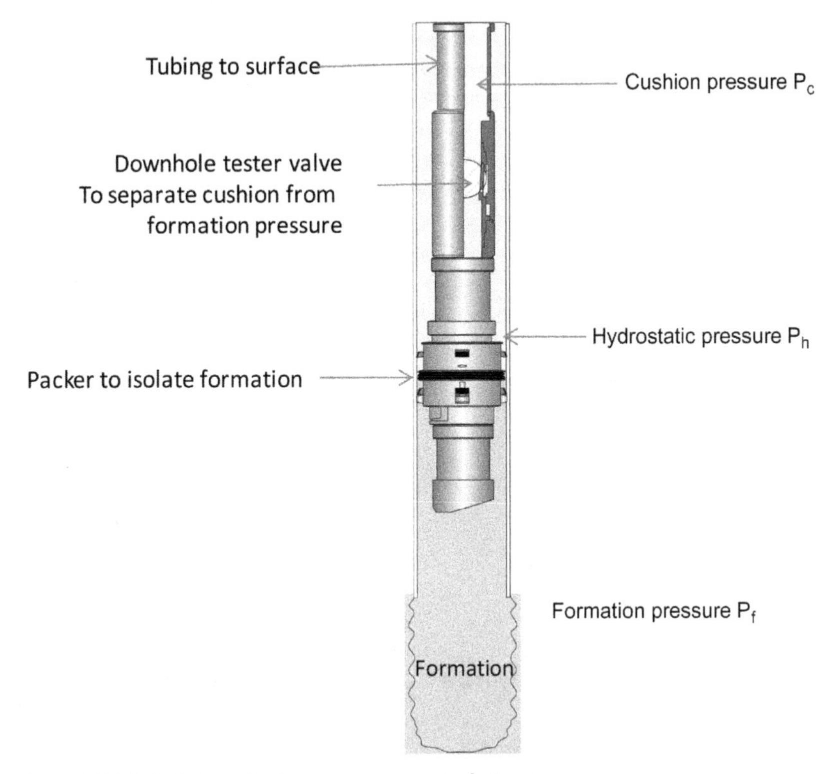

Figure 12.16 Drill stem test string pressure isolations.

accordance with well condition, the type of test being performed, and operating company policy and preference. Design will also be influenced by the well status. There are four basic design categories:

- Open hole: The packer is set inside the open hole. This is the least safe option, as hydrocarbons can by-pass the packer in permeable formations.
- Barefoot: The well is cased to the top reservoir. The reservoir is drilled, but no casing or liner has been run. The test packer is set inside the casing above the open formation.
- Cased hole: The reservoir is drilled, and then casing is set and cemented across the zone of interest. The packer is set in the casing above the production zone. The casing will need to be perforated to enable the well to flow.
- Multilayer well: This requires isolation between producing zones. A single string multipacker DST string is used. This enables different zones to be produced individually or comingled. An alternative is to

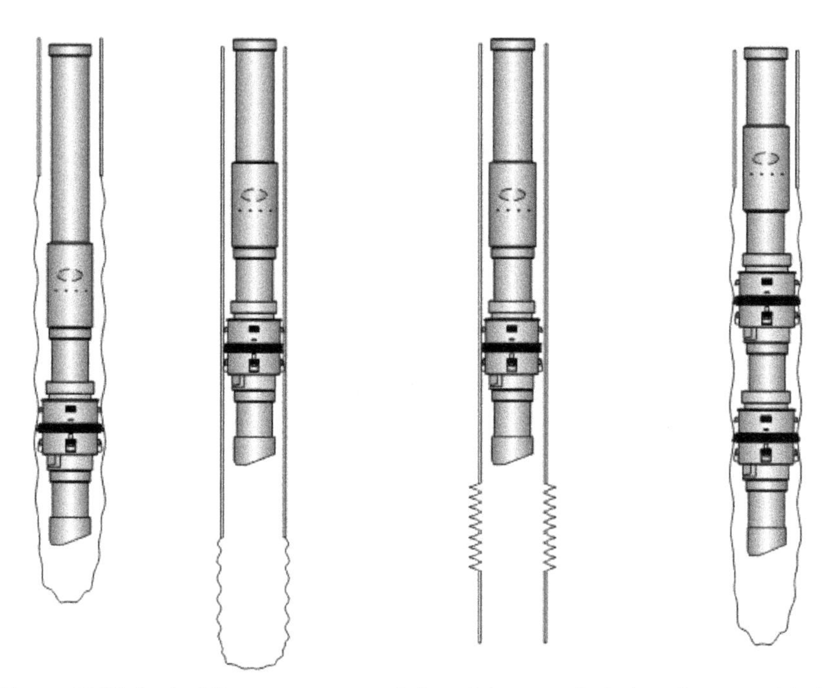

Figure 12.17 Basic drill stem test types, left to right: open hole; barefoot; cased hole; zonal isolation.

test each zone from the bottom up. After a zone is tested, it is plugged and the string moved up to test the next zone (Fig. 12.17).

For testing wells where the reservoir has been cased and cemented, TCP guns are normally run as part of the DST string.

12.6 DRILL STEM TEST COMPONENTS

There is a wide verity of DST tools provided by a number of different vendors. Vendors often use trade names to label tools that carry out identical functions to those of their rivals. The names used here are generic unless otherwise indicated.

12.6.1 Bull-nose or mule shoe

A bull-nose helps guide the bottom of the DST string past possible hang up points such as a liner top or ledges in open hole, and are commonly used on the bottom of TCP guns.

A mule shoe is used on the bottom of a string of pipe. It also acts as a guide. Mule shoes should be internally tapered to aid the passage through tubing intervention tools such as gauges or through tubing perforating guns.

12.6.2 Perforated joint or ported sub

Where TCP guns are run, but where there is no plan to drop the guns postperforation, a perforated joint is used to allow produced fluids to enter the test string tubing. It is positioned above the TCP guns and below the packer.

12.6.2.1 Gauge (bundle) carrier

This sub is used to carry downhole pressure and temperature gauges. Gauges can be positioned above or below the packer. Generally they are more often run above the packer.

- When TCP guns are used, positioning the gauges above the packer protects them from any shock loading associated with gun detonation.
- If the packer cannot be pulled, the gauges (positioned above the safety joint), along with the data they have recorded, can still be recovered. Gauges run above the packer should be positioned below any slip joints.

12.6.3 Debris sub

A debris sub is used to prevent solids settling on the firing head when TCP guns are used.

12.6.4 Gun release sub

In wells where a high rate of production is anticipated, or where through tubing logging access is needed, it will be necessary to drop TCP guns (if run) into a sump below the reservoir. Gun release subs are normally activated by the firing of the gun. Most systems will also feature a mechanical back-up, which is usually wireline activated.

12.6.5 Shock absorbers

Shock absorbers are placed in the string above TCP guns. They absorb the mechanical energy associated with gun detonation, and therefore help protect the packer and downhole gauges against damage. Both lateral and vertical shock absorbers are used.

12.6.6 Packers

Well test packers perform the same basic functions as the production packer used for a completion. They create a mechanical barrier, isolating the reservoir from the surface. Packers may also be required to isolate between zones in a multilayer reservoir, enabling each zone to be tested individually. In addition, some of the downhole tools used with a DST string function in response to changes in annulus pressure; making a sealed tubing/casing annulus essential.

Packer selection is a function of well status (open or cased hole) and reservoir parameters.

Open Hole:

- For open hole testing a weight set packer is normally used. The base of the string is brought to rest at the bottom of the hole. Setting off weight compresses and extrudes the packer seal element, isolating the annulus above the packer from the reservoir. The packer is released with string pick-up.
- Isolation between zones in open hole is normally accomplished using inflatable packers.

Cased Hole:

- Mechanically set retrievable packers are widely used for cased hole well testing. Some of the packers currently available can be set and released several times in a single trip.
- For more hostile conditions (high pressure and/or high temperature) a permanent packer may be required. A seal assembly on the bottom of the DST string stabs into the packer. Because the packer does not have to be released to enable the DST string to be recovered, there is less risk of getting stuck.

12.6.7 Tubing test valve

A tubing test valve enables the tubing string to be integrity tested as it is run into the well. Once the final test has been performed, the tubing test valve is locked open for the remainder of the test. Most tubing test valves are locked open by the application of annulus pressure. There are two basic categories of tubing test valve, manual fill and automatic fill. A manual fill valve requires the tubing to be top filled as it is run. With an auto fill valve, the tubing fills as it is run.

12.6.8 Safety joint

If a retrievable packer is used, it is standard practice to run a safety joint immediately above the packer. Most safety joints are constructed with a coarse left hand thread. Right hand rotation disconnects the DST string at the safety joint, allowing the string above to be retrieved. The remaining stump should have a standard fishing neck to aid the recovery of the packer using standard fishing equipment run on drill pipe.

12.6.9 Hydraulic jars

Jars are normally positioned in the DST string immediately above the safety joint. Impact forces from jarring can be used to assist in the recovery of a stuck packer.

12.6.10 Relief valve and by-pass tool

This tool allows fluid to by-pass the packer when the strung is being run. It is normally positioned a short distance above the packer, and the ports in the tool are in the open position as the string is run. The open ports allow fluid to easily by-pass the packer, and this in turn reduces surging and swabbing. The risk of inducing losses or swabbing in a kick are significantly reduced, which is an important consideration when running into a well with an open reservoir. Once the packer is set, weight down on the string closes the by-pass ports. At the end of the test, tension in the string opens the ports, allowing tubing to annulus communication and equalization.

12.6.11 Tester (shut-in) valves

A tester valve or tester tool (not be confused with a tubing test valve) allows the well to be closed in close to the reservoir. Typically, a well test will consist of a period of production followed by a shut-in. Having the shut-in mechanism close to the reservoir reduces well bore storage effects, and therefore improves data quality obtained during the test. In addition, the tester valve can be used as a mechanical barrier, isolating the reservoir from the surface. At the end of a well test, the tester valve is normally cycled closed before kill weight fluid is circulated in.

12.6.12 Radioactive marker sub

Radioactive marker subs contain a gamma ray (GR) source and are placed in the string to aid depth correlation. The signal from the sub will be clearly visible on GR correlation logs, and will enable the DST string to be placed on depth with a high degree of precision. It is common to place a second marker sub in the production casing. This is particularly beneficial in situations where the formation GR is lacking in features.

12.6.13 Drill collars

Drill collars are run in a DST string to give the setdown weight needed to keep a retrievable packer in place during the test. For example, a 9⅝″ mechanically set packer needs about 20,000 lbs of compression to remain fully set. Most drill collars are manufactured with IF connections and are not gas tight. For high pressure wells or gas wells, heavyweight pipe with premium connections will be required.

12.6.14 Slip joints

Slip joints are needed if the tubing string is fixed at two points (e.g., the packer and the wellhead). Slip joints allow the tubing to expand and contract along the longitudinal axis caused by changes in pressure and temperature. Drill string test slip joints should be nonrotating to allow torque to be applied to the bottom of the string for packer setting and other functions. Drill string test string slip joints generally have limited stroke length, typically between 3 and 5 ft, and some wells will require more than one joint.

12.6.15 Reverse circulating valve: single operation

This tool is used at the end of the well test to reverse hydrocarbons out of tubing. Pressure applied to the tubing or, more commonly, the annulus will shear a rupture disk opening a circulation path between the annulus and tubing. Some models of single use circulating valves have shear pins that hold an inner mandrel in place across circulating ports. Shearing pins enable a mandrel to move, exposing circulation ports. This type of valve is designed for single use only, and is usually run as a back-up to multiuse valves.

12.6.16 Reverse circulating valve: multiple operation

Multioperation reverse circulating valves are operated by either casing pressure or tubing pressure. Most of the major vendors offer both options. Pressure cycles are used to open and close the valve, so keeping track of the number of cycles is important. The benefit of the multioperation valve is the ability to open and close a number of times during the test.

12.6.17 Cross-overs

A DST string will contain many components that have incompatible threads. Cross-overs will be needed. A component as simple as a cross-over can be directly attributable to operational and integrity problems during well tests. Drill string test vendors often contract out the manufacture of cross-overs to local machine shops, where quality assurance is poor. There have been instances of cross-overs being made with little or no internal taper, or with too small an ID. The difficulties for through tubing logging operations are obvious, but of far greater concern is the possibility of a cross-over being made from substandard material. Drill string test strings have failed because of this.

12.6.18 Tubing and drill pipe connections

Standard (5½″) DST tools are normally supplied with 3½″ IF connections if they are to be run with drill pipe. Tools run on premium tubing normally have either a 3½″ Hydril PH-6 thread or 3⅞″ CAS. Both of these connections are robust, allowing repeated make up and break out. Premium connections use integral metal-to-metal seals and are gas tight; a requirement for many operating companies—especially for high pressure/high temperature conditions (Fig. 12.18).

12.6.19 Drill string test string design

Preliminary string design will normally be carried out by well testing vendors working in close cooperation with the operating company reservoir engineers, well test engineers, and completion engineers. It is important the vendor DST specialists are given as much relevant data as possible to enable them to plan the most appropriate string configuration. It has to be accepted that on exploration wells some key parameters will have to be estimates. However, the estimates will be within a range of probability, and the string will normally be designed for a worst case scenario, particularly for pressure. Vital well information supplied to the vendor will include:

- Expected reservoir temperature. This influences:
  - Elastomer selection.
  - Metallurgy.
  - Data collection requirements.
  - Corrected annulus fluid density.
- Surface pressure:
  - Surface equipment pressure rating.

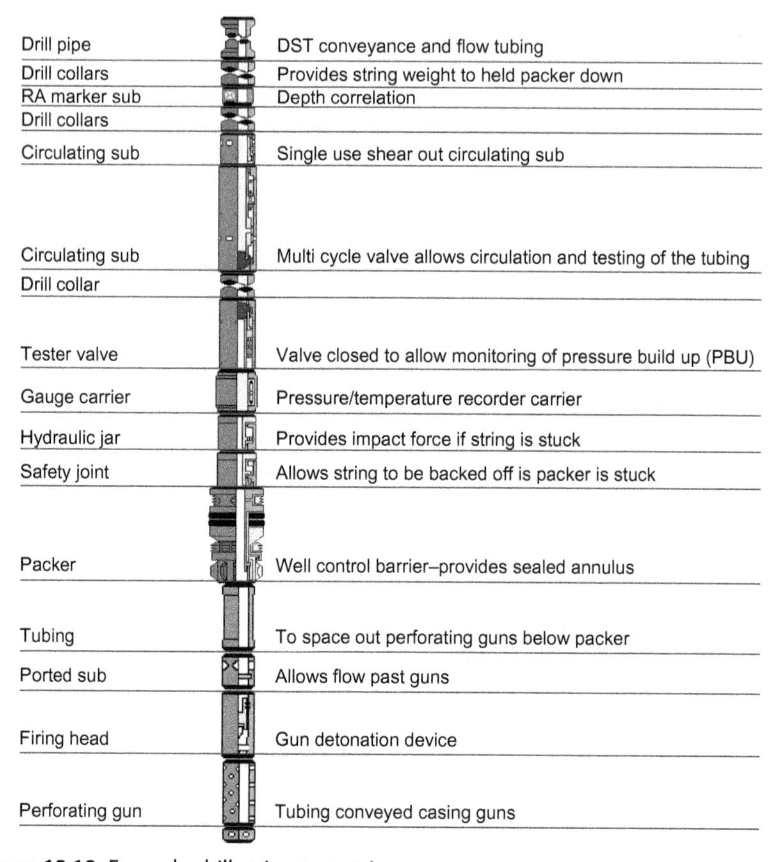

Figure 12.18 Example drill string test string.

- Downhole pressure:
  - Pressure rating for DST string components and tubulars.
  - Packer type.
  - Type and volume of cushion.
  - Data collection requirements.
- Produced fluid properties and rates:
  - Oil, gas, water, H_2S, and CO_2 and solids production.
  - Elastomer selection.
  - DST tool requirements.
  - Surface equipment requirements.
  - Test duration.
- Wellbore fluids:
  - Mud and brine specifications.
  - Elastomer specification.

- Well status prior to the test:
  - Casing and liner configuration.
  - Measured and vertical depth (MD and TVD).
  - Trajectory data.
- Rig data:
  - BOP type and configuration.
  - RTE to wellhead elevation.
  - Air and electrical supply.
  - Deck layout.
  - Drill string details (if being used for the test).
- Facilities data:
  - If produced fluids are being sent to the permanent production facilities downstream of the test spread, the test crew will need to know operating pressure and temperature limits and throughput capacity.

12.7 WELL TESTING OPERATIONS

When a well is tested, the tubing is deliberately placed in an underbalanced condition to initiate flow; the well is deliberately allowed to kick. However, instead of killing the well immediately—as would be the case in an unplanned kick, it is instead flowed through well test surface equipment. At the end of the test, the well must be killed before the DST string can be safely removed.

Well integrity and well control are vital. Supervisors must have a good working knowledge of the purpose of each well test tool and how it functions. They should also have a clear understanding of the function and operation of the surface equipment. Supervisors must have a proper understanding of the barrier requirements and pressure control envelope at each and every step during a well test, and must know how to close in the well in the event of an emergency.

12.7.1 Barriers during a well test

Defining barriers during a well test will depend entirely on the well configuration and the DST string design. Supervisors must think through each step of the operation and must be able to clearly identify both primary and secondary barriers at all stages of the operation.

Some examples of barriers in place during a well test are outlined below. Additional guidance is contained in the chapter of the book detailing barrier policy (Chapter 6).

During the running of a test string, the barriers in place will vary depending on the status of the wellbore. If the DST string is being run into a well that has cemented and tested unperforated casing or liner, then the following barriers would apply:

- Primary barrier. Cemented casing or liner. For the casing (or liner) to function as a barrier, ideally an inflow test will have been carried out in addition to a positive pressure test.
- Secondary barrier.
  - Kill weight fluid.
  - If the DST is being run with underbalanced fluid in the well, the secondary barrier would be the BOP pipe rams.
- Tertiary barrier. Shear rams and blind rams.

If the DST string is being run in a perforated well, or a well with open reservoir:

- Primary barrier: Kill weight fluid (mud or brine).
- Secondary barrier: BOP pipe rams.
- Tertiary barrier: Shear rams and blind rams.

When the well is producing there are two mechanical annulus barriers between the reservoir and surface. These are:

- The packer.
- BOP pipe rams closed on the test string tubing.

In many tests, kill weight fluid will remain in the annulus throughout; however, this is not universal. Some well tests will be conducted with a lighter fluid in the annulus and it will therefore not form a barrier.

During production, flow from the well is controlled at the choke. The well can be closed in by:

- Closing the master valve and flow wing on the STT (primary barrier).
  If the primary barrier fails:
- Close the subsurface safety valve (if fitted)—secondary barrier.
- Close the downhole tester valve—secondary barrier.
- Tertiary barrier—shear seal BOP rams.

After killing the well and during the recovery of the DST string the barriers are:

- Primary barrier—kill weight drilling mud (or brine).
- Secondary barrier—BOP pipe rams.
- Tertiary barrier—shear rams and blind rams.

Operating company policy with regards to barriers during a well test vary. However, most will conform to basic minimum standards and would include:

- Configuring the wellbore/DST string such that there are two independent pressure containment envelopes (two barrier isolation).
- Each of the well barriers will have been tested to ensure it is leak tight (where possible in the direction of flow) at the time of installation and before the well is flowed.
- Where a component has a permissible (acceptable) leak rate, additional sealing elements should be included in the string to ensure isolation.
- Where H_2S is present, or where the test is exploratory in nature, fail-safe valves may be preferred (in place of "normally open" systems).
- Where BOP pipe rams are to be used as a component of the barrier envelope (i.e., closed pipe rams to contain annulus pressure necessary to function DST components) they must be tested before the string is run—even if they are not due a routine test.
- In the event of a barrier failure, DST operations must be halted until the barrier can be properly reinstated and tested.
- Check valve placement, and numbers should comply with API 14C (2001).[2] The aim of these fundamental steps is to prevent any loss of containment whilst the well is "live"—flowing to surface.

12.7.2 Premobilization equipment check

If the equipment arriving at the wellsite is in good working order and is the correct specification for the planned well test, then the risk of a well control incident is reduced. Before mobilizing equipment, engineering staff should ensure that:

- Vendor held maintenance inspection and calibration records for all relevant equipment is available for inspection. Equipment that has time limited certification must have a duration of at least 1½ times the expected test duration (including mobilization transit time) before expiry.
- Function testing and pressure testing of surface and downhole equipment has been carried out.
- Confirm calculations for DST tool rupture disk or shear pin settings. Installation of shear pin/rupture disk should be witnessed by an operating company well test supervisor. Supervisors should also witness the preparation of any DST tools that use a nitrogen charge.

12.7.3 On-site equipment checks

- Ensure all surface equipment is grounded.
- Properly calibrate torque wrenches must be used for making up flange and hub type connectors.
- Before pressure testing any of the surface equipment, confirm the equipment layout conforms to any piping and instrumentation drawing (P&ID) plan provided (Fig. 12.19). The best way to do this is for the operating company supervisor and a senior member of the well testing vendor crew to "walk the line." A copy of the rig up P&ID should be used as reference. Points to look out for:
  - Ensure all the lines are properly restrained in case of failure.
  - Line connectors are compatible.
  - Gauges, sample points, and other ancillary equipment is both accessible and aligned in a way that minimizes the risk to personnel if a connection fails.
  - Valves, manifolds, controls, data, and sample points are all easily accessible and there are clear escape routes.
  - Clearly visible and properly labeled barriers are in place around hazardous areas.

In addition to the P&ID drawing, it is also useful to have a plan for the equipment layout at the location. For offshore installations with limited space (small jack-up rigs), a deck plan becomes essential and can normally only be completed after a site visit by a member of the well testing crew (Fig. 12.20).

12.7.4 Pressure testing surface equipment

- Pressure tests should be monitored and recorded on calibrated instruments.
- Test should be witnessed by an operating company supervisor.
- Test records should be endorsed by the well test vendor and the operating company supervisor, and the record retained.
- Test pressure must be higher than the maximum anticipated pressure expected during the well test. In most instances a 10% excess is acceptable.
- Pressure testing is normally carried out using a water/glycol mix to reduce the risk from hydrates. Gaseous nitrogen (N_2) is also permitted.
- It is good practice to include a low pressure (i.e., 200–300 psi) stability test before increasing to the final test pressure.

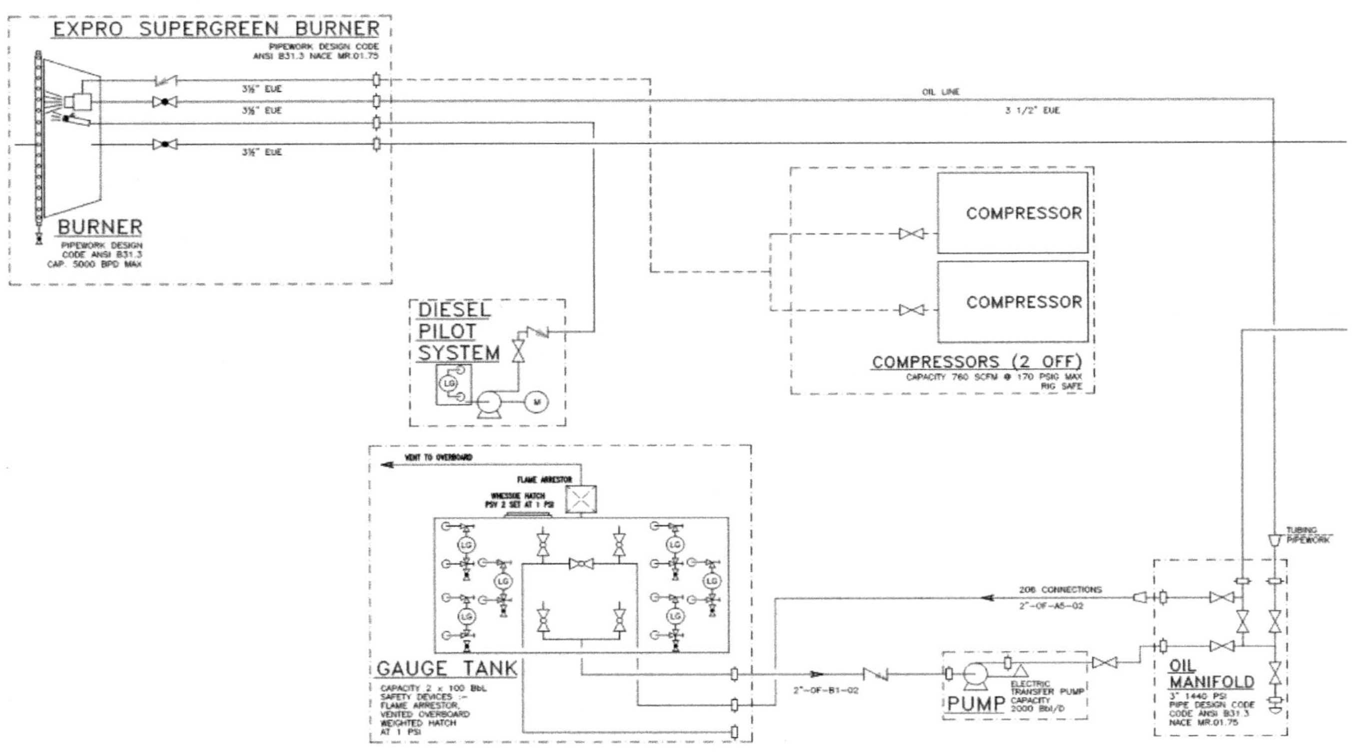

Figure 12.19 Detail from a typical offshore well test P&ID drawing.

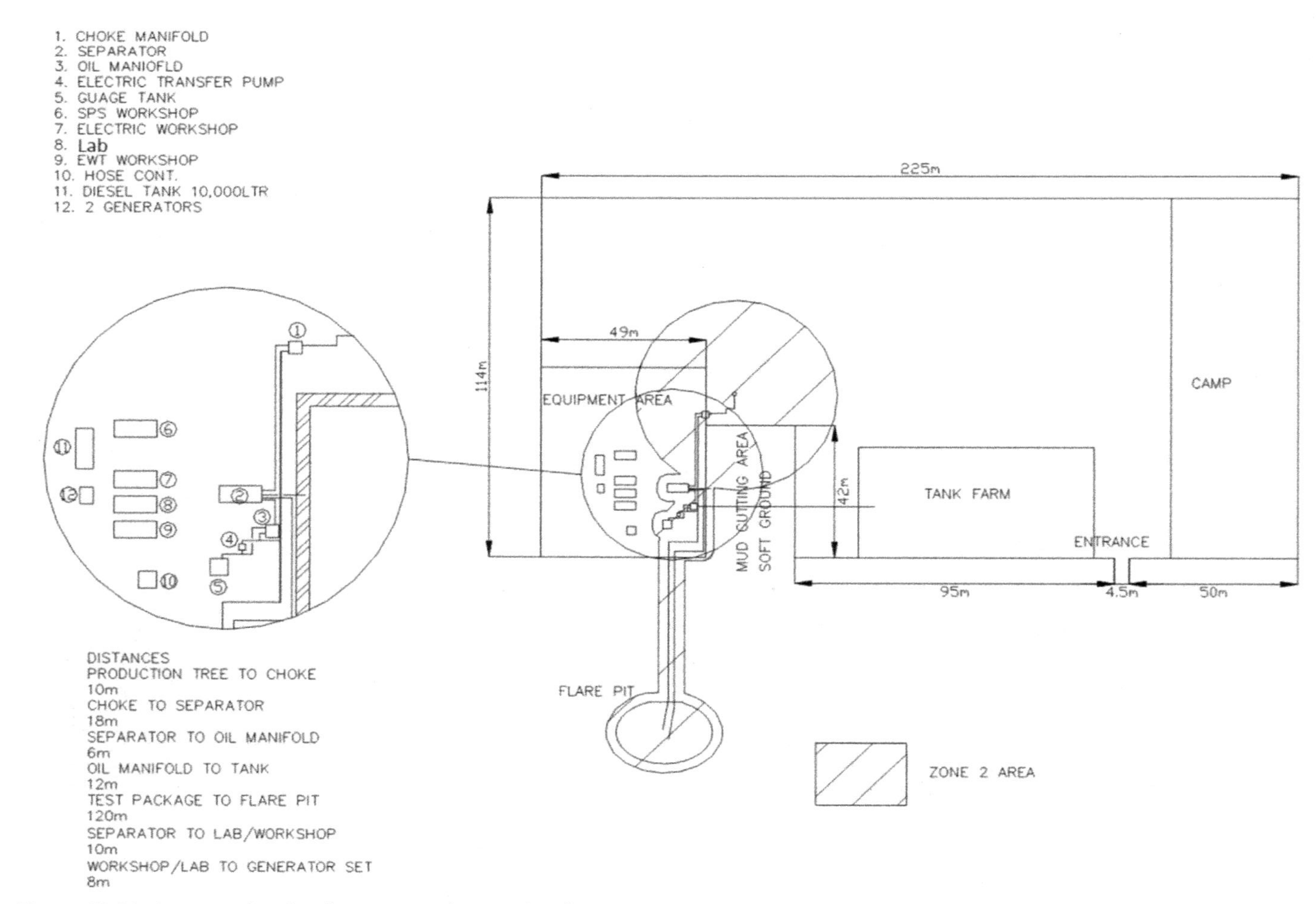

Figure 12.20 An example of and equipment layout plan for a land rig well test.

- Bring the pressure up in stages and hold the final pressure for as long as is required to confirm integrity—normally at least 10 minutes.
  - Test and process equipment exposed to hydrocarbon flow should be purged with nitrogen before flow starts.

12.7.5 Brine and brine preparation

With few exceptions, a DST string will be run with brine already in the well. In some cases the well will be displaced to brine through the DST string, although this is not recommended as mud solids tend to interfere with DST tool functions.

It is good policy for the annulus to be filled with a clean brine containing no more than 0.05% solids by volume. Obtaining that level of cleanliness will normally mean filtering the brine. Filtering normally takes place on location. Good practice will also require an adequate volume of kill weight fluid be held on site—ideally at least twice the casing capacity. Further guidance on brine preparation and brine quality can be found in Chapter 5, Completion, Workover, and Intervention Fluids.

12.7.6 Running the drill string test string

Running a DST string is analogous to running a completion. Many of the same risks will apply—particularly if the string is being run into a well with open reservoir. Supervisors and the drilling crew must be aware of those risks and the measures necessary to control them.

- If TCP guns are being run, an approved perforating safety protocol must be in place and the correct procedures followed when the guns are assembled and armed (Fig. 12.21).
- If a workstring with premium connections is being run, it and the DST tools must not be made up with rig tongs. Specialized tubular make up equipment should be used.
- Running speed should be moderated to reduce surge/swab effects—especially if the reservoir is exposed. The use of a by-pass sub is recommended.
- Some DST tools use pressure cycles to activate open/close functions. Surge pressure might be enough to cause a tool to cycle without the knowledge of the crew. This could lead to a premature opening of a valve and a loss of integrity later in the operation—again, running speed needs to be moderated to reduce surge pressure.

Figure 12.21 Making up tubing conveyed perforation guns prior to running the drill string test string.

- Tubing fluid displacement must be carefully monitored at the trip tank. If the DST string is being run with a closed tubing test valve, then displacement volume will be significant and a small influx could go unnoticed.
- If manual fill valves are run, the tubing should be topped up in accordance with program requirements.
- A kelly cock or an equivalent full opening safety valve (with a crossover to the DST string premium connection) must be on the rig floor at all times.
- A minimum pressure differential between each pressure operated tool is required, usually at least 1000 psi.

Once the string is at depth, the packer can be set. If packer setting is depth critical, and if the string has been run with a radioactive marker sub, a GR/CCL will be run on wireline to confirm the string depth.

Once the packer has been set, tubing and annulus integrity must be confirmed in accordance with the well test program. If a downhole safety valve has been run as part of the string, this will also be (inflow) tested at this point. Supervisors must be absolutely sure that integrity is confirmed before proceeding to flow the well.

12.7.7 Initiating production and the initial flow period

Initiating flow during a DST is critical. It is the time when the possibility of a well control incident is probably the greatest. Before the well is flowed, the wellsite supervisor should ensure that everyone involved attends a prejob meeting. Well test objectives and procedures must be clearly understood by all involved, and roles and responsibilities clearly defined. Good communications between the operating company supervisor, well test engineer, driller, DST operator, and the well test crew are essential.

If the well is to be perforated, initiating flow will normally mean perforating with an underbalance cushion in the tubing. The well may be perforated using wireline. It is more common to perforate with tubing conveyed guns run as part of the DST string, although in some cases the well might have to be perforated before the DST or completion is run.

For open hole completions, the most common method of flow initiation is to run the tubing with a closed tester valve. The pipe above the tester valve is left empty, or partially filled to create an underbalance. After setting the packer and closing the BOP pipe rams around the DST string, pressure is applied to the annulus to open the tester valve. Because the pipe above the tester valve is fully (or partially) evacuated, the reservoir is exposed to an underbalance and the well begins to flow.

12.7.7.1 Cased and perforated wells

Where the well has been cased and cemented, and where an inflow test and positive pressure test have confirmed the integrity of casing and cement, it is common to circulate in an underbalance fluid, usually a clear filtered brine, before the completion or DST is run. Having an underbalance fluid in the well allows flow to begin as soon as the casing is perforated. Alternatively, the tubing can be run with kill weight fluid in place. Once the packer has been set and tested, underbalanced fluid is circulated into the well before the guns are fired.

12.7.7.2 *Open hole and barefoot completion*

If the wellbore is open to the reservoir before the DST string is run, then clearly a kill weight fluid must be in place to control reservoir pressure. Putting the well on production requires the well to be put in an underbalanced condition. There are two options available. Firstly, the tubing could be run with a tubing test valve in the closed position and the pipe above the valve partially or fully evacuated. With the tubing in place, packer set, and the BOP rams closed, opening the valve would allow the well to flow. However, from a well control point of view, running with a closed tester valve has some drawbacks:

- Closed end tubing has a much bigger displacement volume. It is therefore more difficult to detect a small influx.
- No circulation path is immediately available if a well control situation arises, although the test valve can be opened to restore the circulation path.
- No immediate ability to monitor tubing head pressure if there is a kick. The tubing test valve needs to be opened.

An alternative would be to run the pipe open ended and circulate in the underbalanced fluid once the string is at depth, the packer set (and tested), and the BOP pipe rams set (and tested). Underbalance fluid would typically be forward circulated through an open circulating port close to the packer. Overbalance against the formation can be maintained using the choke, but a more common, and arguably safer, method is to circulate in the underbalance fluid above a closed tester valve. Before initiating flow, supervisors must be sure that:

- Packer integrity has been confirmed by pressure testing the annulus, and that the correct amount of weight has been set down on the packer.
- An accurate tally of the number of pressure cycles is kept—this will affect the opening/closing of some DST tools.
- The BOP pipe rams are closed and integrity ensured.
- Fluid pumped volume and annulus return volume should be monitored and measured carefully to stop underbalance cushion fluid from entering the annulus. This is especially important if nitrogen is used and there is no accurate method for knowing where the fluid level is in the tubing.

Once the underbalance cushion is in place, the tester valve can be opened to allow the well to flow. Before flow can begin some checks are necessary:

- There should be a facility to pump kill fluid into the well. In most cases this will mean having a tested high pressure line (Chiksan) hooked up from the cement unit (or a similar high pressure pump) to the kill wing on the flowhead.

- A separate pump must be lined up to the annulus. This allows the application of the pressure necessary for the function of downhole well test tools. It also allows for the pumping of kill fluid at the end of the well test.
- If applicable, a back-up pump can be lined up to the flare cooling system.
- Both the well test engineer and operating company supervisor should make a final check of the lines and manifolds to confirm valve positions and flow direction. On offshore platforms and rigs, flow must be directed to the downwind burner. Confirm that the initial flow (noncombustible fluid) is directed to by-pass the separator.
- If a steam generator is required, this must have been fired up at least 2 hours before flow starts.
- Confirm that there is an adequate supply of hydrate inhibitor (usually methanol), and that the chemical injection pump is connected and ready to operate.
- Confirm that the hydraulic wing valve and the SSV close on command, and that the closure time is within limits (normally 10 second maximum). For operations under Norwegian jurisdiction, the NORSOK requirement is for a 5 second closure.
- If H_2S is expected (or even suspected), then two people, fully kitted out with BA sets must be on stand-by at the rig floor before the first gas reaches surface. H_2S gas detectors must be positioned at strategic locations.
- A valve status board is an extremely simple but effective way of avoiding confusion about valve status and configuration. It needs to be kept at a suitable location, and only works if it is kept up to date and at all times. Properly managed, the status board shows the position (open/closed) of all the relevant valves in the DST string, flowhead, and surface equipment. It also works best if only one nominated person looks after the board and keeps it up-to-date. This is normally the driller on rig supported well tests. If no rig is on location, the well test supervisor is usually the nominated person. To avoid confusion and contradictory instructions, many operations stipulate that all instructions to open or close valves must only be passed via the nominated person.

- A DST vendor representative must be present at all times when the tools are in the well. If TCP guns are in the hole, the TCP vendor representative must be present until the guns have fired and flow has begun.
- For offshore operations only, ensure the air compressors are running and the flare/burner pilot flame is burning before flowing the well.

12.7.8 Opening up and flowing the well

In some locations, the initial flowing of the well will only be permitted during the hours of daylight. Where H_2S is expected, this is usually a mandatory requirement. Whilst there are very sound reasons for only allowing a well to be opened up for the first time in daylight, supervisors need to make sure the crew is not rushing things or taking short-cuts to beat the sunset. It has led to mistakes in the past, particularly in Northern latitudes during winter when daylight is short.

As flow begins:

- Constantly monitor the annulus pressure. This is critical for the following reasons:
  - Annulus pressure is used to control the tester valve function—in most DST strings the tester valve is held open using annulus pressure.
  - Thermal expansion of the fluid in the annulus will exceed tubing collapse/casing burst unless it is monitored and relieved.
  - It is good practice to have two independent monitors of annulus pressure.
- When the well is closed in, the annulus must be topped up and kept full at all times.
- As soon as the first well effluent appears at the surface check for H_2S.
- The separator should be by-passed until flow is free from solids (unless there are facilities to remove solids upstream, or the separator is able to handle solids).
- Throughout the flow period, monitor the surface equipment for vibration and leaks.

12.7.9 Surface sampling

Whilst the well is flowing, samples are taken upstream of the choke (normally at the data header). Additional oil, water, and gas samples are taken at the separator. Although not directly related to well control, sample taking can be hazardous if not carried out properly. Personnel should take samples in accordance with operating company or well test vendor best practice and guidelines. All sample points must be fitted with double valves, and the outer valve used as the working valve when samples are obtained.

12.7.10 Downhole sampling and logging requirements

Wireline (slickline and e-line) is commonly used during well tests. Operations include running and recovery of data acquisition tools, fluid sampling, perforating, and real time logging of the reservoir. Standard wireline pressure control equipment is used and is rigged up on top of the flowhead. Wireline pressure control is covered in detail in Chapter 9, Wireline Operations.

12.7.11 Killing the well

If the completion is in place, there will not normally be a need to kill the well when the test finishes. Instead, the well will be handed over to the facilities operators and put on line. However, if the tests took place through a DST string, the well must be killed to allow the tubing to be recovered. Procedures for killing the well must be clearly laid out in the well test program. In the overwhelming majority of cases, the well will be killed by reverse circulation. The exact sequence of operational steps will depend on the DST tools in the string and their method of operation. A number of options are available, but by way of example a post well test kill might include the following steps:

- If applicable, recover downhole pressure and temperature gauges.
- Close the downhole tester valve.
- Most wells are killed through a reverse circulation valve. The type and operation of the valve is string design and vendor specific. The string might be equipped with a multiple operation, single operation valve, or both.
- Normally at least twice the tubing volume will be reverse circulated with the tubing content sent to the burner or the flare pit.
- Once the well is dead (full circulation complete) and the DST string design allows, the circulating valve is closed and the tester valve locked in the open position.
- Residual hydrocarbon below the tester valve can be bullheaded back to the formation. Bullhead pressure must be below fracture pressure. If

lost circulation material is needed, it can be circulated in before the tester valve is opened.
- After killing, observe the well to confirm it is static—no fluid loss or fluid gain.
- Open the BOP pipe rams in preparation for string recovery.

12.7.12 Pulling the drill string test string

All of the precautions that are normally applied during pipe tripping operations must be applied when the string is pulled. In addition:
- At the end of the well kill, hydrocarbons can be left behind the tail-pipe (below the packer). These will migrate to the surface through the annulus once the packer is released. This should be anticipated and controlled at the choke.
- It is advisable to allow the packer sealing element time to relax. If the string is moved too early swabbing is likely.
- The test program should include contingency measures for dealing with a stuck packer.
- Cleaning up and flowing the well could have improved inflow potential significantly. The degree of drawdown required to swab the well in can be low after a test. Ensure that tripping speed is commensurate with well and fluid properties. Keep the hole full and monitor the trip tank. Similarly, after producing the well, the newly cleaned perforations or open-hole section of the well are more prone to take fluid. Lost circulation material and adequate quantities of kill fluid must be available at the well site.

12.8 EMERGENCIES AND CONTINGENCY PLANS

A well written well test program will include emergency procedures. Some of these will be site specific. Some emergency actions, e.g., what to do if there is a major leak in the surface equipment, are universal.

12.8.1 Production shut-down at the host facility

As described, there are occasions when, production downstream of the separator will be routed through permanent production facilities. This is done to reduce flaring.

When this is done, it is essential to have good lines of communication between well test operators and the host facility control room technicians. In the event of an unplanned plant shut-down, the control room operator must immediately inform the well test crew who will close in the well at the choke. If the shut-down is likely to be prolonged, the flowhead valves should be closed as well.

If the equipment is in place and there are no constraints on flaring, it might be possible to divert flow to the flare, thus allowing the test to continue without serious interruption.

12.8.2 Muster alarm (offshore installation)

Nonessential personnel should go to their muster station. Essential personnel will remain at their post. Who stays and who goes must be made clear at the prejob briefing. On most installations the crew remaining at the job-site will need to contact the installation management and report numbers and well status.

Close in the well at the flowhead and depressurize the surface equipment. The operating company supervisor should try to determine the reason for the alarm, and the probability of the situation escalating.

Depending on the situation, further action may be advisable. If time allows, and it is judged safe to do so:

- Close the downhole tester valve and bleed down any tubing pressure.
- Open the reversing valve and circulate kill weight fluid into the well.
- Close the reversing valve. Bleed off any residual pressure in the tubing and annulus and secure the well.

12.8.3 Abandonment alarm or fire alarm (offshore installation)

The well should be shut-in immediately. This is best done by hitting the ESD button. If the test is being conducted on a drilling rig, it will usually be the driller who will be assigned the responsibility of shutting down production. If the alarm is genuine and it is clear that there is a fire or a major emergency of any sort, anyone can and should activate the ESD. All members of the crew should know where each and every ESD station is located. Go to muster stations.

12.8.4 Loss of containment: surface equipment

The rig floor, well test choke manifold, test separator, and flare must be manned and monitored at all times. If a leak is observed, the affected area must be depressurized as quickly as possible. In most cases, that means closing in the well using the ESD button. As soon as the well is closed in the following people need to be informed:

- Operating company well test supervisor or well test engineer.
- If the well was flowing to the process facilities, then the control room will need to be told that flow has stopped.

At the same time the leak needs to be isolated and the affected equipment depressurized.

12.8.4.1 Leak on the swivel below the flowhead

If wireline is in the hole when the leak develops the following action should be taken:

- If the leak is very small and the well site supervisor judges it safe to do so, pull the wireline toolstring back above the tester valve.

If the wireline is above the tester valve, or if there is no wire in the hole, carry out the following actions:

- Inform the operating company test supervisor. Do not operate the ESD system. Closing in at the surface will increase system pressure and make the leak worse.
- Close in at the well downhole. Usually this will be at the tester valve. If the string was run with a surface controlled downhole safety valve this would be closed.
  - If wire is in the hole above the tester valve, but below the downhole safety valve, close the tester valve. Leave the downhole safety valve open until the wire is above the valve.
- Bleed off the string above the closed valve to the burner.
- Continue to pull the wireline out of hole and back to surface—monitor trip tank levels. Close in at the flowhead.
- Rig down wireline.
- Before repairing any leaks there must be two valves closed upstream of that leak. If the string design does not allow double barrier isolation, a bridge plug can be used. Failing that the well will need to be killed.

12.8.5 Downhole leaks

Downhole leaks will be detected by a change in annulus pressure, rising or falling above that expected from the temperature effects. Tubing to annulus communication means that the primary barrier envelope has failed. Immediate measures must be taken to reinstate the barrier.

Terminate the test and make immediate preparations to kill the well. Inform the operating company well test supervisor straightaway.

Before continuing, the following senior staff should be consulted:

- Operating company well test supervisor (well site).
- Drilling supervisor (well site).
- Vendor well test supervisor (well site).
- Vendor well test manager (base office).
- Operating company supervisory staff at the base office.

12.8.5.1 *Considerations and options*

For a major leak with the annulus pressure decreasing:

- Bullhead kill weight fluid down the string while keeping the annulus full.
- Maintain enough annulus pressure to keep circulating valves open.

For a major leak with annulus pressure increasing, hydrocarbons must be entering the annulus:

- Slowly shut-in the well at the choke manifold to prevent shock loading the casing.
- Line up the kill pumps to reverse circulate.
- Activate single use reversing valve and reverse hydrocarbons out of the DST string with kill weight fluid.

After completing the rest of the programed well kill procedures, pull the test string with care, since the leak may have washed out and weakened a connection.

12.8.6 Downhole test tool failure

Downhole test tool failure will usually mean the end of the test—especially if the valve fails closed. The well will have to be killed and test string pulled:

- The first priority is to make the well safe.
- It may not be possible to bullhead past the failed closed valve. If the string design allows, kill the well by reverse circulation above the failed vale.
- Hydrocarbons will be trapped below the failed valve. Exercise caution when releasing the packer and when the tester valve reaches surface.

12.8.7 Hydrates

There is an elevated risk of hydrates forming during a well test, particularly when the first gas reaches the surface. Wellbore temperature is still relatively low at this point, and in all probability water based completion fluid is still present in the wellbore. Well test programs should include preventative measures which, if properly implemented, will significantly reduce the risk of a hydrate forming. Routine measures to reduce the formation of a hydrate include:

- Using a water/glycol mix to perform pressure tests of surface and downhole equipment.
- Injecting methanol upstream of the choke if there is any risk of gas/water contact, and the temperature is low.
- Avoiding closing in the well until the water based fluids have been removed.

Hydrate formation is rapid. If a hydrate forms upstream of the flowhead, a sudden loss of flowing pressure and rate will be noticed. A hydrate forming at the choke is likely to result in an increase in wellhead pressure combined with a reduction in rate downstream of the choke. Iced up pipework at the choke is a visual indicator of the low temperature associated with a hydrate forming. If a hydrate forms:

- Inform the operating company supervisor, well test supervisor, and the driller.
- If possible, determine the location of the hydrate (downhole or surface).

Hydrates are removed by raising the temperature, lowering pressure, and chemical dispersal, or a combination of these. However, hydrate removal is not without risk—as detailed in Chapter 1, Introduction and Well Control Fundamentals. Wellsite supervisory staff need to carefully plan hydrate removal so as to minimize risk.

REFERENCES

1. McAleese S. *Operational aspects of oil and gas well testing*. Elsevier; 2000.
2. API 14C: 2001 (7th Edition). Recommended practice for analysis, design, installation, and testing of basic surface safety systems for offshore production platforms.

CHAPTER THIRTEEN

Subsea Completion and Intervention Riser Systems

13.1 INTRODUCTION

Drilling, completions, and interventions on subsea wells are carried out from a floating rig or drill-ship that is either anchored or dynamically positioned (DP). Most of the time the rig is on location it will be connected to the seabed through either a subsea drilling riser or production riser. Completion and intervention specialists are not always familiar with subsea production, intervention, and drilling equipment. This is understandable and it can be argued that the operation of complex and specialized subsea equipment is best left in the hands of subsea engineers and specialist vendors. However, during the running or recovery of a subsea completion, and during interventions on subsea wells, completion and intervention specialists will be working with these systems. As such, they must understand how this subsea equipment is configured, deployed, and operated.

13.2 SUBSEA BLOW OUT PREVENTER AND MARINE RISER SYSTEMS

The subsea blow out preventer (BOP) has the same function as its land based equivalent; to close in flow from the well in the event of a kick. In addition, the subsea BOP is sometimes used to suspend the work string should there be a requirement to disconnect the riser. Since it is located on the seabed, a riser is required to deploy the BOP act as a conduit between the rig and the seabed and provide a circulation path between the seabed and the rig (Fig. 13.1).

Well Control for Completions and Interventions.
DOI: https://doi.org/10.1016/B978-0-08-100196-7.00013-0

Figure 13.1 Running a subsea blow out preventer.

13.2.1 System overview

Fig. 13.2 illustrates the marine riser system used to drill and complete subsea wells. The same riser system is also required when intervening in subsea wells equipped with a horizontal tree.

13.2.2 The wellhead connector

The wellhead connector is a hydraulically activated latch and sealing mechanism that is used to connect the BOP stack to the wellhead. The connector is latched to the wellhead high-pressure housing, with 18¾ in. being the most common size in use. The same connector is also used to latch the top of horizontal subsea Christmas trees (SSXTs) (Fig. 13.3).

13.2.3 Subsea blow out preventer stack

Most subsea BOP stacks will have four or more ram type preventers. These are configured on a well-by-well basis, but it is common to include:

- Pipe rams.
- Variable bore rams (VBR). All major BOP manufacturers make VBRs for their subsea BOP stacks. These have the advantage of being able to close over a range of pipe sizes, which is a useful feature when working subsea where changing out a conventional fixed diameter ram would involve recovering and re-running the BOP. Pipe hang off using VBRs is not recommended and limits are lower than for pipe rams.
- Shearing blind rams.
- Casing shear rams.

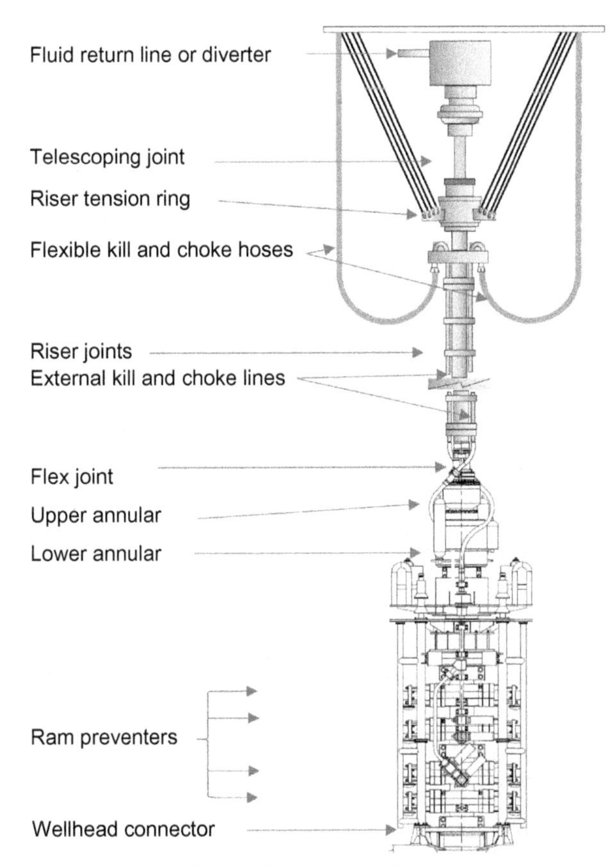

Figure 13.2 "Subsea" BOP and riser. System overview.

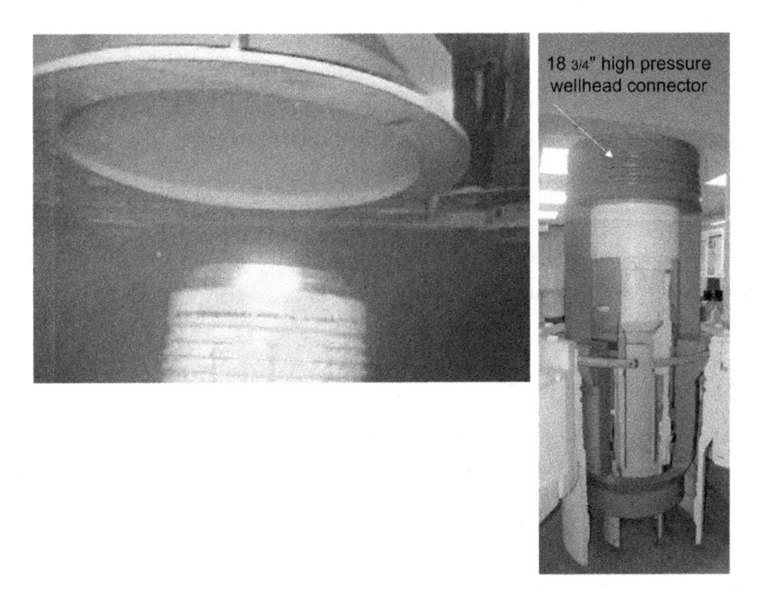

Figure 13.3 Wellhead connector: Remotely operated vehicle (ROV) screenshot of connector landing on wellhead (left); 18¾ in. 10,000 psi wellhead connector (right).

Ram preventers for subsea use, whilst having similarities with their land based equivalent, have additional requirements. These are:

- Fixed diameter pipe rams must be able to support the weight of the tubing string, since it may have to hang from the rams during a riser disconnect.
- Rams for subsea BOPs are fitted with hydraulic locks, such as the Cameron "Wedgelock" system.

Most subsea BOPs have two annular preventers, one located on the BOP stack and a second on the lower marine riser package (LRMP). Subsea preventers must be able to close in 60 seconds or less.[1]

One advantage of annular over ram type preventers is the ability to seal around a range of shapes and sizes. Several manufacturers make annular preventers suitable for subsea use. The Cameron model D and Shaffer annular preventers are essentially the same as the land based model (see chapter 4: Well Control Surface Equipment). Hydril make two annular preventers which are specifically designed for subsea use:

- The Hydril GL preventer: Although designed primarily for subsea use, is used for some land operations. It has a secondary closing chamber which compensates for marine riser hydrostatic pressure in deepwater. The secondary chamber also allows additional closing force on the contractor piston. This may be necessary in some instances, since the preventer is only slightly wellbore pressure assisted.
- The Hydril GX annular preventer: Uses a balanced pressure piston, allowing it to be used in deepwater applications.

13.2.4 Lower marine riser package connector

The LMRP connector is a hydraulically operated latch that allows the LMPR to be disconnected from the BOP stack and the riser lifted clear of the stack. Riser disconnect is either a planned event or an unplanned response to an emergency.

13.2.5 Lower flex joint

The lower flex joint is located above the upper annular preventer and allows a degree lateral movement in the riser. Typically flex joints are limited to approximately 5 degrees of lateral movement from the vertical.

13.2.6 Marine riser

The marine riser provides a conduit between the rig and the BOP stack. It is used to deploy and recover the BOP, and as a circulation path via choke and kill lines that are permanently attached to the exterior of the marine riser.

Each length of riser must be able to support the combined weight of the BOP stack and riser. In deepwater applications, riser hanging weight is reduced by the attachment of buoyant material to the exterior of the riser joint. Recommended practices for riser design and construction are the subject of API RP 2Q.[2]

13.2.7 Telescopic joint or slip joint

The telescopic joint compensates for rig heave and connects the top of the riser to the rig. It is made up from two main components. The outer barrel (lower component) is connected to the riser and remains stationary (relative to the seabed). It is attached to the vessel and supported via the riser tensioning system.

The inner barrel (upper section) has either the fluid return line or the diverter system attached. It is connected to the underside of the rig substructure, and moves up and down with the rig. The ID of the inner member is matched to the riser system ID, with the most common size being 18¾ in.

A pneumatic or hydraulically actuated resilient packing element enclosed in the upper section of the outer barrel seals around the outside of the inner barrel, thus providing riser integrity as the vessel heaves.

13.2.8 Marine riser tensioning system

Without support, the riser would quickly topple over. The marine riser tensioning system keeps tension on the riser whilst allowing for vessel movement (heave).

Most rigs use a wire rope tensioner system consisting of multiple hydraulic cylinders with wireline sheaves. Wire is reeved around the sheaves with one end attached to the outer barrel of the riser slip joint at the tension ring (Fig. 13.4 & 13.5).

The hydraulic cylinders are operated by high-pressure air stored in pressure vessels. Tension on the wire lines is directly proportional to the pressure of the stored air. As the rig heaves upward, fluid is forced out of the hydraulic cylinders, compressing the air. As the rig heaves downward, the hydraulic cylinder strokes in the opposite direction, forced by the compressed air (Fig. 13.5).

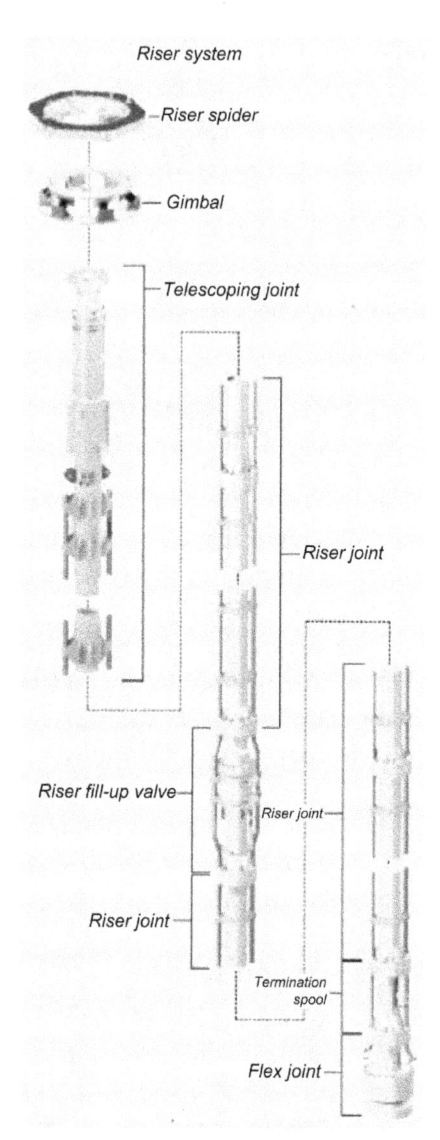

Figure 13.4 Riser components. Source: *Image courtesy of World Oil.*

13.2.9 Kill and choke lines

Kill and choke lines run from the outlets on the subsea BOP to the rig floor choke manifold via external piping on the marine riser. These lines are usually 3 in. nominal diameter or larger, and must equal or exceed the

Figure 13.5 Wire rope attached to the riser tension rig (left). Tensioner cylinders and sheave assembly (right).

Figure 13.6 Riser connection with sealing main bore, kill lines, and choke lines.

pressure rating of the ram preventers. The kill and choke lines are attached to the main riser section and are therefore run simultaneously with the marine riser (Fig 13.4). Most connections are in the form of stabs with a hard-faced pin connecting to a box with integral elastomer seals (Fig. 13.6). The kill and choke lines are connected to the riser as far as the telescoping joint. To allow for rig movement, flexible hoses are used to link the kill and choke lines on the riser to hard piping on the rig.

13.2.10 Choke and kill line "failsafe" valves

Fail-safe (closed) choke and kill line valves are placed on the outlet spools that are positioned between the ram preventers of the BOP stack. The location of each outlet varies, depending on the stack layout, but most contingencies can be dealt with by having one outlet below the rams used for hanging off the tubing string, and a second outlet above the ram.

The choke and kill lines use hydraulically actuated gate valves. Each line has two valves mounted on the stack adjacent to the outlet. They are protected by the outer frame of the BOP stack. As space is limited, it is common to build two gate valves into a single block. On the outboard side of the block a 90 degree elbow is used to change the direction of flow to the vertical whilst keeping the kill and choke lines within the confines of the BOP frame.

In deep water, the hydrostatic head of the control line fluid is sufficient to prevent the power spring in the actuator from closing the valve. To overcome this problem, some valves allow seawater hydrostatic pressure to act on the spring side of the hydraulic piston and assist valve closure. Other valves use system hydraulic pressure to assist valve closure.

13.2.11 Subsea blow out preventer control systems

All the functions on the BOP stack and LMRP are hydraulically operated. Hydraulic fluid under pressure is used to:

- Lock and unlock the wellhead connector and riser connector.
- Open and close the choke and kill valves.
- Open and close the ram preventers and the annular preventers.

The function of the control system is to direct sufficient volumes of hydraulic fluid under pressure to the required BOP stack components. The most common type of BOP stack control system is a hydraulic pilot operated system.

In a piloted hydraulic system, the size of the control umbilical running from surface to the control pod on the BOP is reduced by using a single large diameter hydraulic power supply line. This line is surrounded by several small diameter pilot lines; each pilot line transmits a hydraulic signal to a pilot valve (generally referred to as an SPM).[a] When a hydraulic signal is received from a pilot line, the SPM valve diverts hydraulic power fluid to the specified BOP function. The pilot valves are located inside the control pods on the BOP stack. Hard (small diameter) piping takes the control fluid from the pod to the individual BOP functions. To improve reliability, there are two independent pods, normally referred to (and painted) as blue and yellow. Subsea hydraulic systems are open circuit; fluid is vented into the sea when a valve functions. This speeds up response time and eliminates the need for return lines in the control umbilical. Subsea controls use an environmentally-friendly water based fluid (Fig. 13.7).

[a] SPM: sub-plate mounted.

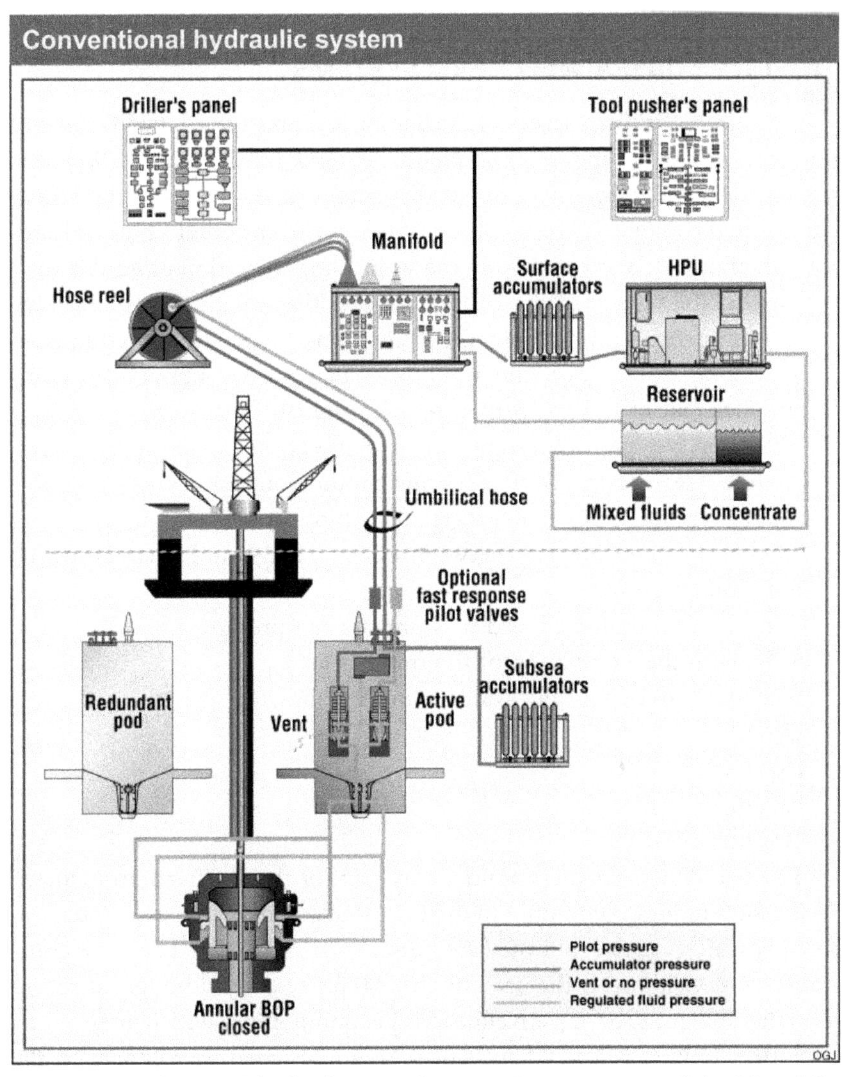

Figure 13.7 Subsea BOP hydraulic control system. *Image courtesy of the Oil and Gas Journal.*

To improve response times, power fluid from the umbilical is used to charge accumulators mounted on the BOP. Stack mounted accumulators typically have a capacity of 1½ times the total volume required to operate all the BOP stack functions.

13.3 SUBSEA WELLHEAD SYSTEMS

A subsea wellhead serves as a structural and pressure containing anchor point for all the drilling and completion systems on the seabed. Wellhead systems incorporate internal profiles that support the various casing strings and isolate the annulus. The wellhead system is also the connection and support when drilling and completing the well. The wellhead must be able to:

- Support the weight of the BOP stack plus the weight of any drill pipe or tubing string hung from the BOP rams. If the well is completed with a horizontal tree, the wellhead will need to support the combined weight of the horizontal tree plus the BOP stack.
- Withstand bending resulting from riser offset.
- Provide a seal for the BOP stack and production tree.
- Support and lock down all casing and tubing strings.
- Provide pressure integrity between casing strings, and between casing and production tubing.
- Contain maximum wellhead pressure and test pressure.
- Assist with the orientation of the subsea tree.

Wellheads are commonly available in the following sizes:

- 13$\frac{3}{8}$ in.
- 16$\frac{3}{4}$ in.
- 18$\frac{3}{4}$ in.
- 21$\frac{1}{4}$ in.

The size designates the nominal bore (ID) of the wellhead, in inches.

Pressure ratings commonly in use are 10,000 and 15,000 psi. Systems at 20,000 psi are available, but they are not in widespread use. The most commonly used wellhead is the 18$\frac{3}{4}$ in. 10,000 psi system, although in recent years the 18$\frac{3}{4}$ in. 15,000 psi wellhead is becoming more widely used.

Subsea wellheads and Christmas trees are designed to conform to the following API standards:

- API 6 A, Specifications for Wellheads and Christmas Tree Equipment.
- API 17D, Specification for Subsea Wellhead and Christmas Tree Equipment.
- API 17D, Recommended Practice for Design and Operation of Subsea Production Systems.
- API RP 17H, Remotely Operated Vehicle (ROV) Interfaces on Subsea Production Systems.
- Wellheads and trees will also have to conform to standards set by national regulatory authorities, for example NORSOK in Norway.

13.3.1 Wellhead components

The main components that comprise a subsea wellhead are:
- Temporary guide base (TGB)
- Permanent guide base (PGB)
- Conductor housing
- Wellhead housing

13.3.2 Conductor housing, temporary guide base, and permanent guide base

Deployment of the TGB, conductor housing, conductor, and the PGB varies depending on several factors including seabed competency, water depth, the type of well, and local experience. If the seabed is unconsolidated, the TGB, conductor housing, conductor pipe, and the PGB are run on drill pipe. On reaching the seabed, the conductor is washed into place until the TGB (or mud mat) comes to rest on the seabed. If the conductor cannot be jetted into place, the TGB is placed on the seabed. A 36 in. hole is then drilled through the guide funnel in the center of the TGB. The conductor, conductor housing, and PGB are then run as a single assembly and the conductor is cemented into place (Fig. 13.8).

13.3.3 Wellhead housing

The wellhead housing is the main high-pressure housing and supports the intermediate casing, production casing strings, and the production tubing (Fig. 13.9).

The wellhead housing is attached to the top joint of structural casing (usually 20 in.). This casing string is cemented into place and gives the required structural integrity to the wellhead. The exterior of the housing is equipped with the latching profile. This profile is used to latch the drilling BOP to the wellhead. Upon completion of the well, the same profile is used to connect the SSXT to the wellhead.

13.4 SUBSEA WELL CONSTRUCTION

Completion and intervention specialists are unlikely to be closely involved with the drilling of the well. However, they should have an appreciation of the pre-completion well construction process and the configuration of the wellhead.

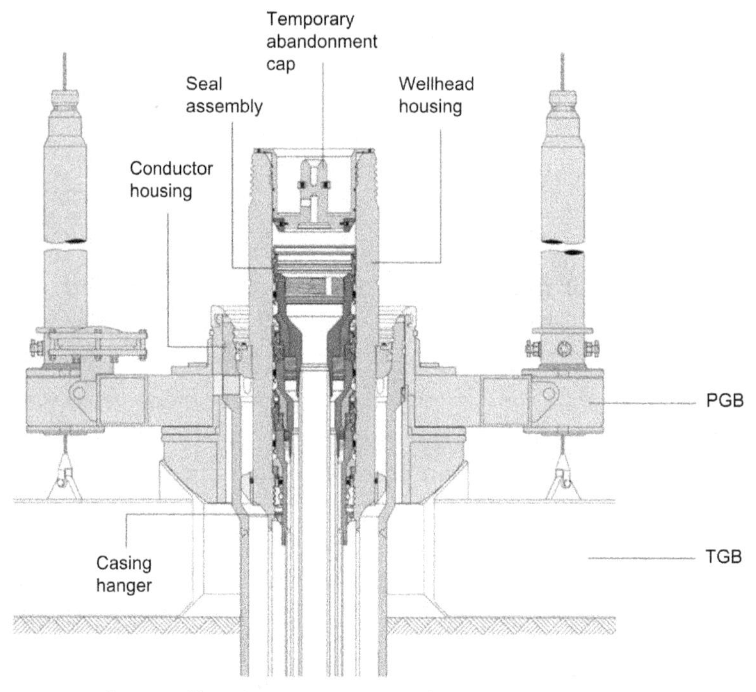

Figure 13.8 Subsea wellhead. *Image courtesy of Drill-Quip.*

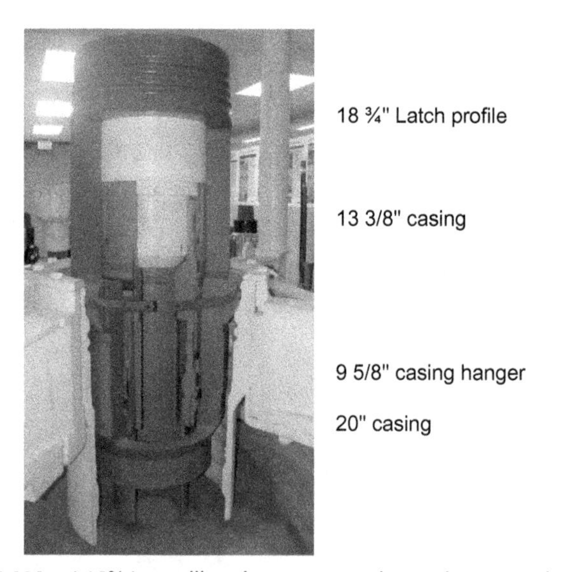

Figure 13.9 10,000 psi 18¾ in. wellhead cutaway to shows the casing hangers.

The sequence outlined below is representative of many subsea drilling operations:

- Place the TGB (or mud mat) on the seabed at the intended well location.
- Drill a 36 in. hole through the guide funnel in the TGB. In most cases, the 36 in. hole will be drilled to about 250–300 ft (80–100 m) below the seabed.
- Run conductor and cement back to the seabed.

Alternatively:

- Run conductor and TGB and jet the conductor into place.

Then:

- Run the PGB
- Drill a 26 in. hole to approximately 2600 ft (800 m)
- Run and cement 20 in. casing
- The top casing joint will be made up to the 18¾ in. high-pressure housing
- Run the drilling BOP and marine riser
- Drill 17½ in. hole to the required depth
- Run and cement 13⅜ in. intermediate casing
- Drill 12¼ in. hole to above top reservoir
- Run and cement 9⅝ in. production casing
- Drill 8½ in. hole to TD
- Run reservoir completion
- Run upper completion

13.5 WELLHEAD INTEGRITY

Although completion and intervention specialists are unlikely to be directly responsible for wellhead and casing design, they should have a basic understanding of the forces acting on the wellhead housing and the casing strings it supports. Detailed casing design and stress analysis is required to fully determine the transmission of force acting through the casing and tubing strings to the wellhead. Stress analysis (casing design) will usually be carried out by the drilling engineer.

Casing stress analysis in subsea wells is complicated by thermal effects on the trapped annulus. In subsea wells, only the inner (A) annulus can be monitored and vented, since there are no valves on the outer (B and C) annulus. A crucial part of any subsea well design is calculating thermally induced fluid expansion and pressure build-up in the outer annulus. As the well is brought on to production, the fluid in the annulus will heat up and expand, increasing pressure in the annulus. If measures are not taken to reduce or relieve this pressure build-up, the burst or collapse of one or more of the casing strings will occur.

13.6 SUBSEA TREES

The functional requirements of a subsea tree are broadly similar to those of a land or platform tree:

- It is a vital component in the well integrity barrier envelope.
- It attaches to the wellhead and directs flow through a series of valves to the flowline.
- It is used to isolate flow from the well.
- Flow is regulated at the choke that is normally attached to the tree downstream of the wing valves.
- It provides access for well intervention operations.

Subsea trees differ from surface trees in several important respects:

- The method of connecting the tree to the wellhead.
- Specialist equipment is required to install the tree.
- All tree functions must be capable of remote operation.
- It is possible for external fluid (seawater) to leak into the wellbore.

There are three types of tree in widespread use throughout the industry: the conventional dual bore vertical tree; the vertical mono-bore tree; and the horizontal tree, sometimes called the spool tree.

13.6.1 Conventional dual bore vertical trees

Conventional dual bore trees have a 5 in. production bore and a 2 in. annulus bore through the body of the tree. Flow from the production bore is isolated using the master valve and/or production wing. Most trees have two master valves (upper and lower). A swab provides access to the production bore for intervention purposes. A smaller secondary bore (2 in. nominal ID) is used to access, monitor, and vent the annulus. Most trees have only a single master valve and a swab valve on the annulus side (Fig. 13.10).

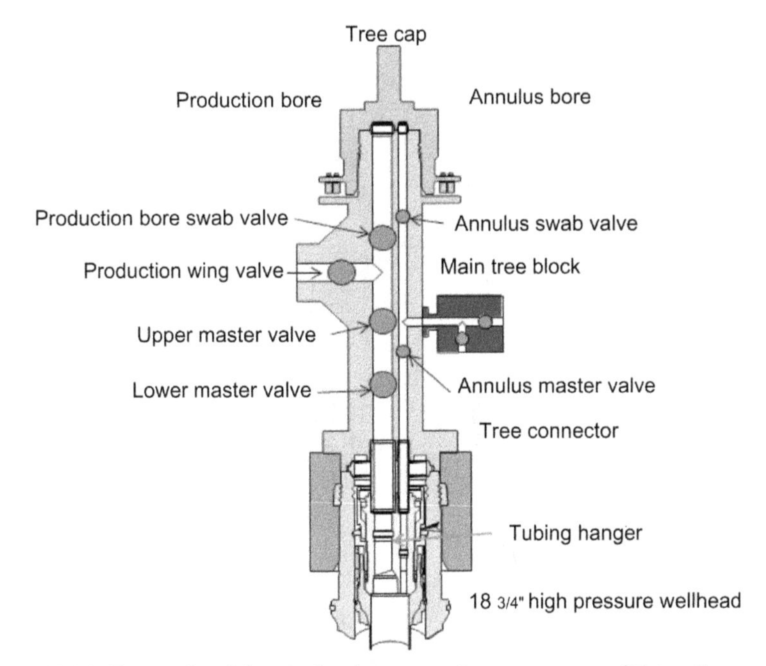

Figure 13.10 Conventional (vertical) subsea tree. *Image courtesy of Vetco Grey.*

External pipework connects the production bore to the annulus bore. Opening cross-over valves in the pipework allows pressure building up in the annulus to be vented into the flowline downstream of the production wing valve.

After running the completion and plugging the well, the tree is run and locked onto the wellhead high pressure housing using a dedicated dual bore tree running riser package. The riser system is also used for well interventions.

13.6.2 Enhanced (large bore) vertical tree

With the "industry standard" 5 in. × 2 in. dual bore vertical tree, tubing sizes larger than 5½ in. could only be run if the completion was reverse tapered below the tubing hanger. Although some 7 in. completions were run below 5 in. × 2 in. vertical trees, the design was far from ideal. A reverse taper can give access problems, and the cross-over below the hanger is subjected to very high stress. Conventional dual bore vertical trees also need a dedicated dual bore riser for tree deployment and well intervention.

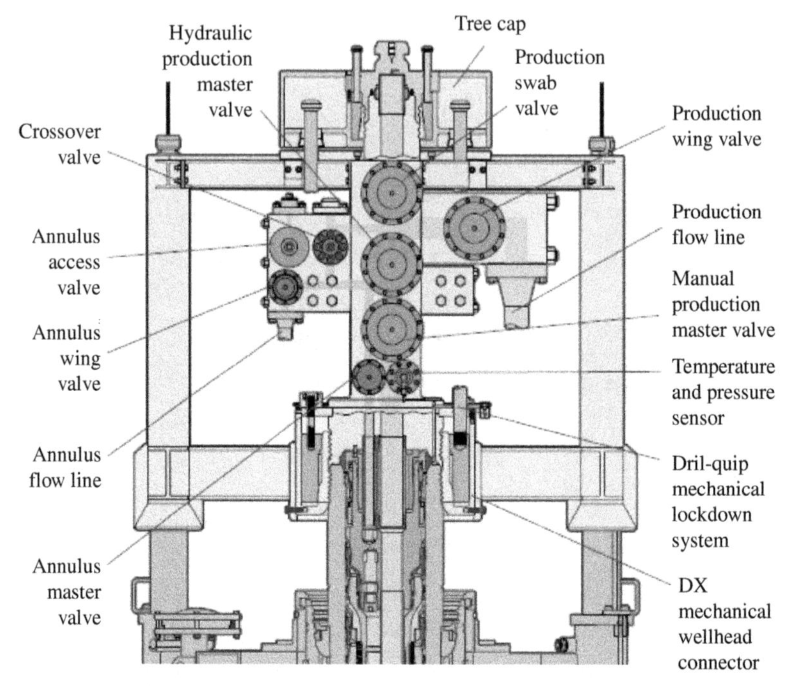

Figure 13.11 Drill-Quip SingleBore subsea tree. *Image courtesy of Drill Quip.*

The introduction of large bore vertical trees has combined most of the advantages of horizontal and vertical trees. A large through bore allows large tubing (typically up to 7 in.) to be run without having to reverse taper below the hanger. A dedicated riser system is not needed; the tree can be run on drill pipe. Intervention access is possible with a mono-bore riser system, available as a rental package when required. The riser system allows circulation with fluid returns through a large bore hose in the control umbilical (Fig. 13.11).

13.6.3 Horizontal subsea trees (spool tree)

Cameron introduced concept of the SpoolTree in 1992, and it was first used in 1994 in the North Sea Gryphon Field development.[3] Since then, other manufacturers have followed with similar products. The horizontal tree differs fundamentally from the conventional tree. When a conventional tree is used, the tubing hanger sits inside the wellhead and the tree is connected to the wellhead after the completion has been run. With a horizontal tree, the tubing hanger sits inside the tree block. The "tree" is in the form of a

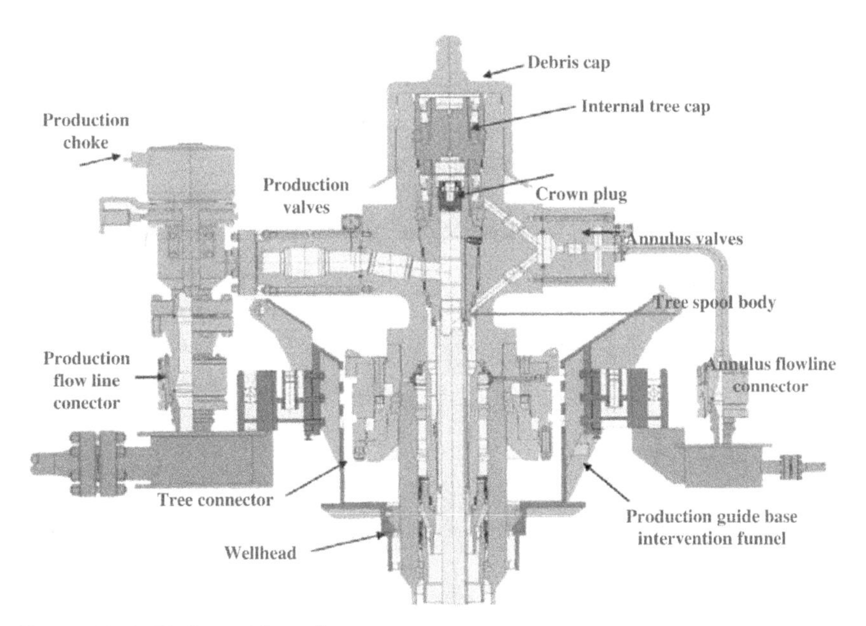

Figure 13.12 Horizontal (spool) tree.

concentric bore spool with a wellhead connector below and a second wellhead connector above. Since the tubing hanger lands inside the tree block, the tree must be installed before the completion tubing is run.

A spool tree is very different in appearance from a conventional tree. All the flow control gate valves are attached to the outside of the spool; there are no valves in the vertical bore (Figs. 13.12 and 13.13).

During the landing of the tubing hanger, a production port on the hanger body is aligned with its counterpart in the spool of the tree and is the main flow conduit when the well is producing. Additional ports within the spool provide access to the production ("A") annulus. Annulus pressure can be vented to the production flowline through a cross-over loop controlled by gate valves (Figs. 13.13 and 13.14).

Tubing hangers for horizontal trees have a single concentric bore and side production outlet. There are additional penetrations for downhole control and instrument lines. The exact configuration will depend on operator requirements and vendor design.

Orientation of the tubing hanger is necessary to align the production outlet in the hanger with the outlet in the spool. Most systems use an orientation helix built into the bottom of the hanger assembly. The helix engages a key slot in the tree spool, giving 180 degree of passive

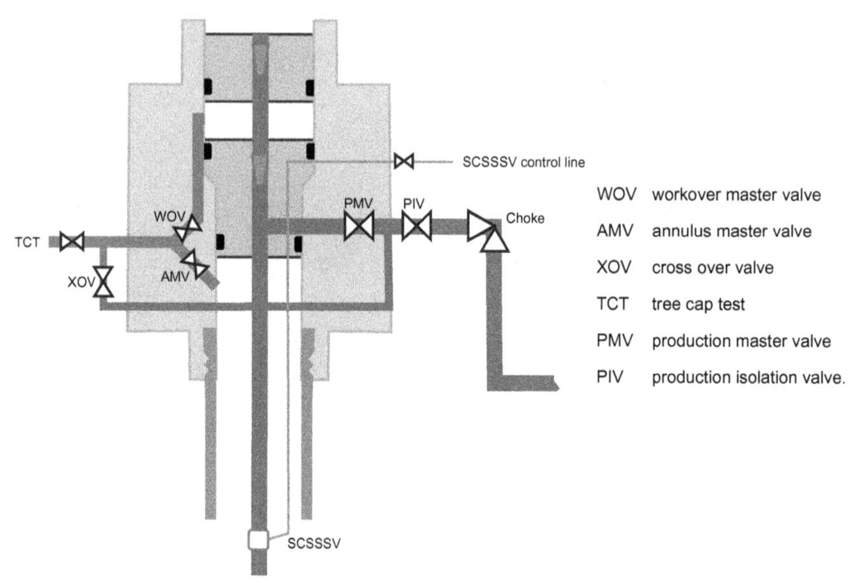

Figure 13.13 Horizontal tree valve configuration.

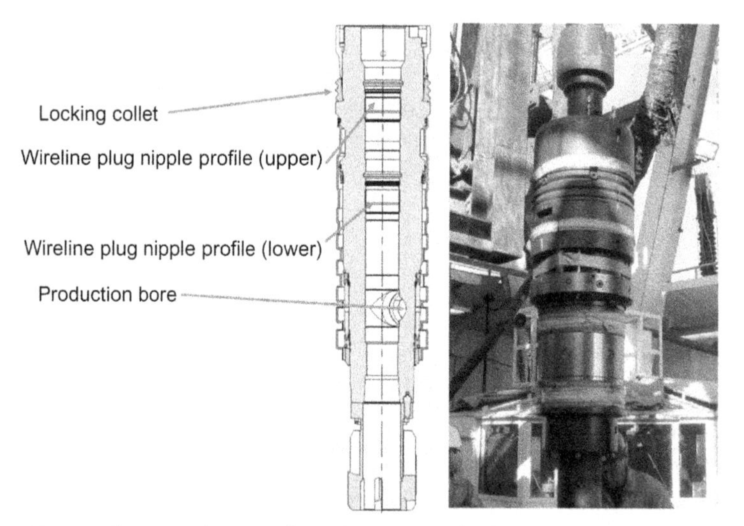

Figure 13.14 A horizontal tree tubing hanger ready for running. *Photo courtesy of Danny Thomas.*

orientation as the hanger is landed. Once landed the hanger is hydraulically locked in place. The main hanger seals are metal-to-metal.

As there are no gate valves in the vertical section of the main production bore, mechanical barriers are needed for isolation of the produced

Figure 13.15 Wireline set plugs of the type used in the internal tree cap and tubing hanger. *Source: Photo from NOV Elmar.*

fluids. An external metal-to-metal seal prevents pressure escaping between the hanger body and the internal diameter of the tree spool.

A wireline conveyed mechanical bridge plug is set in the top of the tubing hanger assembly above the production off-take port. Set in the upper section of the tree spool (above the hanger), a high-pressure internal tree cap forms a second pressure barrier, ensuring conformance with double barrier isolation policy. The internal tree cap also has a wireline set plug to isolate the production bore. It is worth noting that these wireline set plugs form an essential component in the well integrity barrier envelope. To reduce the possibility of a leak, they are run without an integral equalizing device. This can make the plugs difficult to pull, particularly the lower tubing hanger plug (Fig. 13.15).

Horizontal trees were designed to save rig time, reduce capital expenditure, and increase safety. Eliminating the need to remove the tree to re-complete the well certainly saves time, making horizontal trees a good choice for completions that require frequent replacement. As there is no need to invest in customized and dedicated riser systems to run the hanger and tree, capital expenditure is certainly reduced. However, horizontal trees may not be the best solution for wells where frequent interventions are anticipated. The requirement to have to pull and replace the crown plug and tubing hanger plug each time the well is entered is an additional operational expenditure. The manufacturers also maintain that using a spool tree is safer than the conventional alternative, as there is uninterrupted BOP coverage from the time the completion is run until the rig leaves location (Table 13.1).

Table 13.1 Horizontal and vertical tree comparison summary

Type of tree	Advantages	Disadvantages
Horizontal	The tubing can be pulled without having to remove the tree, which is an additional advantage with subsea operations as the need to disconnect flowlines and control umbilical is also eliminated	If the tree needs to be replaced, the completion must be recovered first
	Completion installation, tree deployment, and any subsequent well interventions are performed using standard drilling BOP, marine riser, and (well test) vendor supplied rental equipment—subsea test tree (SSTT)	There are no gate valves in the vertical bore of the tree. Well integrity is reliant on the BOP and SSTT during completion operations and post-completion perforating and well testing. This is arguably less secure than a conventional tree configuration
	The subsea tree provides an integral, precise, and passive hanger orientation system. No BOP modification (orientation pin) is needed	Well interventions are complicated by the need to remove and install the tubing hanger plugs to gain access to the wellbore
	Large bore. Large tubing size if required	The subsea tree must be able to withstand the loading associated with the subsea BOP and marine riser system
	Tree can be installed before drilling the reservoir if desired	Rental costs for SSTT
Vertical (dual bore)	No requirement to recover internal "crown" plugs when re-entering the well for interventions	Tree needs to be recovered when performing a tubing workover. This will mean having to disconnect flow line and control umbilical
		Limited through bore size (5 in. nominal)
		Dedicated LMRP/EDP and riser needed to deploy the tree and reconnect for interventions
Vertical (mono-bore)	Large through bore means large tubing size if required	Restricted flow path for returns if forward circulating—high ECD
	No dedicated riser system required	
	Well control barriers can be configured to allow the tree to be installed after rig departure—potential to save significant amount of rig time	
	No requirement to recover internal "crown" plugs when re-entering well for interventions	

13.7 SUBSEA TREE RISER SYSTEMS

Dual bore vertical trees must be run using a dedicated (vendor supplied) dual bore riser system. The dual bore riser is also used to run the tubing hanger. Mono-bore vertical trees do not require a dedicated riser and are normally run using drill pipe, or in some cases deployed by crane from a support vessel after the rig has moved off location. The tubing hanger is run on a landing string made up from premium tubing.

Horizontal trees do not require a vendor supplied riser, and are normally run from the rig using drill pipe. Unlike the vertical trees, horizontal trees must be run before the completion, since the tubing hanger locks into the tree spool and not the wellhead. The tubing hanger is run on premium tubing with a well test subsea test tree (SSTT), providing barriers to flow until the tubing hanger and crown plugs have been installed.

13.7.1 Vertical tree (dual bore 5 in. × 2 in.)

Conventional dual bore subsea trees are deployed using a dedicated dual bore riser package. Some elements of the riser system are also needed for running and landing the tubing hanger. In recent times, dual bore trees have been largely replaced by mono-bore and horizontal trees. However, the riser package is still a necessity for re-entering existing wells.

When completing a well with a dual bore vertical tree, the tubing hanger is run and landed using the same dual bore riser joints used during the deployment of the tree. They are used in combination with the tubing hanger running and orientation tool (THROT).

13.7.1.1 Tubing hanger running tool

The tubing hanger running tool (THRT) is used to run the tubing hanger into the wellhead and lock it in place. It is connected to the tubing hanger at the rotary table and control line terminations made. The tubing hanger is run on a dual bore riser, usually the same dual bore riser that is later used to run the tree. A control umbilical, run with the riser, operates the hanger lock/unlock function. Downhole instrument and hydraulic functions are also controlled and monitored via the umbilical, enabling the Surface Controlled Sub Surface Safety Valve (SCSSSV) to be kept in the open position as the completion is landed. Having landed the hanger, wireline is normally used to install mechanical barriers in the well to enable the BOP to be recovered and the SSXT run (Fig. 13.16).

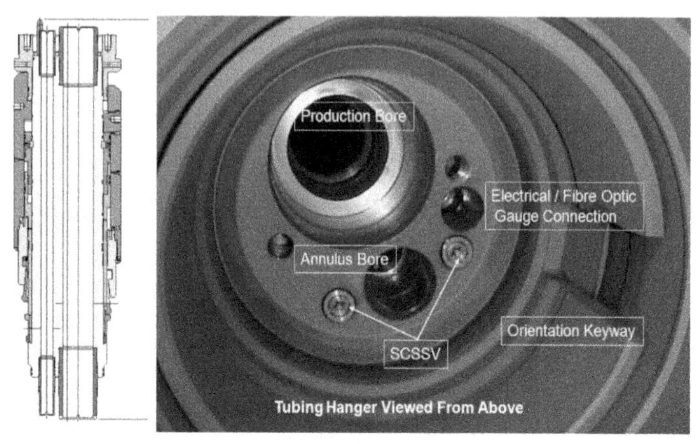

Figure 13.16 (Left) tubing hanger running tool. (Right) the top of a tubing hanger for a dual bore tree.

13.7.1.2 Tubing hanger orientation helix

The tubing hanger must be orientated correctly as it is landed in the wellhead. Failure to do so will prevent the tree from landing properly, as the production and annulus bores will not be aligned correctly. The method most commonly used is to place an orientation helix in the landing string above the THRT (Fig. 13.17). A pin mounted inside the drilling BOP turns the helix to the correct orientation as the hanger is lowered into the wellhead.

13.7.1.3 Completion landing string

A dedicated landing string is used to run the tubing hanger, running tool, and orientation helix. The string is made up of dual bore (5 in. × 2 in.) riser—the same riser that would be used to run the tree. In most operations, the THRT, orientation helix, and BOP adapter are made up into a single assembly, collectively known as the THROT (Tubing Hanger Running and Orientation Tool). The BOP adapter allows the BOP pipe ram or annular BOP to be closed around the landing string to provide an additional well control barrier.

An example of the type of procedure followed when landing a dual bore tubing hanger in a subsea wellhead follows:

1. Run the completion to the point where the tubing hanger will be installed.
2. Make up the tubing hanger.
3. Carry out all necessary control line and instrument line terminations.

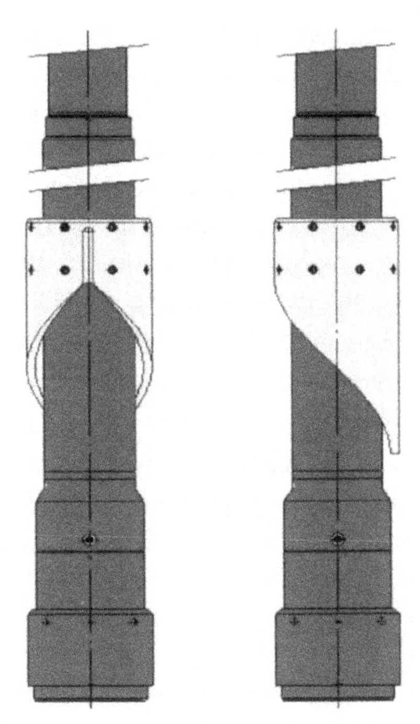

Figure 13.17 Tubing hanger running tool orientation helix.

4. Pick up the THROT and latch the tubing hanger. An over-pull is normally used to confirm the latch.
5. Perform function checks and confirm communication through the THROT/hanger to downhole functions, i.e., the SCSSSV.
6. Configure the control panel confirming each umbilical line is correctly pressured. Isolate the panel from the umbilical reel and disconnect the jumper in preparation for running in.
7. After running the hanger below the table, riser spiders are usually rigged up at the rotary table for riser running.
8. Run the hanger on dual bore riser until above the wellhead.
9. Reconnect the jumper lead between the umbilical spooler and the control panel.
10. Engage the motion compensator and lower the hanger into the wellhead.
11. Lock down the hanger.
12. Over-pull to confirm lock down. Pressure test hanger seal.

Before the THRT can be unlatched and recovered, well control barriers must first be installed. This is usually done by setting wireline conveyed plugs. To conform to a two-barrier policy, a deep-set plug is normally set in the production tubing with a second plug in the tubing hanger nipple profile. A second wireline set plug will be set in the tubing hanger annulus bore. With the well plugged, the riser and THROT can be recovered. Finally, the marine riser and drilling BOP would be recovered in preparation for running the tree (Fig. 13.18).

The subsea tree is run using the LMRP, emergency disconnect package (EDP), and the dual bore riser.

13.7.1.4 Lower marine riser package

The LMRP has a dual bore construction, with each bore sized to match the tree and riser (5 in. × 2 in.). It contains valves for shutting

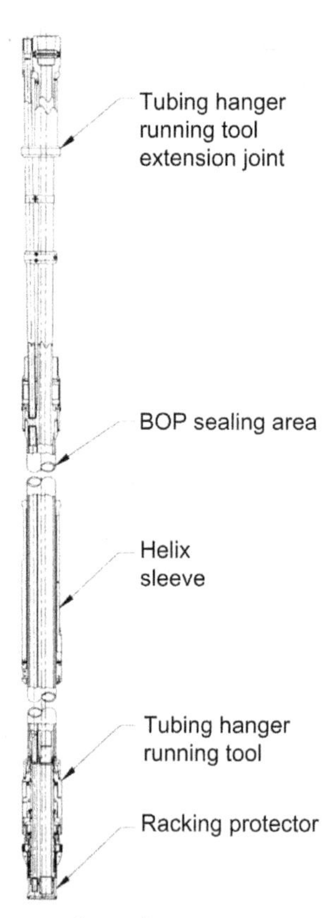

Figure 13.18 Tubing hanger running string.

off flow from the well and allowing the riser to be flushed and tested. It is hydraulically latched to the top of the tree during deployment and interventions, and remains on top of the tree in the event of a riser disconnect (planned or unplanned). There are two deployment options.

1. The LMRP is used to run the SSXT. It is stacked up on top of the tree, usually in the rig moonpool. The EDP is then stacked up above the LMRP. The whole assembly (SSXT/LMRP/EDP) is run down to the wellhead using the dual bore riser system. A hydraulic control umbilical, deployed with the assembly, is used to lock the SSXT connector onto the wellhead and then operate the SSXT/LMRP and EDP. With the tree in place, the mechanical barriers can be removed. Work associated with preparing the well for production can be performed through the riser system. For example, it may be necessary to perforate the well and clean up through a well test system before recovering the riser.
2. The LMRP is used for re-entry into an existing well. The LMRP and EDP are made up and run down to a tree already in place on the seabed. They are latched onto the tree to give access to the well for interventions, or if necessary, remove the tree in preparation for a re-completion.

The LMRP is generally equipped with the following valves:

- Production shear ram capable of cutting coiled tubing and wireline. The shear ram can be configured to fail "as is" or to fail closed. Rams are typically fitted with tungsten inserts and must be able to cut the largest diameter, highest yield coil likely to be used in the well.
- Fail-safe production bore gate valve.
- Wire cutting fail-safe gate valve on the annulus bore.
- Cross-over valves and flow loop, enabling fluid to be circulated down one side of the riser with returns taken on the other. These are used to flush the riser of hydrocarbons after well entry.

During any intervention, the SSXT valve functions are normally controlled from the umbilical controlling the LMRP. Taking control of the SSXT valves usually requires jumper leads connecting the umbilical stab plate to each tree valve. Connections are usually made using an ROV, but in some older systems divers may still be used. ROVs are also used for mechanical valve manipulation where the tree is ROV compatible (Fig. 13.19).

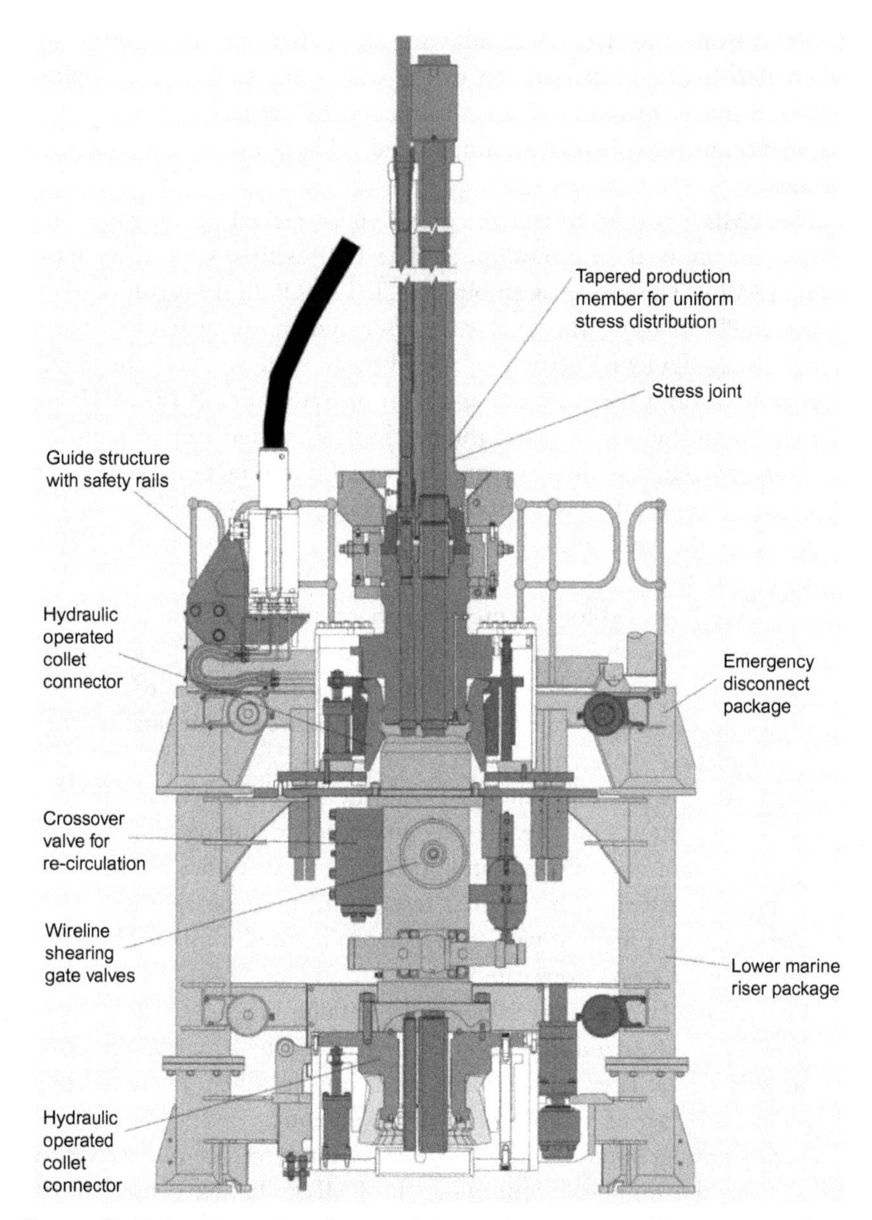

Figure 13.19 Lower marine riser package and emergency disconnect package. *Image courtesy of Cameron.*

13.7.1.5 Emergency disconnect package

The EDP connects the riser to the top of the LMRP. In common with the other components of the system the EDP has dual bores compatible with the riser and LMRP; usually 5 in. × 2 in.

The EDP allows the riser to be disconnected from the LMRP in a controlled manner in the event of bad weather or emergencies. It is designed to allow the riser to be disconnected when the vessel or rig is offset from the wellhead up to a pre-determined angle, sometimes as much as 30 degrees. Connection and disconnection of the EDP is initiated by functioning a hydraulically actuated locking mechanism. The same locking mechanism is used to reconnect the EDP to the LMRP when the weather improves or the emergency is over.

The riser control system must be configured such that a disconnect leaves the well closed-in and in a safe condition.

13.7.1.6 Dual bore riser

A dual bore riser is used to run the EDP/LMRP/SSXT. It acts as a conduit for produced fluids during well testing operations, and gives access to the well bore for intervention tools. At the end of completion or intervention operations the riser is used to retrieve the LRMP and EDP from the seabed. A dual bore riser is assembled from the following components.

13.7.1.6.1 Stress joint

The stress joint is positioned between the EDP and the bottom standard riser joint in the riser. The tubular member is tapered with the thickest wall at the bottom and provides additional bending capacity at the base of the riser.

13.7.1.6.2 Standard riser joint(s)

Standard riser joints are commonly 45 ft. in length, with the tubing and annulus bore clamped together at intervals. The clamps have fastenings to support and protect the control umbilical. Riser joint connections vary from vendor to vendor, but must be able to support the weight of the riser, EDP, LMRP, and subsea tree, as well as any riser stresses associated with vessel movement and ocean currents. The riser has the same working pressure limit as the tree and wellhead.

13.7.1.6.3 Tension joint

The tension joint is used to maintain tension on the riser. The tension joint is a riser joint equipped with a tension rig. The riser tensioner wires are connected to the tension ring and keep a constant, or near constant, tension on the riser as the rig heaves.

13.7.1.6.4 Surface joint

Sometimes referred to as the "landing joint" or "wear joint" the surface joint is a section of dual bore riser, with an outer shroud of casing running almost the full length of the joint. When the Christmas tree lands on the wellhead riser movement stops (relative to the seabed). The rig, however, carries on moving in relation to the riser. The outer shroud protects the riser and umbilical against rig movement where they extend through the rotary table.

13.7.1.6.5 Surface test tree

Surface trees for use with dual bore riser systems have two bores: a main (5 in.) production bore; and a secondary (2 in.) annulus bore. The master valve and wing valves (production and kill) are normally hydraulically actuated and tied into the well test shut-down system. Although the surface tree is supported in the blocks during post-completion well operations, it is the riser tensioners that support most of the riser weight.

Fig. 13.20 illustrates a complete riser system with the LMRP/EDP, stress joint, dual bore riser, and riser tension ring. In this illustration, pup joints are shown between the tension ring and the surface joint. These are used to space out the riser, ensuring the surface test tree (STT) is the correct elevation above the rotary table (Fig. 13.21).

13.7.2 Enhanced vertical tree (mono-bore)

Mono-bore vertical trees do not need to be run on a dedicated riser system. They can be deployed from the rig using drill pipe and a control umbilical. Some operators have run the SSXT from a support vessel after the rig has moved off location. They are run using the vessel crane and a control umbilical. To enable this to happen, the well is typically plugged using a formation isolation valve (FIV) and a tubing hanger plug. Once the tree is installed, the tubing hanger plug is removed using riserless intervention techniques (through water wireline). Pressure cycles are used to open the FIV. This technique makes for significant cost savings, as the rig can leave location as soon as the completion has been landed and the well plugged.

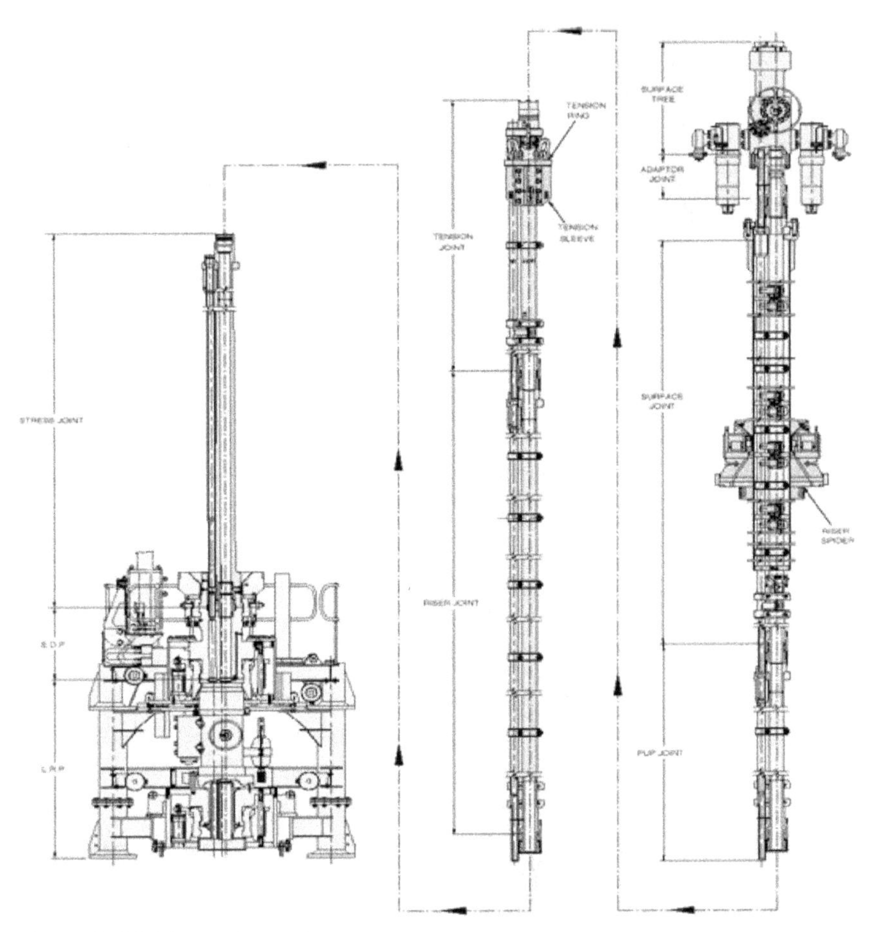

Figure 13.20 The dual bore riser system.

A riser system is required for well interventions. The intervention riser is a rental item and can be adapted for use on any vendor's tree. A widely used system, manufactured by World Oil Machines (WOMS) is operated by Helix.

The Helix WOMS riser has a 7⅜ in. production bore and a 2 in. flexible umbilical connects with the annulus. A foot valve in the EDP retains hydrocarbons in the riser following a disconnect. The LMRP has a hydraulically assisted cutting valve designed to cut coil of up to 2⅞ in. diameter with a yield of 80,000 psi. It also contains fail-safe closure valves (Fig. 13.22).

Figure 13.21 A vertical tree with its LMRP and emergency disconnect package is lowered through the splash zone. *Image courtesy of Cameron.*

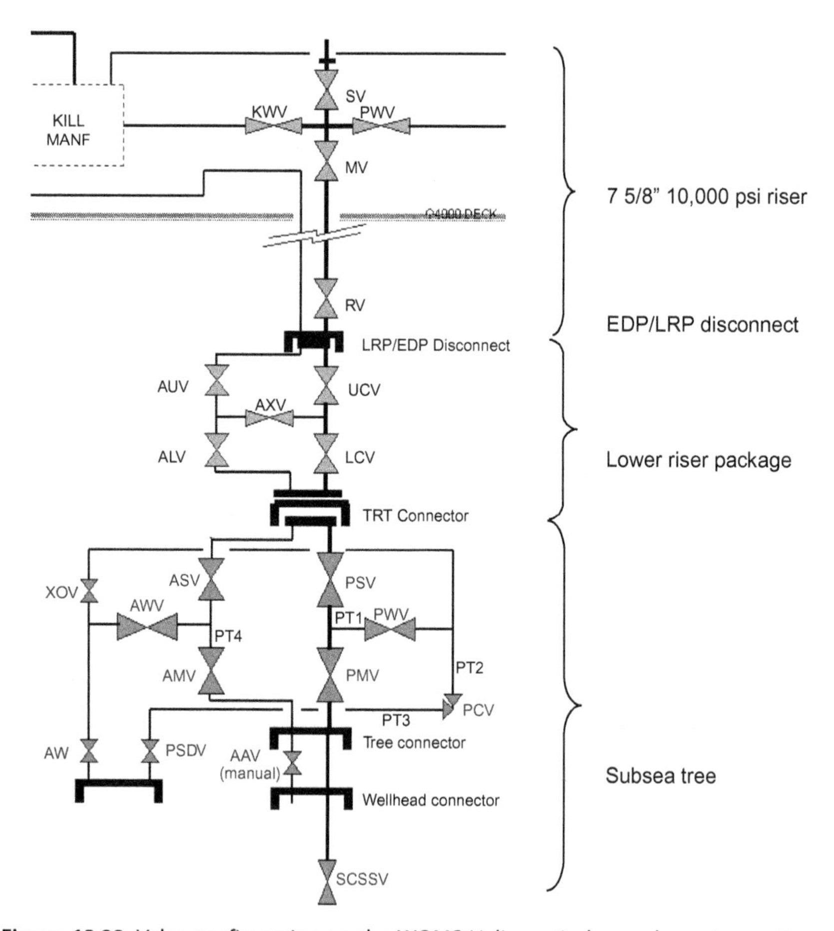

Figure 13.22 Valve configuration on the WOMS Helix vertical monobore riser system.

13.7.3 Horizontal tree riser and re-entry system

Horizontal trees are run on a simple workstring, usually drill pipe. However, as the tubing hanger is located inside the tree spool, the tree must be in place before the completion can be run, and so it is usually run from the rig. It is not uncommon to install the tree before the reservoir is drilled. For example, the well is drilled, cased and cemented to above the top of the reservoir, then suspended using a retrievable packer. The drilling BOP is recovered to the surface, and the horizontal subsea production tree run down to the seabed and latched to the wellhead. After performing any required function and pressure tests, the tree landing string is recovered. The drilling BOP is re-run and latched to the hub on top of the newly installed tree. After recovering the suspension packer, the reservoir is ready to be drilled.

Once the reservoir has been drilled, a wear bushing must be recovered from the tree spool before completion operations begin. After running the completion tubing, the tubing hanger is made up to the final joint of tubing in the rotary table. The THRT, already connected to a SSTT, is picked up and hydraulically latched to the tubing hanger in the rotary table. The completion is then run down through the drilling riser until the hanger lands out in the SSXT. Premium connection tubing is used to assemble the landing string and the control umbilical is clamped externally. Surface controlled, fail-safe, valves in the SSTT provide well control barriers to flow through the production bore. Pipe rams are closed around the SSTT slick joint to provide an annulus barrier (Fig. 13.23).

13.7.3.1 Horizontal tree: tubing hanger running tool

The THRT is used for installing and recovering the hanger and, where applicable, the high-pressure internal tree cap. It is also used to create a pressure tight connection between the tree and the landing sting during interventions by latching into the tree cap, although in some newer enhanced horizontal trees there is no separate tree cap so the running tool latches the hanger profile.

The THRT is connected to and run below a well test SSTT, the test tree providing the well control barriers during post-completion intervention activities.

Hanger hydraulic control functions (hydraulic latch, unlatch, and test) are run from the control umbilical through the running tool to the

Figure 13.23 Running a horizontal tree on drill pipe. *Photo courtesy of Danny Thomas.*

hanger. The control umbilical also operates the hanger and SSTT functions.

13.7.3.2 Subsea test tree

The SSTT is a vital component in the landing/intervention string for any well equipped with a horizontal tree, since it provides the essential well control barriers up until the time the tubing hanger and crown plug are installed. SSTT rental items are supplied by a well test vendor. At the time of writing, two vendors, Schlumberger and Expro, dominate the market.

To perform an intervention on a well equipped with a horizontal tree, the BOP and marine riser are first installed. The THRT is connected to the SSTT and the assembly is run into the riser on production tubing (normally with premium connections). The THRT is hydraulically latched onto a sealing profile in the top of the high-pressure internal tree cap. Wireline pressure control equipment would then be rigged up on top of the riser and wireline used to pull both crown and tubing hanger plugs. With the plugs removed, valves in the SSTT are used to control flow from the well. They become the primary well control barrier until the wireline plugs are replaced.

Fig. 13.24 illustrates the well control barrier configuration of the SSTT after it has latched the high-pressure internal cap. The main points of note are:

- The THRT is locked onto the high-pressure internal tree cap.

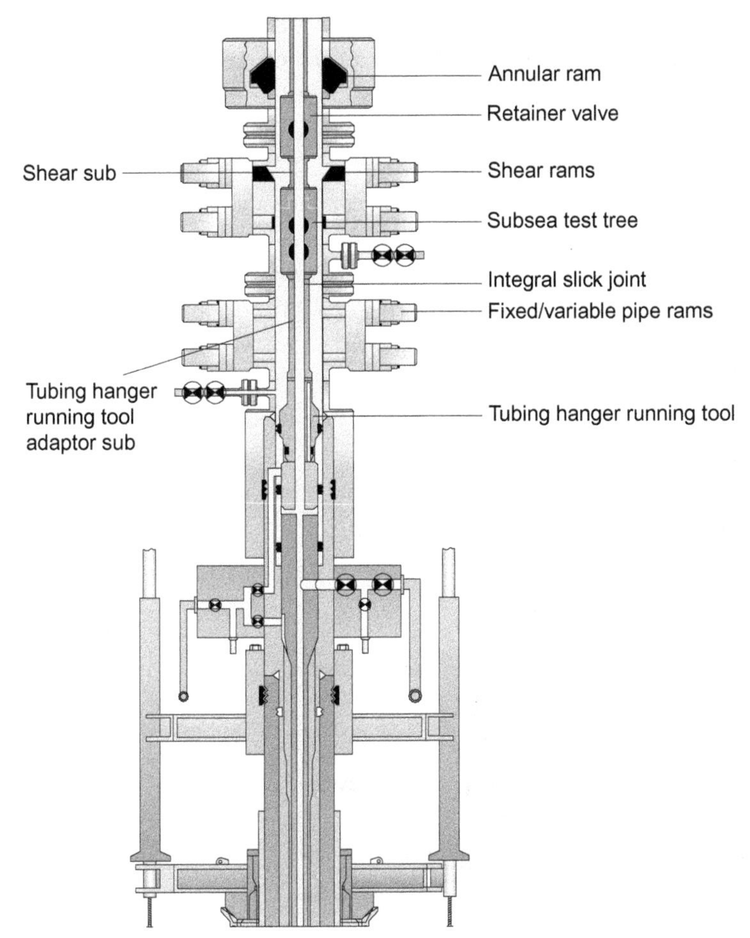

Figure 13.24 A subsea completion and test tree landed inside the drilling blow out preventer.

- BOP pipe rams are closed around the SSTT slick joint. The closed rams are above the outlet for the kill and choke lines. The close rams provide a well control barrier when circulating.
- Valves in the tree can be opened to allow communication with the annulus, and to provide a circulation path.
- There are two valves in the production bore of the SSTT. The lower valve is a ball valve with a wire and coiled tubing cutting capability. The upper valve is either a ball (Expro) or a flapper (Schlumberger); both are fail-safe.
- Above the two fail-safe valves is the disconnect latch.

- Above the disconnect latch is a shear sub. This must be positioned across the BOP shear ram. In the event of an emergency, or a vessel drive off, the shear rams can be closed to disconnect the landing string. Shearing the SSTT severs all the hydraulic lines, causing the SSTT valves to closed (fail safe closure).
- In the foot of the landing string a retainer valve prevents loss of hydrocarbons in the event of a disconnect.
- Some riser systems will have an additional "lubricator valve" positioned above the retainer valve (Fig. 13.25).

13.7.3.3 The landing string

Standard joints of production tubing are used for the landing string. The main selection criteria are:

- The string must be pressure tight. This usually means that only premium connections will be used.

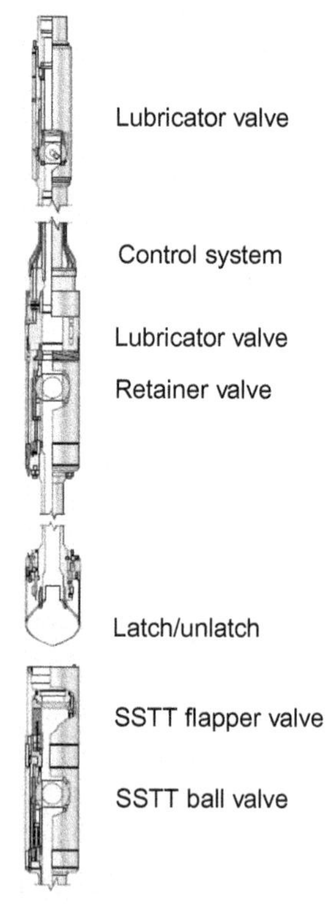

Figure 13.25 Schlumberger SenTREE 7. Subsea test tree (SSTT). *Illustration courtesy of Schlumberger.*

- The string must be able to support the weight of the completion, including allowances for drag and unforeseen events. If the landing string is being used to recover a completion, allowances must be make for any over-pull requirements; unseating packers or pulling swollen and damaged v-packing out of seal bores, for example.
- The string material must be compatible with the produced fluids. This is particularly important if the well is to be produced through the string as part of a post-completion well clean up or test.
- The string ID must be more than the maximum OD of the largest item of equipment to be run into the well, usually the crown plug.

13.7.3.4 Surface test tree

The STT or flowhead provides well control at the surface when completing, testing, or performing live well interventions. A surface tree used with a horizontal tree landing string has a single bore and is usually equipped with hydraulically actuated fail-safe valves. During the running of the completion, the landing string is run to a suitable distance above the subsea tree and set in the slips. The rig blocks are used to lift the coiled tubing lifting frame into the derrick, and bails and elevators are attached to its base. The elevator is then attached to the STT and it is lifted to the vertical before it is attached to the landing string. The whole assembly is then picked up out of the slips before it is lowered to land off in the tree. After landing, tension is applied to support the landing string and the compensators are set, usually at mid stroke. The surface tree should be high enough above the rotary table to allow for rig heave.

The STT actuated valves are controlled from a console positioned at or close to the rig floor, and are also linked into the well test emergency shut-down (ESD) system.

Fig. 13.26 clearly shows the flexible (coflex) flowline and kill line hooked up to a surface tree. The bails and elevators, connected to the base of the lift frame, in turn, are latched to the SST. Well intervention pressure control equipment can be rigged up inside the lift frame using the integral winch. In this photo, a crash frame protects the actuated valves. A swivel, used to facilitate makeup of the SST onto the landing string is clearly visible below the STT and the production tubing (landing string) at the bottom of the photo.

Figure 13.26 A surface tree rigged up.

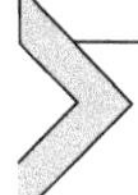

13.8 SUBSEA INTERVENTION AND WORKOVER CONTROL SYSTEMS

The selection of a control system for use with subsea completion and intervention risers depends on several factors, and criteria include:

- Water depth
- Intervention riser (tree) type
- Required response time
- Deployment vessel:
  - Anchored vessel
  - DP vessel

For an anchored vessel, the response time (ESD closure and disconnect timing) is normally expected to be 60–120 s. For DP vessels shutdown response time should be 30–60 s.

Mono-bore and dual bore risers systems with vertical trees use:

- Direct or piloted hydraulic control for water depths up to 3000 ft. (1000 m) and where the rig is anchored.
- For DP rigs and where the water depth is excess of 3000 ft., electro-hydraulic multiplex systems are used.

Horizontal tree THRT/SSTT controls:

- Direct or piloted hydraulic control for water depths up to 2000 ft. (600 m) and where the rig is anchored.
- For DP rigs and where the water depth is excess of 2000 ft., electro-hydraulic multiplex systems are used.

Based on board the rig or vessel will be:

- Hydraulic power unit (HPU). This is the main control panel and is used to function the valves in the intervention riser system and, where relevant, the SSXT valves.
- Jumper hose to connect the HPU to the control umbilical reel. This is disconnected whilst the riser is run and pulled. Pressure is locked into the reel as required.
- When operating with a horizontal tree:
  - THROT/SSTT control umbilical reel.
- When operating with a vertical tree:
  - Reel for either THROT or SSXT/LMRP/EDP.
- ESD control stations (Fig. 13.27).

13.8.1 Direct hydraulic control

Direct hydraulic control is the simplest of the control methods. Each function on the riser package and subsea tree has its own supply line operated from a surface HPU. To open a subsea valve, the control panel-mounted solenoid operated control valve is placed in the open position, allowing hydraulic fluid to flow to the actuator via the umbilical. As the pressure builds up on the actuator piston, the spring is compressed and the valve opens. To close a subsea valve, the control panel-mounted valve is moved to close, isolating the umbilical from the supply header, and allowing the pressure in the umbilical and actuator to vent.

- Topside equipment:
  - HPU
  - Umbilical hose reel
  - Hand operated control valves

Figure 13.27 Workover control system for a direct hydraulic control THROT/SSTT umbilical during a well completion.

- Subsea equipment:
  - Umbilical
  - Umbilical termination unit
  - Hydraulic jumper hoses

Advantages

The main advantage is simplicity and reliability. A simple system is:

- Low cost.
- More reliable than complex systems.
- Easy to maintain. The complex parts of the system are at the surface in the HPU/control panel.

Disadvantages

- Response time is slow and gets slower as water depth increases.
- Large complex umbilical, since each function needs a dedicated hose.
- Unless electrical conductors are integrated into the umbilical, there is no method of monitoring downhole gauges during the landing of the completion.

13.8.2 Piloted hydraulic control

A piloted control system gives better response time, in comparison with direct hydraulic control. A large diameter hose in the umbilical line supplies the seabed control pod and accumulator with hydraulic power. The umbilical also contains multiple hydraulic pilot lines that transmit pressure signals for specific functions. To open a subsea valve, the control panel-mounted valve is opened, allowing hydraulic fluid to flow into the pilot valve via the umbilical, the pilot valve in the control pod will operate allowing hydraulic fluid to flow from the accumulator to the valve actuator, opening the valve. To close a valve, the control panel-mounted valve is moved to vent the pressure from the pilot line. This causes the pod mounted pilot valve to vent pressure from the actuator.

- Topside equipment:
  - HPU
  - Hand operated control valves
- Subsea equipment:
  - Umbilical
  - Umbilical termination unit
  - Hydraulic jumper hoses
  - Subsea control module

13.8.3 Hard-wired electrohydraulic controls

In an electrohydraulic system, hydraulic signals from pilot lines are replaced with electrical signals. The control umbilical carries a multi-conductor electrical cable and a large diameter hydraulic hose to charge the accumulators. In the control module on the seabed, electrically operated solenoid valves are used to control hydraulic flow from the accumulator to individual valve functions.

To open a subsea valve the appropriate switch on the surface control unit is closed, allowing current to flow to the solenoid valve in the control pod. Opening the solenoid valve allows hydraulic fluid from the accumulator to valve actuator. Opening the switch on the panel cuts off power to the solenoid, closing the valve and venting power fluid from the valve actuator and causing it to close.

Topside equipment:

- HPU
- Electrical power unit (EPU)
- Hand operated switches

Subsea equipment:

- Umbilical
- Umbilical termination unit
- Hydraulic jumper hoses
- Subsea control module

Advantages:
- Operated with fast response times in deepwater. No theoretical limit to water depth.
- Independent control of selected function.
- Ability to monitor downhole gauges using electrical conductors in the umbilical.
- Normally a smaller umbilical than that used for direct or piloted hydraulics.

Disadvantages:
- More costly than direct hydraulic systems.
- More difficult to maintain—complex components on the seabed.
- Generally, not as reliable as direct hydraulic control because of the reliance on electrical components in a pressurized salt water environment.

13.8.3.1 Multiplexed electrohydraulic

Multiplex has similarities with hard-wired electrohydraulic systems, but uses multiplex technology to reduce the number of electrical conductors in the umbilical and the complexity of the subsea electrical connections. Using multiplexed digital data many commands can be processed through a single cable. Electrical coding/decoding logic is required at the surface and subsea. Like the hard-wired system, a HPU on the surface supplies hydraulic power to the seabed accumulators via the control umbilical. Control valves used with multiplex systems are normally a latching type valve with pulse-energy solenoids. They remain in the last commanded position when the electric control signal is removed.

Topside equipment:
- HPU
- EPU
- Master control station (MCS)

Subsea equipment:
- Umbilical
- Umbilical termination unit
- Hydraulic jumper hoses
- Subsea control module

Advantages:
- Reduced umbilical complexity (single conductor)
- Rapid response time

Disadvantages:

- Cost and complexity
- Additional surface equipment—MCS (computer)

REFERENCES

1. American Petroleum Institute Recommended Practices. API RP 53. 4th ed.; 2007.
2. API RP 2Q. Recommended practice for design and operation of marine drilling riser systems. 2nd ed.; 1984.
3. Lopez RH, Mansell M, Stewart AA. The spool tree: first application of a new subsea wellhead/tree configuration. *7427-MS OTC conference paper*; 1994.

CHAPTER FOURTEEN

Well Control During Subsea Completion and Workover Operations

Subsea completion workover and intervention operations share the same well control problems that are experienced by those working on land or platform operations. In addition, there are well control concerns that are unique to the specialization of subsea operations. Placing the wellhead on the seabed involves complex control and riser systems for subsea blow out preventer (BOP's) and Christmas trees. The offshore environment is often harsh. Operating in deep water from a vessel that is continually in motion brings problems that do not effect operations on land or platform. Anyone working on a drillship or semisubmersible drilling rig must be aware of these differences.

14.1 SUBSEA WELL CONTROL

Placing the wellhead, BOP, and Christmas tree on the seabed affects the way well control is managed for subsea wells. The main concerns are:

- Kick detection.
- The effect of water depth on fracture pressure.
- The impact a riser disconnect will have on hydrostatic overbalance.
- Kick tolerance is reduced.
- Choke and kill line friction.
- Pressure changes at the choke when gas enters the choke line.
- Trapped gas in the BOP stack.
- Intervention riser complexity.

14.1.1 Kick detection

Kick detection is complicated by vessel motion. Modern semisubmersible rigs have sophisticated heave compensation systems that allow operations

Well Control for Completions and Interventions.
DOI: https://doi.org/10.1016/B978-0-08-100196-7.00014-2

to proceed in rough seas. However, the problem of brine surging in flow lines, and fluid movement in the pits, remains. Fluid levels are more difficult to read accurately when the vessel is moving, and so kick detection becomes more difficult, especially if the volume is small.

14.1.2 Effects of water depth on reservoir fracture pressure

Reservoir fracture pressure is a function of both overburden pressure and reservoir pressure. As water depth increases, the effective overburden (as measured from the rig floor) decreases, and so the pressure required to fracture the formation is less in relation to well depth. Increasing water depth reduces the pressure differential between the brine weight required to control the well (overbalance) and the pressure that will fracture the formation.

This is significant during completion and workover operations, not just in terms of limiting kill weight and equivalent circulating density (ECD), but for setting maximum pump pressure if scale or corrosion inhibitor treatments are necessary.

Fig. 14.1 shows a land well that is 6000 ft (1829 m) deep. This is true vertical depth (TVD) below rotary table (BRT). The well has a formation fracture gradient of 0.79 psi/ft (0.1787 bar/m), giving a formation fracture pressure of 4740 psi (327 bar). A 12 ppg (1.43791 sg) brine gives a

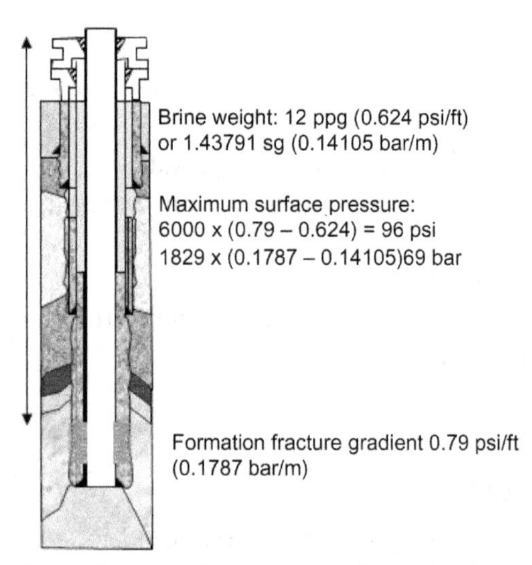

Figure 14.1 Land rig: maximum surface pressure—formation fracture.

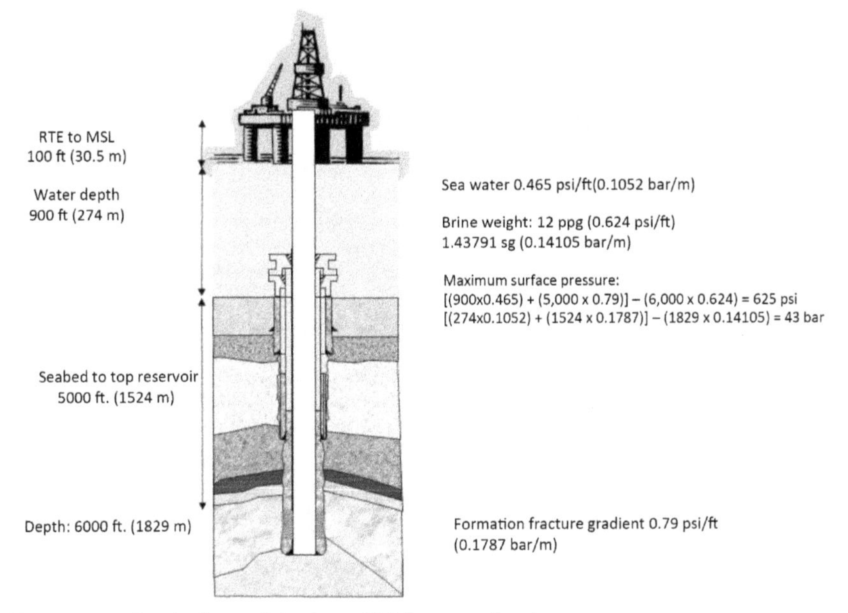

Figure 14.2 Semisubmersible rig at 900 ft water depth.

hydrostatic pressure of 3744 psi (258 bar). The maximum surface pressure that could be applied before formation fracture occurs is $4740 - 3744 = 996$ psi (69 bar).

Fig. 14.2 shows what happens to fracture pressure if a 6000 ft TVD BRT well is drilled from a semisubmersible rig in 900 ft (274 m) of water. Fracture pressure reduces because there is less overburden. The top 100 ft of 0.79 psi per foot fracture gradient formation has been replaced with air. In addition, from 100 ft TVD to the seabed at 1000 ft, seawater at 0.465 psi/ft provides the overburden, replacing the heavier 0.79 psi/ft formation. The reduction in overburden is $(1000 \times 0.79) - (900 \times 0.465) = 371$ psi. The maximum surface pressure that can be applied over a 12 ppg (1.43791 sg) brine is reduced from 996 psi (69 bar) to 625 psi (43 bar).

Fig. 14.3 shows the effect of increasing the water depth to 2800 ft. In theory, the maximum surface pressure that could be applied at the surface would be 7 psi (0.447 bar). In practice, this is too small a margin, as ECD during circulation would more than overcome the fracture limit.

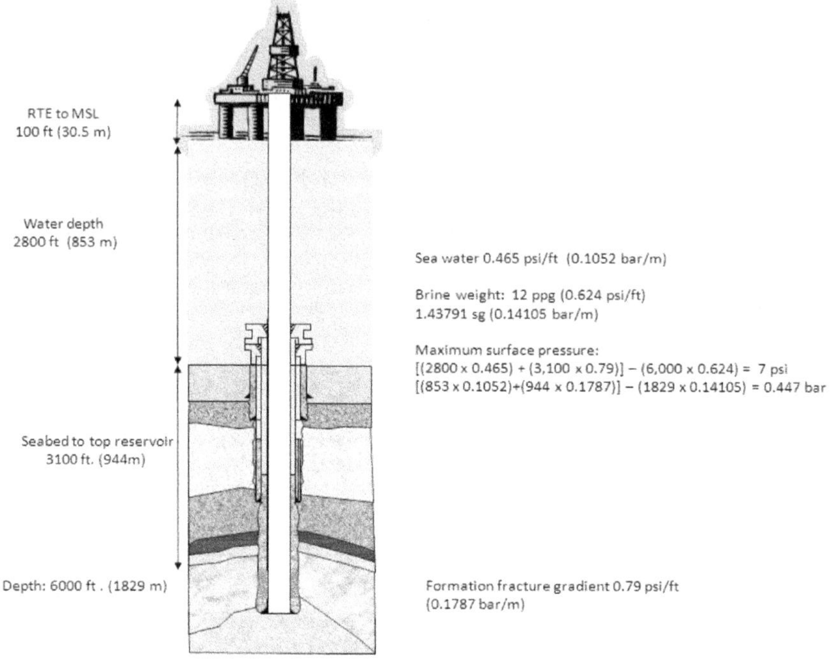

Figure 14.3 Semisubmersible rig at 900 ft water depth.

14.1.3 Riser margin

If a marine riser fails, leaks, or becomes disconnected, the completion or workover fluid gradient in the riser is replaced by a sea water gradient from the seabed, or point of failure, to the rig floor. If the loss of wellbore hydrostatic pressure associated with riser failure is sufficient, the well will flow. Similarly, although not as serious from a well control perspective, is the overbalance condition that can arise when seawater enters the well following a riser disconnect or failure. This happens where the workover fluid being used is at a lower density than seawater. The resultant losses to the formation have the potential to seriously damage the formation and postworkover well productivity.

The riser margin is defined as the minimum incremental brine weight required to maintain a hydrostatic overbalance above reservoir pressure after disconnecting the riser. The minimum brine weight including a riser margin is calculated as follows.

14.1.3.1 Oilfield units

$$\mathrm{BW}_2 = \frac{(\mathrm{BW}_1 \times \mathrm{TDV} \times 0.052) - (L_2 \times \mathrm{SW} \times 0.052)}{0.052 \times [\mathrm{TVD} - (L_1 + L_2)]} \tag{14.1}$$

where:

BW_2 = Brine weight with riser margin (ppg)

BW_1 = Brine weight to overbalance the formation by the required amount (typically 200 psi)

TVD = True vertical depth (ft)

L_2 = Depth of seawater (ft)

SW = Weight of seawater (ppg)

L_1 = Air gap (ft).

14.1.3.2 Metric units

$$BW_2 = \frac{(BW_1 \times TDV \times 0.0981) - (L_2 \times SW \times 0.0981)}{0.0981 \times [TVD - (L_1 + L_2)]}$$

where:

BW_2 = Brine weight with riser margin (sg)

BW_1 = Brine weight to overbalance the formation by the required amount (typically 14 bar)

TVD = True vertical depth (m)

L_2 = Depth of seawater (m)

SW = Weight of seawater (sg)

L_1 = Air gap (m).

Example calculation:

Workover operations are taking place on a well where the water depth is 320 ft (98 m) and the air gap is 80 ft (24 m). A 12 ppg (1.43 sg) brine gives the required 200 psi (14 bar) overbalance against the reservoir pressure at 6000 ft (1829 m) TVD (Fig. 14.4).

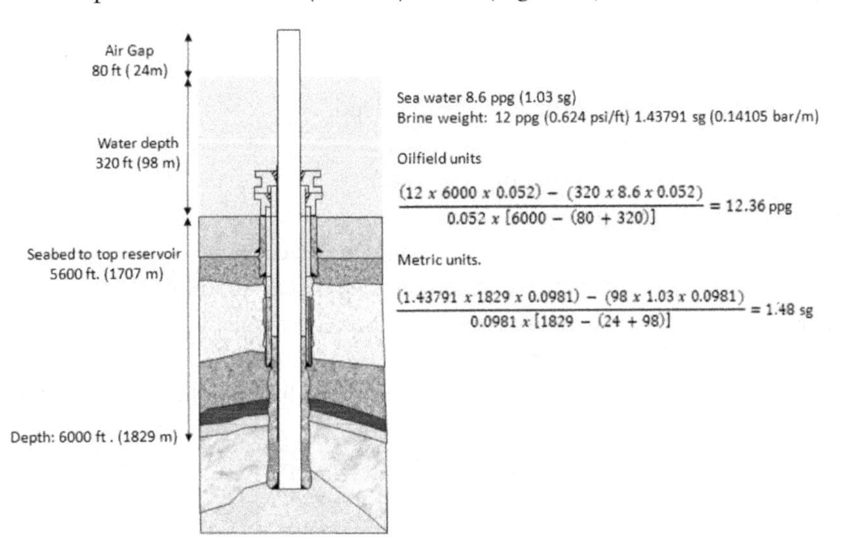

Figure 14.4 Example riser margin calculation.

To account for riser margin, the brine weight would need to be increased from 12 ppg (1.43 sg) to 12.36 ppg (1.48 sg).

14.1.4 Kick tolerance

Since the gap between formation pore pressure and formation fracture pressure narrows as water depth increases, the maximum size of influx that can safely be circulated out of the well without fracturing the formation will be similarly reduced.

14.1.5 Choke line friction

When circulation returns are taken through the choke, a significant pressure loss is seen. This is caused by friction losses in the small diameter choke line between the subsea BOP and the rig floor. This pressure drop is observed during slow circulating rate measurements when circulating through the marine riser and increases with water depth.

Choke line friction loss (CLFL) is measured and recorded, and the data used to enable circulation through the choke without exceeding maximum pressure downhole. If the normal method of bringing pumps to kill speed is followed (i.e., choke manifold pressure maintained equal to SICP until kill rate is achieved), bottom hole pressure will be increased by an amount equal to CLFL. The additional back pressure caused by circulating through the choke could cause losses or fracturing of the formation.

As fracture pressure generally reduces with increasing water depth, and CLFL also increases in direct proportion to water depth, choke loss must be factored in to any well kill calculation.

CLFL are affected by:

- The length of the choke line (water depth).
- Choke line diameter.
- Fluid properties (density and viscosity).
- Circulation rate (pump speed).

Actual choke line pressure loss should be determined at various slow circulation rates and the pressure recorded. Two methods are used to determine actual pressure drop through the choke line.

1. *Calculating choke line friction. Closed system—no open reservoir.*

   This method exposes formation to the full choke line friction during circulation. It is not recommended for wells where the reservoir is exposed. It can be used if the circulation path is isolated from the reservoir, e.g., the circulation path is through a sliding sleeve above a mechanical plug or closed FIV, or in an unperforated well (Fig. 14.5).

a. Static conditions.

b. Record the stand-pipe pressure whilst circulating at different pump rates. Typically, three rates are used, e.g., 20, 30, and 50 strokes per minute (Table 14.1).

c. Close the annular preventer. Open all choke line valves, all the valves and the choke on the choke manifold.

d. Circulate through the choke line at the same three stroke rates (choke fully open). Record the stand pipe pressure.

e. Choke line frictional pressure drop is the difference between slow circulation through the riser (step 2), and slow circulation through the choke line (step 4).

2. *Determining choke line friction (open perforations).*

This method is used where the formation will be exposed to circulating pressure.

a. Static conditions.

b. Record the stand-pipe pressure whilst circulating at different pump rates. The maximum pump rate used should not exceed the highest pump rate that can be tolerated when a kick gets to the choke line.

c. Line up the rig pump to the by-pass line on the choke manifold. Pump down the choke line at the same three rates as used in step (b) whilst taking returns up the riser. Record pressures.

Since pressure loss in the riser annulus is negligible at low circulation rates the choke line friction for each pump rate is equal to the pressure indicated by the choke manifold pressure gauge.

For subsea workover and completion operations, slow pump pressures and choke line friction values should be recorded after any change in the fluid weight or viscosity. If a new value for friction pressure loss has not been recorded, an estimate can be made using the following calculation:

$$\text{New choke friction pressure drop} = \text{Old friction pressure} \times \frac{\text{New fluid weight}}{\text{Old fluid weight}} \tag{14.2}$$

This equation must not be used as a routine replacement for the determination of choke line friction by low pump rates.

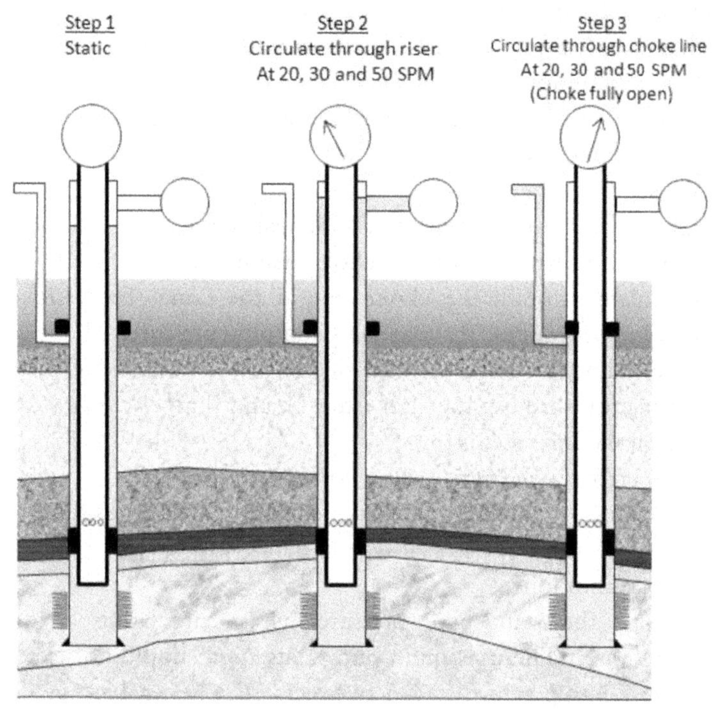

Figure 14.5 Calculating choke line friction losses (CLFL). Method 1.

Table 14.1 Choke frictional pressure drop

SPM	BPM	Stand pipe pressure Riser circulation psi	Stand pipe pressure Choke circulation psi	CFPL psi
20	1.8	300	350	50
30	2.7	500	600	100
50	4.5	1200	1450	250

CLFL are not significant at slow pump rates in shallow water depths. However, in deep water they can be significant and compensating measures need to be taken.

Where CLFL are significant, the most important phase of the operation will be at start of the kill, when the pump is being brought up to the required speed. Conventional kill methods are based on maintaining a constant bottom hole pressure throughout. With the BOP at surface, this is achieved by maintaining a constant casing pressure through manipulation of the choke as the pump is brought up to kill speed. At this stage,

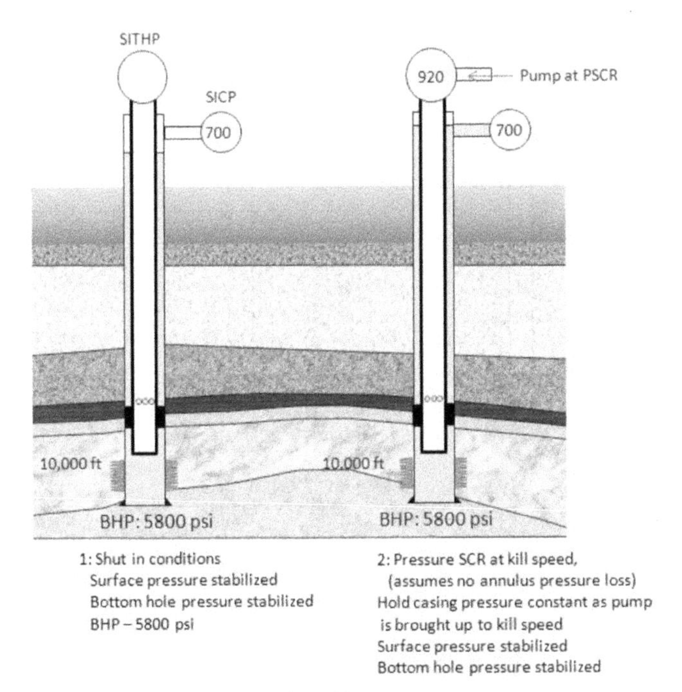

Figure 14.6 Starting circulation—surface blow out preventer.

drill pipe pressure will be at the required initial circulating pressure (Fig. 14.6).

If this start-up procedure is used on a floating rig with significant CLFL, bottom hole pressure would increase by a pressure equal to the friction losses. This overpressure could cause losses to or fracture the formation. Two methods can be implemented to compensate for overpressure (CLFL).

1. As the pump is being brought up to kill speed, the casing pressure is reduced by an amount equal to the CLFL. Once the pump is at kill speed, tubing pressure should be at the required initial circulating pressure. This method can be difficult to implement (Fig. 14.7).
2. The kill line can be used as a pressure monitor, assuming the fluid density in the kill line is the same as that in the wellbore. As the pump is being brought up to kill speed, the kill line pressure at the surface is held constant. There will be no pressure losses in this line, as no brine is being circulated. With the pump at kill speed, tubing pressure will be at the required initial circulating pressure, and casing pressure will have reduced by an amount equal to the friction losses (Fig. 14.8).

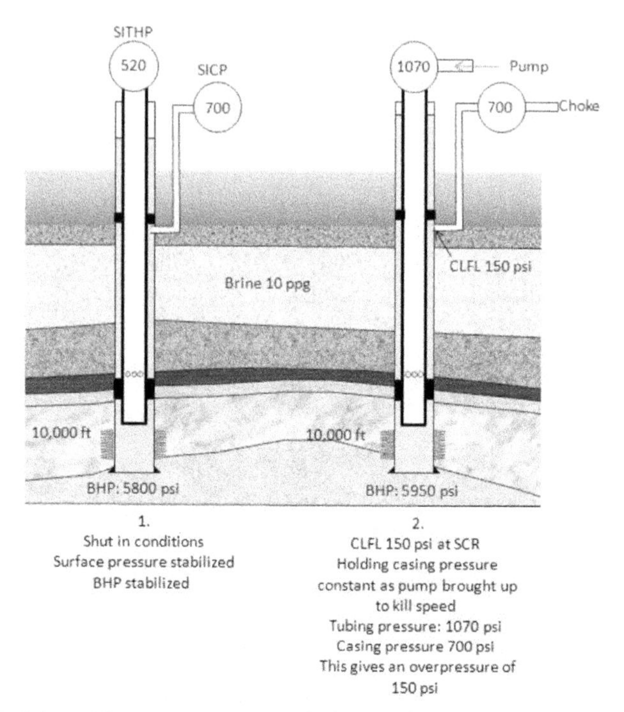

Figure 14.7 Subsea blow out preventer. Choke line friction losses startup. Method 1.

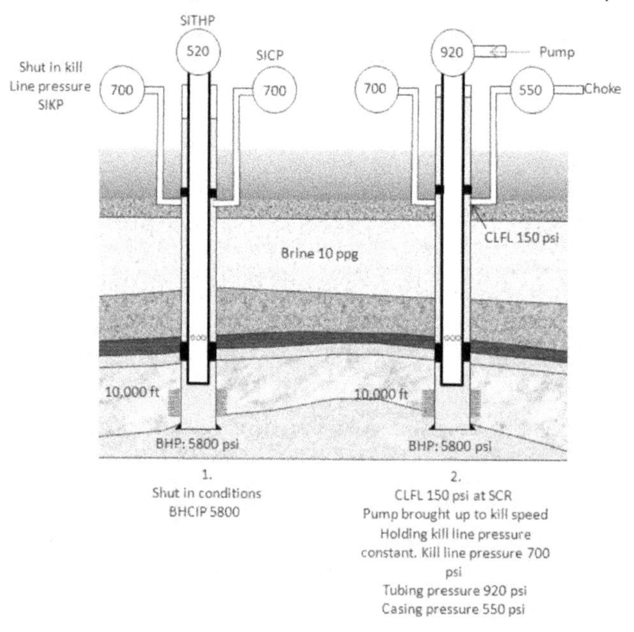

Figure 14.8 Subsea blow out preventer. Choke line friction losses start-up. Method 2.

If CLFL are greater than shut-in casing pressure (SICP), then the well will be subjected to an overpressure. The amount of overpressure is the pressure difference between the CLFL and SICP. With the pump at kill speed the choke device will be fully open. To minimize overpressures, consideration should be given to slower pump rates, if the pump speed is reduced, friction losses will reduce. In some cases, consideration maybe given to circulating through both choke and kill lines. Using both lines increases the flow area, thereby reducing the friction losses. Using this would normally require the addition of a dedicated pressure monitoring system on the stack, since the kill line could not be used as a pressure monitor.

14.1.6 Pressure changes at the choke when gas reaches the choke line

A critical point when circulating a kick out of a subsea well occurs as gas reaches the choke line. At this point the hydrostatic pressure in the choke line begins to drop, as the low density gas replaces higher density fluid (brine). As the capacity of the choke line is significantly less than annular capacity, the change in pressure when the gas/fluid interface reaches the choke line is rapid and must be anticipated. Choke size must be reduced to maintain constant bottom hole pressure and prevent a further influx. A similar situation occurs when kill weight brine reaches the choke line and begins to displace gas. Hydrostatic pressure in the choke line will increase. Again, the pressure change must be anticipated, and adjustments made to the choke.

The ability to maintain a constant bottom hole pressure is influenced by circulation rate. There will be more time to react if circulation rate is reduced in anticipation of gas reaching the choke line. One option is to choose a circulating rate that will allow the gas to be properly handled when it reaches the choke line, and then maintain that rate for the entire circulation.

14.1.7 Handling trapped gas in the blow out preventer stack

During a well kill, gas will accumulate in the upper part of the stack, between the bottom of the closed BOP and the outlet used to circulate out the kick. Trapped gas released into the riser has been the cause of several serious incidents, so it must be removed in a controlled manner. The severity of the problem depends on the size and pressure of the trapped gas. If the well kill required a kill fluid density increase, then once the gas has been removed, the riser will need to be displaced to kill weight fluid. Gas can be removed from the BOP as follows:

- Close the pipe rams below the choke and kill line outlet spool (normally lower or middle rams). A flow path across the top of the closed rams and back up the choke line must be available to isolate the well during the stack venting operation.
- Displace the kill weight brine by pumping a lighter fluid, such a water or base oil, down the kill line and take returns through the choke line. Maintain backpressure on the choke equivalent to the hydrostatic pressure of kill weight fluid (choke pressure should be equivalent to the difference between the hydrostatic pressure of the kill fluid and the lighter fluid used to flush the gas from the stack). A kill line pressure schedule helps. If there is a hydrate risk, the flush fluid must be properly inhibited to prevent the formation of hydrates.
    - *Note: It is important to hold back pressure to keep the BOP pressure from falling below the original pressure and allowing premature expansion of the gas bubble.*
- When the returns are clear fluid, stop pumping down the kill line and close the choke, holding the same backpressure as before.
- Close the kill line.
- Completely open the choke as quickly as possible to bleed off the pressure from the choke line, allowing water and gas to escape from the choke line.
- Once the flow stops, displace the choke line, kill line, and stack with kill fluid, pumping down the kill line side until the kill fluid returns from the choke line.

14.1.8 Riser kill

- Keep the lower pipe rams below the choke and kill lines closed to isolate the wellbore. Close the diverter, then open the BOP upper rams. Pump kill weight fluid down the choke and kill lines (and marine riser boosting line if available) and up the marine riser.
- Once kill fluid is observed at the surface in the marine riser, the well should be static. Close the fail-safe valves on the choke and kill lines, open the lower pipe rams, and monitor the well for flow.

14.2 SHUT-IN PROCEDURE

Shut-in procedures incorporated in most well control manuals only deal with the procedure for having drill pipe across the BOP. Shutting-in the well might have to be handled differently during the running (or pulling) of a completion.

14.2.1 General shut-in procedures: drill pipe

When a kick is detected, the well should be closed in on the annular preventer. After shut-in, drill string reciprocation through the annular preventer is not recommended, as significant wear may occur on the sealing element, especially if a tool joint is reciprocated through the element. On floating rigs, heave can cause some unwanted reciprocation. To avoid this, the workstring should be hung off on a pipe ram as soon as possible. Additionally, a trapped gas bubble will be left in the BOP stack between the annular preventer and the choke line outlet used to circulate the well. This trapped bubble may be hazardous to handle at the surface, depending on its size and pressure. To minimize the risk of annular wear, and the size of the trapped bubble, the following procedure is followed:

1. The well should be shut-in on the annular preventer in the lower marine riser package (LMRP), and operation to hang off the string on a pipe ram begun straight away. Pipe reciprocation should be minimized to reduce trapped gas and annular element wear. It is acceptable to hang off on variable bore rams only if a tool joint will rest on ram blocks, not on the fingers.
2. When circulating out the influx, returns should be taken through the choke or kill line outlet directly beneath the closed pipe rams. Once the influx is circulated from the wellbore, remove the trapped gas from the BOP stack, then fill the riser with kill weight fluid. The diverter controls and overboard line should be manned before the well is opened. When circulation begins up the riser, any residual gas bubble will then be diverted overboard through the diverter lines as necessary.
3. Significant quantities of gas can be trapped in the BOP stack for any of the following reasons:
   - Failure to hang off the drill pipe on the pipe rams.
   - Neglecting to close the well in until gas reaches the BOP stack.

- Pipe rams leak with the annular closed.
- Inability to utilize a choke or kill line outlet directly beneath the hang-off rams.

14.2.2 Shut-In procedure whilst circulating

The annular preventer on the LMRP (upper annular) is normally used as the initial closing element, absorbing damage until the pipe can be supported on the ram preventers. Minimizing reciprocation through the annular preventer should limit any damage. The annular preventer on the stack (lower annular) is still available for well control. The Drilling Representative must always know the relative positions of tool joints in relation to the annular and ram type preventers.

Closing in the well would typically be as follows:

1. Sound alarm.
2. Pick up and position any tool joint clear of the upper annular sealing element and stop the pumps.
3. Close upper annular preventer. Regulate closing pressures so pipe moves freely.
4. Open the upper kill line valves and monitor the SICP while hanging off.
5. Space workstring to ensure that a tool joint won't interfere with closing pipe rams.
6. Close pipe rams with normal operating pressure (approximately 1500 psi).
7. Close ram locking device, if not automatic.
8. Lower workstring slowly until it is supported on closed pipe rams.
9. Bleed well pressure between annular and ram preventers via kill line. Observe well to verify rams are holding. Open annular preventer.
10. Adjust the motion compensator to mid-stroke and support only the string weight above the hang off rams, plus a nominal overpull (15–20,000 lbs).
11. Close the upper kill line valve, open the choke outlet just below the hang-off rams, read and record the SICP.
12. Read and record the shut-in tubing pressure, bleed off any trapped pressure on the annulus, and continue with the kill procedure.

14.2.3 Shut-In procedure while tripping

1. Sound alarm.
2. Install a fully opened safety valve in the workstring. Close the safety valve.
3. Close upper annular preventer. Regulate closing pressure so the pipe moves freely.
4. Open the upper kill line outlet to monitor the SICP.
5. Read and record the SICP.

14.3 SHUT-IN PROCEDURES WHILST RUNNING OR PULLING A COMPLETION

Special consideration must be given to how to shut-in if a kick occurs when running or pulling a completion. There are several situations where closing the annular preventer or pipe rams is ineffective, since they will not seal around some types of completion equipment.

14.3.1 Running sand control screens or slotted liner

If a kick occurs whilst screens or slotted liner are across the stack, closure of the annular (or pipe rams) is ineffective. Cutting the pipe will be the only option. If there are concerns that the screens (or liner) cannot be cut, or the shear rams fail to cut, the only alternative is to drop the completion. Dropping the string intentionally must be planned in advance, and the crew properly briefed.

1. With the string latched in the elevators, lower the tubing until the elevator is close to the rotary table.
2. Set the compensator to mid stroke
3. Close the pipe rams with enough pressure to support the weight of the string.
4. Slack off the elevator and unlatch the tubing.
5. Open the BOP to allow the string to drop.

   This method can only work if the screens clear the BOP once dropped.

14.3.2 Running control lines and instrument cables

Subsea completions frequently use multiple lines (bundled into flat packs) for downhole chemical injection, instruments, and flow control. Electric submersible pumps are occasionally used in subsea wells. Control line bundles or ESP cables will prevent both annular and pipe rams from sealing. It may be possible to get a partial seal if the annular is closed, but this will only likely work on a single nonencapsulated line if wellhead pressure is relatively low—an unlikely combination in a subsea well.

In most circumstances, the only possible response to a kick will be to shear the tubing and close the blind rams.

14.3.3 Pulling spent tubing-conveyed perforating guns

BOP rams will not seal around spent (fired) hollow carrier guns. These will have to be cut and dropped if the well kicks.

14.3.4 Pulling damaged completion tubing

Closing rams of holed or damaged pipe will be ineffective. Shearing the pipe and closing the blind rams is the only safe option if the well kicks.

14.4 PRE-WORKOVER: PLANNED WELL KILL

Most subsea well control manuals and procedures are written by drillers for drillers, and deal only with well control incidents during the drilling phase of the well construction process. Well kill is in response to a kick, and is not a planned part of a workover operation. As such, all the procedures are based on having the drilling BOP in place, and describe only a forward circulation. None of the manuals cover a planned well kill using the well intervention equipment. Moreover, manuals that are written to conform to the International Well Control Forum (IWCF) intervention and completion well control syllabus make only the briefest mention of subsea equipment, and rarely include any procedures. Completion and intervention supervisors who work on floating rigs should be aware of the main differences between a planned well kill carried out on a fixed installation, and one carried out on a floating rig.

14.4.1 Establishing a circulation path

In wells equipped with a horizontal tree, the well kill will be carried out once the tubing hanger running and orientation tool (THROT), subsea test tree (SSTT), and landing string are in place. If the well is killed by forward circulation, returns would be taken through the BOP choke line (see Fig. 14.13). For a reverse circulation, kill fluid would be pumped down the kill line with returns taken back through the wing valve of the surface test tree (STT).

Killing a well equipped with a conventional dual bore vertical tree takes place before the tree is removed, and is carried out through the dual bore intervention riser (see Fig. 14.14). Here, the path for forward circulation would be down the production side of the dual bore riser, with returns taken through the annulus bore, reversing is the opposite.

Well kill methods from a floating rig share many similarities with those used for land and platform operations (as described in Chapter 7: Well Kill). The well can be killed by circulation (forward or reverse), bullhead, or lubricate and bleed. Performing a bullhead kill or lubricate and bleed kill does not differ significantly between a floating and fixed installation. However, for a circulation kill, the flow path is different.

If a horizontal tree is used, returns must pass through a gallery in the tree block. This restricts flow, and adds to the already significant pressure drop through the choke line. Similarly, if a vertical tree is in use, the frictional pressure drop through the 2″ annulus string of the dual bore riser results in a frictional pressure drop of similar magnitude to that recorded for the choke line. Pressure drops associated with forward circulation mean that ECD in subsea wells is normally lower during a reverse circulation. Since water depth is directly related to formation fracture pressure, reverse circulation is usually the preferred choice. A secondary benefit is that the lower density hydrocarbon in the tubing can be easily routed to a well test spread for safe disposal.

14.4.2 Handling hydrocarbons

For any circulating well kill carried out on a floating rig, thought must be given to the handling of hydrocarbons. When performing a reverse circulation kill on a land or fixed platform base, it is common practice to leave the flow-line in place, and route the hydrocarbons into the process facilities. Whilst this option could be considered for a subsea tree, it would only be applicable if very close cooperation between the rig and the host facility were possible. In addition, the requirement to have isolation in place before the well is officially handed over from the host facility to rig further complicates matters. It is far more common to take hydrocarbons back through a temporary well test package on board the rig.

For a reverse circulation well kill, returns from the tubing would be routed through the flow wing of the Surface Test Tree (STT) to the well test choke manifold (Figs. 14.13 and 14.14). A forward circulating kill through a horizontal tree would require returns to be routed from the rig choke line to the well test choke manifold. If a vertical tree is used, the returns would be routed from the STT annulus bore wing valve to the well test choke manifold.

14.4.3 Riser disconnect

Subsea BOP stacks are designed to allow the drilling vessel to disconnect the riser without compromising well integrity. Riser disconnection will either be a planned operation, or a response to an emergency. Planned disconnection is often a response to a poor weather forecast, where sea state (wave height) is expected to be over and above the rig operational limit. Situations requiring an emergency disconnection of the riser include loss of dynamic positioning (drive off), loss of well control, sudden unexpected changes in the weather, or too long a delay in making a "planned" disconnect.

Note: These procedures cover disconnections before the tubing hanger is landed. Disconnection with the landing string in place is covered in the next (intervention) section.

14.4.4 Planned (nonemergency) disconnect

1. Circulate bottoms up. If the string is close to the bottom, it will have to be pulled back a distance equivalent to the length of the riser, to allow the hang off tool to be run from the rig floor to the BOP stack.
2. Figure space out to land hang off tool in the wellhead.
3. Make up hang off tool in the string and run it into the hole.
4. Land hang off tool inside the wellhead with the compensator stroked open. Position compensator in mid-stroke, and adjust the compensator to support the pipe weight above the rams.
5. Close the appropriate pipe rams and locks.
6. Back off the right-hand release sub on the hang off tool.
7. Pick up the workstring above the BOP stack. Check for flow and close the blind shear rams. Close the wedge locks.
8. Displace the riser with sea water. Dump the subsea accumulator bottles, if applicable. Deballast part or all of the air buoyancy tanks on the riser, if necessary. Adjust the riser tensioners to allow for the difference between the weights of the fluid in use and seawater.
9. Pull the remainder of the drill pipe out of the riser. Disconnect the LMRP and pull the riser.

Note: Most of the disconnect procedures found in well control manuals relate only to disconnection during drilling operations. If the need to disconnect during production tubing running (or pulling) operations there will be additional complications.

- The hang off tool used during drilling operations is designed to sit in the wellhead when the wear bushing is in place. During completions and some phases of a workover, the wear bushing will not be in place, and the hang off tool could damage the wellhead hanger bowl. The tree/wellhead vendor should be consulted to determine the best method for hanging off the string.
- If there is a need to pull the string back (to space out) it should be remembered that pulling premium tubing is always going to take longer than pulling back drill pipe in stands, especially if there are control lines in place.
  - Many completions, in particular subsea completions, are equipped with downhole flow control systems and instrumentation that are operated by hydraulic and electric control lines. A planned disconnect would mean having to cut these lines before the hang off tool could be run. Where possible, the running of a completion that is equipped with complex control and instrument cables should only be attempted if the weather is settled and the forecast for the duration of the installation is good.
  - Wells with open reservoirs are vulnerable to swabbing and surging during the running (and later recovery) of the hang off tool, especially if there is a packer in the well. If weather conditions are deteriorating rapidly and the crew is rushing to deploy the hang off tool, it is all too easy to exceed safe tripping speed.

14.4.5 Emergency (unplanned) disconnect

1. Pick up and space out to hang off string on closed pipe ram.
2. Close hang off ram. Closure pressure should be specified in the rig procedure, but is typically 1500 psi. Close wedge locks.
3. Close lower pipe rams for a backup.
4. Adjust compensator to support string weight above rams plus approximately 10,000 lbs (time permitting). Applying tension to the string will ensure it lifts when sheared.
5. Shear the string with full ram operating pressure.
6. Disconnect the LMRP and pull the riser if possible. If not able to pull the riser, move the vessel off location while dragging the riser. Stay clear of shallower water and away from the BOP stack.

Note: As with the planned disconnection, the procedures found on most rigs are limited to drilling operations. An emergency disconnect during a completion operation can be complicated by other factors:

- Shear ram limitations. There may be components in a completion string that are difficult to shear. For example, sand control screens made from high yield material (and with a wash-pipe) could be difficult, even impossible to cut. Shear capability should be confirmed before deploying high yield completion components through a subsea stack.
- Hang off with completion tubing couplings. Drill pipe has large tapered upsets that allow it to be hung off easily within the pipe rams. The different profile of most completion tubing may not make them suitable for hanging in the rams. This is certainly the case if flush joint tubing is across the stack. Moreover, if there are control lines across the rams, this will prevent sealing. Situations arise where cutting and dropping a completion string is the only option available.

14.4.6 Reentering a well following a disconnect

After disconnection, gas can migrate toward the stack. In addition, the cooling effects from circulating before the rig moved off means that the wellbore can warm during the rigs absence, causing a rise in pressure. When reentering the well, a pressure increase should be anticipated, and measures taken to remove the trapped gas. This can be accomplished as follows:

1. After reconnecting the LMRP, open the lower pipe rams (if closed), upper kill line valves, and the lower choke line valves. Check pressure on the choke line for indicators of trapped gas. Attempt to bleed off any pressure.
2. Circulate the proper weight fluid down the kill line while taking returns through the choke. Displace the riser with this same fluid using the workstring.
3. After clearing the well of any trapped gas, open the blind shear rams.
4. Screw into the hang off tool. If the pipe was sheared, dress off the top of the workstring and latch with an overshot.
5. Open the middle pipe rams and monitor the well for flow.
6. Pull the hang off tool (or overshot if pipe was sheared) out of the hole.
7. Trip to the bottom, circulate and condition fluid as required.

14.5 SUBSEA INTERVENTIONS

Consider an intervention on an easily accessible land well. All that is required to carry out a simple slickline operation is a wireline unit, a hoist to rig up some pressure control equipment, and a competent slickline crew consisting of no more than two or three people. They would be perfectly capable of performing a range of operations with little or no assistance from operator company personnel. Depending on the type of intervention, the operation could be completed in a few hours.

Performing the same task on a subsea well needs substantially more time, equipment, and support. A rig would have to be moved to location, and the intervention riser system deployed.

This seemingly simple slickline operation requires a fully crewed semisubmersible rig, and all of the logistical support that goes along with it. In addition, equipment specialists from multiple vendors are needed to prepare, deploy, operate, and recover the subsea intervention access systems. However, once the intervention riser is in place, the actual wireline run will be essentially the same as the land-based equivalent. Similarly, many of the procedures and practices that are used for land and platform operations can be equally as well applied to operations carried out on subsea rigs. There are, however, some fundamental differences. Principally with pressure containment during live well interventions, but also with the way the equipment is rigged up.

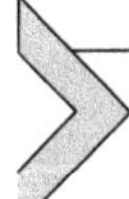

14.6 RIGGING UP USING A COILED TUBING LIFT FRAME

As the name suggests, the lift frame was originally designed to enable coiled tubing to be rigged up on floating rigs. The principle reason for using the lift frame is safety. Rigging up without one is dangerous. As their use became widespread, most operators began to use the lift frame for wireline as well as coil tubing operations. A consequence of the additional safety provided by the lift frame is a more complex (and time consuming) rig up.

A dual bore riser (or landing string) is run from the rig floor. During deployment, it hangs in the rotary table when connections are made.

Consequently, the riser moves up and down with the rig as it heaves. Once the LMRP lands out on the subsea tree or the tubing hanger running tool latches the horizontal tree internal tree cap, the riser is no longer able to move and becomes stationary (relative to the seabed). The rig, however, continues to heave. A person on the rig floor will observe the riser apparently stroking up and down through the rotary table. However, it is not the riser that is moving—it is the rig. Consider a wireline lubricator that has been picked up with one of the rig floor winch lines and is ready to be stabbed (connected) to the top of the riser. The sheave supporting the cable is attached to the derrick structure. This means the suspended lubricator will heave up and down with the same frequency and elevation as the rig. In poor weather conditions, movement is significant. Trying to stab a moving lubricator pin connection into the box on top of a stationary riser is extremely hazardous, and people were injured trying this before coiled tubing lift frames were widely used.

The coil tubing lift frame allows coiled tubing and wireline pressure control equipment to be made up and backed out using a winch that is stationary relative to the riser. This is accomplished by mounting the winch on the top cross-member of the lift frame, then attaching the lift frame to the top of the riser. The lift frame is supported by the compensated traveling block, keeping it under constant tension as the rig heaves. Rigging up the coiled tubing lift frame is usually carried out as part of the riser deployment and proceeds as follows:

- Run the dual bore riser (or landing string) to a point just above land-off, then suspend in the rotary table.
- Latch the elevators around the lifting sub on the top cross-member of the CT lift frame (Fig. 14.9).
- Pick up on the blocks until the lift frame is hanging vertically in the derrick, and swivel the frame to the correct orientation.
- Attach bails and elevator to the bottom of the lower cross member.
- If the riser (or landing string) has not been landed the next step will be to pick up the STT (flowhead) in the elevator hanging off the bottom of the lift frame (Fig. 14.10).
- The joint below the flowhead is made up into the joint in the rotary table (dual riser joint or landing string joint).
- Activate the motion compensator (normally mid stroke), and lower and land out the riser or landing string.
  - On a well equipped with a vertical tree, the riser is lowered until the connector on the base of the LMRP latches the subsea tree.

Figure 14.9 Elevators latched onto the top of the lift frame.

  - On a well equipped with a horizontal tree, the landing string is lowered until the THROT latches the internal high pressure cap profile.
- The riser (or landing string) needs to be spaced out so the bottom of the flowhead does not contact the rotary table as the rig heaves; there needs to be sufficient "stick up" of riser (or landing string) above the rotary table. For most operations where bad weather is common and wave height is significant, at least 15–20 ft of stick up is suggested.
- Once the riser or landing string has landed out on (in) the tree rig, movement is absorbed by the action of the motion compensator. This allows the lift frame and riser below it to be kept at a constant tension (Figs 14.11 and 14.12).
- Well intervention equipment is then rigged up inside the CT lift frame using the winch fitted to the top cross-member of the frame.

For some operations, the riser (or landing string) will be in place before the coiled tubing lift frame is rigged up.

14.7 WELLBORE ACCESS: HORIZONTAL TREES

When working on a well equipped with a horizontal tree, the retreivable crown and tubing hanger plugs must be removed to gain access to the wellbore. These plugs are nearly always pulled using wireline,

Figure 14.10 Picking the Surface Test Tree. The bails and elevator seen in the picture are connected to the base of the lift frame.

Figure 14.11 Making up a dual ram blow out preventer inside the coiled tubing lift frame.

although coiled tubing could be used. Once the first (crown) plug has been removed, there will only be one mechanical barrier in place between the reservoir and the surface. One or both lubricator valves in the Subsea Test Tree (SSTT) must be used to isolate the wellbore and provide double barrier isolation when venting and opening the wireline lubricator.

Figure 14.12 Wireline pressure control rig up with the coiled tubing lift frame.

Once the tubing hanger plug has been removed, there are no barriers to flow in the bore of the tree. Isolation of the wellbore is achieved by closing at least two of the valves in the SSTT, almost always the lower and upper lubricator valves. Where toolstring length allows, the gate valves on the STT will be shut when breaking out the wireline lubricator.

All subsequent intervention work, whether it be wireline, coiled tubing, flow testing, or pumping (stimulation) will use the valves in the SSTT to isolate the well, and where possible the valves on the STT would be used as well.

Fig. 14.13 is representative of the valve configuration used when a SSTT is positioned across a drilling BOP and the tubing hanger running tool latched to the high pressure internal tree cap (or tubing hanger). The main points to note are:

- No valves in the bore of the horizontal tree. Isolation is achieved by closing the lubricator valves (marked as UBV and LBV).

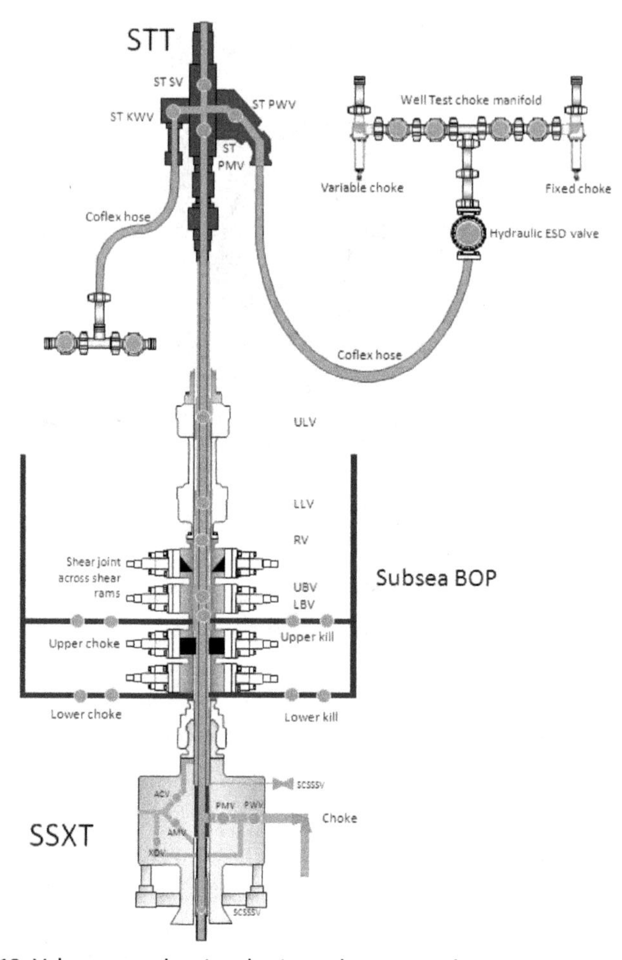

Figure 14.13 Valve status drawing: horizontal tree completion.

- BOP pipe rams are closed around a slick joint. Annulus pressure can be monitored, and circulation managed through the lower kill and choke line. These lines can also be used to test the pipe rams.
- The riser annulus can be circulated through the upper kill and choke line.
- The STT is hooked up to the well-test choke, and the kill line is connected to a fluid pump, usually the cement unit.
- For most operations, the SSTT and tubing hanger running tool functions are controlled from a single hydraulic panel, normally supplied by the SSTT vendor.

When intervention operations are complete, the tubing hanger and crown plugs must be reinstated before the THROT and SSTT can be unlatched and recovered to surface on the landing string.

14.8 WELLBORE ACCESS: VERTICAL TREES

As the valves in a vertical tree are conventionally arranged in the main production and secondary annulus bore, intervention access is relatively simple. The tree valves are opened and closed as necessary, using the hydraulic panel at the surface. As well as the ability to close in the well using the tree valves, valves in the LMRP and on the STT can also be used if additional barriers are required.

Fig. 14.14 illustrates a typical valve configuration when using the dual bore riser system in conjunction with a vertical tree. The main points to note are:

- The circulating (cross-over loop) on the LMRP. This is used to flush hydrocarbons out of riser prior to disconnecting.
- A dual bore STT allows access to either the main production bore or the annulus bore.

14.9 WELL CONTROL DURING SUBSEA INTERVENTION OPERATIONS

Interventions carried out from floating rigs are rigorously planned and should be the subject of an extremely detailed and comprehensive work program. The costly and complex nature of subsea interventions demands this.

14.9.1 Roles and responsibilities

It is common for intervention activities on land wells or offshore platforms to be supervised by well intervention specialists. In many of those cases, the supervisor will be the person ultimately responsible for the safe conduct of operations at the wellsite. When a semisubmersible rig or drillship is used to support well intervention activities, the senior operating company person on board will almost always be the drilling representative. In the past, there have been incidents resulting from lack of clarity

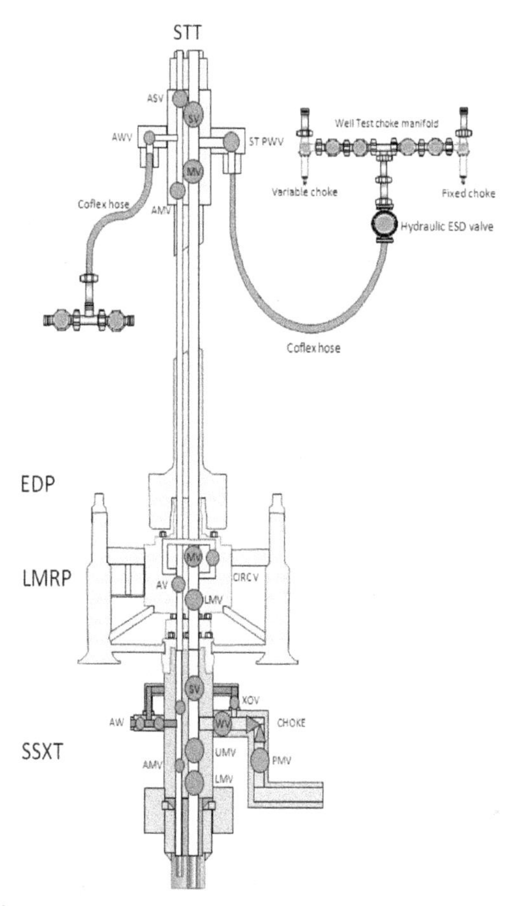

Figure 14.14 Valve status drawing: vertical tree access riser system.

regarding roles and responsibilities—well intervention contractors have been given conflicting instructions by well intervention specialists, and by drilling representatives. Although the drilling representative will normally be the person responsible for well control, during intervention operations the detailed technical expertise often rests with the well intervention specialist. Roles and responsibilities must be clearly defined. Outlined below are some important considerations when defining roles and responsibilities at the rig location.

14.9.1.1 Drill crew

- Horizontal tree:

  The drill crew will be responsible for the deployment and operation of the drilling BOP and marine riser. The drill crew will also assist with the deployment of the THROT and SSTT, running of the landing string, and rigging up of the STT and coiled tubing lift frame. Whilst the rig crew will normally report to the drilling representative, detailed knowledge relating to the deployment of the SSTT rests with the vendor and well intervention specialists.

- Vertical tree:

  The drill crew will assist with the deployment of the LMRP/EDP and marine riser. They will also assist with the rigging up of the coiled tubing lift frame. Although the drill crew report to the drilling representative, the expertise needed to deploy the riser system remains with the vendor and well intervention specialists.

14.9.1.2 Subsea test tree vendor

The SSTT vendor crew is responsible for the preparation, deployment, and operation of the SSTT. The same crew will be responsible for securing the well and disconnecting the SSTT prior to the drill crew disconnecting the marine riser at the LMRP should the need arise—planned or emergency. However, the instructions to disconnect will normally come from the drilling representative. The SSTT, tubing hanger running tool, and subsea Christmas tree (SSXT) are normally operated from the same control panel, supplied by the SSTT vendor. It is vital that the SSTT and tubing hanger running tool deployment is planned and coordinated by both the test tree vendor and the production tree vendor. Similarly, the space out of the SSTT within the drilling BOP requires meticulous planning and cooperation between the drilling contractor and the SSTT vendor.

14.9.1.3 Well testing vendor

Having a temporary well test package on board the rig during well intervention operations is commonplace. If a well test package is required, the crew will be responsible for the preparation rig up and pressure testing of all their equipment. They will also be responsible for function testing the ESD system before hydrocarbons are produced to the surface.

14.9.1.4 Subsea Christmas tree vendor

- Horizontal tree.

  The SSXT tree vendor supplies the THROT used to connect the landing string (and SSTT) to the high pressure internal cap inside the

SSXT. Where the THROT is owned by the operating company, it will still be prepared and deployed by the manufacturer. The tubing hanger running tool and SSTT are controlled from the SSTT panel.

- Vertical tree.

  If a vertical tree is used, the tree vendor will be responsible for the preparation, deployment and operation of the LMRP/EDP and dual riser. The crew will also be responsible for securing the well and disconnecting the riser following instructions to do so.

14.9.1.5 Wireline crew (slickline and logging)

The wireline crew will be responsible for the preparation, rigging up, and testing of the wireline pressure control equipment. They will also be responsible for the break out, make up, and retesting of the lubricator between runs. Loss of lubricator containment (leaks) will, in the first instance, be dealt with by the wireline crew.

14.9.1.6 Coiled tubing crew

The coiled tubing crew will be responsible for the preparation, rigging up, and testing of the coiled tubing equipment, including the pressure control equipment. They are also responsible for the preparation, maintenance, and operation of the coiled tubing lift frame. They will also be responsible for the break out, make up, and retesting of the injector head between runs. Loss of containment (leaks) will, in the first instance, be dealt with by the coiled tubing crew.

14.9.1.7 Stimulation crew and stimulation vessel

Pumping equipment may be located on the rig if the treatment volume is small. The pumping of a large volume treatment will require a "frac boat."

If a frac boat is to be used, it is standard practice to send the treating iron to the rig where it is rigged up and pressure tested ahead of the vessels arrival. Coordination of vessel and rig is normally the responsibility of a well intervention specialist acting in a supervisory role.

14.9.2 Riser and landing string preparation

The riser system (vertical tree) or landing string (horizontal tree) form an integral part of the well control envelope during interventions on subsea wells. There is considerable variance between systems depending on vendor, water depth, rig facilities, and operating company requirements.

Intervention specialists cannot be expected to have an intimate knowledge of every system. That said, subsea operations are normally planned well in advance, so specialists will have adequate time to learn about the riser system that will be used on any operation. Vendor expertise is invaluable when at the rig site. Although there is considerable variation between different intervention riser systems, all have some basic features in common, and the important checks performed before and during use are industry wide.

14.9.2.1 Horizontal tree system

- After locking the tubing hanger running tool on to the internal high pressure tree cap (intervention operation), or after landing the tubing hanger (completion operation), the BOP pipe rams should be closed on the slick joint, and circulation through the lower kill and choke line confirmed.
- Function testing and pressure testing of the SSTT valves should be performed in accordance with vendor recommendations and operational requirements. The test sequence and pressure requirements should be properly documented in the program.
- The valves in the STT must be function tested and pressure tested.
- Pressure test the well test choke (if relevant).
- Confirm annulus monitoring and access capability via the lower kill line and tree annulus master valve.
- Confirm function of the SCSSSV from the rig control panel.

14.9.2.2 Vertical tree system

- After landing, the tree/LMRP interface must be tested. Test pressure will be determined by the rating of the equipment and operational requirements.
- Function testing is normally carried out to confirm operation of the LMRP production bore valves and annulus bore valves. Where valves have visual indicators a remotely operated vehicle (ROV) will be deployed to observe valve functions.
- Circulation through the riser is confirmed by opening the cross-over valve in the LMRP.
- Pressure test are required to confirm riser integrity and LMRP valve integrity.
- The STT will be function tested and pressure tested in accordance with the work program.

Table 14.2 Control panel (and umbilical) configuration for landing a tubing hanger running and orientation tool in a horizontal tree

Line no	Description	UMBILICALREEL	LOCK INPRESSURE
3	Secondary Reference	OPEN	5000 PSI
7	THRT Latch	OPEN	3000 PSI
1	THRT Unlatch	OPEN	0 PSI
8	TH Lock Monitor	OPEN	0 PSI
9	TH Lock	OPEN	0 PSI
2	TH Unlock	OPEN	1500 PSI
10	Vent/Test Return	OPEN	0 PSI
4	TH Soft Land	OPEN	3000 PSI
6	SCSSSV 1	OPEN	5000 PSI
5	SCSSSV 2 (Not required: Spare line)	CLOSED	N/A
E	Retainer Valve Open/SSTT Relatch	OPEN	3000 PSI
1U	Upper Lubricator Valve Open	OPEN	1000 PSI
2U	Upper Lubricator Valve Close	OPEN	0 PSI
3U	Upper Lubricator Valve Chemical Injection	CLOSED	0 PSI
1L	Lower Lubricator Valve Open	OPEN	1000 PSI
2L	Lower Lubricator Valve Close	OPEN	0 PSI
3L	Lower Lubricator Valve Chemical Injection	CLOSED	0 PSI
C	SSTT Unlatch/Retainer Valve Close	OPEN	0 PSI
F	SSTT Upper Open	OPEN	3000 PSI
D	SSTT Lower Open	OPEN	3000 PSI
A	SSTT Assist Close	OPEN	0 PSI
B	SSTT Chemical Injection	CLOSED	0 PSI

Table 14.3 Valve status table: an example for a subsea completion program

Subsea tree (SSXT)						
Valve	**PMV**	**PIV**	**XOV**	**WOV**	**AMV**	
Status	Open	Close	Open	Open	Closed Inflow tested	
Subsurface test tree (SSTT)						
Valve	**LLV**	**ULV**	**Retainer**	**LBV**	**UBV**	**TRSSSV**
Status	Closed Inflow tested 50% THP glycol water	Closed Tested 6000 psi Glycol water	Open	Open	Open	Closed Inflow tested glycol water mix
Surface test tree (STT)						
Valve	**PMV**	**KWV**	**SV**	**PWV**	**Choke**	**Tree cap**
Status	Open	Closed	Closed Tested 6000 psi	Open	Closed	Installed and tested 6000 psi

There are many different intervention riser systems in use. That, coupled with the diverse nature of operations that are carried out, necessitates a detailed and comprehensive completion or intervention program. It should have clear, unambiguous step-by-step instructions itemizing all the function checks and pressure tests necessary to run, test, and operate the intervention riser system. Many programs incorporate a series of tables that serve as check lists.

An intervention supervisor, working with the vendors, can use these tables to confirm valve status and control umbilical pressure at each step of an operation.

Illustrated below are two examples. The first was used to ensure the control umbilical was correctly configured before landing a tubing hanger in a horizontal tree (Table 14.2).

The second example shows the valve configuration prior to a wireline run into a well equipped with a horizontal tree (Table 14.3).

Valve status schematics (Figs. 14.13 and 14.14) or a valve status board (Fig. 14.15) are a useful method of keeping track of valve status as a job progresses.

If a valve status board is used, it is good practice to have all valve movements reported to a single central location. For example, on some rigs the valve status record is controlled by the driller with the status schematic located in the dog house. All instructions to open and close the valves in the subsea tree/riser/surface tree are relayed through the driller to the personnel operating the control panel. The driller then updates the status of the board, after getting confirmation that the valve movement is

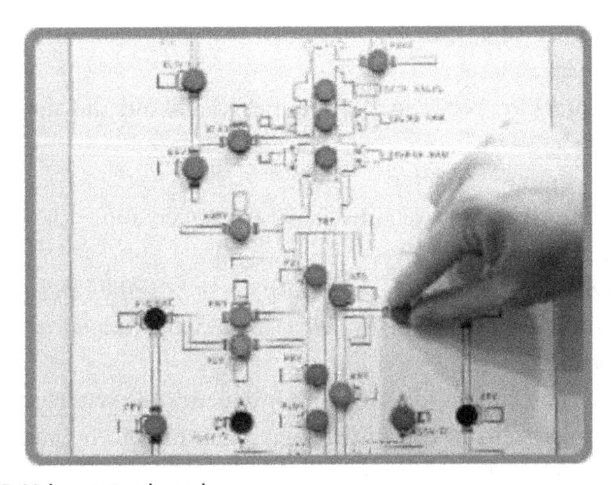

Figure 14.15 Valve status board.

complete. This simple method removes ambiguity and doubt regarding the valve status, and reduces the risk of well control incidents caused through poor communication between the interested parties.

14.10 INTERVENTION RISER DISCONNECT

An ability to disconnect the riser during completion and intervention operations is vital. As with drilling operations, disconnecting may be a planned response or a result of an emergency. Completion and intervention supervisors should understand the disconnect sequence, and the consequences of having intervention tools (wireline or coiled tubing) in the well when a disconnect is planned, or an emergency disconnect occurs.

14.10.1 Riser disconnect: horizontal tree

Before the marine riser can be disconnected, the landing string needs to be disconnected at the SSTT.

14.10.1.1 Planned disconnection

An example of a planned disconnect is listed here.

1. Pull out of the hole with any wireline or coiled tubing.
2. Depressurize then rig down pressure control equipment.
3. Lower SSTT ball valve closes.
4. SSTT Flapper valve closes.
5. Retainer valve closes.
6. Trapped pressure between the retainer valve and the flapper valve is vented.
7. Latching dogs are retracted.
8. The landing string and retainer valve are pulled above the BOP.
9. Blind rams are closed.

Note: Different systems may operate in a slightly different sequence. Supervisors must acquaint themselves with the operating features of the system in use before work begins.

Response time for the disconnection depends on water depth and the type of control system in use (direct hydraulic, piloted hydraulic, or electrohydraulic). The entire disconnection sequence should take no more than 15–20 seconds from initiation to disconnection. The SSTT remains in place, straddling the drilling BOP. Well integrity is provided by the two closed SSTT valves, and the BOP pipe rams shut on the slick joint. Once the disconnection has been made, the landing string is pulled back to the surface, or picked up enough to allow the drilling riser to be disconnected. The closed retainer valve prevents any loss of hydrocarbon in the landing string when separation occurs. If wireline or coiled tubing is in the well, and time permits, it would be recovered from the well and the surface pressure control equipment rigged down (lubricator or injector head) before the SSTT is disconnected.

Figs. 14.16 and 14.17 illustrate a planned disconnection.

Note: The illustrations are based on the "Schlumberger SenTREE 7" subsea test tree. The Expro test tree performs the same function, but uses two ball valves instead of a ball and flapper.

14.10.1.2 Planned disconnection with intervention tools in the well (coiled tubing or wireline)

If there is no time to recover wireline or coiled tubing from the well, a planned disconnection can still be made. The ball valve in the SSTT is designed to cut slickline, braided cables (e-line), and some diameters of coiled tubing (normally up to and including 2″). The coil cutting capabilities of the tree valves must be confirmed with the vendor before any intervention programs are implemented. If in doubt, tests should be

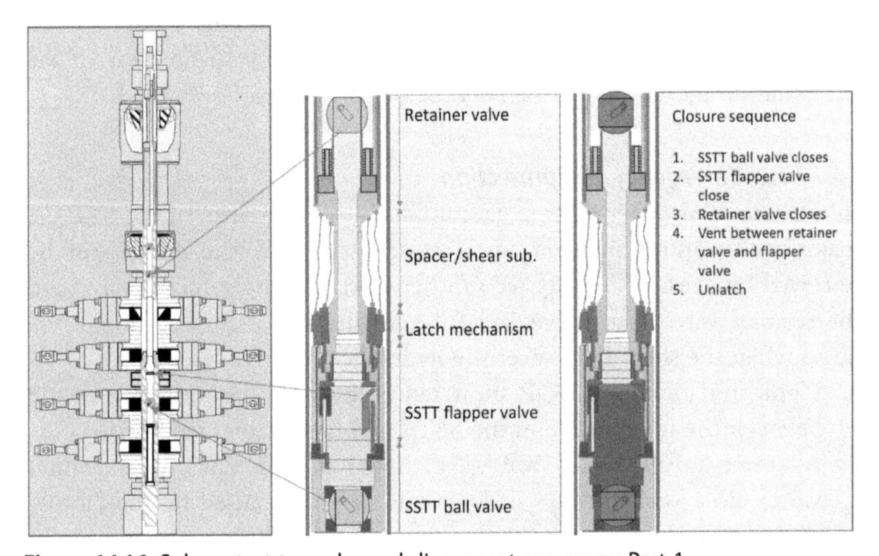

Figure 14.16 Subsea test tree planned disconnect sequence: Part 1.

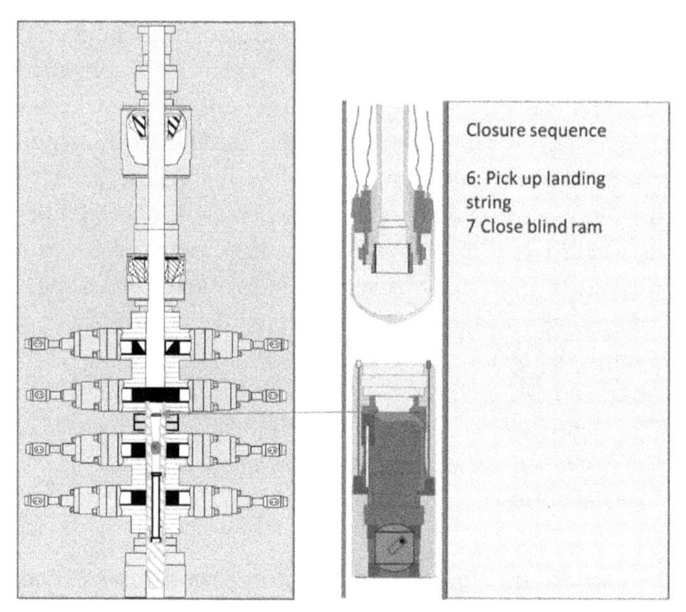

Figure 14.17 Subsea test tree planned disconnect sequence: Part 2.

carried out on the coil to be used by the SSTT vendor, to confirm cutting capability. The coiled tubing vendor cannot be allowed to change the specification of the coil without first consulting the operating company engineer and the SSTT vendor.

If there is a need to disconnect whilst intervention pressure control equipment is still rigged up in the coiled tubing lift frame, there needs to be enough height in the derrick to pick the landing string clear of the BOP and up inside the marine riser, before it in turn is unlatched.

14.10.1.3 Emergency disconnection

In the case of an emergency, e.g., complete loss of control of the SSTT, a major well control incident, or a vessel drive off, the string can be retrieved by shearing the spacer sub between the upper SSTT valve and the retainer valve. The valves are fail-safe, and if not already closed will do so when the shear rams sever the hydraulic control lines.

1. Upon initiation, the BOP shear rams close and shear the spacer sub between the upper valve in the SSTT and the retainer valve.
2. All three valves (lower ball valve, upper flapper valve, and retainer valve) close with the loss of hydraulic pressure caused by the cutting of the control lines.

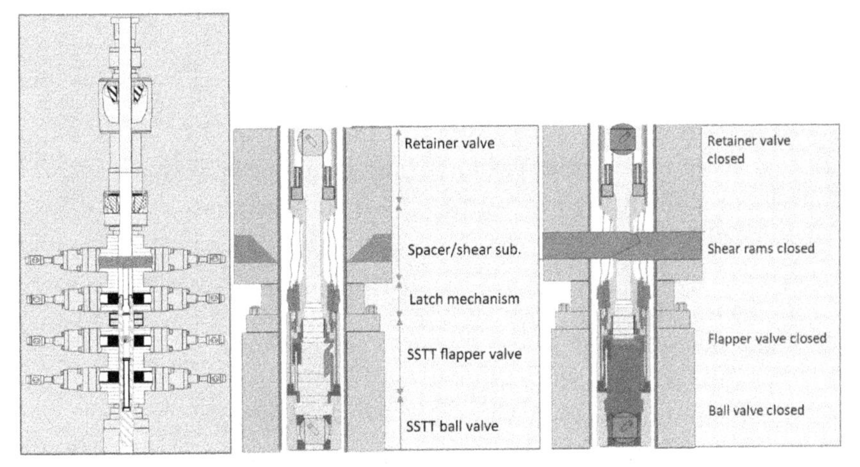

Figure 14.18 Emergency disconnect sequence.

3. The landing string can be picked up above the shear rams.
4. The drilling marine riser is disconnected (Fig. 14.18).

14.10.2 Riser disconnect: vertical tree

The workover control panel provides the means of operating the SSXT, the LMRP, and the emergency disconnect package (EDP). Disconnecting from a vertical tree when rigged up using a dual bore riser, EDP, and LMRP will vary, depending on the type of control system that is in use and how the control system is configured. Most systems have different levels of shut-down and disconnect. For example, the first stage would be a closure of the STT, followed by a shut in of the tree and LMRP valves, followed by a disconnect.

The disconnection sequence described here is representative of many systems, but is by no means universal. Before arriving on the rig, completion and intervention supervisors should familiarize themselves with the intervention riser package.

14.10.2.1 Controlled disconnect

In a nonemergency situation, the riser disconnect sequence might be as follows:

1. Recover any wireline or coiled tubing from the well. If the tools can be picked up above the surface tree valves, then the following valves would be closed.

   a. STT-PWV
   b. STT-PMV
   c. STT-AWV
   d. STT-AMV.
2. Rig down intervention pressure control equipment.
3. Close the following valves on the subsea production tree.
   a. SSXT-PWV
   b. SSXT-PMV
   c. SSTT-AMV
4. Close the SCSSSV.
5. Open the XOV on the LMRP.
6. Open the surface tree AWV, AMV, PWV, and PMV.
7. Open the LMRP upper gate valves on the annulus and production bore.
8. Flush the riser by pumping down the annulus bore, taking returns through the production bore.
9. Once clean returns are observed, close in the LMRP gate valves and XOV.
10. Close the surface tree valves.
11. Unlatch the EDP from the LMRP.
12. Pick up on the riser tensioners until the EDP is clear of the LMRP.
13. Move rig off position so the EDP and riser are not directly above the well (Fig. 14.19).

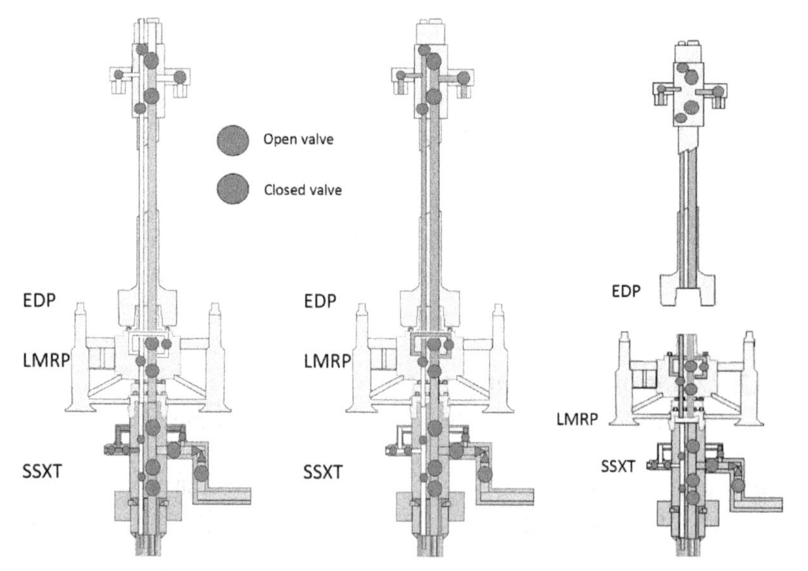

Figure 14.19 Dual bore riser. Planned disconnect sequence.

14.10.2.2 Emergency disconnect

Most systems are designed to shut in the well and disconnect the riser in an ordered sequence. The example shown here is based on a Cameron system that uses piloted hydraulics to shut down and disconnect from the well in three stages. The intervals between each stage are designed to allow the shutdown to take place safely without damage to subsea equipment.

ESD 1: Surface tree shut in.

1. Immediate closure of the surface tree production wing valve and annulus wing valve.

ESD 2:

1. LMRP production gate valve closes.
2. LMRP annulus gate valve closes.

   The gate valves in the LMRP should be able to cut coiled tubing and wireline. Many systems use subsea accumulators to assist valve closure. Before any intervention is carried out, shear tests must be carried out to confirm that the LMRP valves are able to cut any coil or wire that it is planned to use.

   After a short delay of about 15 seconds:

3. Production wing valve closes (if not already closed).
4. LMRP XOV fail-safe opens.
5. SSXT XOV valve closes.

   After a further 30 seconds:

6. SSXT-PMV closes.
7. SSXT-PSV closes.
8. SSXT-AMV closes.
9. ASV closes.

   The 30 second time delay between closure of the SSXT XOV and the closure of the production master and production swab should be sufficient to allow any cut coiled tubing to drop below the tree valves.

   After a further 60 seconds:

10. EDP unlocks.
11. Riser tensioners lift the riser.
12. Rig moves away from the subsea tree.

The whole disconnect sequence should take approximately two minutes.

Most modern intervention systems can be shut in and disconnected using an ROV if the control system fails.

14.11 ADDITIONAL WELL CONTROL AND WELL INTEGRITY CONSIDERATIONS FOR SUBSEA INTERVENTION OPERATIONS

In some respect, interventions carried out on land and platform wells are performed for the same reasons and use the same techniques as those carried out on subsea wells. Any difference lies with the equipment needed to access the well, and some additional complications that arise out of operating in a subsea environment.

14.11.1 Hydrate prevention

Hydrates are far more likely to form during subsea interventions. Temperature at the seabed is often much lower that at the surface, and large volumes of water are used when pressure testing the landing string above the SSTT or riser above the SSXT. Inhibition of test fluids (with glycol) is essential if hydrates are to be prevented.

14.11.2 Wireline operations

Most of the emergency well control procedures detailed in Chapter 9, Wireline Operations, of this book are equally as applicable when operating on a floating rig. There are some exceptions and additional considerations:

- Ensure that hoses used to operate the BOP's, grease head, and stuffing box are of sufficient length.
- The wireline lubricator will be much higher above the rotary table when compared with a land or platform well, making access reliant on hoists.

14.11.3 Riser or landing string leak

There is little that can be done about a leak below the BOP whilst wire is in the hole. Fortunately, this is a rare occurrence.

If the wire is above the seabed, the subsea valves can be closed and the landing string depressurized. If the toolstring is below the seabed, then the wire must be cut. Cut the wire with the wire cutting valve in the LMRP or SSTT then secure the well.

14.12 COILED TUBING OPERATIONS

Most of the emergency procedures and well control procedures detailed in Chapter 10, Coiled Tubing Well Control, are equally as applicable when operating on a floating rig. Additional considerations for coiled tubing operations on a floating rig are as follows.

14.12.1 Rig heave

On floating rigs, the coiled tubing is continually flexing where it meets the guide arch (gooseneck). This is because the injector head is stationary (relative to the seabed), whereas the rig (and therefore the reel) is in motion as the rig heaves. The following heave limits are recommended by most vendors.

1. Maximum heave during rig up of coiled tubing unit, change out of bottom hole assemblies, and rigging down of coiled tubing unit—no greater than 5 ft (1.5 m).
2. Maximum heave during coiled tubing operations, RIH, POOH, or any other work in the well. No greater than 10 ft (3 m).
3. If the coiled tubing is going to be stationary in the well during bad weather conditions, a loop should be made in the coil by clamping the coil at the level-wind head on the reel, and using the injector pick up on the pipe about 5–10 ft (1.5–3 m). This will prevent excessive cycling of the coiled tubing at the reel (Fig. 14.20).

14.12.2 Production shut-down

It is unlikely that a well would be produced to the host facility during a coiled tubing operation. However, production through a temporary test spread is common. In the event of a shut-down at the test spread, the coil operators should:

1. Stop fluid pump (or nitrogen converter). The actuated wing valve on the STT and the ESD valve upstream of the well test choke should close automatically.
2. If scale milling or sand washing operations are under way, pick up to get the nozzle clear of open perforations, and above any fill that could settle and stick the pipe when circulation stops.

   If there is time, pull back the distance equivalent to the space between the cutting rams in the LMRP (or cutting ball valve in the SSTT) and the SCSSSV. If the situation deteriorates and a riser

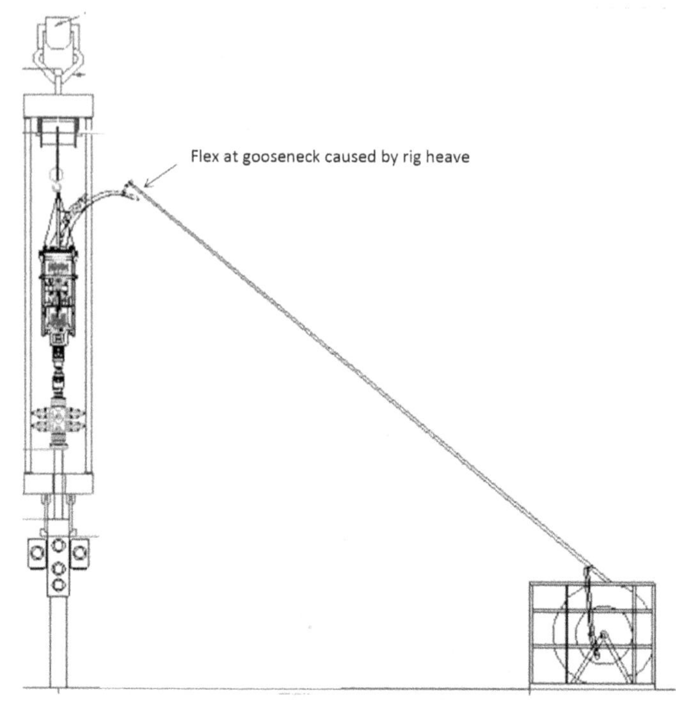

Figure 14.20 Beware of rig heave fatiguing the coil.

disconnect is imminent, the coiled tubing will be cut and dropped. Picking up should ensure it drops below SCSSSV, which can then be closed, as will the remaining open LRMP or SSTT valves

If the coiled tubing BOP is connected to an accumulator, confirm there is enough pressure in the accumulator bottles to shut the BOP, and then shut down the power pack. If the power pack is needed to function the BOP leave it running.

14.12.3 Injector replacement

If the injector fails and it cannot be repaired whilst remaining rigged up, the procedure outlined in Chapter 10, Coiled Tubing Well Control, can be followed providing the injector head can be lifted clear of the cut pipe whilst within the confines of the coiled tubing lift frame.

14.12.4 Riser or landing string leak

When only a short section of pipe is in the hole:

1. If it is thought safe to do so (as judged by the supervisor on location), pull out of the hole to above the LMRP or SSTT.

2. Close in at the SSXT/LMRP (or SSTT).
3. Bleed off pressure and flush riser (or bleed off landing string).

The next step will be determined by the location of the leak, but it will probably be necessary to rig down the coiled tubing and pull the riser (or landing string).

When the pipe is deep in the well and pulling back to surface would expose people and the rig to unnecessary risk:

1. Pull sufficient coiled tubing out of the hole to ensure that the string will drop below SSXT when the cutting valves are closed.
2. Cut coiled tubing using the cutter valve in the LMRP or SSTT.
3. Close the SSXT or SSTT valves to secure the well.
4. Depressurize the riser above the LMRP.
5. Flush riser to remove hydrocarbons.
6. Replace or repair riser or landing string.
7. Commence fishing operations.

Note: If the leak is small, then pulling out of the hole can be attempted. This will be at the discretion of the supervisory staff on site. Pulling out of the hole should only be attempted if:

- There is no H_2S in the leaking hydrocarbon.
- There are coil cutting rams in the LMRP/SSTT.
- The facility to pump into the tubing (coiled tubing/production tubing annulus) is already in place, and can be used without delay.

If these conditions are met then proceed as follows:

1. Pump water (or water glycol mix if there is a risk of hydrates) slowly down the tubing, ensuring coiled tubing collapse pressure is not reached.
2. Pull out of the hole. Monitor the leak continuously. Be prepared to cut the pipe and close in the well if it worsens.
3. As soon as the pipe is back in the riser, close in the well and bleed off.
4. Repair the leak.

14.12.5 Collapsed pipe

If the tubing is being pulled out of the hole, and a collapse is pulled up to the stripper causing a leak at surface:

1. Immediately run the tubing back in the well a sufficient distance to make sure that round pipe is in contact with the stuffing box and across the BOP.

2. Immediately reduce the wellhead pressure by all safe means possible. Either flow the well through a choke at a higher rate, or stop the annular fluid injection if reverse circulating.
3. Increase the coiled tubing internal pressure by attempting to circulate.
4. Once pressure conditions inside and outside the coiled tubing have been optimized, the well will need to be killed.

If it is not possible to run down to place undamaged pipe across the stripper and there is a leak at the stripper:

1. Close the pipe rams. This might reduce or even stop the leak.
2. Clear all nonessential personnel from the vicinity.
3. If the leak is small and the escaping fluids are not hazardous, it might be possible to kill the well without cutting the tubing. This will be at the discretion of the supervisory staff.

If it is not possible to control the well, the tubing must be cut and dropped and the blind rams closed, followed by the STT valves. Be prepared to close the shearing valves (LMRP or SSTT) at the seabed if necessary.

Once the well has been killed:

1. Arrange for clamps to be fitted to the coiled tubing above the injector head.
2. Release the pressure from the stuffing box and open the pipe rams.
3. Cut the tubing at the gooseneck.
4. Reclamp the tubing above the injector head and cut off in 30 ft (10 m) sections (or as appropriate to the space available in the lift frame).
5. Continue pulling and cutting tubing until the tubing pulled to surface can be pulled by the injector head.
6. Once undamaged (not collapsed) tubing is above the injector chains (plus some excess) close the slips and pipe rams. Remove the clamps.
7. Using a dual roll on connector, join the end of the pipe protruding from above the injector to the end on the reel. Take up any slack.
8. Open the pipe and slip rams.
9. Continue to pull out of the hole.

14.12.6 Releasing stuck coiled tubing using a chemical or explosive cutter

The procedure used in the coiled tubing chapter cannot be applied on a floating rig, because there is not enough room in the coiled tubing lift frame to rig up wireline lubricator above the injector head. There is little option other than to cut the pipe using the cutting valves in the LMRP or SSTT.

14.12.7 Tubing parted downhole

The method used for pulling parted tubing back out of the hole (cycling the swab valve to see if pipe is still across it) is not applicable for subsea operations.

In deep water, it might be possible to pick up above the LMRP or SSTT, based on weight of the string in the well, before closing in at the seabed and bleeding off the riser.

In shallow water, the risk of pulling back based on pipe weight alone is that the pipe will be pulled out of the stripper, leading to a loss of containment and a hydrocarbon spill. If in doubt, the well must be killed before an attempt is made to recover the pipe.

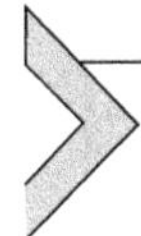

14.13 STIMULATION OPERATIONS: WORKING WITH FRAC BOATS

Many rigs and platforms do not have enough deck space to carry out large volume pumped treatments of their wells. In these circumstances, a frac boat will be used. The important considerations when using a frac boat are:

- In most cases, all of the treating iron will be sent to the rig well in advance of the vessel's arrival. This allows time for the crew to rig up and test all the lines. Once the boat arrives on location, Coflex hoses are picked up from the stern of the vessel and connected to the hard piping on the rig.
- Communications. All the pumps, mixing, and monitoring equipment are on the boat. Before any pumping can begin, it is essential that good lines of communication are in place between the vessel and the supervisory staff on the rig (Fig. 14.21).

14.14 WELL TESTING OPERATIONS

Many of the emergency procedures and well control procedures detailed in Chapter 12, Well Control During Well Test Operations, of this book are equally applicable when operating on a floating rig. There are, however, some exceptions and additional considerations.

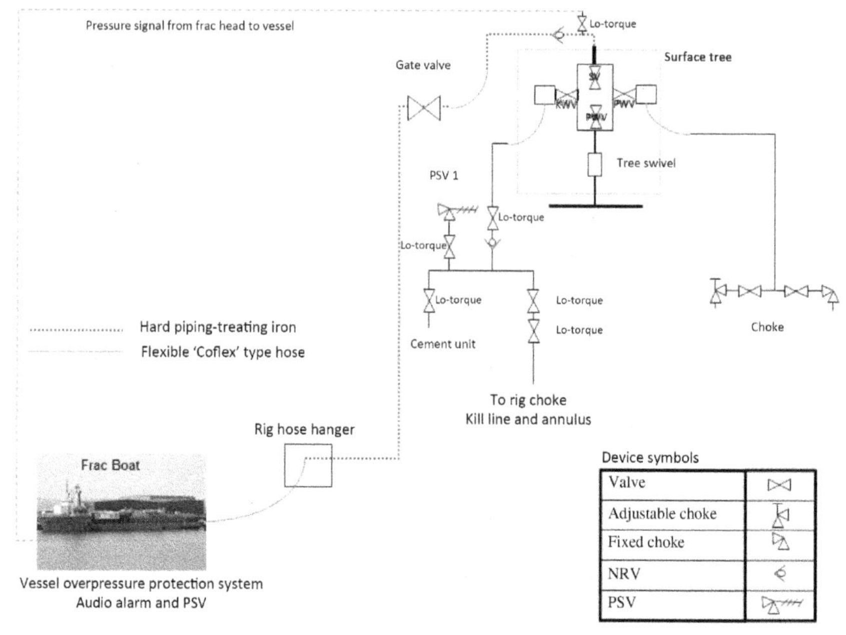

Figure 14.21 Example P&ID drawing for a frac boat supported stimulation operation.

14.14.1 Leak between the subsea test tree and the surface test tree

If there is a leak in the landing string between the SSTT and the STT oil and gas will be seen at surface and there will be a trip tank gain.

1. The driller should instruct the well test crew (who operate the SSTT) to immediately close in the SSTT. If there is a remote ESD station at the rig floor, the driller can initiate the close in.

   Note: Closure of the SSTT will result in the loss of the tool string if wireline is in the hole. If the wire blows out of the stuffing box, the STT actuated master valve should be closed.

2. Close the diverter element to divert the leak through overboard lines.
3. Inform the operating company Drilling Representative and the Well Test Supervisor.

CHAPTER FIFTEEN

Subsea Wireline Lubricator Interventions

The concept of placing a wireline lubricator on the seabed and conducting wireline interventions from a small mono-hulled support vessel was conceived over 40 years ago. In the 1970s BP, ADMA, and CFP carried out trials using a Flopetrol designed subsea well intervention lubricator. The trials took place at a depth of 65 ft in the Zakum Field; offshore Abu Dhabi.[1] In 1984, a joint venture between BP and Camco further developed the subsea intervention lubricator (SIL) to access subsea wells producing to the Buchan Alpha, a converted semisubmersible rig. The Camco intervention lubricator was subsequently deployed from the vessel Stena Seawell[a] (Fig. 15.1) and used on wells throughout the North Sea; it continues to be used up to the present day. In recent years, a new generation of subsea intervention vessels has entered the market (Fig. 15.2).

A major advantage of the subsea lubricator system is the speed of deployment. Dynamically positioned mono-hull vessels can travel to locations more quickly than a semisubmersible rig, and once on location the subsea lubricator is deployed in much less time than a conventional riser.

15.1 MONO-HULL INTERVENTION VESSELS

There are several mono-hull vessels currently designed and equipped to deploy intervention equipment—both lightweight flexible risers and subsea lubricator systems. These vessels have some common features:

- Category 2 or 3 dynamic positioning.[b]

[a] Now the Helix Seawell

[b] Based on IMO (International Maritime Organization) publication 645[7] the Classification Societies have issued rules for Dynamic Positioned Ships described as Class 1, Class 2, and Class 3.

Class 1 has no redundancy. Loss of position may occur in the event of a single fault.

Equipment Class 2 has redundancy, so that no single fault in an active system will cause the system to fail. Loss of position should not occur from a single fault of an active component or system such as generators, thrusters, switchboards, remote controlled valves, etc., but may occur after failure of a static component such as cables, pipes, manual valves, etc.

Well Control for Completions and Interventions.
DOI: https://doi.org/10.1016/B978-0-08-100196-7.00015-4

Figure 15.1 The original intervention vessel and still operating today: Helix Seawell. *Photograph courtesy of Helix Well Ops.*

Figure 15.2 The next generation Well Enhancer. Source: *Photograph courtesy of Helix Well Ops.*

- Derrick for deployment of the subsea lubricator (or lightweight flexible riser).
- Multiple moonpools for the deployment of the Subsea Lubricator, diving bells (where diver support is required during interventions), and remotely operated vehicle (ROV) deployment.
- Work class and observation class ROV capability.

- Work crane (some vessels have cranes with active heave compensation capable of retrieving and deploying equipment to and from the seafloor).
- Most intervention vessels double as dive support ships with full facilities for saturation diving.

15.2 THE DERRICK

The derrick deploys and recovers the subsea lubricator. On some vessels, the derrick is also used to deploy a lightweight riser system.

In most vessels, the derrick is positioned close to midships over a moonpool. Equipment is stacked up in the derrick (above closed moonpool doors) in preparation for deployment. Within the derrick are hoists and winches for the subsea equipment assembly, deployment, and recovery (Figs. 15.3 and 15.4).

15.3 SUBSEA INTERVENTION LUBRICATOR SYSTEMS

The SIL allows live well slickline, braided cable, and e-line operations to be carried out on subsea wells without the need to deploy a riser. Most systems have a working pressure of 10,000 psi, but unlike their land-based equivalent, subsea systems are designed to withstand external (seawater hydrostatic) as well as internal pressure.

Figure 15.3 The Helix Seawell derrick. Source: *Image courtesy of Helix Well Ops.*

Figure 15.4 Opening the moonpool doors on the Helix Seawell. *Photograph courtesy of Helix Well Ops.*

There are minor differences in the subsea lubricator systems available from vendors. However, the main components and the principle of operations are broadly similar, and the main components will be recognizable to anyone who has worked with conventional surface based wireline pressure control systems (Fig. 15.5).

15.3.1 Grease head or liquid seal stuffing box

In common with surface wireline pressure control equipment, a seal around the wireline prevents hydrocarbon leaks when the tree valves are open and the lubricator is exposed to well pressure. Unlike surface systems, the seal must be able to prevent leaks into the lubricator when the hydrostatic pressure of seawater is more than wellhead pressure.

For braided cable and e-line operations, a grease head seals around the wire. Subsea grease heads are normally configured with more flow tubes than on the surface. For example, the 7 1/16″ Helix SIL is configured with eight flow tubes. A double elastomer stuffing box/pack-off is located below the flow tubes. This pack-off has a grease injection point between the two elastomer seals. Packing the void space between the seals makes a gas-tight seal (Fig. 15.6).

For slickline operations, a combination of liquid seal and conventional stuffing box packing are normally used. A tool catcher is located immediately below the grease head or liquid seal.

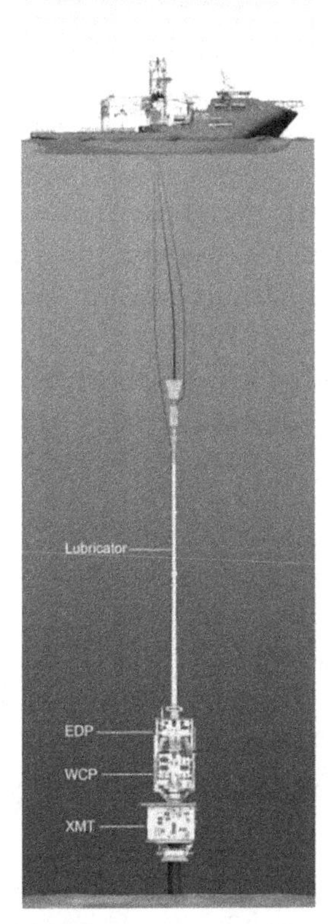

Figure 15.5 Through water wireline in a subsea well.

Figure 15.6 Liquid seal head (left) and grease head (right). *Photograph courtesy of Helix Well Ops.*

15.3.2 Grease head latch mechanism

The grease head is lowered from the surface using one of the derrick winches, and is guided into place at the top of the lubricator using guide wires, an ROV, or in some cases, a diver. After landing out on the lubricator, a hydraulic latch mechanism is activated from the surface control panel. This secures and seals the grease head. The latch mechanism can also be operated using the ROV manipulator arms if the control umbilical malfunctions. The Helix latch mechanism uses the acronym open water latch (OWL) system.

A wire cutting ball valve is positioned immediately below the latch mechanism (Fig. 15.7).

Figure 15.7 The Helix "OWL" latch mechanism. *Photograph courtesy of Helix Well Ops.*

15.3.3 The lubricator

Subsea lubricators are available in a range of configurations. For example, Helix can provide a 5⅛″ system for use on conventional 5″ × 2″ vertical subsea trees. Interventions on large bore vertical trees and horizontal trees are carried out using 7⅜″ and 7¹⁄₁₆″ subsea lubricators. Lubricator length can be adjusted based on requirements.

15.3.4 Lower base section or well control package

Within the lower base section (LBS) are the wireline blow out preventer (BOP) rams and gate valves required to isolate well pressure and close in the well in the event of an emergency. A stab plate for the control umbilical and associated accumulators for the control system are also mounted on the base section. Critical well control functions can be ROV operated in the event of control umbilical malfunction.

The exact valve configuration varies depending on the lubricator size and design, but at minimum the base section will contain three independent methods of shutting in the well.

- Gate Valve: The gate valve (in the base section of the lubricator package) can cut wire. This valve is one of the two mechanical barriers when installing and removing a tool string from the lubricator during operation on horizontal (spool) trees. When operating on vertical trees, the tree valves (swab and master) can be used in preference to the valves in the base section.
- Blind Ram: Wireline BOP rams in the base section are fitted with blind seals of the type used with slickline. The rams are not usually configured for e-line/braided cable, as the double stuffing box in the grease head is designed to seal on stationary wire.
- Shear/Seal Rams: Are the lower rams in the base section, and are designed to cut the heaviest cable in use. Once cut, wire is pulled above the blind ram and gate valve, enabling them both to be closed for two valve isolation (Fig. 15.8).

15.3.5 Control system

A single control umbilical runs from the vessel to a hydraulic stab plate on the lubricator base section. It has multiple functions:

- Direct hydraulic control of all required Christmas tree valves via the tree running tool (TRT).
- Direct hydraulic control of the SIL valves.

Figure 15.8 Lower base section. *Photograph courtesy of Helix Well Ops.*

- Hydraulic control of the grease head latch mechanism.
- Pressure test line.
- Flushing line; to remove hydrocarbons following well entry.
- Lubricator to tree connector latch/unlatch.
- Grease supply for the grease injection head, liquid seal, and pack-off.

A smaller "jumper" line with hydraulic supply for the ball valve in the latch mechanism and grease for the liquid seal head runs from the main umbilical stab plate, vertically up the side of the lubricator (Fig. 15.9).

15.3.6 Main control panel

The hydraulic control panel is normally located on the vessel's main working deck, close to the derrick and moonpool, and is used to deploy, test, and operate the subsea lubricator system. Video monitors relay images from the ROV and divers' helmet cameras (if divers have been deployed). This gives the panel operator visual confirmation of seabed operations as they are performed. The control system includes failsafe closure of all the critical valves from accumulators located in the base section (Fig. 15.10).

Figure 15.9 The main control umbilical connected to the stab plate on the lower base section. *Photograph courtesy of Helix Well Ops.*

15.3.7 Tree bore selectivity

When working on conventional dual bore vertical trees, the lubricator can be configured to access either the main 5″ production bore or the secondary (2″) annulus bore. On the Helix lubricator, a QMU (quick makeup connection) spool is configured for the chosen bore before deployment. The lubricator must be recovered to surface to carry out a bore change (Fig. 15.11).

15.3.8 Tree running tool

The TRT connects the subsea lubricator to the subsea production tree. Since each different model and make of tree requires a dedicated TRT, this piece of equipment is normally owned and supplied by the operator or the tree vendor.

Figure 15.10 Lubricator control panel.

Figure 15.11 Helix quick makeup connection bore selection spool. *Photographs courtesy of Helix Well Ops.*

A dedicated cross-over is used to connect the TRT to the lubricator. Hydraulic control of the subsea Christmas tree (SSXT) valves is achieved by connecting jumper hoses from the umbilical stab plate to the TRT. Stabs on the base of the TRT enable the tree valve functions to be controlled from the vessel main control panel once the tree is latched. Connection of the jumper hoses is usually achieved using an ROV, but in some older systems divers may have to be used (Figs. 15.12 and 15.13).

Figure 15.12 The tree running tool being made up to the lubricator.

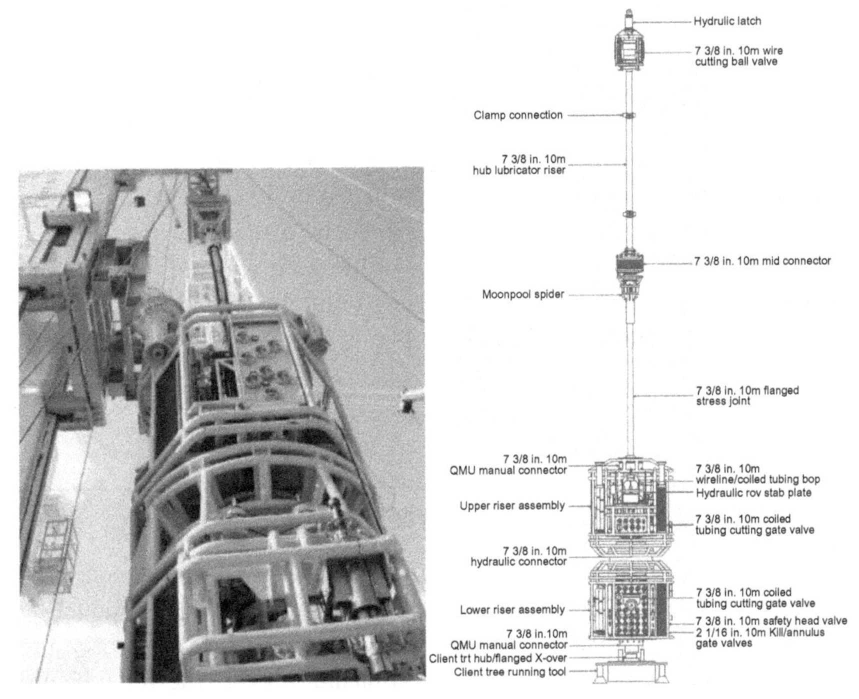

Figure 15.13 7 3/8 in. Subsea intervention lubricator. Source: *Image and photograph courtesy of Helix Well Ops.*

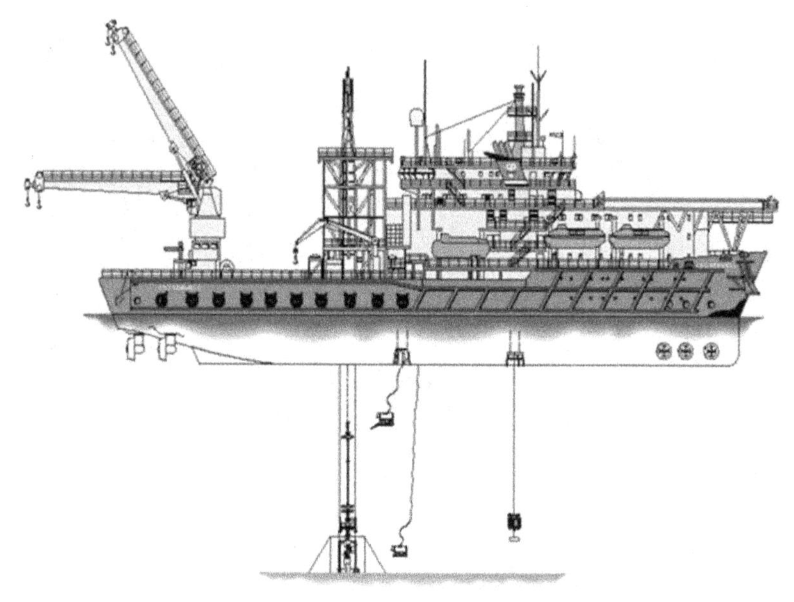

Figure 15.14 A mono-hull vessel on location with the subsea intervention lubricator rigged up on the seabed.

15.4 OPERATIONS WITH THE SUBSEA LUBRICATOR

The types of wireline operation carried out using the subsea lubricator are essential the same as those performed on the surface. Lubricator height limits some activities, and the approach to operations is generally more conservative, as the consequences of breaking wire are more serious than on an equivalent surface operation. Fishing wire using the subsea lubricator is very problematic (Fig. 15.14).

15.5 LUBRICATOR DEPLOYMENT

Once the vessel arrives on location, and DP trials have been satisfactorily performed, control of the well passes from the host production facility to the vessel. Handover must be properly documented to ensure the well is securely isolated from the host facilities before the vessel begins operations. If divers are used, additional isolations may be required.

Before the lubricator can be deployed preparatory operations are required. For example:

- ROV inspection of the wellhead and tree to confirm there are no leaks or damage that could impede the planned operation.
- Removal of trawl protection.
- Deployment of guide wires.
- Recovery of the tree cap.

Stack up and preparation of the lubricator is carried out in the derrick above the closed moonpool doors. This operation requires the cooperation of the operating company subsea engineer, responsible for the preparation and configuration of the TRT, and the vessel subsea engineer responsible for the preparation and configuration of the subsea lubricator system. When the TRT and lubricator have been assembled and all the necessary function checks and pressure tests carried out, it will be lowered to the seabed. Some systems (5⅛″) are deployed as a single assembly, whilst the larger 7″ lubricators may have to be installed as two lifts.

The lubricator or lubricator assembly sections are lowered using the main winch in the derrick. In most operations, guide wires are rigged up from the vessel moonpool to the guide posts on the subsea tree. These both guide the assembly and keep it in the correct orientation as it is lowered. Once the lubricator is latched on to the tree, function tests are carried out. Test sequence and requirements are specific to the operation, and are normally the responsibility of the vessel subsea engineers.

15.6 WIRELINE WELL ENTRY

Once the lubricator has been installed and vital systems function tested, wireline operations can begin.

Rigging up and deployment of wireline tool string is normally carried out as follows:

- Spool slack wire off the wireline unit and feed the wire through the wireline top sheave.
- Suspend the top sheave in the derrick—attached to the motion compensator.

Figure 15.15 The "Witches Hat"—used for assembling the wireline tool string. *Photograph courtesy of Helix Well Ops.*

- Feed wire through the grease head (or liquid seal head) and make up cable head or rope socket.
- Pick up the grease head and suspend in derrick below the top sheave.
- Make up the tool string (from the bottom up) using the "Witches Hat" (Fig. 15.15).
- Connect the tool string to the rope socket or cable head.
- Run the tool string down through open water to the lubricator—the tool string is guided into the top of the lubricator using the ROV, or a diver.
- A derrick winch is used to lower the grease head into place on top of the lubricator, where it is hydraulically latched into place. Once the grease head has been latched, the lubricator is ready for pressure testing.

Pressure test requirements will be specific to each operation. The objective is to confirm pressure integrity of the lubricator package before opening the valves in the tree (or pulling crown and tubing hanger plugs in a horizontal tree) (Table 15.1).

Vessel motion (heave) can be significant. Without heave compensation, wireline operations would be difficult and in some cases impossible. Where guidewires are deployed they act as a reference point for a very effective passive motion compensator. If no guidewires are deployed an active heave system is used.

Table 15.1 Example of a subsea intervention lubricator test schedule

Test description	Test pressure	Duration
Initial well entry		
Pressure test SIL from grease head to gate valve	500/5000 psi	5–10 min
Equalize the pressure above and below the gate valve and open up the gate valve		
Equalize the SIL with the CITHP and open up the blind rams		
From the HCP pump down flushing supply line and equalize against the pressure across the TRSSSV		
Apply 6500 psi operating pressure to TRSSSV control line	6500 psi	Continuous
Flow back 3–4 bbls to the stock tank to confirm TRSSSV is Open		
Pressure test the wellhead connector gasket (WHCST) to 1.1 × CITHP	1.1 × CITHP	15 min
Inflow test the wellhead connector gasket (WHCST)	Inflow	15 min
Subsequent runs		
Pressure test SIL from s/box to gate valve	CITHP + 500 psi	10 min
Equalize the pressure above and below the gate valve and open up the gate valve		
Equalize the SIL with the CITHP and open up the blind rams		

15.6.1 Wireline operations on wells equipped with vertical trees

After pressure testing the lubricator and equalizing with well pressure, the gate valves in the lubricator package and the SSXT are opened and the wire is run into the well. Upon completion of the wireline run, the tool string is pulled back into the lubricator and the lubricator gate valves closed. Depending on the functionality and operational requirements, the gate valves in the SSXT may also be closed.

Before the grease head can be unlatched and brought back to the vessel, the lubricator must be depressurized and valve integrity confirmed. After depressurization, the lubricator must be flushed to remove any hydrocarbons. The contaminated fluids are either flared or stored for later disposal. Once the lubricator has been flushed, the grease head is unlatched and recovered to surface followed by the wireline tool string.

15.6.2 Wireline operations on wells equipped with horizontal trees

After deployment and testing of the lubricator package, the wireline crown plug in the internal tree cap and the wireline tubing hanger plug must be removed before wire can be run below the tree. With the internal plugs removed, there are no barriers to flow in the bore of the tree. Well integrity is therefore entirely reliant on the gate valves in the base section of the lubricator package. To conform to two barrier policy, the lubricator package must have at least two functioning and tested gate valves.

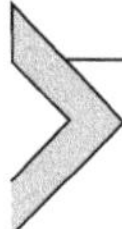

15.7 WIRELINE OPERATIONS: WELL CONTROL PROCEDURES

Once the wireline operator begins to run into the well, the monitoring manipulation and operation of the tool string are carried out in the same way as a surface operation. However, the response to unplanned events is different.

15.7.1 Leaking pressure control equipment

Prior to arrival on location, modeling is carried out to define the vessel operating envelope, accounting for water depth, field environment, and the specification of the subsea tree and wellhead. Allowances are also made for wellhead pressure and wave height.

The lubricator is assembled using API flange connections with BX gaskets. A policy of two barrier isolation is applied, and stringent test procedures both on surface and post-deployment reduce the risk of leaks from the subsea package. Nevertheless, the possibility of a leak must be acknowledged and control procedures put in place:

- Continual ROV monitoring is required to detect leaks—a small leak will not be detectable by monitoring pressure at surface.
- If a leak is detected, the operating company supervisor should immediately be informed. If the leak is minor, a decision to recover the tool string to the surface might be the best option. If the leak is thought to be an environmental risk, then the well must be closed in. Proceed as follows.

15.7.2 Horizontal tree

- Ensure the tool string is not across the gate valves in the lubricator base section. If it is, pick it up to clear the valves.
- Close in the gate valve(s) and blind rams in the base section. If wire is across the valves when they are closed it will probably be ejected from the grease head. Close in the wire cutting ball valve at the top of the lubricator.
  - *Note: Although there is a ball check built in to the grease head to prevent the escape of hydrocarbons, these devices do not always work. Well control measures should be carried out based on the assumption that there will be a hydrocarbon escape as the wire clears the grease head.*
- Flush the lubricator with a glycol/water mix to reduce the size of the spill and prevent the formation of hydrates.
- Close the subsurface safety valve and bleed down the well pressure.

When the well has been made safe, the procedure for repairing the leak will depend on the location of the leak. If it is below the gate valve in the lower lubricator package, the only barrier to flow will be the subsurface safety valve. A well kill might be necessary to enable the lubricator to be removed for repair.

15.7.3 Vertical tree operations

If a leak occurs on the lubricator whilst wire is in the hole and the leak is too severe to allow operations to continue:

- Ensure the toolstring is not across the tree or lubricator valves.
- Close in the well using the wire cutting gate valves and blind rams in the lower lubricator package.
- The wire will almost certainly be ejected from the grease head. Close in the wire cutting valve in top of the lubricator package.
- Close in the well using the vertical tree master valve and swab valve.
- Depressurize the lubricator and flush with a glycol/water mix.
- The lubricator is recovered to the surface and repairs made.

15.7.4 Tool string stuck in the well

For land and platform operations, the usual response to a stuck tool string is to drop a cutter bar, attempting to cut the wire at the rope socket (or cable head), or as deep in the well as possible. Dropping a cutter bar is possible with the subsea lubricator, depending on configuration. Wire has been successfully recovered in the past, although the tool string had to be abandoned owing to lubricator length limitations.

15.7.5 Wire parts above the grease head

If wire parts above the grease head, the severity of the problem is dependent on the location of the tool string in the well when failure occurs. Well integrity is of immediate concern.

- If the toolstring is below the mud-line and sufficiently far above hold up depth, then the wire will be pulled down through the grease head. As the wire disappears through the grease head there is the potential for a leak, unless the ball check in the grease head functions.
  - Close the wire cutting ball valve in the top of the lubricator.
  - Close the wire cutting gate valve(s) in the lower lubricator section.
- Depressurize and flush the lubricator.
- If the tool string is still latched in the tool-catcher, it should be possible to close in the well, depressurize and flush the lubricator, and recover the tool string with the grease head.
- A worst case would be wire parting at the surface, with the tool string stuck across the tree valves. The tool string should be spaced (configured) to allow the lubricator gate valves to close if the tree valves cannot be closed.

15.7.6 Wire parts downhole

Wire parting downhole is managed in much the same way as a land or platform operation. Wire breakage will be indicated by weight loss at surface. It may be possible to estimate how much wire remains in the well.

- Pull the wire from the well. As it approaches the surface, well pressure will begin to eject the wire from the well.
- As soon as the wire is ejected from the well, the ball valve in the top of the lubricator must be closed and the lubricator depressurized and flushed.

Note: Divers should be moved clear of the lubricator as the wire is ejected. The ROV should move to a position where the grease head can be observed, enabling the well to be closed in as soon as the end of the wire clears the lubricator.

15.7.7 Hydrate formation

Hydrates are more likely to occur during subsea operations; low temperatures and water are present. Hydrates are best managed by careful preventative measures—routine pressure testing and flushing of the lubricator with a glycol/water mix. The percentage of glycol required is temperature and system pressure dependent.

Should a hydrate form in the lubricator, the system will need to be depressurized and flushed with methanol to disperse the hydrate.

15.7.8 Vessel drive off (unplanned)

A failure of the vessel dynamic positioning or propulsion system would cause an unexpected drive off the location. Depending on wind and current, this can happen very quickly, leaving no time to recover wireline tools to the lubricator, or to recover the lubricator to the vessel. The following sequence of events will occur as the vessel moves away from the well location:

- Main control umbilical will disconnect from the stab plate on the LBS/LRA section via a shear pin—once the umbilical shears, all in-line SIL gate valves will failsafe close. Any tree valves that are failsafe will also close.
- Guide wires shear release. *Note: If the guide wires wrap around the riser section then they will snap it at the weak point.*

Any wireline in the hole will be cut by the wire cutting valves in the lubricator. It is sequenced to close before the tree valves.

When control of the vessel is regained, the control umbilical is connected to the lubricator stab plate using the vessel ROV and control of the lubricator and tree functions resumed.

15.7.9 Planned disconnect

In a situation where rapidly deteriorating weather conditions force the suspension of intervention operations, but where there is not enough time to recover the lubricator, a planned disconnect can be carried out.

- Stop any pumping operations and recover any wireline tools in the well to the lubricator.
- Tree and lubricator valves would be closed and the lubricator flushed.
- Recover tool string to surface.

- It may be possible to leave the control umbilical (and guidewires) latched by spooling out additional slack. If not:
  - Unlock the control umbilical from the stab plate and spool back to the vessel.
  - Shear and recover the guide wires.

REFERENCE

1. Low-cost wireline and logging operations on a satellite well using a subsea wireline lubricator deployed from a dynamically positioned monohull vessel. D.G. Clarke, A.S. Warne. OTC – 5726 -MS. Offshore Technology Conference. Houston, 2–5th May 1988.

INDEX

Note: Page numbers followed by "*f*" and "*t*" refer to figures and tables, respectively.

E

H

I

S

U

V

W

CPSIA information can be obtained
at www.ICGtesting.com
Printed in the USA
LVHW011705160719
624282LV00016B/1143/P